Y0-AAK-294

PASSIVE AND ACTIVE SOLAR HEATING TECHNOLOGY

MICHAEL MELTZER

Environmental Science and Engineering Program
University of California at Los Angeles

Prentice-Hall, Inc., Englewood Cliffs, NJ 07632

Library of Congress Cataloging in Publication Data

Meltzer, Michael.
Passive and active solar heating technology.

Includes bibliographies and index.
1. Solar heating. 2. Solar energy—Passive systems. I. Title.
TH7413.M45 1985 697'.78 84-26615
ISBN 0-13-653114-8

Editorial/production supervision: *Mary Carnis*
Cover design: *Lundgren Graphics, LTD.*
Manufacturing buyer: *Anthony Caruso*

Cover map courtesy of Solar Energy Research Institute, Golden, Colorado. Solar Energy Research Institute is operated for the U.S. Department of Energy.

To Chris, because she cares so much.

Printed in the United States of America

10 9 8 7 6 5 4 3 2 1

ISBN 0-13-653114-8 01

Prentice-Hall International UK Limited, *London*
Prentice-Hall of Australia Pty. Limited, *Sydney*
Editora Prentice-Hall do Brasil, Ltda., *Rio de Janeiro*
Prentice-Hall Canada Inc., *Toronto*
Prentice-Hall Hispanoamericana, S.A., *Mexico*
Prentice-Hall of India Private Limited, *New Delhi*
Prentice-Hall of Japan, Inc., *Tokyo*
Prentice-Hall of Southeast Asia Pte. Ltd., *Singapore*
Whitehall Books Limited, *Wellington, New Zealand*

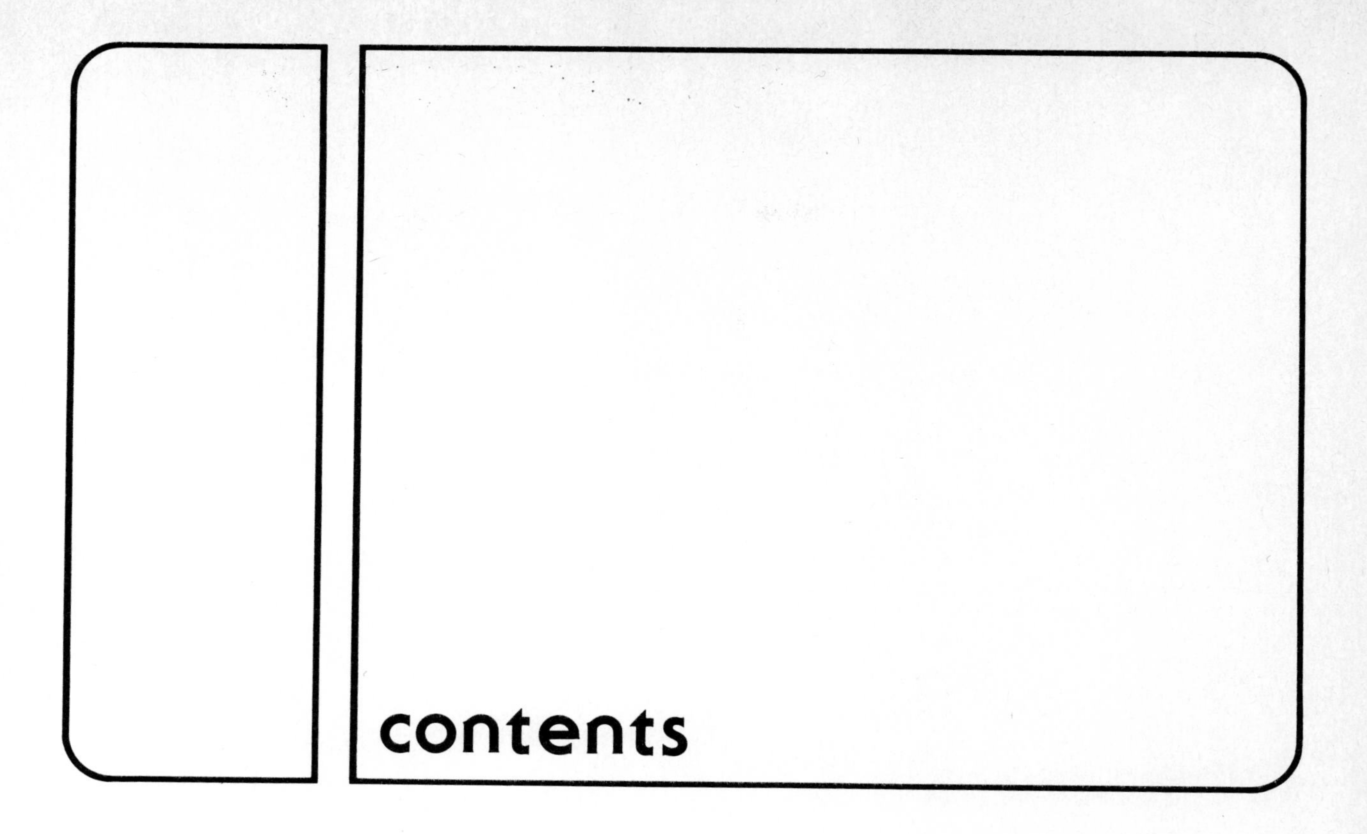

contents

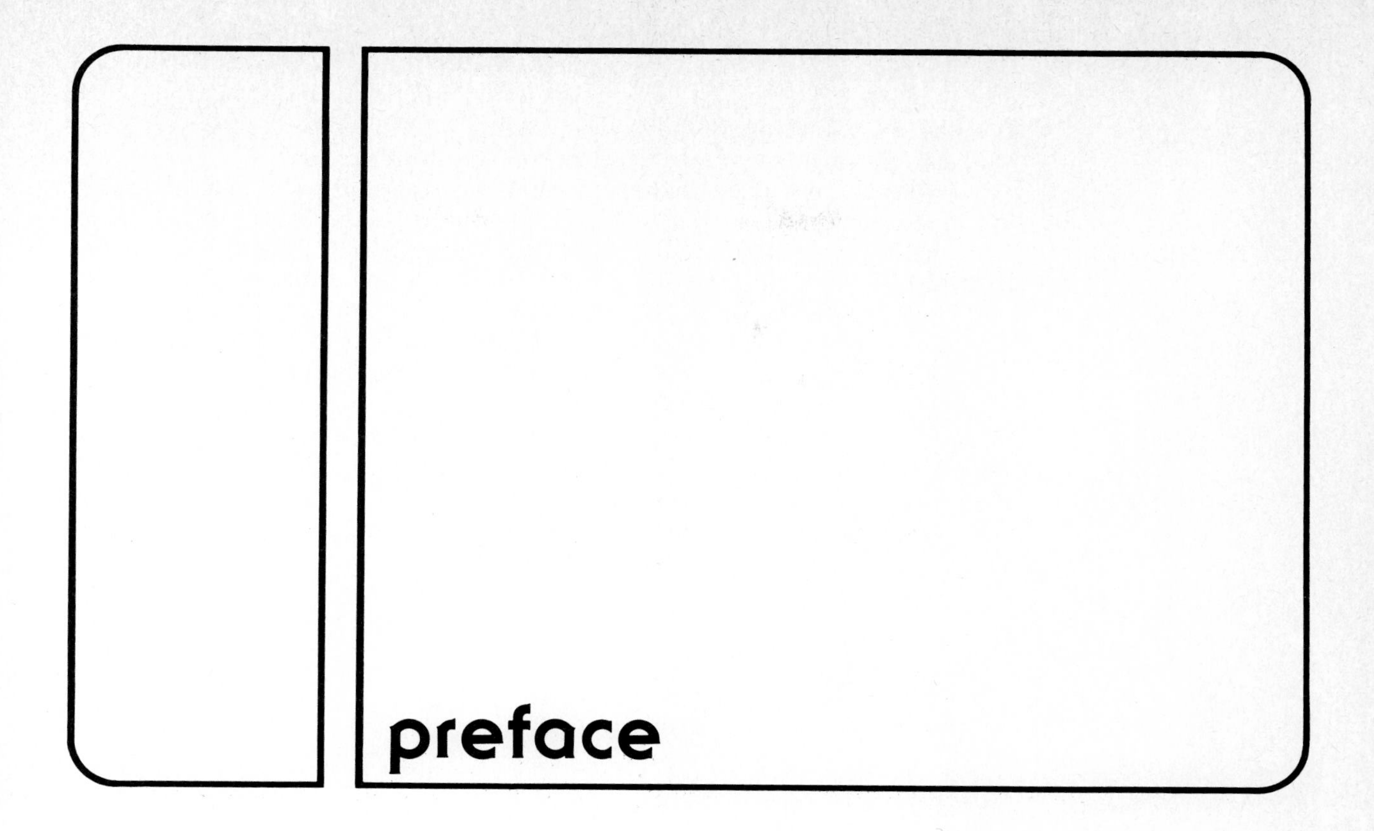

preface

During the five years in which I have taught solar energy classes, the most recurrent problem has been how to find a good, comprehensive textbook. Books available tended to be either too elementary for a college course, or aimed for upper division classes. Other books had the problem of focusing on only one aspect of solar energy, such as domestic water heating.

My aim in preparing this text is to offer a comprehensive source of information about solar heating technology. The book is intended to be used by students in an introductory level course, as well as by solar contractors, owner-builders, or simply readers with an interest in solar heating. I examine in depth both passive heating approaches, such as direct-gain, Trombe wall, and sunspace systems, as well as active designs for heating of buildings, water supplies, and swimming pools. Each system is described and analyzed with the aid of numerous illustrations and tables, after which methods of sizing and design are studied.

Also dealt with in detail are methods of making buildings energy efficient, for conservation of available energy is a basic requirement of a successful solar system. In addition, plans for owner-built solar collectors are included in the appendices. These collectors were built by students in the laboratory sections of my classes.

Finally, there is an appendix on computer-aided solar design that includes a listing of some of the software available.

Most of the text is of a qualitative rather than quantitative nature. Some computations are necessary, however, in order to design systems. The mathematics that is included is kept simple, and operations are fully explained, so that readers with or without technical backgrounds can benefit.

There is sufficient data in the tables to design useful working solar systems as well as to perform economic comparisons of different systems.

I want to give deep thanks to a number of people who helped tremendously during the two and a half years of writing this book. Among them are Christina Meltzer, Margie Lewis, and Joyce Crommet for their patience and dedication in editing and typing the manuscript; Joe Armstrong, for his insights and critiques; Ruth Cohen; Connie Westcott for her excellent illustrations; and Mary Carnis, Chuck Iossi, and the other editors at Prentice-Hall.

I also want to thank SolarVision, Inc. for several things. Their staff was interested and helpful on a number of occasions in answering pertinent questions. Their publications—especially their monthly magazine, *Solar Age*, and their *Solar Products Specifications Guide*—were excellent, important sources of background information for the book.

I hope that you enjoy reading this book as much as I have enjoyed writing it.

Michael Meltzer

Willits, California

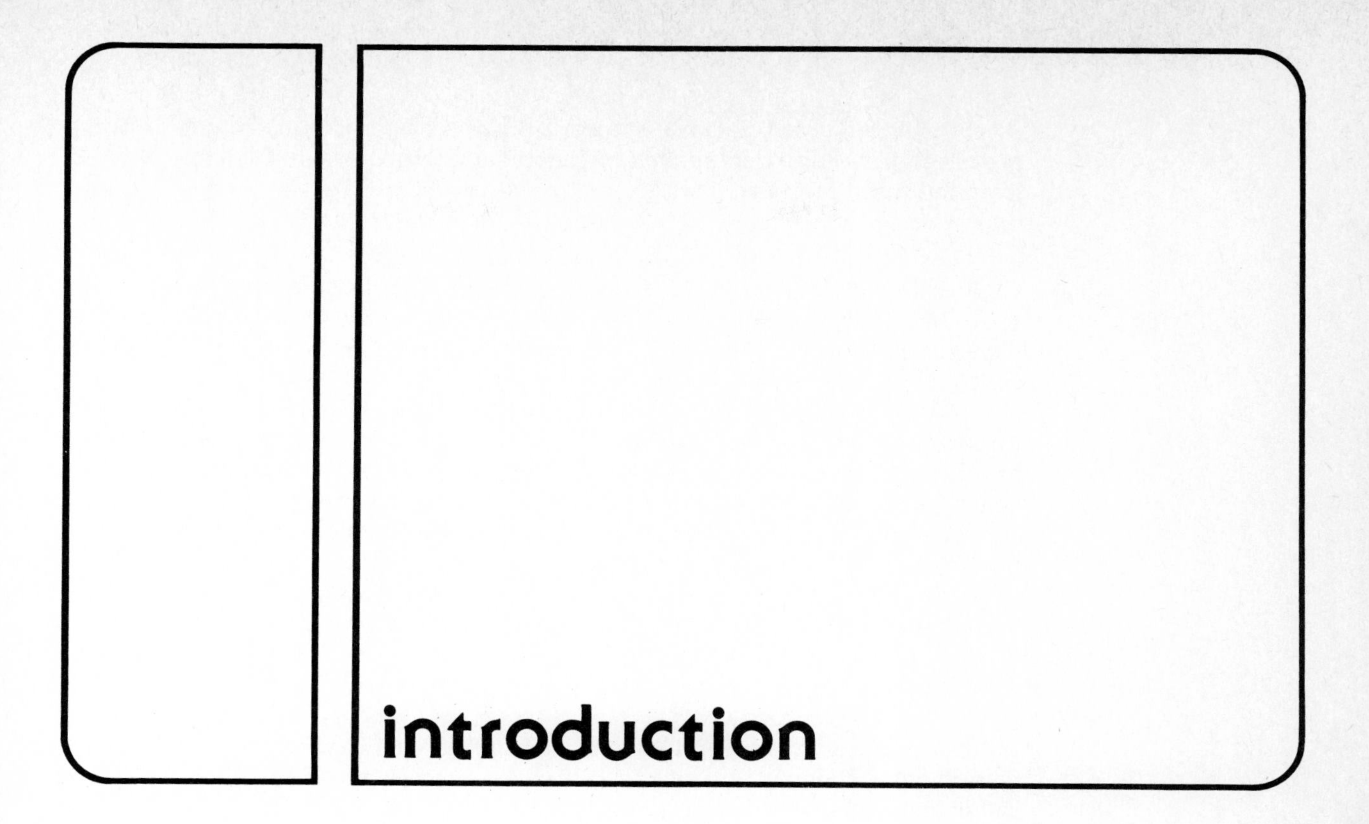

introduction

It is interesting that although our society has been skillful in locating and learning to exploit a variety of energy resources, we have been much less skillful in predicting the social and environmental implications of our energy use patterns. As we entered the Petroleum Age earlier this century, it was not generally foreseen what effect the massive use of petroleum fuels would have on our city's air quality, nor was the effect on our economy foreseen of having to rely so heavily on other nations for our fuel supplies. When nuclear power first came on the scene, it was heralded as a safe source of energy "too cheap to meter." It has been a surprise to many that nuclear generating plants have proved neither safe nor inexpensive to maintain.

Policy decisions concerning energy exploitation matters are too often based on simple economic computations such as the number of kilowatt-hours obtained per dollar of investment, rather than examining the implications of that exploitation in a more global context. Our present type of energy use harms the environment more than any other human activity.[1] It tears up the landscape, it raises the temperature of our atmosphere, and it results in the release of countless tons of pollutants and radioactive by-products. It is the opinion of many of us that in the long run, it will be environmental factors rather than financial considerations that will prove to be most important to the well-being of the organisms that live on this planet.

The people on planet earth are on the verge of a great energy transition. Like it or not, we will soon have to turn from petroleum as our main source of power to other resources. Reports differ widely as to just how many years the earth's oil wells will continue to supply us with so much energy, but whether the figure is 20 years or one century, it is clear that eventually we must seek new types of power.

In recent years, a number of options have been suggested for fulfilling our energy needs once sufficient petroleum is no longer available. Let us examine some of them.

Coal

Coal is the most abundant fossil fuel in the world and has been used as a source of energy for millenia. It was not until the industrial revolution in the eighteenth century, however, that coal began being exploited on a massive scale.

Eclipsed for many years by petroleum, coal is once again gaining in importance as an energy source. Besides being used as fuel in electric power generating plants and for heating of buildings, coal can be converted into other hydrocarbons, such as gasoline. The United States is particularly well stocked with coal, for 19 percent of all the world's deposits are within her borders.

If all the coal in the world could be profitably mined, production would not peak until two or three centuries from now.[2] In actuality, however, it is too expensive to mine much of the world's deposits. They lie in beds that are either too thin or too deep to bother with.

The environmental effects of massive coal use can be very grim. Mine tailings pollute streams and have already affected 11,000 miles of America's waterways. Underground mines can cause subsidence of the surfaces above them and pose severe health threats to those who work in them. Surface strip mining techniques lead to perhaps even more serious problems. To power a 1000-megawatt plant for one year, 20 *miles* of a coal seam 3 feet deep by 225 feet wide is required. Considerable earth must often be removed to get to the seam. Reclamation of the excavated areas is difficult and expensive and has often not even been attempted. In Appalachia, for instance, American coal companies have abandoned 20,000 miles of unrestored strip mine benches.[3] Even rejuvenated land is frequently inferior agriculturally to the way it was in its original state. Where land could originally support crops, after strip mining and reclamation it is often good only for pasturing. Much of our country's coal reserves are in regions with little or no rainfall, and in these areas, reclamation and reconstitution of the topsoil and plants that were there is nearly impossible.

Still another disadvantage of massive coal use is that most varieties of it are rather polluting when burned. Combustion produces ash, sulfur oxides, poisonous metals, and known carcinogens. More mercury is released than in any other industrial process on earth.[4]

Oil Shale and Tar Sands

Mineral deposits known as oil shales and tar sands contain heavy, thick, and gritty oil of the same chemical family as petroleum. The largest and best explored deposits are in northern Canada in the province of Alberta. Some studies estimate that the Canadian fields contain 100 billion barrels of recoverable crude oil, which sounds like a great deal, but at current consumption rates would provide the United States with only 15 to 20 years of power.

Petroleum is difficult to extract from oil-bearing shale and sand. The process requires huge amounts of water and energy. World deposits could total over 100 trillion barrels, but most of it is economically, energetically, and/or environmentally impractical to mine. Many of the deposits lie in areas, for instance, without sufficient water for the extraction processes.

The most useful application for shale oil and tar sands will very likely be in the petrochemical industry, where they could be used as "feedstock" for the manufacture of synthetic rubber, fertilizers, paint, detergents, and possibly even food.

Nuclear Energy

In the 1960s nuclear energy was widely supported as a safe successor to fossil fuel power. The atom seemed to offer an abundant, nonpolluting, and cheap source of energy for many years to come. Why, then, the radical switch in popular opinion? To answer this question, let us look at how a reactor uses the process of nuclear fission to produce electric power.

The majority of a nuclear power plant is very similar to a fossil fuel plant. In both types of facilities, heat from an energy source boils water to produce steam, which in turn drives a turbine. The turbine turns a generator, and an electric current is set up. The difference between the two types of plants, however, is in the primary energy source. A fossil fuel plant produces its heat from burning coal or petroleum. In a nuclear plant, radioactive material in the reactor core is the source of thermal energy.

The radioactive fuel in a nuclear reactor is packed into long, thin "fuel rods," typically 12 feet long and ½ inch in diameter. This shape is necessary so that heat generated can be quickly removed from the vicinity of the rods, or else they would melt. As the material in the rods fissions, it releases thermal energy, which is used to vaporize water into steam and operate the turbines.

A reactor's energy comes not only from the process of fissioning, which accounts for 93 percent of the total heat generated, but also from radioactive decay of the fuel into daughter products, which accounts for the remaining 7 percent of the thermal energy. The fission process can be regulated through the use of control rods, but radioactive decay cannot be. If the reactor core were not constantly being cooled by flowing water or gas, the heat from decay would grow so intense that the core would melt its way through the floor and down into the earth. This "meltdown" reaction has almost occurred several times, notably at Brown's Ferry in Alabama and at Three Mile Island in Pennsylvania.

The radioactivity of nuclear fuels cannot be "turned off" after spent fuel is taken out of the reactor. It remains toxic and dangerous for thousands or millions of years, depending on the fuel. Countries have devised many schemes for disposal of the spent rods, but none are foolproof. Some have turned into disasters.

It used to be believed that the way to dispose of "radwastes" was to sink them to the bottom of the sea. Between 1946 and 1970, 47,500 55-gallon concrete-lined steel drums of waste were dumped into the ocean 30 miles off of San Francisco. Many of these drums (according to some estimates, one in three) are now leaking. There is also evidence that certain sea life around the barrels are showing abnormal rates of mutation.[5]

Deep salt mines were thought for a time to be a good solution to the disposal problem. But when these repositories too developed leaks, and when some of the radioactive wastes found their way into the groundwater system, this idea lost much of its credibility.

Spent fuel rods are considered "high-level" nuclear wastes. If current nuclear growth projections are accurate, the United States could produce 80,000 10-foot by 1-foot canisters of such wastes over the next quarter of a century. "Low-level" wastes include such items as irradiated components of decommissioned power plants, contaminated clothing, and fission by-products diluted in liquid or gaseous solutions. The Environmental Protection Agency estimates that by the year 2000, these wastes will amount to 1 billion cubic feet, which is enough to pave a four-lane coast-to-coast highway 1 foot deep.[6]

Fusion Energy

It has often been stated that nuclear fusion will be a pollution-free, almost limitless source of power. But a look at the reality of the situation produces quite a different view.

The fusion reactors that will be built in the next quarter century will in all probability be of the type that fuses together two hydrogen isotopes, deuterium and tritium. But this type of reactor will not fulfill the promise of limitless power, for tritium is one of the rarest elements on earth. It must be fabricated by bombarding lithium with neutrons, and lithium is not much more abundant than uranium.

Two deuterium nuclei can be fused together to liberate energy, but this reaction is much more difficult to achieve. So even though there is plenty of deuterium available in seawater, it will probably be many years before it can be used as the sole fusion fuel.

Another common fallacy is that fusion reactors are pollution-free. Fusion reactions produce intense neutron fluxes that make the materials surrounding them highly radioactive. Reactor parts might have to be replaced fairly frequently due to damage from the high-energy neutrons. One large reactor could produce as much as 250 tons of radioactive waste material each year.

Renewable Resources

The energy resources that we have been discussing depend on fuel and mineral deposits that can be depleted. Uranium, coal, and petroleum are all in limited supply. They might last 50 years, a century or two—but they will ultimately be gone if we continue using them at present rates.

There are other energy options, though, that will be around as long as the solar system and life on this planet exist. The power of moving water, of wind, of biomass, and of course, of the sun itself are resources of this type. All have enormous potential to light the lamps, heat the houses, and run the machinery of the world in the future. Besides being enormously abundant, they do not emit poison into the air, they are not radioactive, and they require no mining, with its attendant ecological problems.

Solar Energy. One of the most notable of these alternative resources is solar energy. It has been much in the news and widely discussed in recent years, and it is apparent to many of us that it offers a bright alternative to reliance on depletable, traditional energy sources. We have not even scraped the surface of this resource's vast reservoir of power. All the energy that our society uses—for transportation, industrial processes, and residential applications—does not even equal the amount of solar energy striking only the roofs of our buildings.[7]

But for many people, it is not enough simply to talk about the possible benefits of solar energy to our society. Many of us want to learn to *do* something with solar energy. Examined in this text are methods of utilizing the enormous power of the sun in one of its most direct applications: to provide heat for the needs of residential and other small buildings.

The technology of solar heating has been developed to the point where it is now practical and economical to build solar structures. The materials and the know-how are available not only to solarize new construction, but to retrofit systems into existing buildings as well. Some situations call for systems of the passive type that do not use fans, pumps, or other externally powered equipment; some call for active systems that do use this equipment; and all call for buildings that are tightly constructed, well insulated, and energy efficient, so that every Btu of sunlight can be well utilized. All of these areas are dealt with in detail in the following pages.

It is the belief of the author that the more that people know about the techniques of harnessing solar energy, the sooner the dream of allowing the sun to be one of our major sources of power can be realized. If the sun's energy were employed only to heat our buildings and their water supplies, it would cut our country's dependence on fossil fuels and other energy sources by over 20 percent.[8] This could, among other things, greatly reduce our sometimes vassal-like reliance on the Middle East's oil supplies, and ease the political and economic troubles that have resulted from that reliance.

The sun is not a complicated resource to harness. It is within the ability of most of us to design as well as build a variety of systems for our homes. If we believe in the promise and importance of solar energy, it is our responsibility, whether we be present or future homeowners, builders, students, or teachers, to learn more about this resource and to spread its use in whatever ways we can. You who are reading this text can help make a solar society a reality.

NOTES

1. Denis Hayes, *Rays of Hope: The Transition to a Post-petroleum World*, W. W. Norton & Company, Inc., New York, 1977, p. 24.

2. Ibid., p. 39.

3. Ibid., p. 40.

4. Ibid.

5. Helen Caldicott, *Nuclear Madness*, Bantam Books, Inc., New York, 1981.

6. Hayes, p. 54.

7. Ibid., p. 60.

8. Donald Rapp, *Solar Energy*, Prentice-Hall, Inc., Englewood Cliffs, NJ, 1981, pp. 13–18.

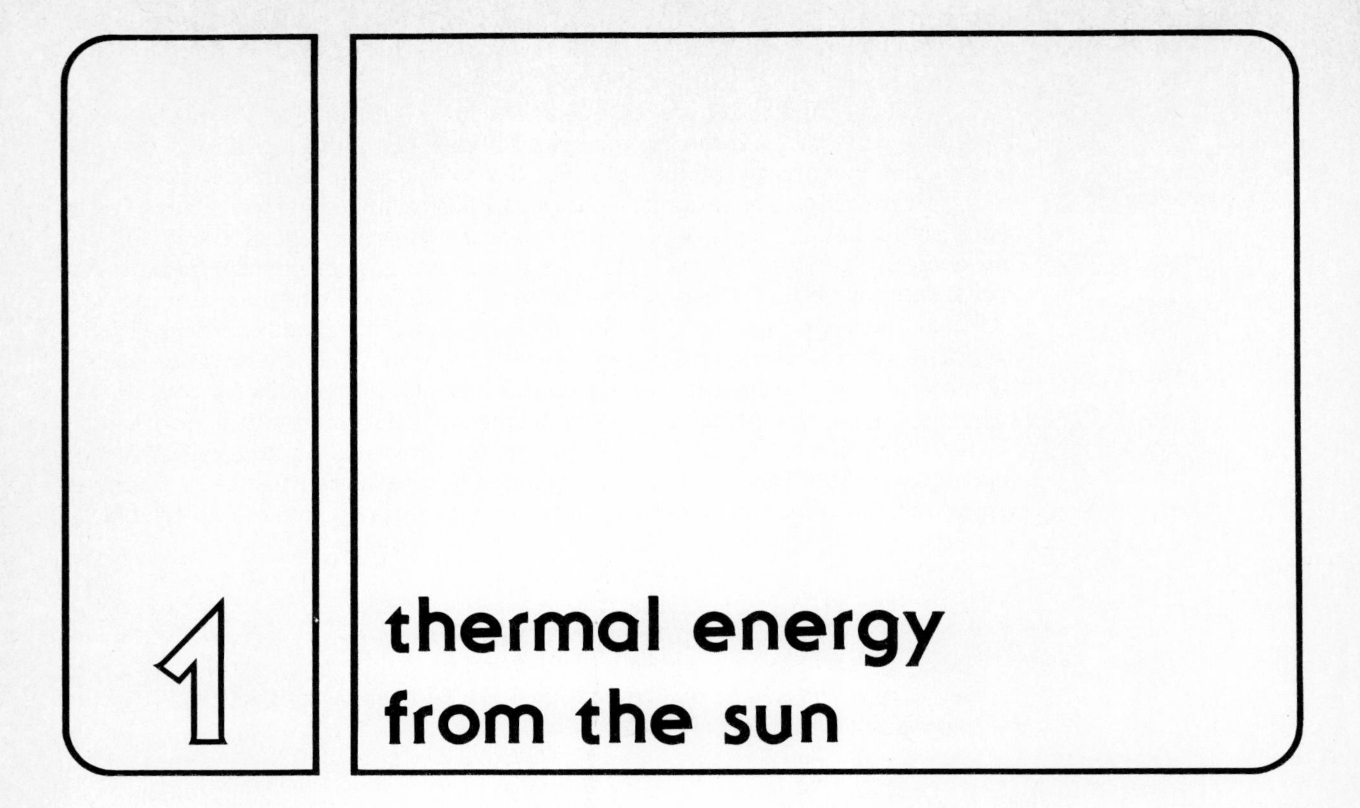

1 thermal energy from the sun

THE NEAREST STAR

The Source of Our Sun's Energy

Our sun, the nearest star, is a ball of gas 865,000 miles in diameter that produces enough energy each second to meet the present needs of all the earth's commercial activities for 2 million years.[1] It is often thought that this energy originates from burning the hydrogen gases that compose most of the sun's volume, but combustion could not begin to produce the enormous quantity of energy being radiated from the solar surface.

The true source of solar energy is from nuclear reactions similar to those that occur in a hydrogen bomb explosion. In the core of the sun, where temperatures reach 12 to 18 million degrees Kelvin (22 to 32 million degrees Fahrenheit) and pressures reach 1 billion atmospheres, 600 million tons of hydrogen atoms are fused into helium every second, in the process releasing 3.7×10^{23} British thermal units (Btu) of energy.

The fusion of hydrogen into helium results in a continually growing helium core at the sun's center. The hydrogen "fuel" for these thermonuclear reactions will one day be used up, but there is no immediate cause for alarm. That day is still 5 billion years away.

The Solar Spectrum

The sun emits energy in the form of *electromagnetic radiation*, also called *electromagnetic waves.* These are not like ocean waves, which require the presence of water, nor are they like sound waves, which would not exist without air or some other medium in which to travel. Electromagnetic waves propagate in the vacuum of space and are composed of electric and magnetic fields. Those of certain wavelengths (from 0.0004 to 0.0007 millimeter) are picked up by the "antennas" in the retinas of our eyes and are called *visible light.* The longest wavelengths in this range appear red, while the shortest appear violet.

Visible light makes up only a small part of the total spectrum of electromagnetic waves emitted by the sun. Those with wavelengths of 0.001 millimeter (slightly longer than those of visible light) to 1 millimeter are termed *infrared radiation* and are sometimes referred to as "heat waves," because our skin feels warm when exposed to them. Those with waves 1 millimeter to many kilometers in length are classified as *radio* waves, for they are the type emitted by radio transmitters.[2]

Waves with lengths shorter than those of visible light make up the ultraviolet, X-ray, and gamma-ray region of the spectrum. Ultraviolet (UV) radiation is responsible for tanning the skin, and in high enough intensities can cause blindness and skin cancer. X-rays and gamma rays are of extremely high energy and penetrate deeply into most substances.

Figure 1-1 depicts more precisely the range of wavelengths of each type of radiation, and includes a graph of the intensity at each wavelength. Notice that the sun's energy peaks in the very narrow band of visible-light wavelengths. In fact, almost half of the sun's total energy is emitted within this band.[3] Thus surfaces designed to collect solar energy should at the very least be good absorbers of visible light if they are to work efficiently.

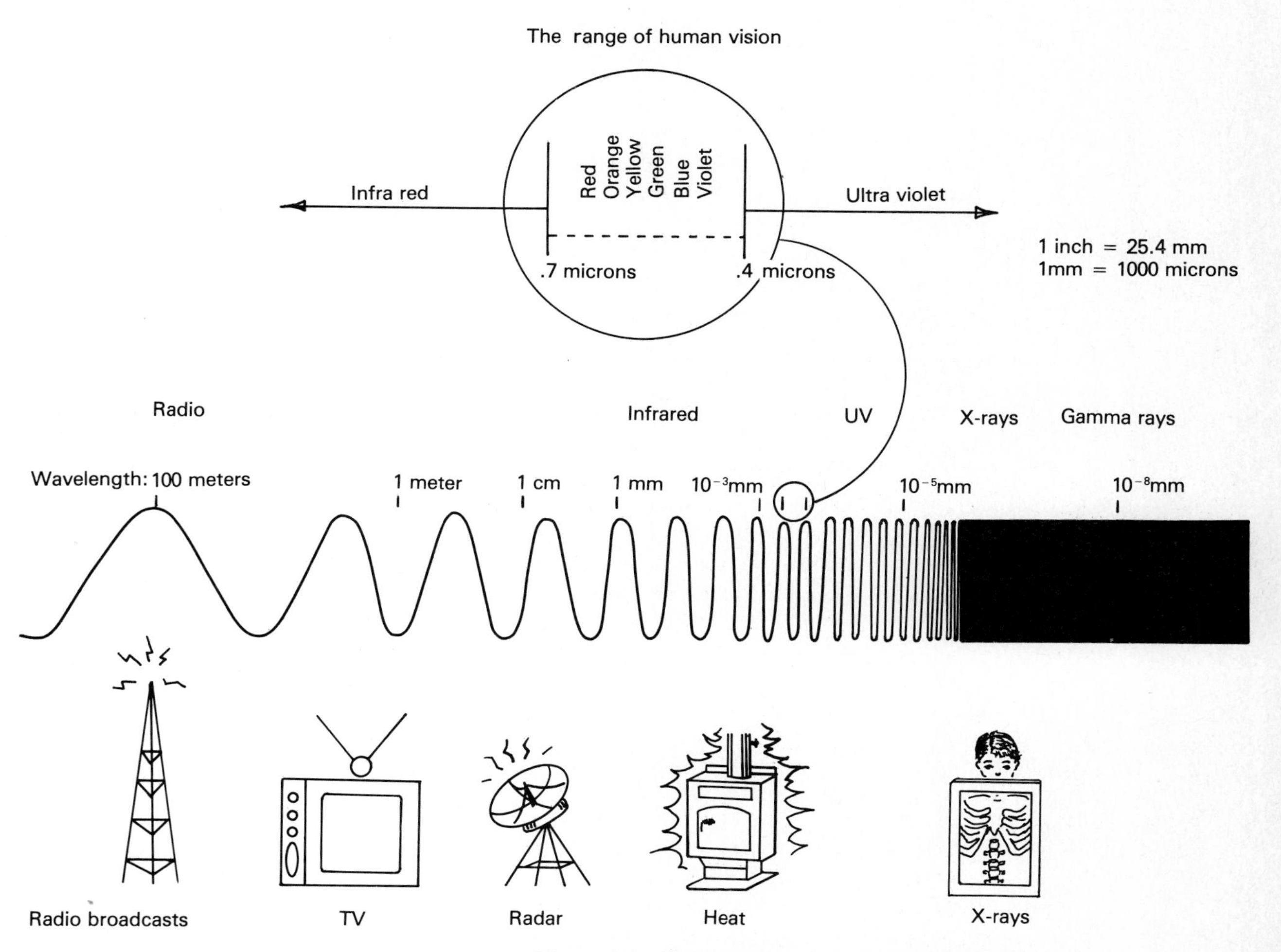

Figure 1-1 Wavelengths of the solar spectrum.

THE SUN'S INTERACTION WITH THE EARTH

The Filtering Atmosphere

The sun's energy must pass through the earth's atmosphere in order to reach its surface, and in so doing it is partially depleted. Short-wavelength ultraviolet waves—the most dangerous kind of UV radiation—are absorbed by ozone gas[4] in the stratosphere (part of the earth's upper atmosphere). As the solar beam penetrates deeper into the atmosphere, much of it is reflected from the upper surfaces of clouds. Some energy is also scattered by individual molecules and dust particles.[5]

Water vapor and carbon dioxide in the lower atmosphere (the "troposphere") are especially good absorbers of infrared radiation. Thus humid air transmits less total radiation than does dry air of the same temperature, even though both air masses might appear to be perfectly clear.

During periods of clear weather, the sunlight that reaches the earth's surface has roughly three-fourths of the amount of energy that it had at the top of the atmosphere. At heavily overcast times, the intensity can drop to less than 20 percent of its clear-sky value.

Why There Are Seasons

To understand why there are seasons, it is helpful to study Figure 1-2, depicting the earth in its journey around the sun. It is seen from the figure that during the whole of its orbit, the earth's axis always points in the same direction, toward the North Star Polaris. Thus the North Pole and the Northern Hemisphere are tilted toward the sun during half of the year (from March until September) and away the rest of the time. Imagine that solar energy is made up of many individual "sunbeams."[6] As illustrated in Figure 1-3, a greater number of sunbeams strike a square foot of Northern Hemisphere surface when it is tilted toward the sun than when it is tilted away. The *intensity* of the sunlight striking the earth is thus greater between March and September, so it is no surprise that the Northern Hemisphere experiences its summer during this period. In the Southern Hemisphere, summer arrives six months later, for the most direct sunlight is received between September and March.

Another consequence of the earth's tilt is that during the summer months in either hemisphere, sunlight strikes the atmosphere at an angle fairly close to perpendicular. In winter, the angle is much more oblique,

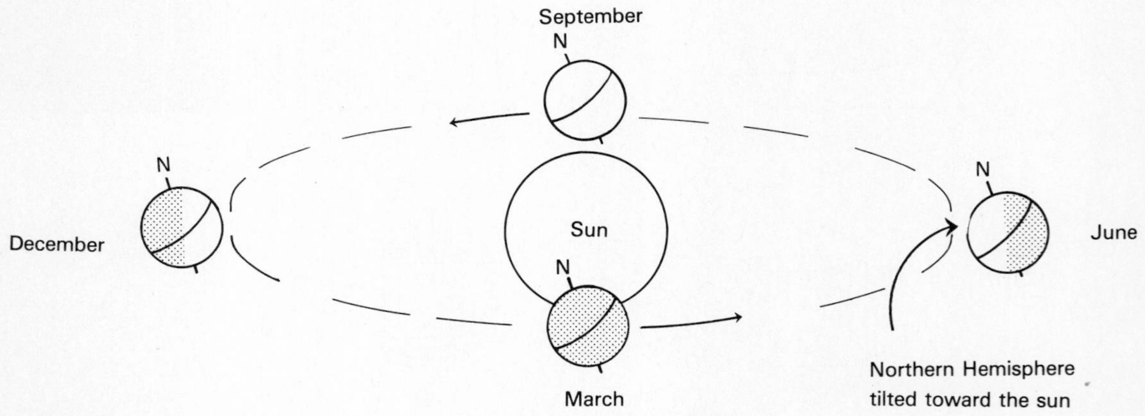

Figure 1-2 The earth's tilt.

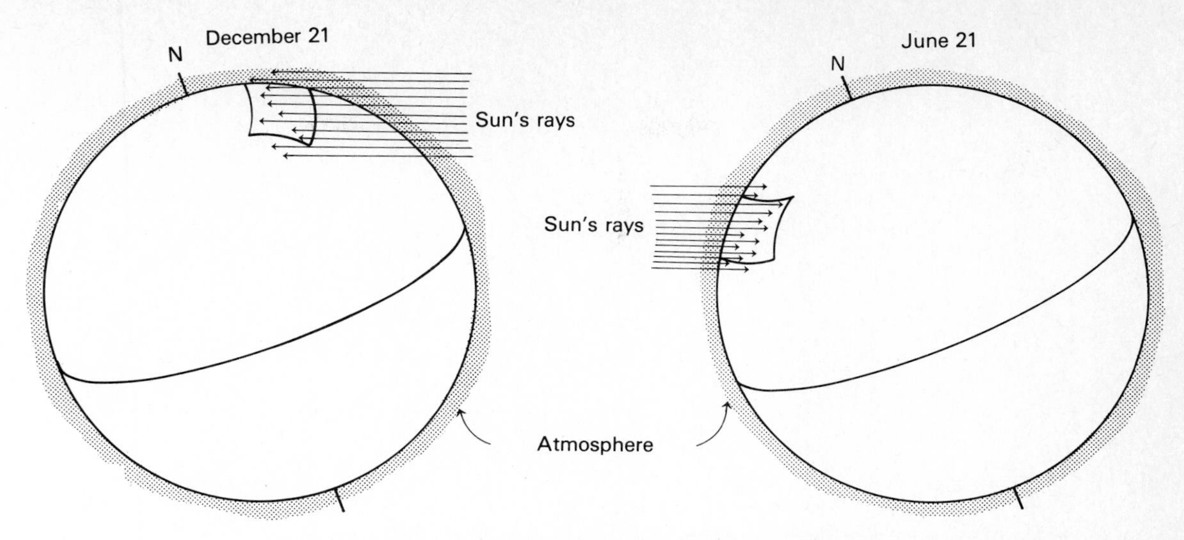

Figure 1-3 Intensity per square foot. More insolation strikes a square foot of Northern Hemisphere on June 21 than on December 21 (effects of cloud cover are not considered). Because of the earth's tilt, sunlight in December is spread out over a larger area in winter and is thus less intense. In addition, winter sunlight must travel a greater distance through the atmosphere to reach the earth's surface, and thus loses more of its energy through absorption and reflection than in summer.

causing sunlight to travel a greater distance through the atmosphere before it strikes the surface. More of its energy is thus filtered out, adding further to summer-winter temperature differences (see Figure 1-3).

Latitudinal Variations in the Sun's Position

The *latitude* of a certain location is its angular distance north or south of the equator. Figure 1-4 illustrates the concept of latitude. Latitudes in the continental United States and Canada range from 25° north (in Florida) to close to the North Pole (in the islands of northern Canada). The geographic North Pole's latitude is 90° north.

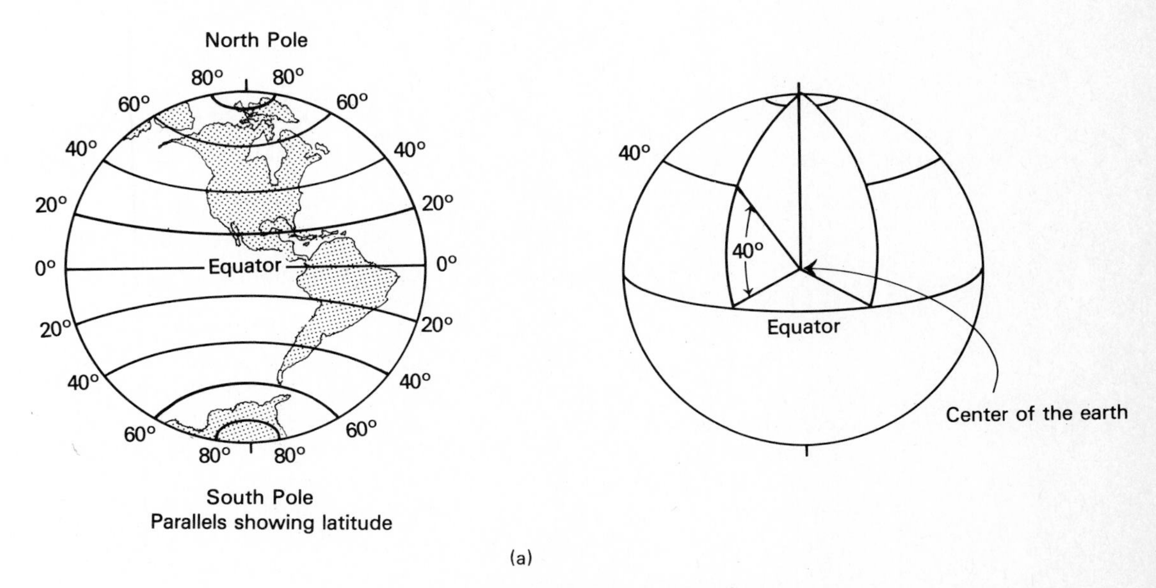

Figure 1-4 (a) Latitude lines.

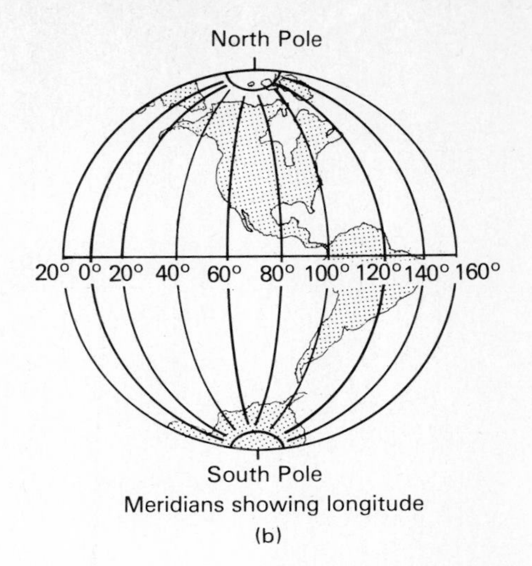

Meridians showing longitude

(b)

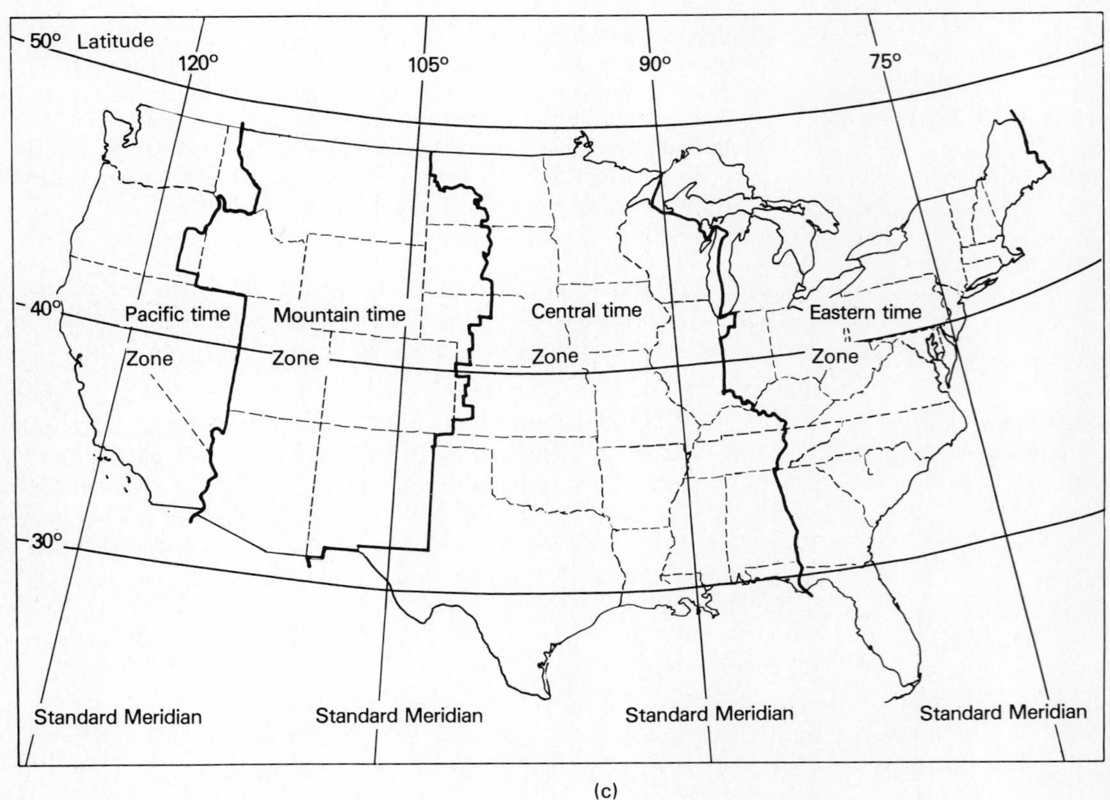

(c)

Figure 1-4 (cont.) (b) Longitude lines. (c) Time zones.

From the point of view of a person on the earth, the sun's daily path through the sky moves slowly northward between December 21 and June 21, reaching its extreme on June 21, the summer solstice. On this day the sun is positioned directly above a latitude line 23½° north of the equator—the Tropic of Cancer (see Figure 1-5). From June 21 through December 21, the sun's path moves south, until on December 21 it is above the Tropic of Capricorn, 23½° south of the equator. On March 21 and September 21, the days of the vernal and fall equinoxes, the sun is directly above the equator.

As the sun's path moves northward, its height in the sky, as viewed

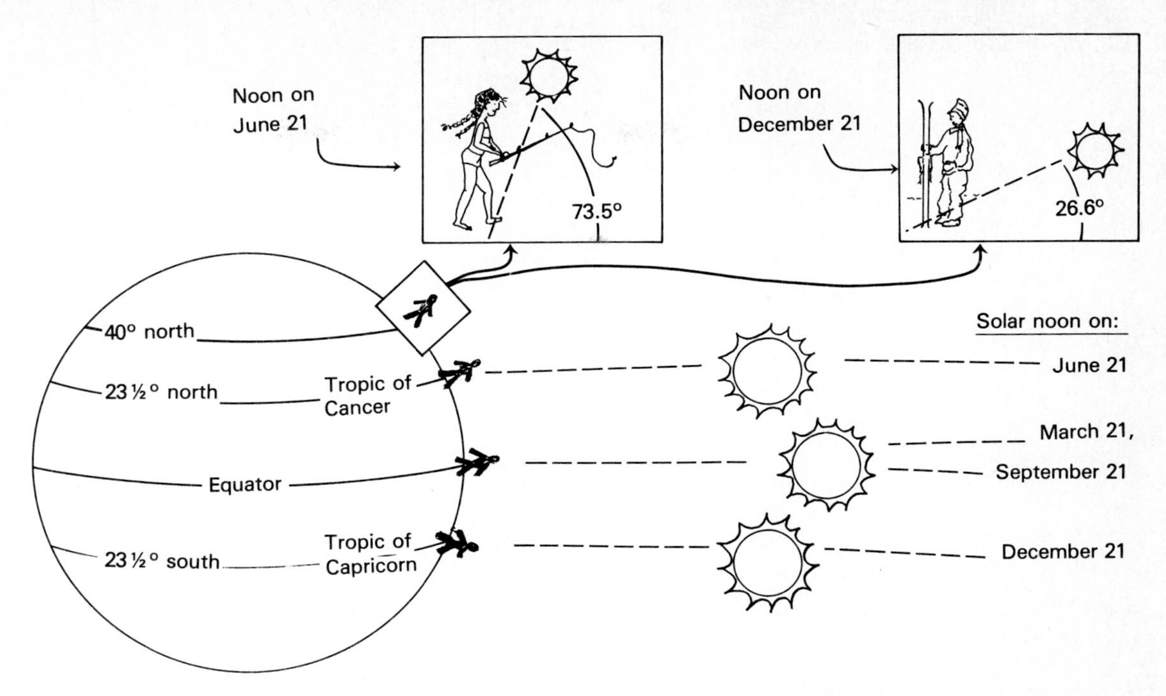

Figure 1-5 Position of sun at various times of the year.

from the United States or Canada, gets greater. At noon on December 21, for instance, the altitude of the sun at a latitude 40° north of the equator is 26.6° above the horizon, while on June 21 it is 73.5° (see Figure 1-5). The tilt angles of solar collecting surfaces should be designed to face the sun as directly as is practical during the time of year when its energy is most needed. This is one of the basic principles of solar energy system design, and will be examined in detail in later chapters.

In the United States and Canada, the sun's rays strike the earth most directly on June 21, the time of the solstice. It might thus be wondered why the hottest days of the year usually occur later in summer. The answer is that the earth's surface—especially its ocean surface—responds gradually to increases in solar intensity and thus so does the air in contact with the surface. In both the North Atlantic and North Pacific regions, for instance, air temperatures do not reach their annual maximum until August on the average.[7]

Solar Time versus Clock Time

For the purposes of solar energy calculations, the designer is interested in a system of time that accurately predicts the position of the sun in the sky at any location on the earth. This system is called *solar time*, and it differs somewhat from the time given us by our clocks. For in any particular time zone, all watches are adjusted to the same setting, whereas the position of the sun appears slightly different from different parts of the zone.

The sun's position in the sky varies not only according to the location from which it is viewed, but also at what time of year it is viewed. Because of the earth's changing velocity as it orbits in its elliptical path around the sun, the position of the sun at noon, standard clock time, is different depending on what date it is. Whereas it might be at its zenith (its highest point) at noon on September 1, it will not be at its zenith at noon the next month.

When it is stated later in this text that surfaces designed to collect solar energy should be oriented so as to receive as much sunlight as possible between 9 A.M. and 3 P.M., it is solar time that is meant. Other time references in this book are also concerned with solar time. To convert from solar time to clock time, it is necessary first to correct for the location of the viewer within the time zone, and then to correct for the time of the year.

Each time zone has a longitude line in it that is termed the *standard time meridian*. For the Eastern Time Zone it is 75°W. For Central, Mountain, and Pacific Time, it is 90°W, 105°W, and 120°W. The Yukon Zone's meridian is 135°W, and that for Alaska and Hawaii is 150°W. The map in Figure 1-4 depicts the various time zones and standard meridians, and may also be used to estimate the longitude of the viewer's location.

The longitude of the time zone's standard meridian is subtracted from that of the viewer, and the difference is multiplied by 4. (This is because the sun appears to move around the earth at the rate of 1° every 4 minutes.) The longitude correction is added to the solar time. The calculation will yield a positive correction if the viewer is west of the standard meridian, and a negative one if he or she is east of it.

As an example, if the viewer is in Crescent City, California (longitude 124°W), he or she would be in the Pacific Time Zone, whose standard meridian is the 120°W longitude line. The longitude correction is calculated as follows:

$$\begin{aligned} \text{longitude correction} &= 4(\text{longitude of viewer} - \text{longitude of meridian}) \\ &= 4(124 - 120) \\ &= +16\text{-minute correction} \end{aligned}$$

Adding this to a standard time of 10 A.M. yields a time of 10:16 A.M.

If the viewer is instead in La Mesa, California (longitude 117°W), the correction is

$$\begin{aligned} 4(117 - 120) &= 4(-3) \\ &= -12\text{-minute correction} \end{aligned}$$

Adding this to a standard time of 10 A.M. yields a time of 9:48 A.M.:

$$\begin{aligned} 10{:}00 \text{ A.M.} + (-12 \text{ minutes}) &= 10{:}00 - 12 \text{ minutes} \\ &= 9{:}48 \text{ A.M.} \end{aligned}$$

To make the time-of-year correction, Table 1-1, the *equation of time*, must be consulted. The table tells us, for instance, that on January 1, 3.6 minutes is added to the solar time in converting it to standard time. On October 21, 15.3 minutes is subtracted from the solar time.

The conversion method is summarized in the following equation:

$$\begin{aligned} \text{standard clock time} = {} & \text{solar time} \\ & + 4\,(\text{longitude of viewer} - \text{longitude of standard meridian}) \\ & + \text{equation of time} \end{aligned}$$

Example Calculation. A solar collection system in New Haven, Connecticut, is designed to receive full sunlight from 9 A.M. to 3 P.M., solar time. During which hours relative to clock time should the collectors receive sunlight? The date is November 1.

New Haven is at a longitude of about 73°W and is in the Eastern Time Zone, whose standard meridian is 75°W. The equation of time for November 1 is −16.3 minutes. Thus at 9 A.M. solar time on November 1, the standard clock time is calculated as follows:

TABLE 1-1 The Equation of Time

	Day of Month		
	1st	*11th*	*21st*
January	+ 3.6	+ 7.7	+ 11.1
February	+ 13.4	+ 14.3	+ 13.7
March	+ 12.7	+ 10.3	+ 7.5
April	+ 4.1	+ 1.1	− 1.3
May	− 2.9	− 3.9	− 3.7
June	− 2.6	− 0.8	+ 1.4
July	+ 3.6	+ 5.4	+ 6.3
August	+ 6.2	+ 5.4	+ 3.4
September	+ 0.4	− 3.0	− 6.4
October	− 10.0	− 12.9	− 15.2
November	− 16.3	− 15.9	− 14.3
December	− 11.1	− 7.1	− 2.2

$$\begin{aligned}\text{standard clock time} &= 9{:}00\ \text{A.M.} + 4(73^\circ - 75^\circ) - 16.3\\ &= 9{:}00\ \text{A.M.} - 8 - 16.3\\ &= 8{:}36\ \text{A.M.}\end{aligned}$$

3 P.M. solar time is converted similarly:

$$\begin{aligned}\text{standard clock time} &= 3{:}00\ \text{P.M.} - 8 - 16.3\\ &= 2{:}36\ \text{P.M.}\end{aligned}$$

THE SCIENCE OF HEAT

Heat, Thermal Energy, and Temperature

The design of solar energy systems is intimately connected with the process of storing heat and transmitting it to locations where it can be of use. *Heat* is the energy that flows into a body by contact with, or radiation from, a warmer body. The bodies can be either solid, gaseous, or liquid. Heat is stored in the form of *thermal energy*. What happens on an atomic level when a body stores thermal energy is that the random motions of the molecules and atoms become more vigorous. These motions include vibrations within the crystal structure of the material as well as random flows of microscopic particles (see Figure 1-6).

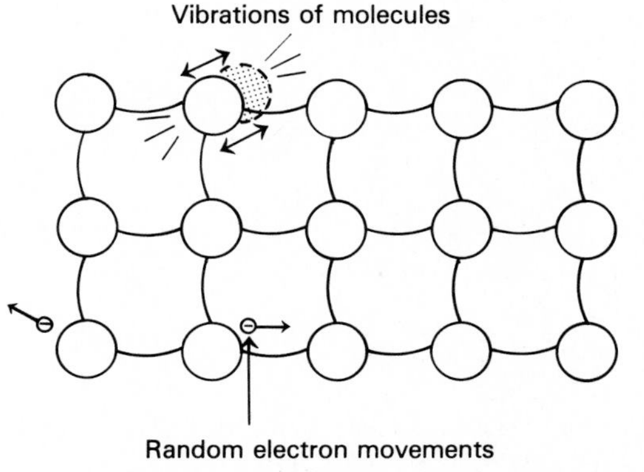

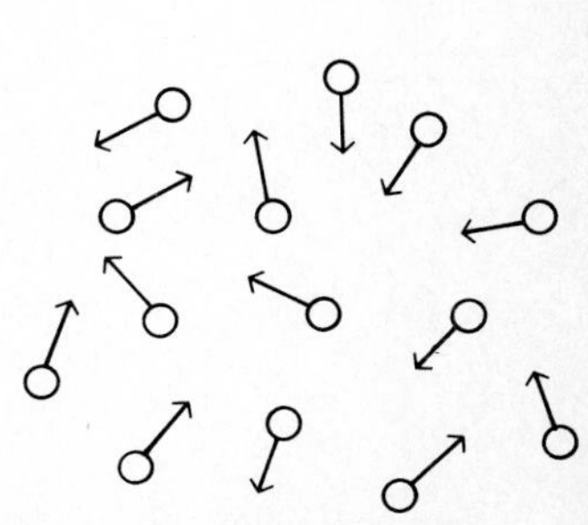

Figure 1-6 Thermal energy. (a) Solid. (b) Gas.

The *temperature* of a body is a measure of the average energy of motion ("kinetic" energy) of its atomic and subatomic components. The molecules in a 100°F glass of water, for instance, move faster on the average than those in 90°F water. Heat flows from a body of higher temperature into one with a lower temperature. When our hands touch a hot stove, energy travels from it into them, increasing the vibratory motions of the skin molecules to the point where it becomes painful. When a sun-warmed floor is in contact with cooler air, thermal energy flows into the air until temperatures are equalized.

The temperature scales most commonly used in everyday life are the Fahrenheit and Celsius scales, whose characteristics are summarized in Figure 1-7. A scale often used in physics or chemistry experiments is one that denotes temperature in degrees Kelvin. The lowest point on this scale is *absolute zero*, at which molecular motion inside the body almost completely ceases.[8] At this point, the body could be said to have zero thermal energy. Absolute zero in degrees Kelvin is equal on the Celsius scale to −273° and on the Fahrenheit scale to −460°.

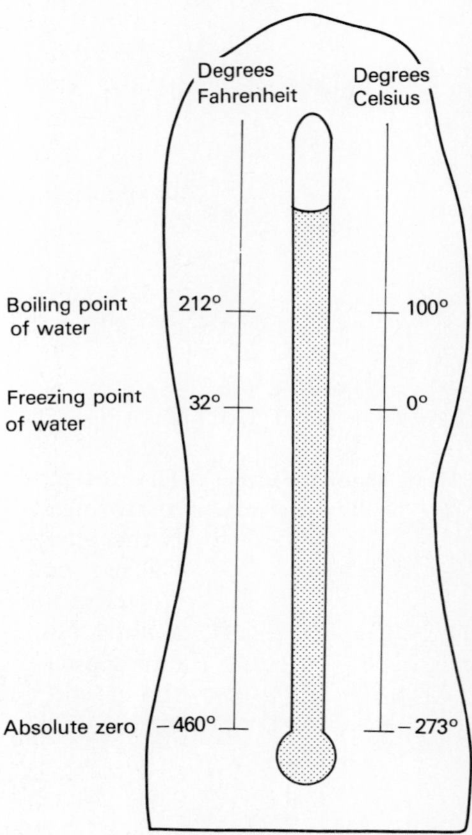

Figure 1-7 Fahrenheit and Celsius scales. To convert from Fahrenheit to Celsius, subtract 32° and divide by 1.8. To convert from Celsius to Fahrenheit, multiply by 1.8 and add 32°.

Thermal Energy Transport

It is important in solar energy work to know how thermal energy travels from one place to another, so that its flow can be directed into areas where it will be utilized and stored. As an example of thermal energy transport, let us follow the path of sunlight as it enters a building and is incident on the interior surfaces.

Transmission, Reflection, and Absorption. Like the earth's atmosphere, window glass transmits only a part of the solar spectrum—its visible portion as well as a small amount of ultraviolet and infrared.[9] The energy that does get transmitted impinges upon interior walls, floors, and other surfaces. When it strikes one of these surfaces, some of it is absorbed, while the remainder is reflected away to be absorbed eventually by other surfaces in the room or lost out the window.

The percentage of sunlight that is absorbed depends largely on the surface's color. Flat black surfaces absorb 95 percent of normally (perpendicularly) incident sunlight, while white surfaces only 20 to 30 percent.[10] The more energy absorbed, the less is reflected back to the eye and the darker the surface appears.

As solar radiation is absorbed, it is converted into thermal energy and causes a temperature rise at the surface of the material. This thermal energy is manifested in the form of vibrations in the molecules on the surface.

Conduction. Thermal energy is transmitted through the interior of *solid* substances largely by the process of *conduction.* To understand this mechanism, imagine a table full of billiard balls arranged in a regular pattern and connected to each other by springs (see Figure 1-8). The balls represent molecules in the crystal structure of a solid, and the springs represent the bonds that hold the molecules together. When a molecule on the edge of the array is jiggled, it passes its energy onto neighboring particles until all are in motion. It is in this manner that thermal energy is conducted throughout the body.

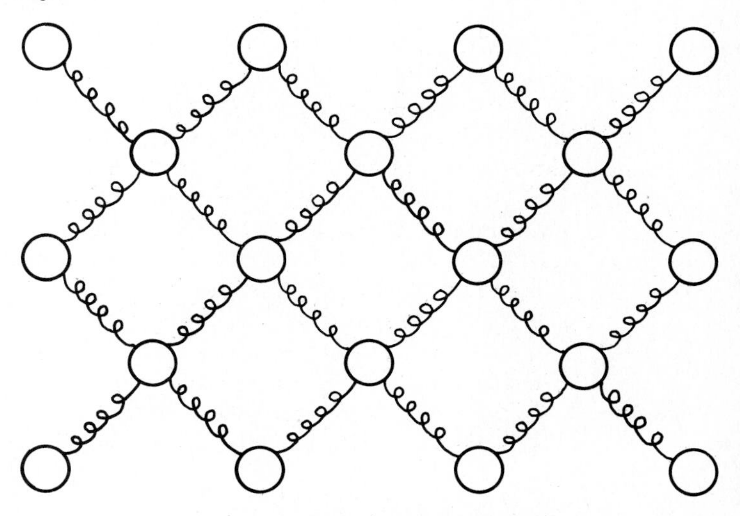

Figure 1-8 Model of a crystal structure.

The speed with which a body conducts heat depends on the nature of its molecular and atomic bonding. Metallic substances in which there are free electrons that can leap from atom to atom conduct thermal energy better than most other materials, and it is for this reason that they are used for the absorbing plates in many solar collectors. Concrete and brick are moderate conductors, wood is fairly poor, and materials such as Styrofoam or glass wool (fiberglass) are very poor and for that reason are termed *insulations.* When designing a solar system, insulation is used to contain thermal energy within a certain area, while better conductors are employed to transmit the heat (see Figure 1-9).

As the temperature of sun-warmed surfaces in a room increases, thermal energy is not only conducted deeper into the material below the surface, but is also conducted into the air molecules that come in contact with the surface. Air is a very poor conductor of heat, as are most fluids (liquids or gases). This is because the bonding between their molecules is weak, and the molecules themselves are often much farther apart than in solids. But fluids have a way of transmitting energy that solids do not have—the process of *convection.*

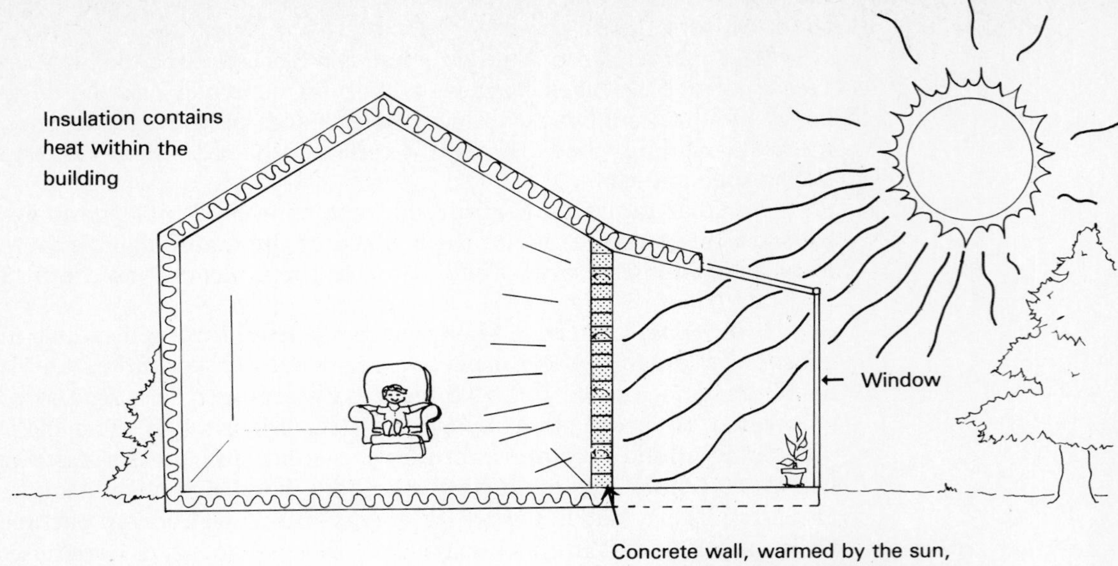

Figure 1-9 Application of conductors and insulators.

Convection. To understand this process, let us examine what happens to the air heated by contact with a warm floor. As its temperature rises, the random motions of the air molecules become more vigorous. They tend to push outward against surrounding air molecules harder than do those molecules that are pushing on them. In more formal terms, the pressure in the heated air increases, causing it to expand. As it expands, the number of molecules contained in each cubic foot decreases, and thus so does the *density*, the total mass per unit volume.

Air near the ceiling has not yet been heated and thus is of a higher density than that at floor level. This is an unstable situation, for the heavier ceiling air will sink downward, pushing low-density warm air up and out of the way (see Figure 1-10a). The heavier air will then become heated and a

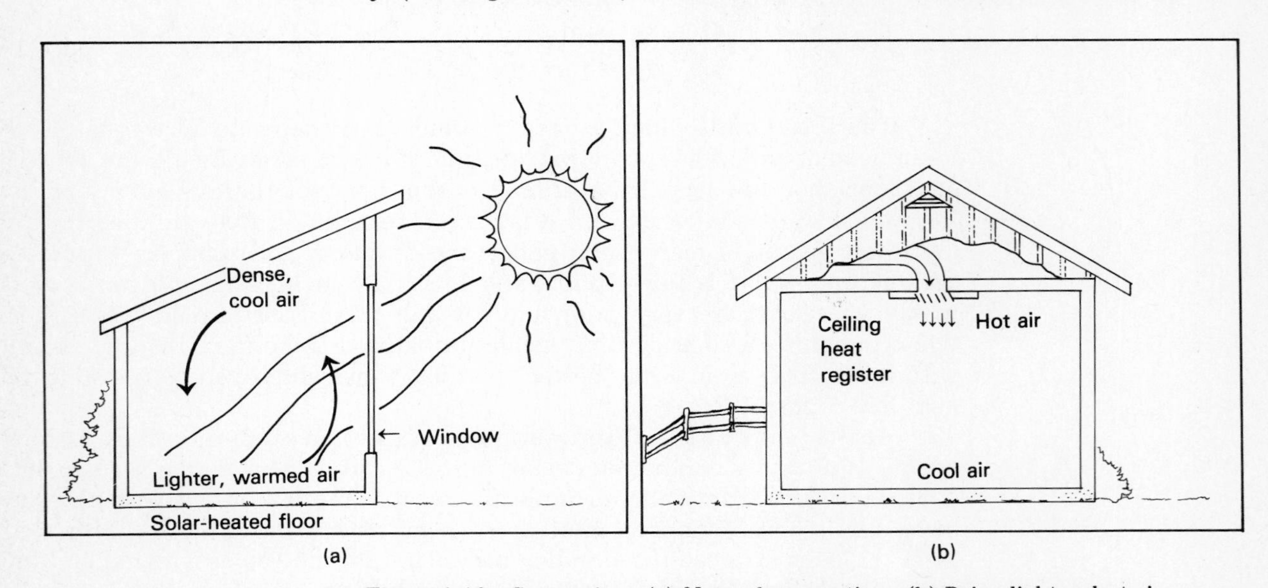

Figure 1-10 Convection. (a) Natural convection. (b) Being lighter, hot air from the ceiling heat register will remain near the ceiling unless driven down by blowers (an example of forced convection).

cyclical flow of rising and falling air will be initiated. Through this cyclical flow, thermal energy is transmitted throughout the room. Fluid flows generated by differences in temperature between one location and another are called *natural convection* currents.

Convection is a powerful method of heat transmission and will frequently carry energy faster than conduction is able to in solids. Convection currents are encountered often in daily life. Those in the atmosphere result in winds. Those in a heated pot of water help circulate the warmth. It is also the process of convection that brings high-temperature water in a water heater to the top of the tank and cooler temperature liquid to the bottom.

Differential heating of a fluid does not always result in natural convection. For instance, if the ceiling rather than the floor of a room is heated, a stable layer of warm air will be created above a cooler layer (see Figure 1-10b). To drive the warm air down toward the lower part of the space where it can be utilized, fans might have to be employed. Thermal energy transfer in a fluid resulting from flows powered by external mechanisms such as fans is termed *forced convection.*

Radiation and Reradiation. Radiative energy transmission from the sun to the earth was discussed earlier in this chapter. It is not only the sun that emits electromagnetic waves, however—*all* objects give off some sort of radiation. The reason for this is related to oscillations of the molecules and subatomic particles within the objects. To understand the process better, imagine an analogous case involving water waves: A buoy in a lake is moved up and down, creating waves that propagate out from it (see Figure 1-11). If the frequency of the buoy's oscillations is increased, more waves propagate out from it each second, although the length of each wave is decreased. Because it takes more energy to oscillate the buoy at a rapid frequency than at a slower one, the amount of energy transmitted outward also increases as the oscillation frequency does.

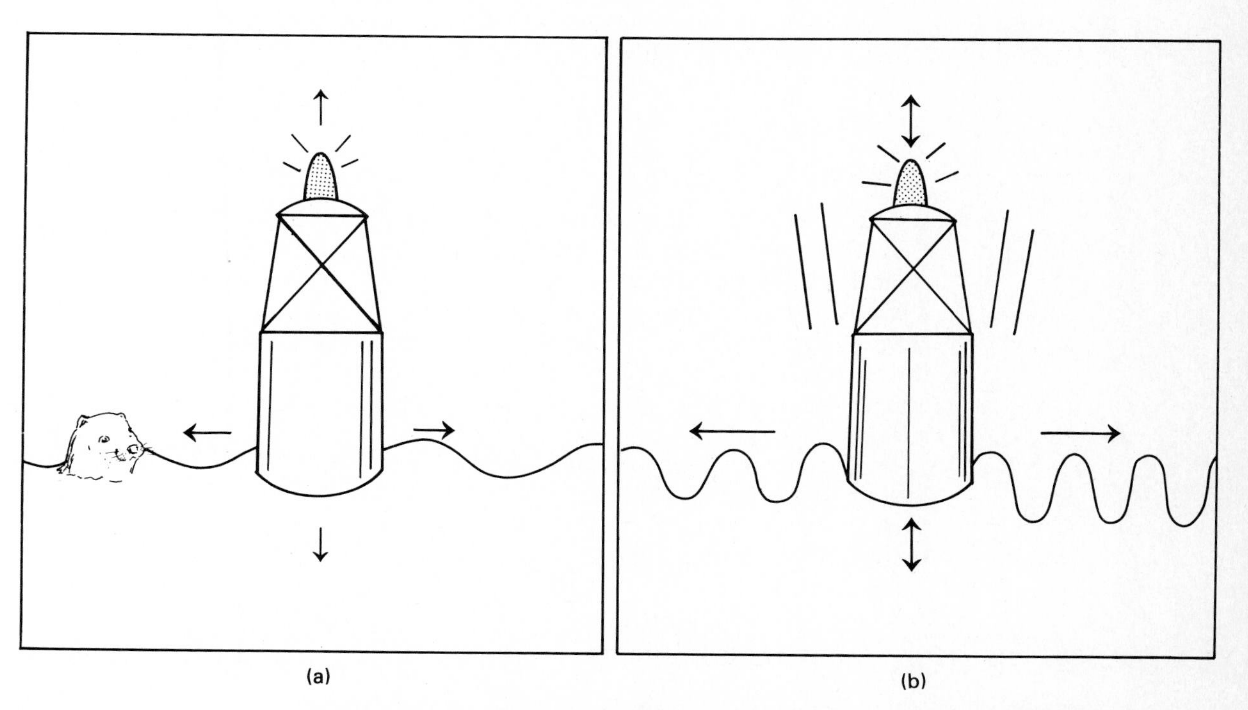

Figure 1-11 Oscillation frequency and wavelength. (a) Slow oscillation. (b) Fast oscillation. More waves are produced per second, with shorter wavelengths. Energy flow outward from buoy increases in intensity.

Now substitute subatomic charged particles for the buoy and electromagnetic waves for water waves. As the particles vibrate, electromagnetic waves are given off. The more rapid the oscillations, the shorter the wavelength and the higher the energy of the emitted radiation.

Since the temperature of a material is a measure of the energy of motion of its subatomic particles, it is also related to the frequencies of their oscillations. The higher the temperature, the more rapid the oscillations, and thus the emitted radiation is more energetic and of shorter wavelength. Room temperature (70°F) bodies emit radiation with wavelengths in the infrared region. A 500°F (260°C) wood stove gives off shorter, higher-energy infrared waves that when absorbed by our skin are experienced as heat. If the wood stove temperature is raised to 932°F (500°C), it produces waves of such short length that they are in the range visible to the eye. The stove, in other words, glows "red hot" (see Figure 1-12).

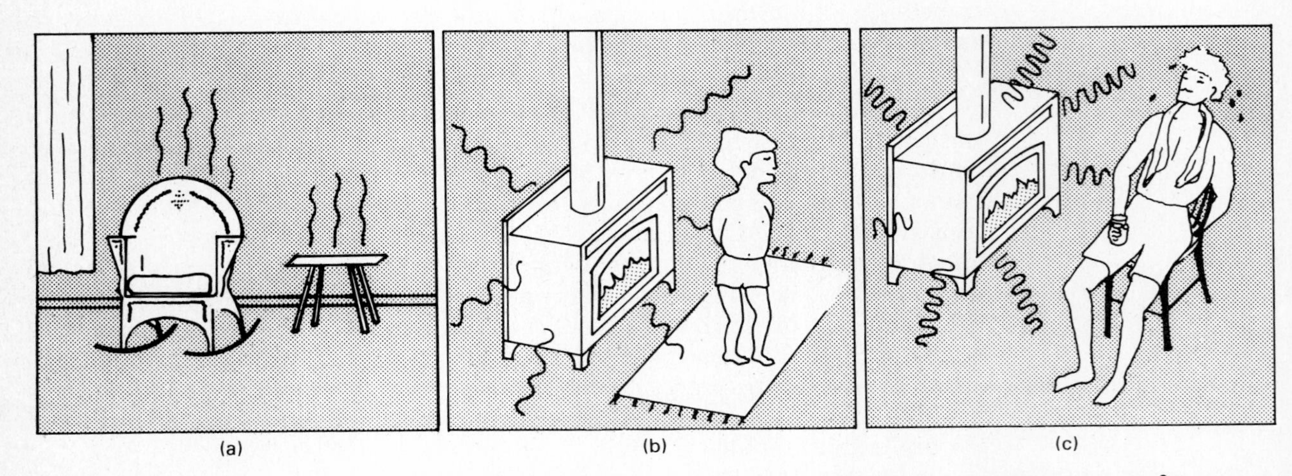

Figure 1-12 Temperature and wavelength. (a) Room-temperature (70°F) objects emit waves in the infrared range. (b) A 500°F wood stove gives off shorter-wavelength infrared that is experienced as heat. (c) A 932°F stove gives off wavelengths so short, they are in the visible range. The stove is "red hot."

It might be supposed, since all objects emit electromagnetic radiation, that they constantly lose energy, causing their temperatures to decrease steadily. The reason they usually do not is because the objects are continually *absorbing* energy emitted from other bodies around them and an equilibrium situation is set up. After solar energy is absorbed by a room's surfaces, it too is eventually released either through reradiation or by conduction or convection. It is then picked up by other surfaces, only to be released from them after a time.

The open space in a room is usually thought of as containing nothing but air. In reality, it is permeated by radiation of many different frequencies, traveling in all directions. The following quote expresses a view of the universe that reflects this fact:

> Our first impression of the universe is one of matter and void—when actually the universe is a dense sea of radiation in which occasional concentrates are suspended.[11]

According to this view, our earth, the other planets, and our sun are nothing more than "occasional concentrates" floating in an enormous radiation-filled ocean!

The Greenhouse Effect

The visible portion of the solar spectrum is able to pass freely through glass as well as through many plastics. After entering a building in this way and being absorbed by interior surfaces, radiation is reemitted in infrared wavelengths by the surfaces. Glass, and to some extent transparent plastic, is *opaque* to most of these wavelengths, however, so the energy is trapped within the structure. This causes interior temperatures to rise until as much energy is escaping through cracks and by conduction and convection in the skin of the building as is being absorbed by the building's interior surfaces. It is by this process that florists' greenhouses are able to attain temperatures far above those outside. It is also by using this process that passively solar-heated buildings are able to stay warm (see Figure 1-13).

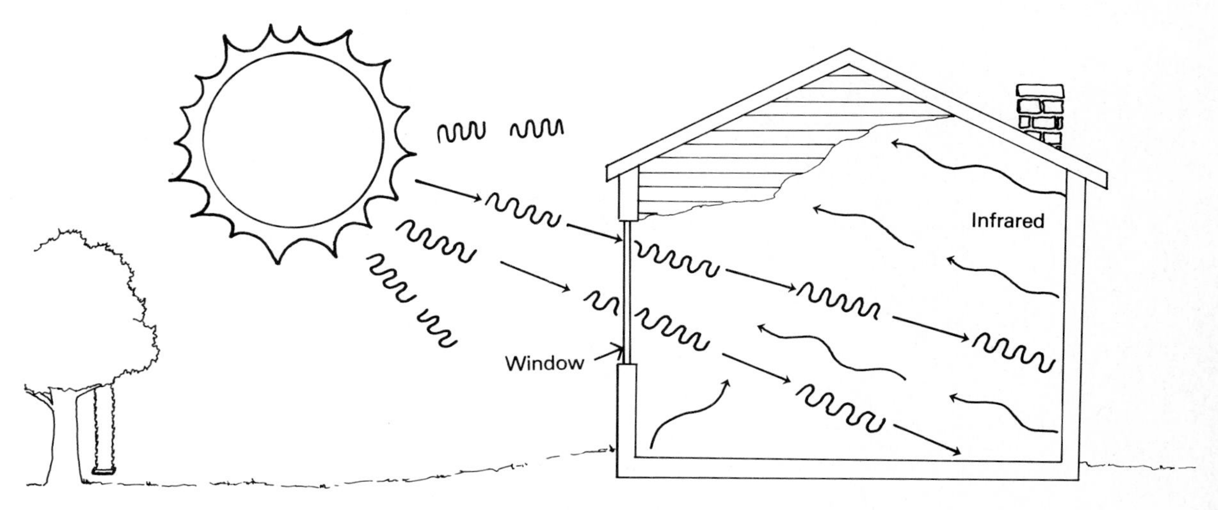

Figure 1-13 The greenhouse effect. Sunlight enters through the window and warms the room's surfaces. They emit long-wavelength radiation that cannot escape through the glass.

NOTES

1. The sun produces 3.7×10^{23} Btu of energy every second through the process of nuclear fusion. The earth's people use about 1.83×10^{17} Btu per year in their commercial activities. 1.83×10^{17} Btu per year $\times$ 2 million years $\simeq 3.7 \times 10^{23}$ Btu.

2. George O. Abell, *Drama of the Universe*, Holt, Rinehart and Winston, New York, 1978, p. 106; John A. Day and Gilbert L. Sternes, *Climate and Weather*, Addison-Wesley Publishing Company, Inc., Reading, MA, 1970, p. 142.

3. Day and Sternes, p. 144.

4. Ozone gas, made up of molecules composed of three oxygen atoms (normally, oxygen combines in two-atom molecules) is an extremely good absorber of the high-energy ultraviolet rays that can cause cancer and eye damage. It is possible that humankind's activities on the earth can destroy the ozone layer. If this were to happen, catastrophic changes for life on earth would probably occur.

5. It is because air molecules selectively scatter the shorter violet and blue wavelengths of the solar spectrum throughout the sky better than they scatter the longer red waves that the sky appears blue.

6. This is not such a bad model, as electromagnetic radiation can for many purposes be thought of as composed of millions of discrete "photons," or parcels of energy.

7. William L. Donn, *Meteorology*, 4th ed., McGraw-Hill Book Company, New York, 1975, pp. 41–420.

8. The word "almost" is used because there are some nonthermal motions that do not cease.

9. The specifics of what happens to nontransmitted light are examined in Chapter 4.

10. J. Douglas Balcomb et al., *Passive Solar Design Handbook*, Vol. 2, Los Alamos National Laboratory, Los Alamos, NM, for U.S. Department of Energy, 1980, p. 75.

11. Paul G. Hewitt, *Conceptual Physics—A New Introduction to Your Environment*, Little, Brown and Company, Boston, 1977, p. 384.

SUGGESTED READING

ABELL, GEORGE O., *Realm of the Universe.* New York: Holt, Rinehart and Winston, 1976. How the sun produces its energy, and more information on seasonal variations in insolation.

HEWITT, PAUL G., *Conceptual Physics—A New Introduction to Your Environment.* Boston: Little, Brown and Company, 1977. Excellent, lucid explanations of thermodynamic principles.

REVIEW QUESTIONS

1. How is the sun's energy produced?
2. What is the difference between infrared radiation and visible light? Between X-rays and visible light?
3. Do you think that there is more or less ultraviolet radiation intensity striking the moon's surface than that of the earth? Explain your answer.
4. What is the Tropic of Capricorn? What does it have to do with the sun?
5. What is the difference between heat, thermal energy, and temperature?
6. Define *conduction.* Can it occur in gases and liquids as well as in solids?
7. Define *convection.* What is the difference between natural and forced convection?
8. How is the radiation emitted by an object related to its temperature?
9. What is the greenhouse effect? Why is it of concern to solar energy system designers?

PROBLEMS

1. What effect could destruction of the atmosphere's ozone layer have on the inhabitants of earth?
2. If the earth's tilt were increased so that the North Pole tilted more toward the sun in June and more away from the sun in December, how would the severity of the seasons be affected? It might be helpful to

draw a picture illustrating what parts of the planet would be in shadow during June and December and what parts would receive sunlight.

3. A solar collection system in Kansas City, Missouri, is designed to receive full sunlight from 9 A.M. to 3 P.M., solar time. During which hours relative to clock time should the collectors receive sunlight? The date is December 1.

4. Repeat the calculation for a system in San Diego, California, on January 1.

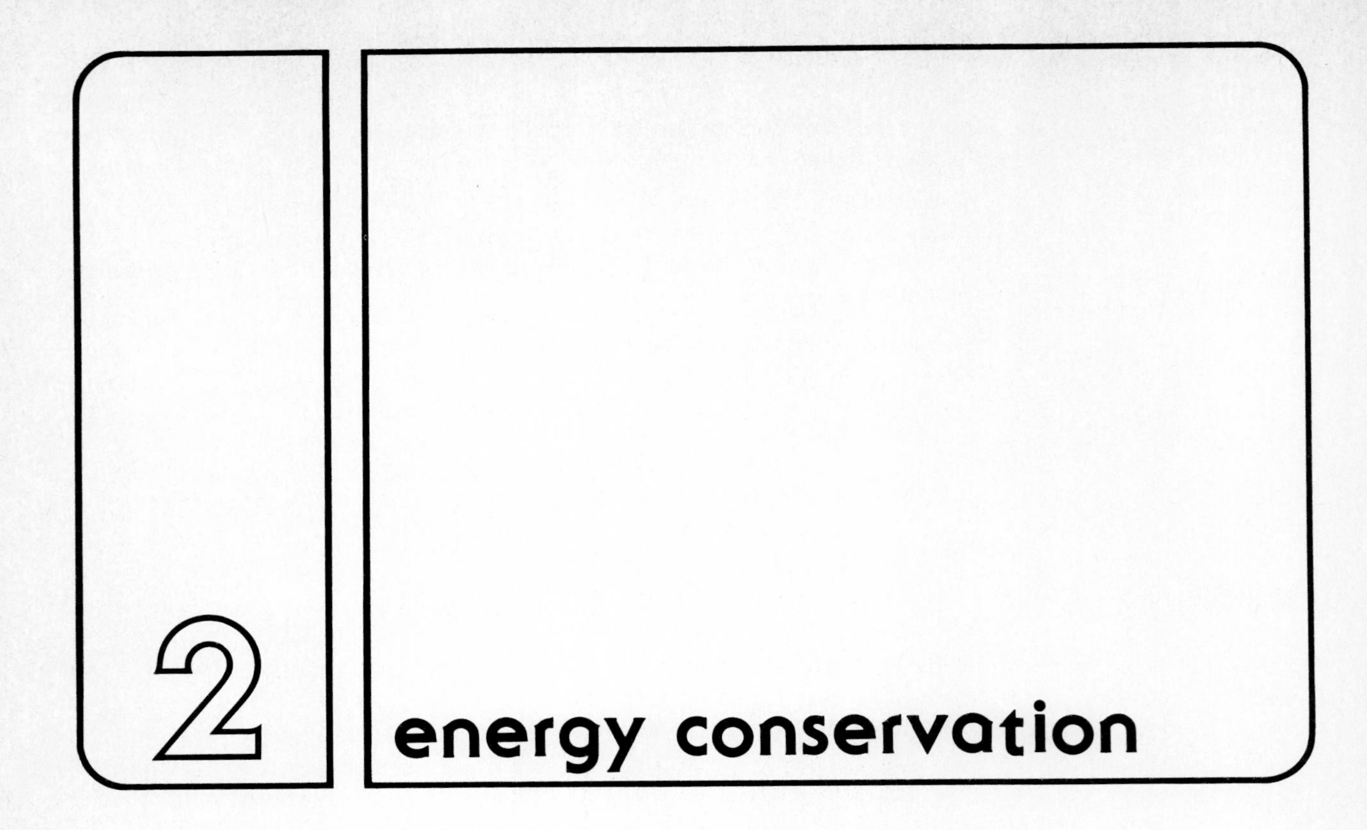

2 energy conservation

The foundation of an effective solar system is a well-constructed, energy-efficient building. A drafty, poorly insulated structure requires far too large and expensive a solar energy system to keep the temperature at a comfortable level. A sound method of system design is first to maximize the building's energy conservation potential, and only after this has been done go on to the planning of the solar collection components. Four critical areas of energy conservation that must be examined are (1) the thermostat setting, (2) infiltration control, (3) insulation methods, and (4) energy-saving appliances.

THERMOSTAT SETTING

It is common practice throughout the United States and Canada to set a building's heating system thermostat to a certain temperature, usually between 68 and 72°F, and leave it at that setting 24 hours per day. This is a custom that remains with us from the days of cheap fuel, but it is an expensive practice to maintain in today's world. A dramatic savings in heating bills results if the thermostat is set no higher than 68°F during the day and *lowered* each night half an hour before going to bed to 60°F. In a study of houses in Boston, Denver, and Nashville[1] it was found that lowering the temperature setting for 9 hours every night reduced the annual heating requirements by 15 to 18 percent, which is quite a significant amount for so simple a conservation measure.

The next two sections deal with two methods of reducing thermal energy losses from a building: by preventing infiltration of cold air from the outside environment and by adequately insulating all components of the *building envelope*, the outer skin of the building that encloses all its conditioned spaces.

INFILTRATION CONTROL

Cold air infiltration in the average home accounts for, according to many estimates, 50 percent or more of the dollars spent on heating.[2] Air leaks are of two types: those around operable parts of the building envelope such as windows and doors, and those in stationary or nearly stationary places such as seams, cracks, and building envelope penetrations by vents and plumbing. The first types of leaks are controlled using various forms of *weather stripping*; the second by *caulking*. In new construction, infiltration control measures should be performed whenever possible *before* insulation is applied to the building envelope, for it is easier to find leaks before they are hidden by fiberglass rolls or polyurethane batts.

Weather Stripping

Although there are over two dozen types of weather stripping on the market today,[3] only a few will perform well for a given situation. The wrong type will either do little to stop infiltration, will wear out quickly, or might even worsen drafts by impairing operation of a door or window.

In a survey of 13 American and Canadian experts in the field of weatherization and energy conservation,[4] it was found that only about half the weather-stripping categories on the market were even recommended for use. The categories recommended fall under the general headings of *compression gaskets, tension strips*, and *specialty designs.*

Compression Gaskets. Compression gaskets are materials designed to be squeezed between two surfaces, such as a door and its jamb. The simplest types are strips made of felt, foam, or sponge rubber. Sponge rubber is the most durable of these materials. Although many types of weather-stripping are sold with an adhesive backing, longer service lives will result if, in addition to gluing, they are nailed, stapled, or screwed in place.

Gasket and flange compression gaskets are composed of a long, flat strip (a flange) to which is attached a thicker strip, sometimes tear-shaped in cross section (the gasket). This and other weather-stripping designs are illustrated in Figure 2-1. The flange is secured to the surface, while the gasket strip provides the weather-seal. The assembly is generally made of rubber or vinyl, although the flange is sometimes reinforced with a metal or plastic clamp (see Figure 2-1a).

Brushlike sets of bristles attached to a metal backing, such as depicted in Figure 2-1b, make up another popular type of weather-stripping design. Sliding glass doors and windows are often sold with bristle weather stripping preinstalled.

Tension Strips. Tension strips provide a weather-seal by virtue of the springiness of the metal or plastic from which they are constructed. They have V- or L-shaped cross sections that are squeezed shut when the door or window to which they are attached is closed (see Figure 2-1c). They are especially suitable for the sides of doors, but will not seal well against surfaces that are warped or not flat.

Specialty Designs. Specialty designs are those with a particular use in mind. Among the most popular products in this category are *door sweeps,* typically constructed of a flexible flap of vinyl attached to a rigid metal,

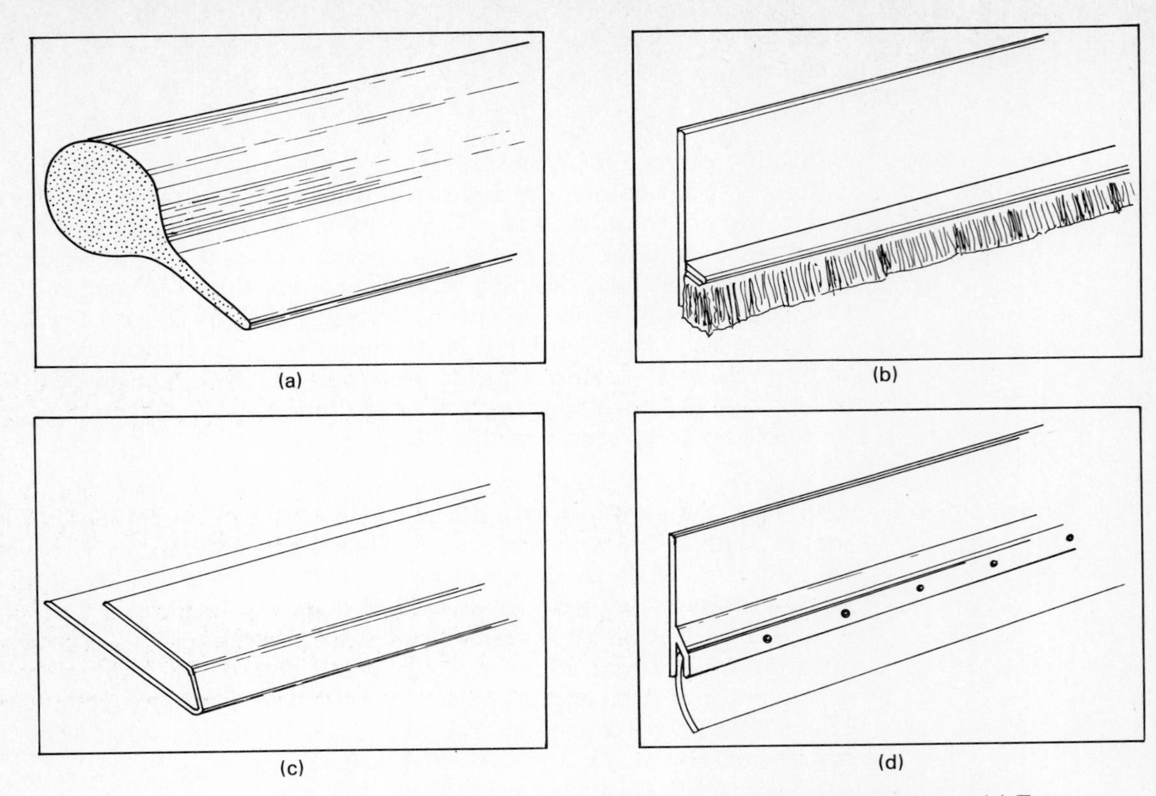

Figure 2-1 Weather stripping. (a) Gasket and flange. (b) Bristles. (c) Tension strips. (d) Door sweep.

wood, or plastic flange screwed or nailed onto the bottom of a door. Door sweeps are generally usable on most doors (even garage doors), although they sometimes bind when swept over thick carpets.

Choosing Weather Stripping. The sealing device that four out of five of the weatherization experts interviewed recommended for the sides of standard doors were tension strips. Most recommended the more durable metal ones, although a few suggested plastic varieties. Door sweeps were strong favorites for swinging door bottoms. Bristle compression gaskets were advocated for sliding doors.

Tension strips were also the favorites for metal casement windows, although they tied with closed cell (vinyl) foam as the best choice for wood casements. Foam conforms better to slight warping and irregularities in the window assembly. Vinyl foam was by far the favorite on nonoperable windows, for in this application there is no movement to cause rubbing and abrasion that can quickly wear out a foam strip.

Jalousie windows (depicted in Figure 2-2) were considered next to impossible to weather-strip adequately, no matter what product is used. For a tight house, it would be best to replace jalousies with other windows. Pet doors and mail slots are also difficult to seal. Commercially sold pet doors with magnetic sealing strips tend to be superior to the homemade variety. Mail slots are perhaps better caulked shut and replaced by external mailboxes.

Caulking The purposes of caulking gaps in the building envelope are to improve the energy conservation and comfort levels of the house by blocking air infiltration as well as to prevent moisture leaks that can damage framing, insula-

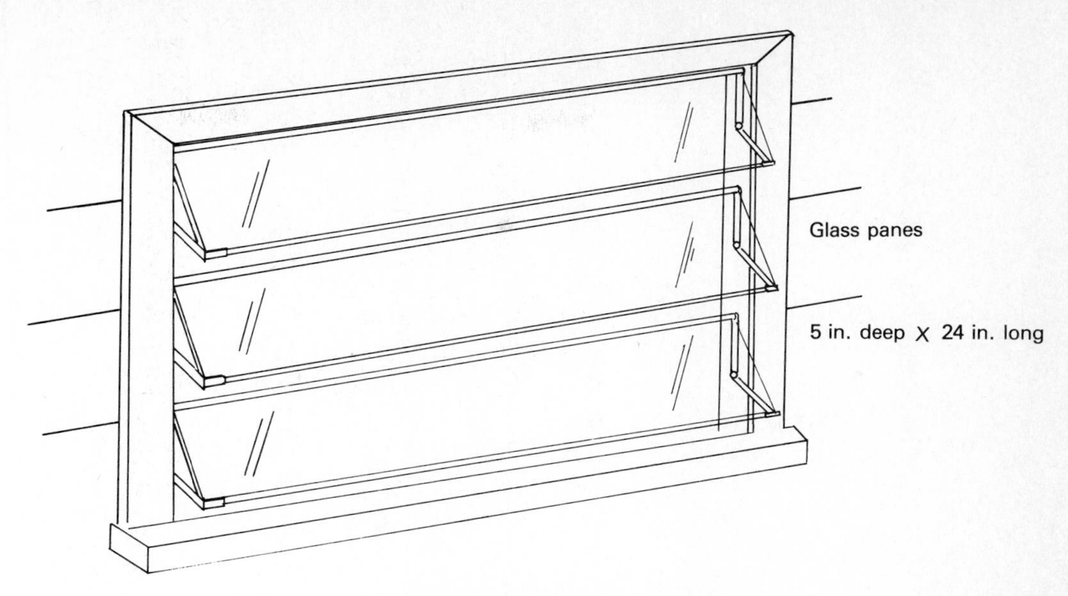

Figure 2-2 Jalousie windows.

tion, and paneling. Spots that typically need to be filled with caulk are: seams around doors and windows, building envelope penetrations for pipes and electrical conduits, seams between foundation and sill plates, and the junction area between the walls and roof.

Caulking compounds range in price (and quality) from $1.29-per-tube oil-based variety to $7-per-tube silicones.[5] The more extreme the conditions the caulk will be subjected to, the higher in quality it must be.

The amount of money spent on caulking a building under construction is a tiny fraction of the total cost of the structure, even when the more expensive caulks are used. Yet improper caulking methods or the installation of cheap, short-lived products can lead to huge infiltrative heat losses and thousands of dollars in extra heating expenses over a 10- or 20-year period. It can also lead to unwanted labor and material expenses whenever recaulking is necessary. It simply does not make sense to try to save money by using poor-quality caulks.

In determining the type of caulk suited to a particular job, there are two very important considerations:

1. Is the seam into which the caulk will be installed completely static, or is there some movement in the building materials around it? If the seam is between two different types of materials, such as metal and wood, there will almost certainly be relative movement due to their different expansion rates.
2. Is the seam on the interior or the exterior of the building?

Caulking for Completely Static Locations. If the seam is totally static and on the interior of the building, *acrylic latex* caulks will provide long life (10 or more years) at a modest price. They are easy to apply, although metal surfaces often need to be primed or painted first. Unlike oil-based caulks, they can be cleaned up with water. Because their flexible surfaces can be punctured, they should be installed in relatively protected locations. Brand names of acrylic latexes as well as other sealants are listed in Table 2-1.

TABLE 2-1 Characteristics of Various Caulks[a]

Type	*Brand Names*	*Approximate 1984 Prices (cents) per Linear Foot*	*Expected Lifetime (years) If Installed According to Instructions and Recommendations*	*Comments*
		CAULKING FOR STATIC SEAMS		
		Interior Locations		
Acrylic latex	DAP Acrylic Latex with Silicone Sears High Performance Acrylic Latex Gibson-Homans Eternaflex 1045 Macklanburg-Duncan Acrylic Latex	8-10	3-10	Dependable, relatively inexpensive interior caulk. Primer recommended on metal surfaces.
Latex	Weldwood 3 Year Latex Macklanburg-Duncan All-Purpose Adhesive and Caulk H.B. Fuller Latex	7-9	3-5	These caulks fall between acrylic latex and oil-based sealants in both performance and price range. Primer recommended on metal surfaces.
Oil-based	Gibson-Homans Draftite DAP Architectural Grade Macklanburg-Duncan Speed Load	4-7	1-5	Does not usually perform as well as the other caulks listed. Adhesion diminishes once the oil base has dried up. Tendency to bleed and stain surrounding surfaces.
		Exterior Locations		
Butyl rubber	H.B. Fuller Butyl Rubber Sealant DAP Butyl-Flex Gibson-Homan Butyloid Macklanburg-Duncan Butyl Rubber Sealant	11-14	4-10	Painting of caulk is recommended in exposed, exterior locations.

Urethane foam	Geocel Expanding-Foam Sealant Fomofill Urethane Foam Sealant Macklanburg-Duncan Insulating Foam Sealant	1-2	10-20	Low price per linear foot. Can fill large cavities and cracks that other materials would have difficulty sealing. Provides insulation (R-5 to R-6 per inch) as well as a weather seal. Does not adhere to polyethylene vapor barriers. Must be covered with trim or some other protection in exterior locations.
		CAULKING FOR NONSTATIC SEAMS		
Polyurethane	Sears Polyurethane Base Kenseal Vulkem Polyurethane Sealant Pecora Pool-and-Patio Sealer	16-22	20-30	Excellent on concrete and other masonry surfaces. Does not adhere to polyethylene vapor barriers.
Silicone	Pecora Silicone Architectural Sealant Pecora Silicone Glazing Sealant Dow Corning No. 999 Glazing Sealant and No. 8644 Silicon Rubber Sealant	20-28	20-30	Good performance on glass, ceramic, and metal surfaces. Primer recommended on concrete and wood. Ultraviolet degradation.

[a] *New Shelter* (published by Rodale Press, Emmaus, PA), February 1983, p. 42. Prices have been updated by author to 1984 levels.

Oil-based and plain latex caulk are somewhat cheaper per tube than acrylic latexes, but have an expected lifetime of only 3 to 5 years. Oil-based sealants are the least expensive, although they are rather messy to work with and clean up. They also tend to lose adhesion to the joint after the oil dries up.

Butyl rubber caulks are good choices for static seams on building exteriors. They adhere well to metal and masonry and are fairly resistant to degradation from exposure to rain and other moisture. Because they are easily punctured and pick up dirt readily, they should be applied to joints that will be protected by the building's siding, trim, or some other means.

Urethane foam is also of use in static, exterior applications. It is especially good for filling large seams that other materials will not span. In addition to providing a weather-seal, urethane foam has insulating properties of about R-5 per inch of thickness. It will not weather well if exposed to direct sunlight and so should be covered with wood trim or some other protection.

Caulking for Nonstatic Locations. If there is a certain amount of relative movement in the building materials around the seam, elastic caulk such as polyurethane or silicone should be used. Polyurethane is cheaper, although careful surface preparation is required before it is applied. Primers are recommended for good adhesion to most surfaces.[6] Polyurethanes are especially effective on masonry but will not adhere to plastics such as polyethylene (a material that is often used for building envelope vapor barriers). Polyurethane caulk should not be exposed to direct sunlight.

Silicone sealants *can* be exposed to sunlight and outside weather conditions. Besides having excellent resistance to ultraviolet degradation and oxidation, silicones are able to withstand extreme temperatures, from −65 to 350°F. This type of caulk adheres well to most substrates, although a primer is necessary if it is to be applied to porous surfaces such as concrete. Silicone works well in large, nonstatic joints and is one of the few sealants that can be used in the rigorous environment of solar collectors.

If applied according to instructions, the expected lifetime of both polyurethane and silicone sealants is 20 to 30 years. Once they have cured, both will withstand severe heat and cold, although polyurethane must be applied at temperatures above freezing. Silicones can be applied at temperatures as low as 0°F.

Infiltrative Trouble Spots

In addition to the locations discussed above that must be caulked or weatherstripped, there are a number of other places that are often missed.

Trim of all types—baseboards, wall and ceiling trim, and trim around window and door frames—frequently hides cracks that let in large volumes of cold air. In new construction the gaps should be securely caulked before trim is installed. Do not depend on the wood to stop leaks. In existing buildings, where the trim has already been nailed in place, seal the cracks between trim and wall with clear caulk.

Junctions between bathroom and kitchen fixtures and the building envelope must also be carefully examined. Infiltration behind sinks, cabinets, and bathtubs is very common.

Cold air flow around electrical outlet boxes can be stopped by installing foam rubber gaskets behind the faceplates. If the house has a basement or heated crawl space, its doors and windows must be thoroughly weatherized. The seams between the top of the basement wall or the foundation and the sill plates, between the sill plates and the headers, and between the headers and the subfloor and floor joists should be caulked, for if cold air is freely able to enter a basement or heated crawl space, warm rising air from the floors above can draw it into the living spaces.

Cracks under windowsills are sometimes channels for the infiltration of cold air. Perhaps because they are out of sight, those cracks are often not painted carefully, which leaves them wide open. Another trouble spot is at the edge of wall-to-wall carpeting. Builders sometimes rely on the carpet itself to stop infiltration rather than adequately caulking the cracks between wall and floor.

Air infiltration through a particular crack is frequently not bad enough to detect by simply placing your hand over it. Many small leaks, however, can add up to a large energy loss. A simple way of detecting small leaks is to hang a sheet of toilet paper from a coat hanger or pencil. Very slight air currents will move the paper.

If condensation forms between the year-round windows and the storm windows of a building, it is usually an indication that the primary windows need weatherization, for the moisture is carried by living-space air that works its way around the primary window and condenses when it reaches the cold storm window.

An Infiltration Control Checklist

The checklist shown in Table 2-2 summarizes the common infiltration locations just discussed and how to fix them. It can be used both in new construction and as a retrofit guide in existing buildings.

TABLE 2-2 Infiltration Control Checklist

Location	*Prevent Infiltration with:*
Operable components of the building envelope	Weatherstripping. Types recommended:
Sides of doors.	Tension strips (preferably the metal varieties).
Bottoms of swinging doors.	Door sweeps.
Sliding doors and windows.	Bristle compression gaskets.
Casement windows.	Tension strips or closed cell (vinyl) foam.
Jalousie windows.	Very hard to weather-strip adequately. Consider replacing them with another design.
Static or nearly static building envelope penetrations	Caulking. Types recommended:
Totally static seams on the interior of the building envelope. (*Note:* If the two surfaces bounding the seam are of different materials, the seam will not be totally static.)	Use acrylic latex caulk. Baseboards and ceiling and wall trim are generally not tight enough to stop infiltration completely. Caulk of some type should be installed before the trim is put in place (or in existing building, cracks between trim and wall should be caulked.)
Static seams on the building exterior, such as trim-protected gaps between boards or panels of siding.	Use butyl rubber or urethane foam caulk, protected from the weather and from abrasion by the building's siding, trim, or some other means. Urethane foam will span wide seams that other caulks cannot.
Nonstatic seams	Caulking. Types recommended:
Seams around doors and windows, between foundation and sill plate, or around building envelope penetrations for pipes and electrical spots; around or behind bathtubs, sinks, and electrical outlet boxes.	Use polyurethane or silicone. Polyurethane is cheaper but must be protected from direct sunlight, whereas silicone stands up well under intense UV bombardment as well as extremes of temperature. The seam between a nonoperable window and the surrounding framing can also be weatherproofed at the time of installation with closed-cell vinyl foam.

Air Quality in Low Infiltration Buildings

As buildings are tightened up and allow less and less outside air to enter in winter, a new problem must be dealt with: how to handle the increased concentrations of moisture, polluting gases, and smells that in more leaky buildings simply flowed out through the cracks? Foam insulations and particleboard sometimes give off formaldehyde vapors, for instance. Radioactive radon gas, constantly being emitted from certain soils, can also accumulate.

Some people have suggested that the problem can be solved by simply allowing buildings to leak more. But this will allow far too much heat to be lost, especially during cold and windy periods. It is recommended that in buildings where the rate of air exchange between inside and outside is less than 0.5 air change per hour, air-to-air heat exchangers be installed (see Table 2-9 for determining the rate of air change).

Air-to-air heat exchangers contain blowers that expel interior air while bringing in fresh outside air, transferring heat from one flow to the other before the fresh air is allowed into the room. Typical heat exchangers for domestic use are about the size of window air conditioners but draw far less power (about as much as one light bulb) and cost $300 to $800. They admit carefully controlled amounts of fresh air into the house, something that "allowing a building to leak a bit more" does not do. Characteristics of a number of air-to-air heat exchangers on the market are listed in Table 2-3.

INSULATION METHODS

To be thermally efficient, a structure's building envelope must not only be devoid of infiltrative leaks, it must also be adequately insulated, for otherwise large conductive losses through it will result. There are three points to keep in mind when insulating a building. It is important that:

TABLE 2-3 Air-to-Air Heat Exchangers

Name	*Supplier*	*Dimensions (in.)*	*Weight (lb.)*	*1984 Prices*	*Comments*
Wall- or Ceiling-Mounted Exchangers					
Econofresher GV-120	Berner International Corp. 12 Sixth Road Woburn, MA 01801 (617) 933-2180	21½ × 12 × 8¼	21	$425	Rotary type; recovers latent as well as sensible heat
Lossnay VL-1500-Z	Mitsubishi Electric Sales America, Inc. 3030 East Victoria Street Compton, CA 90221 (213) 537-7132	22 × 15 × 12	29	$280	Fixed-plate design, which has the advantage of simplicity; no moving parts except the blowers
Exchangers Designed for Attic, Basement, Crawl Space, or Utility Room Mounting					
Z-Duct 79 M.4-RU	Des Champ Laboratories P.O. Box 448 East Hanover, NJ 07936 (201) 884-1460	62 × 16 × 14	75	$685	Fixed-plate type
VanEE R200	Conservation Energy Systems, Inc. P.O. Box 8280 Saskatoon, Saskatchewan Canada S7K 6C6 (306) 665-6030	48 × 15 × 22	70	$795	Fixed-plate type

1. The right type of insulation for a given situation be chosen
2. A sufficient thickness of it be used
3. It be installed in an effective, long-lasting manner

The most common method for rating a material's insulating potential is its *thermal resistance*, often called its *R-value*. The higher the resistance, the slower heat travels through the substance, and the better an insulator it is. Thermal resistance is defined as the reciprocal of the thermal conductance, which expresses, in English units, the Btu of heat that flow through a 1-square-foot section of a certain material each hour, when there is a 1 degree temperature difference between its surfaces.

The R-value of a material depends on both its composition and thickness. Table 2-4 lists resistances of a number of common substances, for a variety of thicknesses.

TABLE 2-4 Insulation Statistics

		Statistics for Insulation with a Thermal Resistance of R-19	
Type of insulation	*Rated R-Value per Inch*[a]	*Price (dollars) per Square Foot*[b]	*Thickness (in.)*
Batts and rolls: fiberglass and rock wool	3.2	0.28–0.41	6
Loose fill			
Fiberglass	2.2		9
Poured in place		0.30–0.38	
Blown in wall[c]		0.98–1.40	
Blown in attic[c]		0.31–0.56	
Rock wool	3.1		6
Poured in place		0.30–0.38	
Blown in wall[c]		0.98–1.40	
Blown in attic[c]		0.31–0.56	
Cellulose	3.2		6
Blown in wall		1.20–1.80	
Blown in attic		0.27–0.52	
Rigid boards			
Expanded polystyrene ("beadboard")	4.0	0.66–0.90	5
Extruded polystyrene ("blueboard")	5.0	1.14–1.52	4
Urethane and isocyanurate	7.2	0.83–1.58	2½
Sprays and foams			
Cellulose	3.5	2.09–4.37	5½
Urethane	6.2	1.90–3.04	3

[a]*New Shelter* (published by Rodale Press, Emmaus, PA), November/December 1982, pp. 32–33. Figures are results of FTC standard tests. Actual values vary slightly above and below these values due to differences in manufacturing.

[b]1984 prices.

[c]Blown, sprayed, or poured-in-place prices include the cost of installation by a contractor.

Types of Insulation Available

Building insulation is sold in the form of flexible batts and rolls, loose fill that can be blown or poured into the area needed, rigid boards that are nailed in place, or foams and sprays that can fill irregular, hard-to-get-to places. A structure may use one kind of insulation throughout or a variety of types in different parts of the building envelope.

Flexible Batts and Rolls. The most common type of batt or roll insulation is *fiberglass*, so named because it is composed of numerous glass filaments held together in a wooly mass. The many small air pockets in the mass are what make the material a good insulator. *Rock wool*,[7] a fibrous material made from molten rock or slag by passing steam through it, has properties very similar to those of fiberglass and has also been used in many buildings.

Batt and roll insulation is readily installed between the studs, rafters, and joists of standard frame buildings. It will not burn at the temperatures it is likely to encounter, although its foil or paper backing can catch fire. It is strongly recommended that installers wear respirators (throwaway types are inexpensive and easy to obtain in building supply stores), eye protection, gloves, and full-coverage clothing when handling fiberglass or rock wool, for the small particles given off are irritating to lungs, eyes, and skin. Continued exposure to loose glass fibers can result in severe lung damage and possibly cancer.

Loose Fill. The advantage of loose-fill insulation is that it can be poured or blown into nearly inaccessible or unusually shaped locations unsuitable for batt or roll insulation. Loose fill is often ideal for retrofits, for it can be installed in wall cavities by simply drilling a small hole in the wall that can then be covered with trim. It can also be installed in horizontal spaces such as attics far more quickly than batts or rolls can be.

Blown-in insulation is usually installed by an experienced contractor, although equipment can be rented by the homeowner. Care must be taken to blow it into cavities at a sufficiently high density, or settling is likely to occur over time. This will lead to voids and cold spots in the wall.

Loose-fill insulation does not provide a vapor barrier. If moisture gets absorbed by it, an impaired insulating value as well as rotting in the surrounding wood can result. It is thus advisable to use a polyethylene or building paper vapor barrier between the loose fill and the interior paneling.

Fiberglass and rock wool are widely available in loose-fill form. Because they are highly fire resistant, they can be packed into hot areas such as those around chimneys and stovepipes. The same health precautions should be taken during installation as with roll and batt insulation. Blown fiberglass has a lower R-value than other types of loose fill (R-2.2 per inch) and can get hung up on wires and nails rather than evenly distributing itself throughout a cavity.

Cellulose insulation has the highest thermal resistance per inch (up to R-3.7) of all loose fills. It does not cause skin or lung irritations and resists air infiltration better than does rock wool or fiberglass. It can be blown through smaller holes than those required for fiberglass, although it requires thorough chemical treatment to make it fire resistant. Only cellulose with a class A fire rating should be installed.

Rigid Boards. Rigid board insulation is more resistant to air infiltration than is loose fill or batts and rolls. It also has higher R-values per inch of thickness. It is often sold prebonded to heat-reflective foil materials that increase the insulating value if oriented so that they face an air gap. The foil also provides a vapor and infiltration barrier. Some board insulations are sold prebonded to interior and exterior finish wall surfaces such as gypsum board or plywood. These modular wall section "sandwiches" are not inexpensive, but they dramatically reduce building construction time.[8]

Board insulations are difficult to install in unusually shaped cavities and corners, for they must be custom cut to fit the location. Most board insulations are combustible and emit toxic fumes when burned. A fire barrier such as ½-inch or thicker gypsum board must be installed between the building interior and the insulation.

Extruded polystyrene is a material that, because of its high moisture resistance and compressive strength, may be applied to the exterior surfaces of foundations and basement walls in direct contact with the ground. When installed above ground, an abrasion and ultraviolet-resistant covering such as stucco or wood is installed on top of it (see the "Outsulation" section in Chapter 5).

Extruded polystyrene is sometimes called "blueboard," and should not be confused with less weather resistant *expanded* polystyrene ("beadboard"). "Blueboard" is often sold through marine supply companies, for it is employed to insulate boats. Dow manufactures a widely used type of blueboard. Some extruded polystyrenes on the market contain fluorocarbons. It is not yet known whether these will escape from the styrene over time and be released to the environment.

Urethane and *isocyanurate*[9] rigid foams have the highest thermal resistance per inch of thickness (R-7.1 to 7.7) of any commonly available insulation. Brand names include Thermax, Trymer, High-R, and others.

Rigid foams sometimes give off formaldehyde vapors, especially when subjected to high temperatures. If there is a good vapor barrier as well as a solid wall between the insulation and living space, however, these vapors should not be able to enter the building's interior and pose a potential threat to the health of its occupants.

Fiberglass is also manufactured in board form. Because it is more flexible than plastic foams it is easier to apply it to slightly irregular surfaces or gentle curves.

Sprays and Foams. Like loose fill, sprayed and foamed-in-place insulation can be installed in hard to get to or irregularly shaped locations. It is of great use in many retrofit situations. Because it adheres to the surfaces around it, it provides a continuous airtight seal in framing gaps or penetrations in the building envelope, such as around plumbing vents.

Sprays and foams are usually installed by insulation contractors with specialized equipment. Cellulose foam should have a class A fire rating if it is to be used. Urethane foam must have a fire-resistant surface such as gypsum board between it and the living area and should never be used near stove pipes and chimneys.

Insulation Comparisons. Table 2-4 lists characteristics of the insulation material discussed above. The price per square foot and thickness statistics refer to insulation of sufficient depth to provide a thermal resistance of R-19. In choosing an insulation, it should be remembered that although cost per square foot is certainly an important consideration, its other characteristics must also be carefully studied. For instance, R-19 polyurethane foam costs considerably more per square foot than do R-19 fiberglass batts, but the foam is only 2½ inches thick, whereas the batts are 6 inches.[10] A wall made of 2 × 4 studs is of sufficient thickness to contain the foam, whereas if fiberglass is used, more expensive 2 × 6 or 2 × 8 studding must be installed.

In addition, foam is self-supporting, whereas batts and rolls are not. The advantages of this are especially apparent in open beam ceilings. Foam insulation can simply be nailed to the roof sheathing and the shingles installed on top; batt or roll insulation, however, requires a support grid of wood to which it is nailed and possibly another layer of sheathing on top of it. This means increased building material and labor costs that would probably outweigh the savings from buying cheaper insulation.

Degree-Day Calculations and Thermal Resistance Selection

In addition to deciding on the types of insulation to be used, it is necessary to determine the thickness needed for different parts of the building envelope. The colder the climate, the thicker it should be. A common measure of climate severity is the system of *degree-days.* The daily degree-days that accrue for a particular location are calculated by subtracting the average temperature for the day from 65°F.[11] For instance, if the average temperature of a building site is 45°F on October 13, the number of degree days that accrue are equal to 65°F minus 45°F, or 20 degree-days. If the average temperature is 65°F or higher, no degree-days accrue.

To arrive at a monthly or annual degree-day figure for a certain location, all the daily degree-days that have accrued are added up. An example computation is listed below.

Date		*Average Temperature (°F)*	*Daily Degree-Days*
February	1	30	35
	2	28	37
	3	29	36
	4	36	29
	5	37	28
	6	42	23
	7	45	20
	8	49	16
	9	44	21
	10	43	22
	11	41	24
	12	32	33
	13	26	39
	14	27	38
	15	24	41
	16	22	43
	17	18	47
	18	19	46
	19	26	39
	20	30	35
	21	33	32
	22	35	30
	23	37	28
	24	32	33
	25	35	30
	26	36	29
	27	30	35
	28	32	33
		Total	902

Table III-1 of Appendix III contains monthly and annual degree-day figures for various locations averaged over many years. Degree-day figures for a particular location not listed in the table are often available from local farm advisory boards or building inspectors.

The reason that 65°F is chosen for the base temperature from which degree-days are calculated is that when external temperatures are 65°F or higher, supplemental building heat is usually not needed, for the thermal energy given off by appliances, lights, and people is usually enough to keep internal temperatures at about 70°F.

Because of the present tendency toward lower thermostat settings and better insulated, tighter construction, some researchers are now compiling degree-day tables with base temperatures lower than 65°F.[12]

Thermal Resistance Selection. After degree-day figures for a building site are obtained, proper thermal resistances for the insulation in different parts of the building envelope can be determined using the data in Table 2-5. Necessary thickness of insulation is calculated by dividing the R-values just obtained by the R-value per inch of the insulation chosen (the R-value per inch is listed in Table 2-4). For instance, if a certain wall requires R-19 insulation and the type chosen is urethane foam, Table 2-4 informs us that its thermal resistance is R-7.2 per inch. Dividing 19 by 7.2 yields a necessary insulation thickness of about 2½ inches.

TABLE 2-5 Recommended Thermal Resistance Levels[a]

Heating Degree-Days of the Building Site	*Roof or Ceiling*	*Exterior Wall*	*Floor over Unheated Spaces*	*Foundations and Below-Grade Walls*
Less than 2500	R-11	R-11	—	—
2500–4000	R-19	R-11	R-8	R-5
4000–5500	R-30	R-19	R-11	R-10
5500–7000	R-40	R-25	R-19	R-15
Above 7000	R-50	R-30	R-25	R-19

[a]These R-values are for the insulation itself, rather than the total R-values for the building envelope component, which includes sheathing, siding, framing, and paneling as well as insulation.

The Do's and Don'ts of Installation

A problem that often occurs from improperly installed insulation is heat loss due to "voids"—empty spaces in which the insulation either has not been put or where it has fallen away. Areas where insulation is frequently not installed are those near the corners of exterior walls, junctions between interior and exterior walls, hollow spaces over and around windows and doors, areas just above the foundation, and spaces between stories on exterior walls. It is often because these places are difficult to reach or because it is time consuming to custom-fit pieces of insulation into irregularly shaped spaces that they are frequently neglected. In new construction it is cost-effective not to overlook these errors. Once the house is built, though, only the larger voids are usually economical to fix.

Joists that rim the perimeter of the building just above the foundation or between stories can be especially severe sources of heat loss if left uninsulated. Voids around stovepipes and chimneys are also trouble spots. Framing is usually kept at least 2 inches away from the masonry or metal surface for fire safety reasons. The heated air near these surfaces will carry many Btu of thermal energy away unless a nonflammable insulation such as fiberglass is packed into the gap between the heated surface and the framing. Because fiberglass is so porous, it must be stuffed quite tightly into place to stop the passage of air. Unfaced fiberglass must be used, for kraft or metalized paper can catch fire.

Recessed ceiling-light fixtures are also potential sources of fire if insulation is piled on top of them. Excess heat cannot escape from such fixtures, and temperatures can reach dangerous levels. If uninsulated, however, recessed fixtures can be sources of heat loss. The best solution to the problem is to replace them with surface-mounted light fixtures.

The most frequent problem that occurs with loose-fill insulation is that it settles over time, leaving a void near the top of the space. Older buildings were often insulated with vermiculite, a rather heavy loose fill with low thermal resistance compared to insulation popular today. Vermiculite settling frequently led to continuous voids along the tops of the walls in those buildings. Fiberglass, rock wool, and cellulose loose fill are also subject to settling problems if blown into the spaces at too low a density.

Loose-fill insulation installed in attics near vents or fans is sometimes blown around by air currents. Nailing window screen over the insulation in trouble areas will keep it in place and still allow evaporation of moisture.

Batt and roll insulation is designed to be held in place mainly by friction between it and the framing members on each side. The paper or foil backing can help secure the insulation if stapled to the stud, but if it is the only means of support the batts might eventually pull loose and fall away, leaving large voids. Insulation installed between floor joists or attic rafters is especially prone to these problems when not supported in some way from below. If the batts do not fit tightly between framing members (as will be the case if the members are not standard distances apart), additional support from cord or wire zigzagged across the bottom of the batts should be used.

The kraft paper or reflective foil side of an insulation batt is a vapor barrier and it, rather than the fiber side of the insulation, should face the building's interior. Otherwise, moisture in the living-space air is likely to work its way through the paneling or wallboard and into the insulation, im-

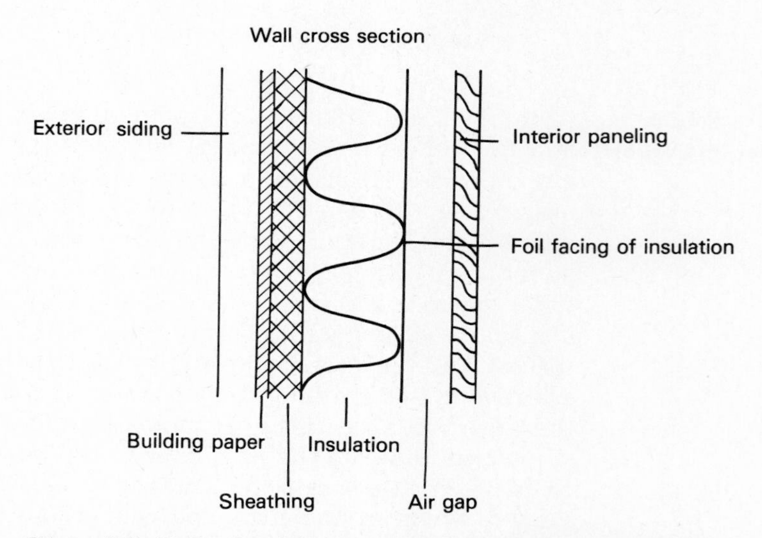

Figure 2-3 Reflective foil insulation facing must be adjacent to an air gap to be effective.

pairing its effectiveness and leading to rot in the surrounding wood. Because the vapor barrier sometimes rips or pulls away from the framing member to which it is stapled, it is advantageous to install an additional vapor/infiltration barrier such as polyethylene or building paper to the building envelope between the insulation and the living space.

Foil-faced insulations can provide higher thermal resistances than those with kraft paper faces, for the foil will reflect considerable infrared radiation back toward the living space. For it to do this, though, it is necessary to mount the batts so that there is an air gap between the facing and the interior paneling (see Figure 2-3).

ENERGY-SAVING APPLIANCES

Water Heaters

The hot water usage in a typical home is about 450 gallons per week.[13] If cold water enters the heater at 50°F and is raised to 140°F, a conventional electric heater will expend about 5100 kilowatt-hours of energy annually heating the water and another 1200 kilowatt-hours making up for standby losses. At 8 cents per kilowatt-hour, the annual water heating bill will run over $500. There are a number of inexpensive ways, however, that this cost can be reduced.

A 1½-inch-thick fiberglass blanket around the water heater will shave 10 percent off the annual heating cost. Changing the fixtures in bathroom and kitchen to the water-saving variety, reducing the tank temperature from 140°F to 130°F, and further insulating the tank to a thermal resistance of about R-14 will cut *44 percent*, or over $220 from the annual cost. The payback period for these improvements (the amount of time it takes for the monetary savings from the improvements to equal the actual costs of the improvements) is very short—often less than a year.

Another option is to install *tankless* water heaters—those that raise the temperature of the water on demand.[14] These types have been used in Europe and other parts of the world for many years with good results. Their big advantage is that they eliminate standby losses, which amount to a cost savings of around $80 per year compared to conventional water heaters, or $40 per year compared to the newer highly insulated models. Typical tankless heater costs range from $200 to $500, while conventional heaters cost $200 to $300.[15] Heaters of all types should be located within the heated (also called the "conditioned") space of the house so that the thermal energy that they lose helps in warming the space.

Other Appliances

Although refrigerators differ widely in their energy usages, manual defrost models are generally far more economical to run than those with automatic defrost.

Gas clothes dryers with pilotless ignition use 45 percent less energy than do models with pilot lights.[16] Another useful feature of some dryers is a sensing mechanism that turns the dryer off as soon as the clothes are dry, rather than running it a prescribed amount of time. Heat exchangers are now available for the exhaust vents of dryers of any type. The exchangers direct some of the thermal energy produced into the conditioned areas of the house rather than allowing it to escape up the flue.

It is common in many homes for one-fifth of all electric power consumed to be used for lighting.[17] Fluorescent fixtures use far less power than incandescents, but many people consider fluorescent light cold and unpleasant and the flutterings that usually go with it irritating. In addition, fluorescents need special fixtures and starter circuitry that can be rather expensive. "Warm-white" fluorescent lamps give off a somewhat gentler light than standard fluorescents. Combinations of fluorescent and incandescent lamps produce softer light with much less flutter.

Because of natural convection processes, the warmest air in a room tends to rise to the highest point, where it is of little help in heating the building. Ceiling fans are very effective in destratifying room air, especially in buildings with high ceilings. In the author's home, which has a ceiling that is 19 feet above the ground floor, typical wintertime temperatures before a fan was installed were 10 to 15°F warmer on the second floor than on the first, but only 2°F warmer after it was installed. The fan, an Emerson Premium, is 4 feet in diameter and draws only 50 to 60 watts while operating.

Several types of new light sources are now available that also use very little energy but that can be screwed into ordinary incandescent fixtures. One is the Philips Norelco SL-18 lamp, which produces as much light as a 75-watt incandescent bulb while using only 18 watts of power. It is expected to have a 7500-hour lifetime, about 10 times that of a normal light bulb.[18] Similar products include Westinghouse's Econ-Nova, General Electric's Halarc, and Interelectric Corporation's U-Lite. These products are not cheap (most cost around $25). If their lifetimes are as long as expected, however, they should pay for themselves after about 4000 hours of use, or if lit an average of 6 hours a night, after less than 2 years.[19]

SUPERHOUSES

There is another level of energy-conserving construction that surpasses those described in previous sections and that is becoming increasingly popular in areas with very cold wintertime climates. Some salient features of these "superhouses" are as follows:

1. All components of the building envelope are insulated to thermal resistances several times greater than in conventional buildings. Roof and wall resistances, for instance, range from R-30 to R-60.
2. Triple glazing (and sometimes quadruple glazing) is used in the windows.
3. Extra care is taken to seal the house against infiltration. Continuous polyethylene vapor barriers are installed throughout the building envelope.[20] Entry vestibules and airtight furnace rooms are typical.

One method often used in constructing superhomes is to make components of the building envelope double-layered. Walls, for instance, will have an inner load-bearing layer framed in a standard way with 2 × 4's and including fiberglass rolls between the studs. There will also be an outer wall framed more lightly (for instance, with 2 × 3's on wider centers) and also filled with insulation. Spacers keep the two walls several inches apart and more insulation is placed between them (see Figure 2-4).

A continuous polyethylene infiltration/vapor barrier is installed between the two halves of each double wall. The sheets extend at least 1½ feet past the top and bottom plates of the wall and are sealed to the roof and floor vapor barriers. The sheets also extend past the sides of the wall and are folded around the building's corners. The building envelope is thus continuously sealed over all of its area.

Performance figures for superinsulated homes are very impressive. Of 11 such homes built in Saskatoon, Saskatchewan, heating expenses for the

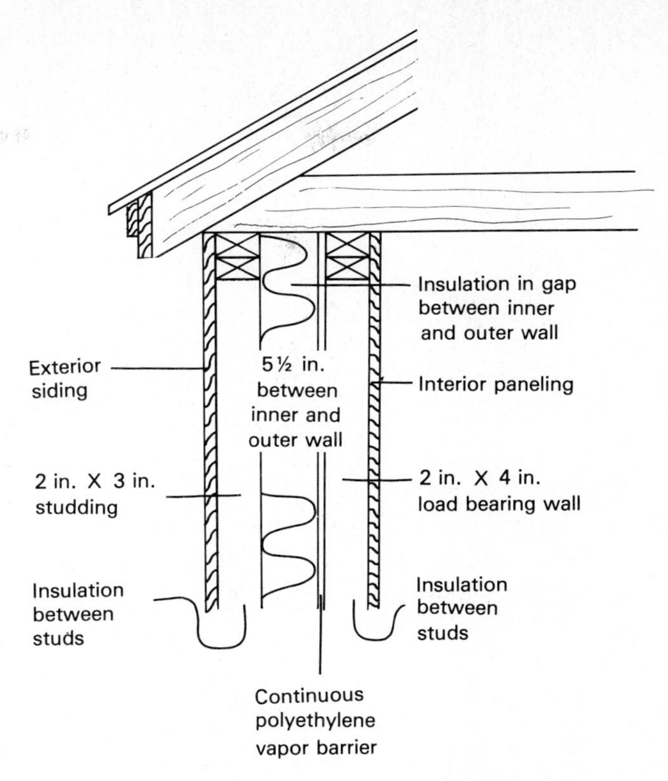

Figure 2-4 Double walls for superhomes.

entire year ranged from only \$59 to \$143.[21] Part of the energy savings was due to passive solar technology and part to the enormous thermal efficiency of the homes. The annual space-heating requirements of an average pre-1975 Saskachewan home ranged from 95 to 142 million Btu. Usages for the super-insulated residences studied were between 31 and 75 million Btu. The extra cost of the energy-conserving features (including the passive solar system) averaged \$5000 per house, only 6% of the total cost of the home.

CALCULATING THE BUILDING HEAT LOSS

Knowing the rate of thermal energy loss through the various components of a building envelope is advantageous for a number of different reasons. The performance of the infiltration control and insulation measures can be evaluated and trouble spots that need more attention can be located. Also, estimates of the cost of heating the building can be arrived at. Finally, heat loss data enable a designer to *size* a passive or active solar energy system for the building.

Thermal energy loss from a structure depends on four factors: the thermal resistance of the components of the building envelope, the surface area of each component, the tightness of construction of the building, and the exterior climate. The first three factors are functions of the building itself,[22] while the fourth depends on its location.

Computing Thermal Resistances

The thermal resistance of a wall or other building envelope component is equal to the sum of the R-values of each of the wall's layers—siding, sheathing, insulation, and interior paneling, as well as air spaces and the thin air

TABLE 2-6 Thermal Resistances of Various Building Materials

	R-Value
(a) Siding and Paneling	
Wood siding, boards 1 in. × 8 in.	0.79
Wood siding, beveled and lapped	
0.5 in. × 8 in.	0.81
0.75 in. × 10 in.	1.05
Aluminum or steel, hollow-backed	0.61
Asphalt roll siding	0.15
Shingles, asbestos cement	0.21
Shingles, wood, 16 in., 7.5 in. exposure	0.87
Gypsum board or plasterboard	
0.375 in.	0.32
0.5 in.	0.45
0.625 in.	0.56
Plywood	1.25 per inch
Particleboard, medium density	1.06 per inch
(b) Masonry Materials	
Brick	
Common	0.20 per inch
Face	0.11 per inch
Concrete blocks, hollow, filled with sand and gravel aggregate	
4-in. thickness	0.71
8-in. thickness	1.11
12-in. thickness	1.28
Concrete, normal density (135-160 lb/ft^3)	0.08 per inch
Limestone or sandstone	0.08 per inch
Plaster—sand, cement aggregate	0.20 per inch
Ceramic paver tile, 3/8 in.	0.03
(c) Roofing	
Shingles	
Asbestos-cement	0.21
Asphalt	0.44
Wood	0.94
Asphalt roll roofing	0.15
Built-up roofing	0.33
Slate	0.05
(d) Flooring and Floor Coverings	
Hardwood (oak, maple), 0.75 in.	0.68
Softwood (fir, pine)	1.25 per inch
Carpet and rubber pad	1.23
Carpet and fibrous pad	2.08
Linoleum and vinyl	0.05
Wood subfloor, 0.75 in.	0.94
(e) Membranes	
Building felt (permeable)	0.06
Two layers of mopped 15-lb felt	0.12
Plastic films	Negligible

TABLE 2-6 (cont.)

(f) Framing Factors

Envelope Component and Framing Type[a]	*Composite R-Value Before Framing Factor Is Taken into Account*						
	3.3	*5.0*	*10*	*12.5*	*16.7*	*25*	*33.3*
Floors							
Joists 16 in. o.c.	0.94	0.95	1.00	1.03	1.10	1.15	1.23
Joists 24 in. o.c.	0.96	0.97	1.00	1.02	1.07	1.10	1.15
Beams 48 in. o.c.	0.95	0.96	1.00	1.02	1.08	1.12	1.19
Ceilings							
Joists 16 in. o.c.							
2 in. × 4 in.	—	—	—	—	1.22	1.29	1.42
2 in. × 6 in.	—	—	—	—	1.13	1.18	1.27
Joists 24 in. o.c.							
2 in. × 6 in.	—	—	—	—	1.08	1.12	1.18
2 in. × 8 in.	—	—	—	—	1.05	1.08	1.13
Walls							
Studs 16 in. o.c.							
2 in. × 4 in.	0.93	0.97	1.07	1.13	1.29	—	—
2 in. × 6 in.	0.91	0.94	1.02	1.05	1.17	1.24	1.37
Studs 24 in. o.c.							
2 in. × 4 in.	0.95	0.98	1.06	1.10	1.22	—	—
2 in. × 6 in.	0.93	0.95	1.01	1.04	1.13	1.19	1.29

[a]o.c., on centers.

Source: Parts (a)-(e) are excerpted, with permission, from *ASHRAE Handbook—1981 Fundamentals*, American Society of Heating, Refrigerating and Air-Conditioning Engineers, Inc., Atlanta, GA. Part (f) is from *Energy Conservation Design Manual for New Residential Buildings*, State of California, 1978.

films on interior and exterior surfaces—divided by a *framing factor* that depends on the size and spacing of the framing members. Thermal resistances of different building materials and air layers, as well as framing factors, are listed in Tables 2-4, 2-6, and 2-7. The R-values of each building material is given per inch of thickness and must be multiplied by its actual thickness.

TABLE 2-7 Thermal Resistances of Air Layers

	R-Value			*R-Value*
Exterior air films	0.17	Interior air films		
		Position of Surface	*Direction of Heat Flow*	
		Horizontal (ceiling, roof)	Upward	0.61
		45° slope (roof)	Upward	0.62
		Vertical (wall)	Horizontal	0.68
		45° slope (roof)	Downward	0.76
		Horizontal (floor)	Downward	0.92

TABLE 2-7 (cont.)

Air Spaces Inside Building Components (for example, an air space between a layer of insulation and paneling)

Position of Air Space	Direction of Heat Flow	R-Values 0.5-in. Air Space	0.75-in. Air Space	1.5-in. Air Space	3.5-in. Air Space
Horizontal (ceiling, roof)	Upward	2.05	2.21	2.40	2.66
45° slope (roof)	Upward	2.44	2.75	2.74	2.95
Vertical (wall)	Horizontal	2.54	3.46	3.55	3.40
Horizontal (floor)	Downward	2.55	3.59	5.90	9.27

Source: Excerpted from *ASHRAE Handbook—1981 Fundamentals,* American Society of Heating, Refrigerating and Air-Conditioning Engineers, Inc., Atlanta, GA, pp. 23.12–23.13. The air spaces considered were ones bounded on one side by bright aluminum foil (such as is used in insulation facing) and on the other side by building materials such as wood, paper, masonry, or nonmetallic paints. Mean temperature in the air space, 50°F; temperature difference, 10°F.

Example Calculation. Calculate the thermal resistance of the wall in Figure 2-5.

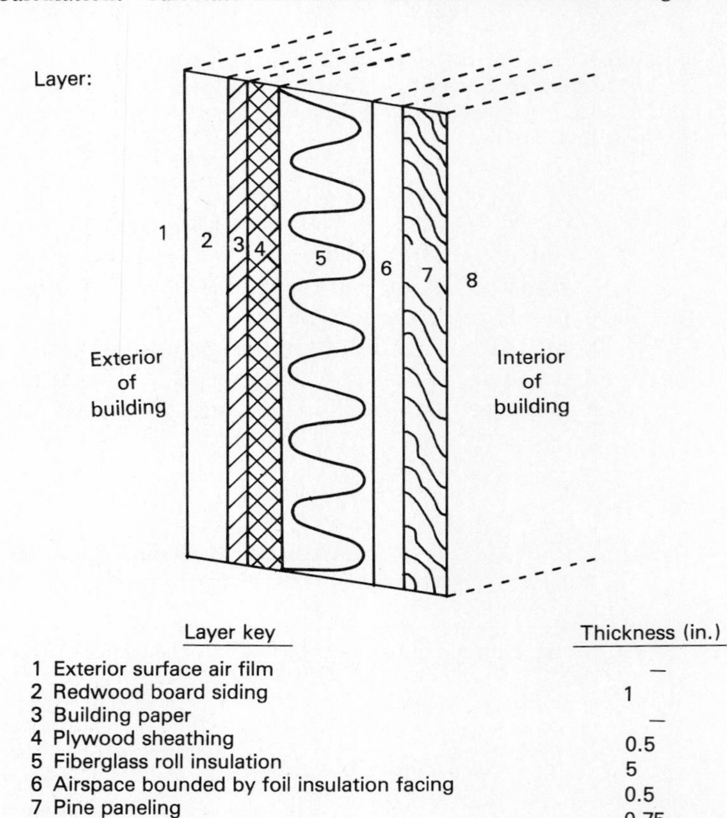

Figure 2-5 Typical frame wall cross section.

The Calculation:

Layer	*Material*	*Thickness (in.)*	*R-Value*
1	Exterior surface air film	—	0.17
2	Redwood board siding	1	0.79
3	Building paper	—	0.06
4	Plywood sheathing	0.5	0.62
5	Fiberglass roll insulation	5	16
6	Air space bounded by foil insulation facing	0.5	2.54
7	Pine paneling	0.75	0.59
8	Interior surface air film	—	0.68
		Total	21.45

The framing factor (obtained from Table 2-6) is 1.17. Dividing 21.45 by this yields a composite thermal resistance for the wall of R-18.33.

Example Calculation. Compute the thermal resistance of the pitched roof in Figure 2.6.[23] 2 × 4 framing on 16-inch centers is used.

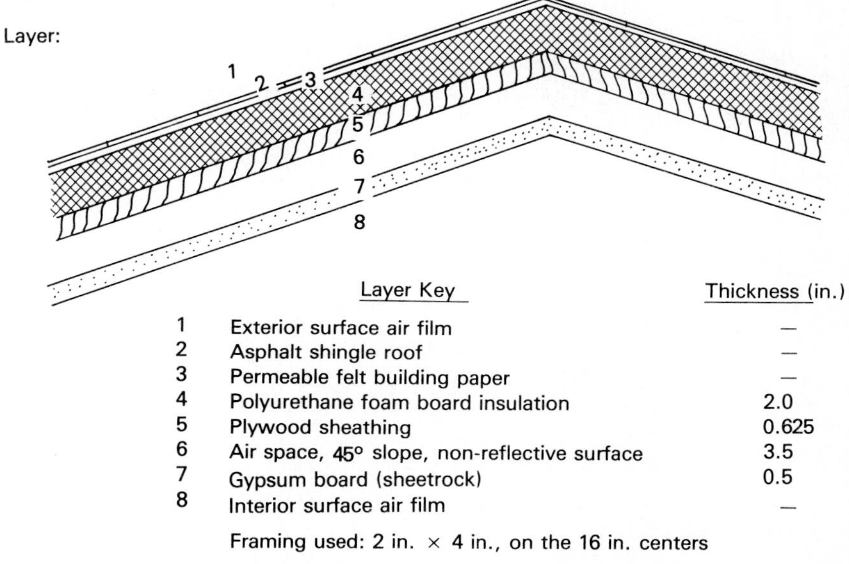

Figure 2-6 Typical pitched roof cross section.

The Calculation:

Layer	*Material*	*Thickness (in.)*	*R-Value*
1	Exterior surface air film	—	0.17
2	Asphalt shingle roof		0.44
3	Permeable felt building paper	—	0.06
4	Polyurethane foam board insulation	2	14.4
5	Plywood sheathing	0.625	0.77
6	Air space, 45° slope, nonreflective surface	3.5	0.96
7	Gypsum board (Sheetrock)	0.5	0.45
8	Interior surface air film	—	0.62
		Total	17.87

Framing factor (Table 2-6) = 1.22

$17.87 \div 1.22 = 14.65$

Composite thermal resistance = R-14.65

A number of composite R-values for the various building envelope components of standardly constructed buildings are listed in Table 2-8. These tabulations can be tremendous timesavers, for if components of the building being analyzed fit the descriptions in the table, their composite thermal resistances can simply be read from the table rather than calculated.

TABLE 2-8 Composite R-Values for Components of Typical Buildings (Not Including Insulation)

FRAME CONSTRUCTION

(a) Walls

	Interior Finish Material			
Exterior Siding Material[a]	*Gypsum Board (½ in.)*	*Wood (¾ in.)*	*Thin Plywood or Composition Board (¼ in.)*	*Plaster (¾ in., metal lath)*
Wood siding, shingles, or panels (¾–1 in. thick)	3.5	4.0	3.3	3.4
Face brick veneer	3.2	3.6	3.0	3.0
Stucco, asphalt roll siding	2.8	3.3	2.7	2.7

(b) Ceilings

	Interior Finish Material			
Exterior Siding Material	*Gypsum Board (½ in.)*	*Wood (¾ in.)*	*Thin Plywood or Composition Board (¼ in.)*	*Plaster (¾ in., metal lath)*
None, such as would be the case if an attic were above the ceiling	1.7	2.2	1.6	1.6
Wood floor (¾ in.) with or without floor tile or linoleum[b]	4.4	4.9	4.3	4.3

(c) Flat Roofs

	Interior Finish Material[c]			
Exterior Surface Material	*Gypsum Board (½ in.)*	*Wood (¾ in.)*	*Thin Plywood or Composition Board (¼ in.)*	*Plaster (¾ in., metal lath)*
Built-up roofing[d]	4.5	5.0	4.4	4.4

(d) Pitched Roofs

	Interior Finish Material[e]			
Exterior Surface Material	*Gypsum Board (½ in.)*	*Wood (¾ in.)*	*Thin Plywood or Composition Board (¼ in.)*	*Plaster (¾ in., metal lath)*
Composition shingles, roll roofing, built-up roofing, wood shakes or shingles[f]	2.0	2.5	1.8	1.8

TABLE 2-8 (cont.)

FRAME CONSTRUCTION

(e) Floors

Exterior Surface Material	*Interior Finish Material*	
	Wood (3/4 in.) with or without Floor Tile or Linoleum[g]	*Wall to Wall Carpeting over Plywood (1 1/8 in.)*
None, as would be the case if crawl space were below the floor	3.8	5.3
Wood (3/4 in.)	5.6	7.1
Gypsum board (1/2 in.)	5.1	6.6

MASONRY CONSTRUCTION

(f) Solid Brick Walls (8-in. thickness)

Exterior Siding Material	*Interior Finish Material: None, or Plaster Applied Directly to Wall*
None	2.2

(g) Concrete Block Walls (12-in. thickness)

Exterior Siding Material	*Interior Finish Material: None, or Plaster Applied Directly to Wall*
None	2.3
Face brick or stone (4 in.)	2.7

ALL TYPES OF CONSTRUCTION

(h) Windows

Single-glazed	0.9
Double-glazed, 0.5-in. air space	2.0
Triple-glazed, 0.5-in. air spaces	3.2

(i) Doors

Solid-core flush door (2 1/4 in. thick)	3.7
Panel door (1 3/4 in. thick) with 1 1/8 in. panels	2.6

[a] All walls are plywood sheathed under the exterior siding.

[b] The R-values were calculated for the following cases: (1) hardwood floor (3/4 in.) over wood subfloor (25/32 in.), and (2) tile or linoleum, over insulation board (3/8 in.), hardboard (1/4 in.), and wood subfloor (25/32 in.). Remember, this is the floor in the room *above* the room being examined.

[c] For the ceiling of the room below the roof.

[d] Over 2-in. wood decking.

[e] For ceiling of room below the roof. Ceiling material is applied directly to the roof rafters.

[f] It is assumed that insulation will be installed in the rafter space.

[g] R-values were calculated for the same cases as in note b.

The composite R-values in Table 2-8 are for building envelope components that are *not yet insulated.* To get the total thermal resistance, add the R-value of the insulation to the value from the table. If the insulation has a reflective foil facing, and this facing will be adjacent to an air space after the insulation is installed, add an additional value of 1.4 to the total thermal resistance, for an air space bounded by a reflective material has greater insulating value than that of one bounded by nonreflective materials (1.4 is an average over the cases considered in Table 2-8 of the effect of adding a reflective boundary to an air space).

Example Calculation. The composite R-value of a 1⅛-inch plywood floor, covered with wall-to-wall carpeting, and situated above a room with a wood ceiling is 7.1. If R-19 insulation were added to the air space between floor and ceiling, the total thermal resistance would be R-26.1. If the insulation were installed in such a way that there was still an air space on one side of it, and if the reflective foil facing of the insulation bounded this space, the total thermal resistance would be R-27.5.

The Building Load Coefficient

After R-values have been obtained, the next step in determining heat losses is to calculate the *building load coefficient* (BLC). The BLC expresses the amount of thermal energy needed per day to increase a building's temperature 1 degree Fahrenheit and hold it at that temperature. Tables in the passive energy section of this book employ the BLC value in sizing solar collection systems. Monthly and yearly heat losses for the building, useful in sizing active solar collector arrays as well as conventional heating systems, may also be calculated using the BLC. The building load coefficient is the sum of the load coefficients for each component of the building envelope. The load coefficients depend on the R-value and dimensions of the particular part of the building, with the exception of the coefficient for infiltration, which is related to air density and air changes per hour as well as building dimensions. The various load coefficients are defined in the following formulas.[24]

Load coefficients for walls, floor, roof, ceiling, and doors:

$$L_{\text{wall}}, L_{\text{floor}}, \text{etc.} = 24 \times \frac{\text{area of component}}{\text{R-value of component}}$$

The effective area of a wall is calculated by computing its total area and subtracting that of all windows and doors. Similarly, skylight areas must be subtracted from total roof areas.

R-values are normally related to heat flow per hour. The factor 24 in the load coefficient equations is included to convert the units to heat flow per day.

If the ground floor is over a heated crawl space or basement, it is not part of the building envelope and its load coefficient is not calculated. That of the basement or crawl-space floor and walls is, however.

Load coefficient for heated basement or crawl space:

$$L_{\text{basement}} = 256 \times \frac{\text{total length of walls}}{\text{R-value of wall insulation} + 8}$$

This formula is also applicable to earth-bermed walls.

If the house is built on a floor slab foundation, heat losses through the bottom of the building envelope are computed using yet another formula:

Load coefficient for a slab on grade:

$$L_{slab} = 100 \times \frac{\text{length of slab's perimeter}}{\text{R-value of perimeter insulation} + 5}$$

Windows not designed as part of a solar collection system are especially large sources of heat loss. The magnitude of the loss depends to a great extent on the number of glazings and is computed as follows:

Load coefficient for nonsolar windows:[25]

$$L_{window} = 26 \times \frac{\text{nonsolar window area}}{\text{number of glazings}}$$

Windows that are integral parts of a solar collection system are designed to be net energy gainers and so are not included in the heat loss calculations. In passive solar systems, the nonsolar windows are typically those not facing within 30° of south. In purely active systems, none of the windows are usually designed as parts of solar collection systems.

Load coefficient due to air infiltration:

$$\begin{aligned} L_{infiltration} = 0.432 &\times \text{(average number of air changes in the building per hour)} \\ &\times \text{(air density ratio)} \\ &\times \text{(ceiling height)} \\ &\times \text{(total area of all floors)} \end{aligned}$$

Building air changes per hour may be estimated using Table 2-9; air density ratios are dependent on the elevation of the site, as illustrated in Table 2-10.[26]

TABLE 2-9 Air Changes per Hour

Description of Building	*Average Air Changes per Hour*
Unweather-stripped doors and windows	1.5[a]
Fully weather-stripped doors and windows	0.9
Fully weather-stripped doors and windows plus continuous polyethylene vapor barrier throughout building envelope. Also, entry vestibule at main door of building, as well as an airtight furnace room for the backup heater.	0.5

[a]Can be considerably higher if doors and windows do not fit snugly within their frames and casements when closed.

TABLE 2-10 Air Density Ratios

Elevation of Site (ft)	*Air Density Ratio*
Sea level	1.0
1000	0.96
2000	0.93
3000	0.89
4000	0.86
5000	0.83
6000	0.80
7000	0.77

As mentioned above, adding up the load coefficients for each component of the building envelope yields the BLC, the building load coefficient.

Monthly Heat Loss To complete the heat loss calculation, the BLC is multiplied by the number of degree-days for the month:

$$\text{monthly heat loss} = \text{BLC} \times \text{degree-days in that month}$$

For heating purposes the most important heat loss is generally that which occurs in the coldest month during which the building is used. For buildings occupied throughout the year, January is generally the coldest month. It is assumed that if the solar system is designed to handle the coldest time of the year, it will certainly handle the more temperate times.

Example Calculation. Compute the BLC for a house with the following R-values and areas for its building envelope components:

Envelope Component	*R-Value*	*Area* (ft^2)
Walls	22	906
Roof	30	1200
Floor (slab on grade)	10 (perimeter insulation)	864
Windows (nonsolar)	—	150
Doors	2	42

The slab's perimeter is 120 feet. The windows are double-glazed. Both windows and doors are weather-stripped. The building is at an elevation of 2000 feet, has an average ceiling height of 8 feet, and has an upstairs area of 572 square feet, in addition to the ground floor area mentioned above.

The coefficient equations are as follows:

$$L_{wall} = 24 \times \frac{906}{22} = 988 \text{ Btu/°F-day}$$

$$L_{roof} = 24 \times \frac{1200}{30} = 960 \text{ Btu/°F-day}$$

$$L_{slab} = 100 \times \frac{120}{15} = 800 \text{ Btu/°F-day}$$

$$L_{window} = 26 \times \frac{150}{2} = 1950 \text{ Btu/°F-day}$$

$$L_{door} = 24 \times \frac{42}{2} = 504 \text{ Btu/°F-day}$$

The air density ratio from Table 2-10 is 0.93, and the total floor area is 864 ft^2 + 572 ft^2 = 1436 ft^2. From Table 2-3, the number of air changes per hour is about 0.9.

$$L_{infiltration} = 0.432 \times 0.9 \times 0.93 \times 8 \times 1436 = 4154 \text{ Btu/°F-day}$$

$$\begin{aligned} \text{BLC} &= \text{sum of load coefficients} \\ &= 988 + 960 + 800 + 1950 + 504 + 4154 \\ &= 9356 \text{ Btu/°F-day} \end{aligned}$$

As can be seen from the example, the largest heat loss by far is in the area of infiltration (44 percent of the total loss). The future homeowner or builder might thus consider the benefits of more rigorous anti-infiltration techniques, such as installation of the continuous polyethylene vapor barriers that were discussed in the "Superhouses" section of this chapter.

The methods described above give a fairly good estimate of heating load, usually within 10 percent of the more detailed ASHRAE (American Society of Heating, Refrigerating and Air-Conditioning Engineers) method. For a more precise heat load calculation, refer to ASHRAE's *Handbook of Fundamentals.*

NOTES

1. Dan Lewis and Joe Kohler, "Passive Principles: Conservation First," *Solar Age,* Vol. 6, No. 9, September 1981, p. 34.

2. New Shelter's Product Testing Workshop, "Caulk Talk," *New Shelter* (Rodale Press), Vol. 4, No. 2, February 1983, p. 40.

3. Dave Sellers and Frederic S. Langa, "Weatherstripping: Expert Choices," *New Shelter* (Rodale Press), Vol. 4, No. 1, January 1983, p. 34.

4. Ibid., pp. 35–42.

5. New Shelter's Product Testing Workshop, p. 42.

6. Ibid., p. 41.

7. Also called mineral wool or slag wool.

8. Rigid insulation can add considerable structural strength when securely bonded to and sandwiched between plywood sheets, although local building codes should be checked if these modules are to be used.

9. Isocyanurate is a form of polyurethane often used in high-temperature applications such as in solar collectors.

10. Calculated from statistics in Bob Flower, "Buying the Right Kind," *New Shelter* (Rodale Press), Vol. 3, No. 9, November/December 1982.

11. More specifically, we are dealing with *heating* degree-days, with base 65°F. Also used are cooling degree-days, or different bases.

12. Los Alamos National Laboratory's *Passive Solar Design Handbook,* for instance, tabulates degree-day statistics for various locations using a variety of base temperatures.

13. Lewis and Kohler, p. 36.

14. Some companies that distribute tankless water heaters include ITS Corporation (Englewood, CO), Tankless Heater Corp. (Greenwich, CT), Chronomite Laboratories (Carson, CA), and Controlled Energy Corp. (Waitsfield, VT).

15. Joe Kohler, "Wrap-Up on Tankless Water Heaters," *Solar Age,* Vol. 7, No. 9, September 1982, pp. 30–31. Prices for tankless heaters are those of Chronomite Laboratories, Carson, CA (1984 prices).

16. *PG&E Progress,* March 1983, p. 2.

17. Ibid., p. 5.

18. Ibid.

19. This figure was calculated based on an electricity cost of $0.08 per kilowatt-hour.

20. These barriers guard against air leaks as well as moisture buildup that can damage the insulation and framing.

21. Jane Meyer and Craig Sieben, "Super Saskatoon," *Solar Age*, Vol. 7, No. 1, January 1982, pp. 26–27.

22. This is not entirely true. One of the terms involved in calculating the thermal resistance of a building envelope is the resistance of a surface air film, which is dependent on the local average wind speed.

23. The example is in part excerpted from the *ASHRAE Handbook—1981 Fundamentals*, American Society of Heating, Refrigerating and Air-Conditioning Engineers, Inc., Atlanta, GA, pp. 23–25.

24. J. Douglas Balcomb et al., *Passive Solar Design Handbook*, Vol. 2, Los Alamos National Laboratory, Los Alamos, NM, for U.S. Department of Energy, 1980, pp. 32–34.

25. In deriving this formula, the approximation was made that the thermal resistance of the window equals the number of glazings divided by 1.1.

26. Note that ceiling height times floor area gives an estimate of the building's volume. Multiplying this by number of air changes yields the volume of air that escapes from the house each hour.

SUGGESTED READING

ASHRAE Handbook—1981 Fundamentals. Atlanta, GA: American Society of Heating, Refrigerating and Air-Conditioning Engineers, Inc., 1981. Extensive tables on building materials, heat load, and other engineering data. A basic reference book for the solar designer.

"The Draft Dodgers, Part Two: Caulk Talk," *New Shelter* (Rodale Press), Vol. 4, No. 2, February 1983.

FLOWER, BOB, "Buying the Right Kind," *New Shelter* (Rodale Press), Vol. 3, No. 9, November/December 1982. Insulation data.

MEYER, JANE, AND SIEBEN, CRAIG, "Super Saskatoon," *Solar Age*, Vol. 7, No. 1, January 1982. Superinsulation techniques.

SHURCLIFF, WILLIAM A., "Air-to-Air Heat Exchangers for Houses," *Solar Age*, Vol. 7, No. 3, March 1982.

REVIEW QUESTIONS

1. In which locations in a building is weather stripping installed? In which locations is caulking applied?
2. What are some locations suitable for tension strip weather stripping? When are tension strips not recommended?
3. Which caulks are suitable for exterior seams in which there is some expansion and contraction of the building materials around the seam?
4. What is the function of air-to-air heat exchangers? In what situation are they recommended?
5. Define *thermal resistance* (R-value).
6. Compare the characteristics of fiberglass roll insulation with those of rigid boards.
7. Compare polyurethane and extruded polystyrene board insulations.

8. What are tankless water heaters? What is their advantage over conventional heaters?

9. Why are entry vestibules often installed in superhouses?

PROBLEMS

1. The average daily temperatures (in °F) recorded at a building site for a 10-day period are: 50, 53, 55, 49, 52, 66, 61, 54, 45, and 42. How many degree-days accrue during this period?

2. What is the recommended thickness of fiberglass roll insulation for the roof of a building in an area that experiences 4124 degree-days annually?

3. What is the minimum thickness of extruded polystyrene board insulation that should be used in the exterior walls of a building in an area that experiences 3300 degree-days annually?

4. Calculate the thermal resistance of a wall with the following layers:

Layer	*Material Used*	*Thickness (in.)*
Exterior siding	Plywood	0.75
Vapor barrier	Building paper	—
Insulation	Fiberglass rolls	3
Air space, bounded by foil insulation facing on one side		0.5
Interior paneling	Redwood barrels	0.75

   The framing is 2 × 4's on 16-inch centers. Remember to take into account exterior and interior surface air films.

5. Compute the thermal resistance of a floor with the following layers:

Layer	*Material Used*	*Thickness (in.)*
Finish flooring	Fir	0.75
Subfloor	Plywood	0.5
Air space, bounded by foil insulation facing on one side		0.5
Insulation	Fiberglass batts	3.5

   2 × 8 joists on 16-inch centers are used. A crawl space is under the floor.

6. Use Table 2-8 to find the R-values of the following components of the building envelope:
   (a) A wood-shingled wall with gypsum board interior paneling.
   (b) Linoleum over a wooden subfloor with a crawl space under the floor.
   (c) A composition-shingled pitched roof, with 0.25-inch plywood interior paneling.

7. If R-19 insulation is added to the wall, floor, and roof examined in Problem 6, how will their composite R-values change?

8. Calculate the load coefficients for the wall, window, and door depicted in Figure 2-7. The wall has a thermal resistance of R-21. A solid-core flush door is used. The window is double-glazed and is not used as a solar aperture.

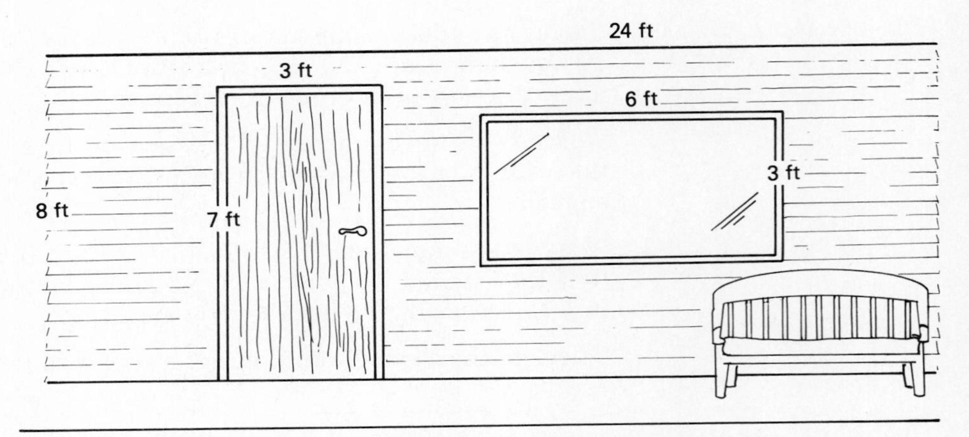

Figure 2-7 The thermal resistance of this wall is R-21. The window is double-glazed and is not a solar aperture. A solid-core flush door is used.

9. Calculate the load coefficient for a slab on grade 36 feet × 24 feet in size. R-12 perimeter insulation is used.

10. Compute $L_{infiltration}$ for a house in Rochester, New York, with 0.5 air change per hour, 11-foot ceilings, and 2000 square feet of floor area. The elevation of the building site may be obtained from Table III-1.

11. The BLC of a house in Greensboro, North Carolina, is 8000 Btu/°F-day. What is its average January heat loss? Refer to Table III-1 for degree-day information.

12. Calculate the BLC of a house with the following characteristics:

Envelope Components	*R-Value*	*Area* (ft^2)
Walls	24	1210
Roof	33	1600
Floor (over unheated crawl space)	13	1370
Nonsolar windows		220
Doors	2.6	63

Windows are double-glazed, and both they and the doors are weather-stripped. The location is Columbus, Ohio (see Table III-1 for elevation). The house has an upstairs area of 300 ft^2, in addition to the ground floor area mentioned above.

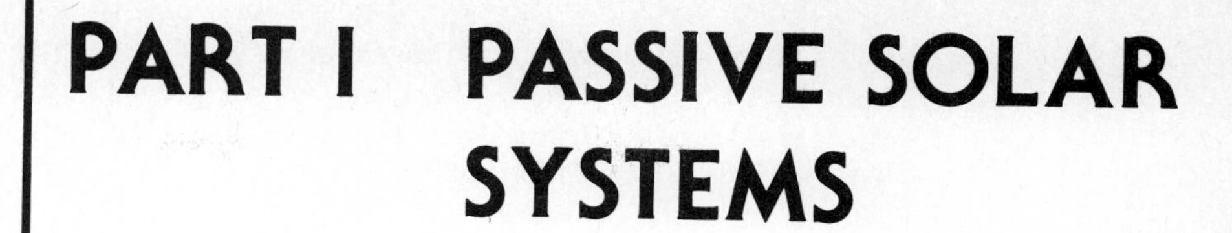

PART I PASSIVE SOLAR SYSTEMS

3 the passively heated building

PASSIVE SYSTEMS VERSUS ACTIVE SYSTEMS

Solar energy systems for heating the interiors of buildings and their water supplies can be divided into two basic types:

1. *Passive systems* are those that collect and distribute heat employing only natural means of thermal energy transmission, such as conduction, natural convection, and radiation.
2. *Active systems* are those that use mechanical means of heat collection and transport, such as air blowers and water pumps.

Typical passive buildings have most of their windows on the south side to allow maximum sunlight to enter during the cold months, overhangs to shade the windows and prevent overheating during the warm months, masonry and concrete floors and walls to absorb and store the thermal energy admitted into the house, and careful site preparation and building orientation to minimize winter shading from trees and other buildings.

Typical active systems use arrays of flat-plate collectors, usually mounted on the roof, to capture solar energy, insulated gravel bins or hot water tanks next to or under the house to store the energy, and blowers, pumps, valves, piping or ducts to distribute it.

There are advantages and disadvantages to both passive and active systems. Passive systems are frequently more economical to install and operate than are active space-heating systems, since the walls, windows, and floors that all buildings need can also function as basic components of the solar system. Passive buildings do not require fans, pumps, or electronic controls for their operation and thus need less maintenance, use no electric power, and last longer.

Active systems' advantages lie largely in their flexibility. The building and its site do not have to be so carefully designed for the system to function successfully. Since the solar collector array can be placed anywhere on or near the house, it is easier to avoid shading problems from buildings and trees. In addition, active systems can usually be retrofitted into existing buildings, whereas passive systems often cannot be.

The interior temperature of a building with an active system can be kept at an almost constant level by using thermostatically controlled heat distribution components. Temperatures in passive homes tend to fluctuate more, although evidence collected by J. Douglas Balcomb of Los Alamos National Laboratory[1] indicates that since the interior surfaces of passive homes are also energy-storing surfaces and thus are at higher temperatures by the end of a sunny day than are those in other houses, they will bathe the buildings' occupants in radiant heat. Because of this, the comfort level of a passive building in many people's opinions equals or exceeds that of other structures (see Figure 3-1).

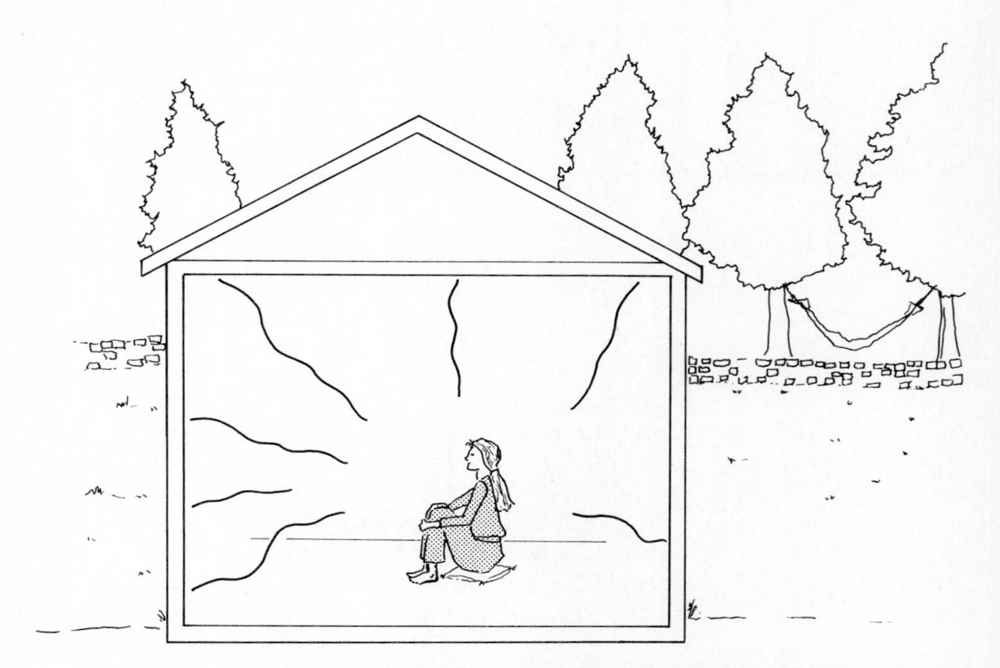

Figure 3-1 Since interior surfaces of passive homes are energy-storing surfaces and at a higher temperature than surfaces in other houses, they bathe the building's occupants in radiant heat. This significantly raises the comfort level.

The purpose of this section is not to prove that one type of system is bad and the other good. Both passive and active designs have their place (see Table 3-1). When planning a solarized building, it is best first to make sure that it is well insulated and weather-stripped, for neither passive nor active systems will perform well in a drafty, underinsulated structure (weatherization of buildings is examined in Chapter 2). Next, the building should be designed to take advantage of whatever passive gains are possible, since passive space-heating elements are usually more economical than are active components. Finally, active systems should be considered for supplying a portion of the remaining thermal energy required for space heating and domestic water heating.

TABLE 3-1 Comparison of Passive and Active Systems

Passive Systems	*Active Systems*
System components are multifunctional, serving as solar energy collectors and thermal energy storage units as well as structural components of the building.	System components usually have only a single function.
No energy source (other than the sun) is required for the system to operate.	Electrical energy is required.
Systems are simple and require minimum maintenance.	Systems are more complex and require more maintenance.
Shading of south wall by nearby obstructions or by the building itself will severely impair performance.	Because collector arrays can be mounted on the roof or away from the house in a nearby location, shading problems can usually be avoided.
It is difficult to retrofit passive systems into existing structures.	Retrofitting active systems into existing structures is usually possible.

Chapters 3 through 8 deal with the design of various types of passive systems. Chapters 9 through 12 examine active systems. All of these chapters should be read carefully by the homeowner or the solar contractor before planning a system.

THE THREE FUNCTIONS OF A PASSIVE BUILDING

The economy and simplicity of a passive solar system lie to a great extent in the threefold purpose for which it is designed. Besides enclosing a living or work space, the passively heated building serves as a *collector of solar radiation* and a *storehouse of thermal energy*. Windows in a passive building, for instance, serve to admit solar energy as well as providing daylight and a view of the outside. A tile floor or masonry wall stores a portion of this energy for use after the sun has set. Roof eaves, if sized and oriented correctly, shade the windows during hot summer days but allow sunlight to fall on them during the colder months.

Figure 3-2 is an illustration of a passively heated house in Bucks County, Pennsylvania.[2] Sunlight admitted through its south-facing windows and skylight strikes the masonry walls and floors within, where it is absorbed as thermal energy and stored for later use. Natural convection processes help carry heat upward from the living room and kitchen on the south side of the house, where most of the solar energy is collected, to the raised bedrooms along the north wall. In summer, awnings and hatches shade south-facing windows and skylights, preventing overheating. The house is designed to have almost half of its space-heating needs met by solar energy.

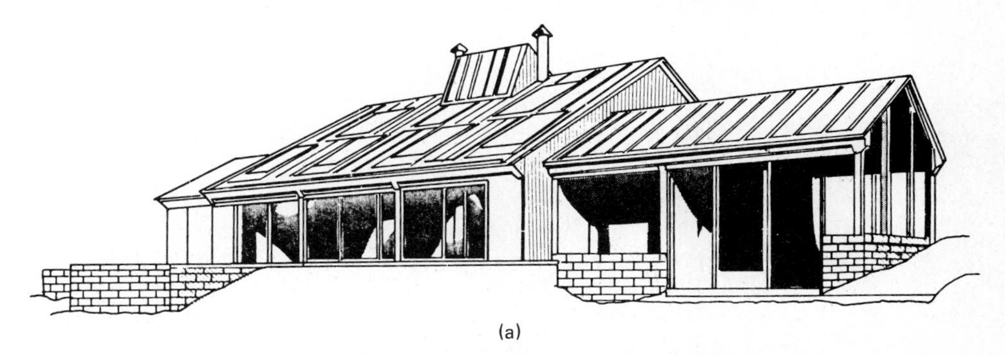

(a)

Figure 3-2 (a) Building using direct-gain passive heating.

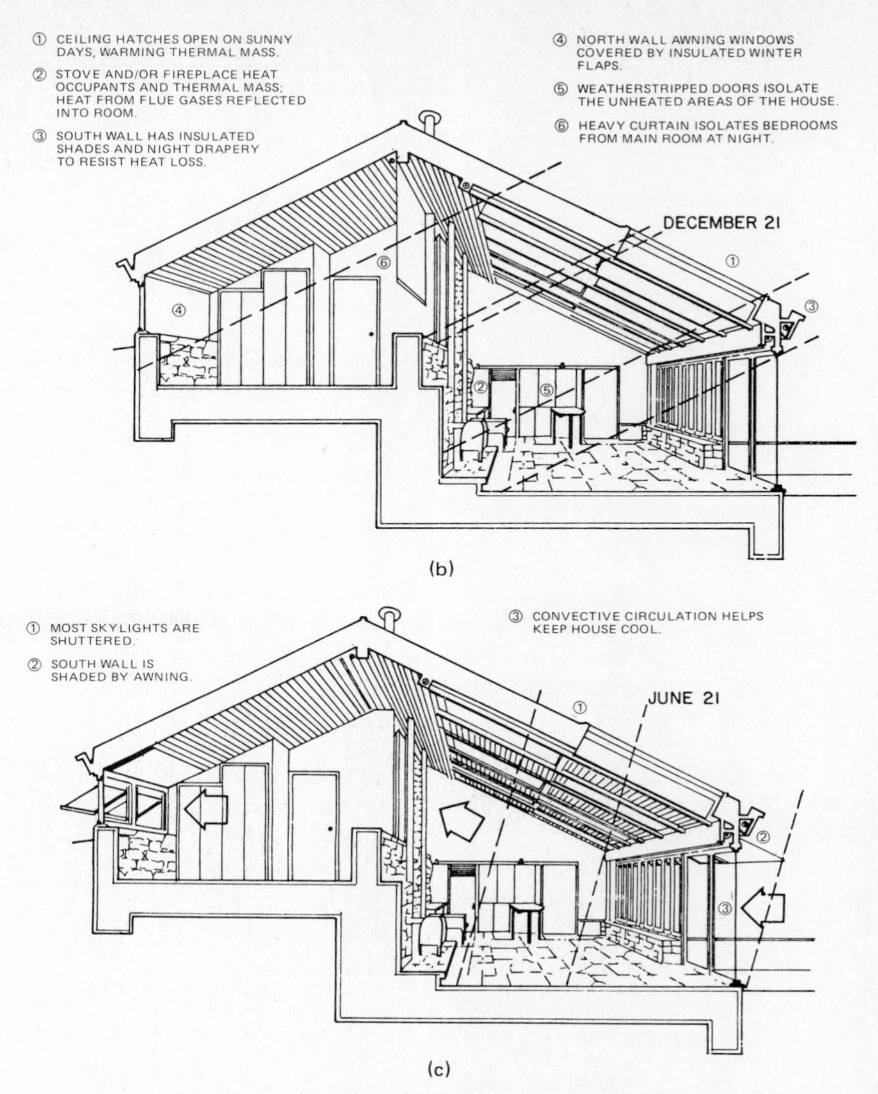
① CEILING HATCHES OPEN ON SUNNY DAYS, WARMING THERMAL MASS.
② STOVE AND/OR FIREPLACE HEAT OCCUPANTS AND THERMAL MASS; HEAT FROM FLUE GASES REFLECTED INTO ROOM.
③ SOUTH WALL HAS INSULATED SHADES AND NIGHT DRAPERY TO RESIST HEAT LOSS.
④ NORTH WALL AWNING WINDOWS COVERED BY INSULATED WINTER FLAPS.
⑤ WEATHERSTRIPPED DOORS ISOLATE THE UNHEATED AREAS OF THE HOUSE.
⑥ HEAVY CURTAIN ISOLATES BEDROOMS FROM MAIN ROOM AT NIGHT.
DECEMBER 21
(b)
① MOST SKYLIGHTS ARE SHUTTERED.
② SOUTH WALL IS SHADED BY AWNING.
③ CONVECTIVE CIRCULATION HELPS KEEP HOUSE COOL.
JUNE 21
(c)

Figure 3-2 (cont.) (b) Winter mode operation. (c) Summer mode operation. (From *Solar Heating and Cooling of Residential Buildings, Sizing, Installation and Operation of Systems*, Solar Energy Applications Laboratory, Colorado State University, for U. S. Department of Commerce, 1980, pp. 3-6, 3-7.)

DIRECT-GAIN, SOLAR WALL, AND SUNSPACE SYSTEMS

The Bucks County house that was just examined uses a *direct-gain* solar system or, in other words, one in which sun is admitted into the living area itself, where its energy is absorbed and stored in the walls, floors, and other surfaces of the rooms. Direct-gain systems are typified by large expanses of glass in the south wall and roof and floors and walls made of massive materials such as concrete, adobe, or brick.

Solar wall systems employ a solar energy collection surface placed between the glazing and the living space. The collecting surface is usually part of a thermal storage mass such as a masonry or water-filled wall (see Figure 3-3).

Sunspaces are enclosed areas whose main purpose is to supply thermal energy to an adjacent building (see Figure 3-4). Solar energy is collected and temporarily stored by the sunspace's floor and/or walls, which are often con-

Figure 3-3 Solar wall. Sunlight passes through the glazing, across an air space, and warms the massive wall. The wall gradually gives off its heat to the living area throughout the evening and night. (Courtesy of North Carolina State University, School of Engineering.)

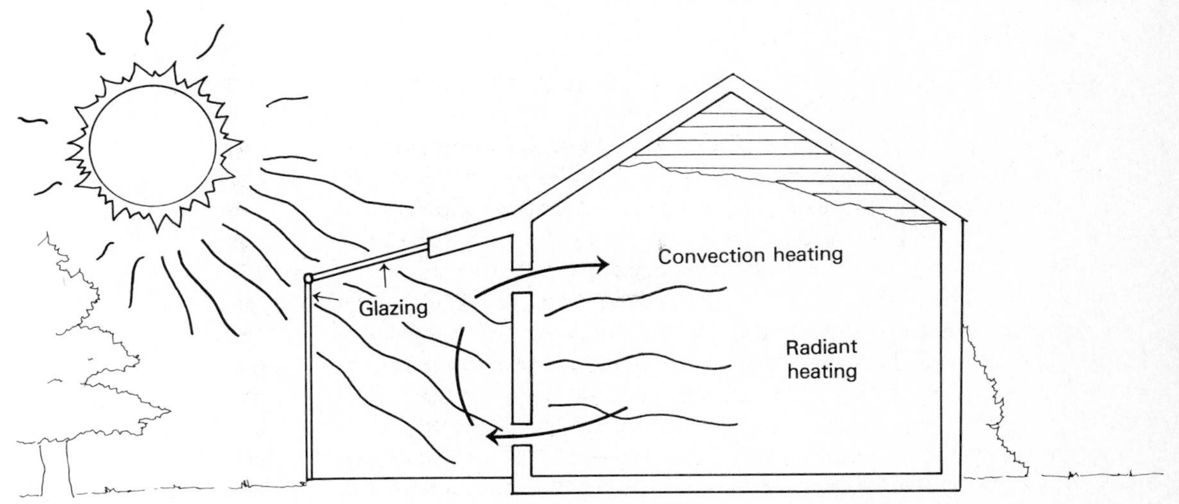

Figure 3-4 Sunspace. Energy is collected by the sunspace's walls and floor. As with solar walls, thermal energy is transferred to the living space both by convective air currents and by radiant heating.

structed of masonry to improve their thermal storage capacities. Secondary functions of the sunspace include use as a greenhouse, as additional living space, or as an enclosure for a spa or swimming pool.

SHADING CONSIDERATIONS

Shading from nearby trees, buildings, or hills can easily spoil an otherwise carefully thought out house design. Problems such as these can often be avoided by carefully inspecting the construction site before plans are drawn up. Instruments such as the Solar Card or the Solar Site Selector[3] are designed to provide a shading assessment of the building site for any month of the year (see Figure 3-5). The Solar Card is very reasonable in cost ($12.95 in 1984) and provides a fairly good shadow assessment. It is simple to use, but a compass or another way of finding true south is generally necessary. A building contractor or architect, on the other hand, might prefer the more

(a)

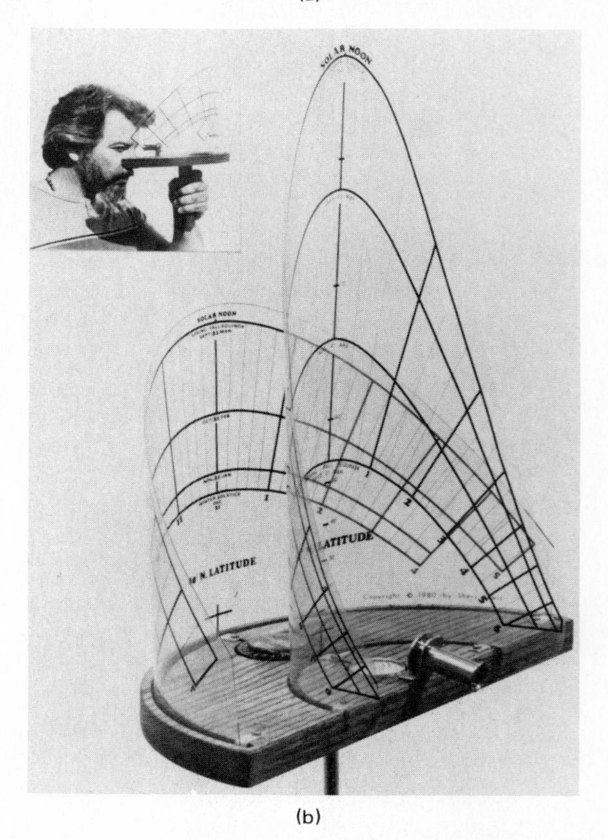

(b)

Figure 3-5 (a) Use of the solar card. (b) Solar Site Selector. [(b), Courtesy of Lewis and Associates, 105 Rockwood Drive, Grass Valley, CA 95945. The man in the picture is David Wright, environmental architect.]

professional and precise Solar Site Selector (its cost was $89.50 in 1984). Built into it is a bubble level and compass, and it is especially good for plotting early morning and late afternoon sun paths as well as those in the middle of the day.[4]

A reasonable shadow assessment can also be made using a compass, a protractor, and a plumb bob (a fishing weight or other small weight can be substituted for the plumb bob). The plumb bob is hung from the center of the protractor's straight side, as illustrated in Figure 3-6.

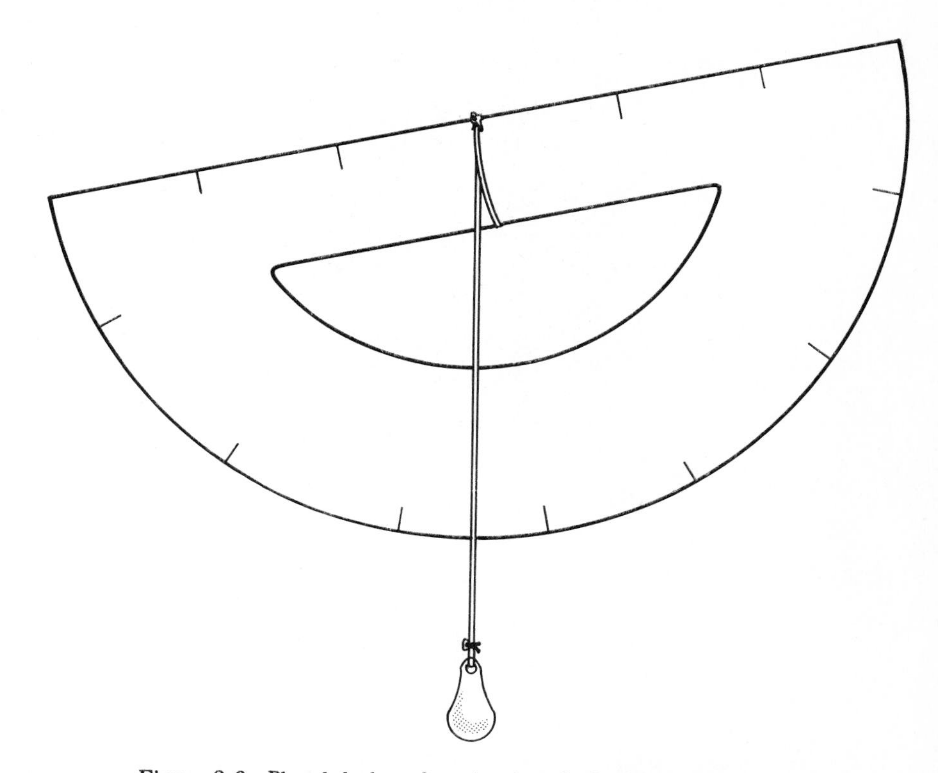

Figure 3-6 Plumb bob and protractor shadow assessment tool.

To do the assessment, two plots need to be made: one of the building site's skyline (the boundary between sky and horizon), and one of the sun's path through the sky. When the two plots are compared, shading problems can be predicted.

Make both plots on one piece of graph paper. Altitude angles of the sun or skyline are marked on the vertical axis, and azimuth angles on the horizontal axis. (Azimuth measures the angle between a certain direction and that of true south. When the sun is directly to the east, for example, its azimuth is 90° east.) Figure 3-7 depicts a sample plot.

To plot the skyline, it is necessary to know the direction of true south. This can be determined using a compass and the isogonic chart in Figure 3-8, which aids in changing magnetic south to true, or geographic south.

Using the protractor and plumb bob and standing at a location where a solar collection surface is planned to be installed, determine the altitude of the skyline for azimuth angles between 90° east and 90° west, taking a reading every 15°. To take a reading, sight along the flat edge of the protractor, as in Figure 3-9, aiming at the skyline. Obstructions such as buildings or hills raise the altitude of the skyline. Plot all readings on the graph paper, and

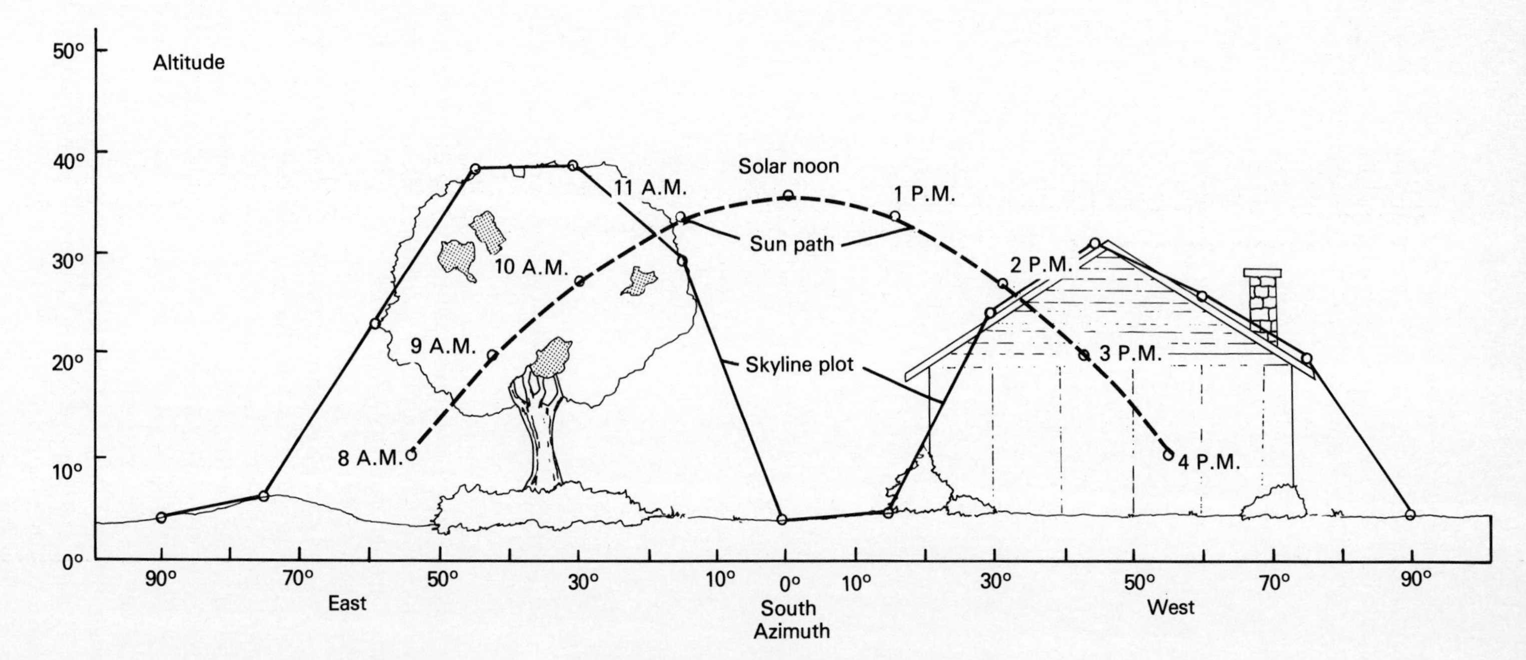

Figure 3-7 Sun path and skyline plot.

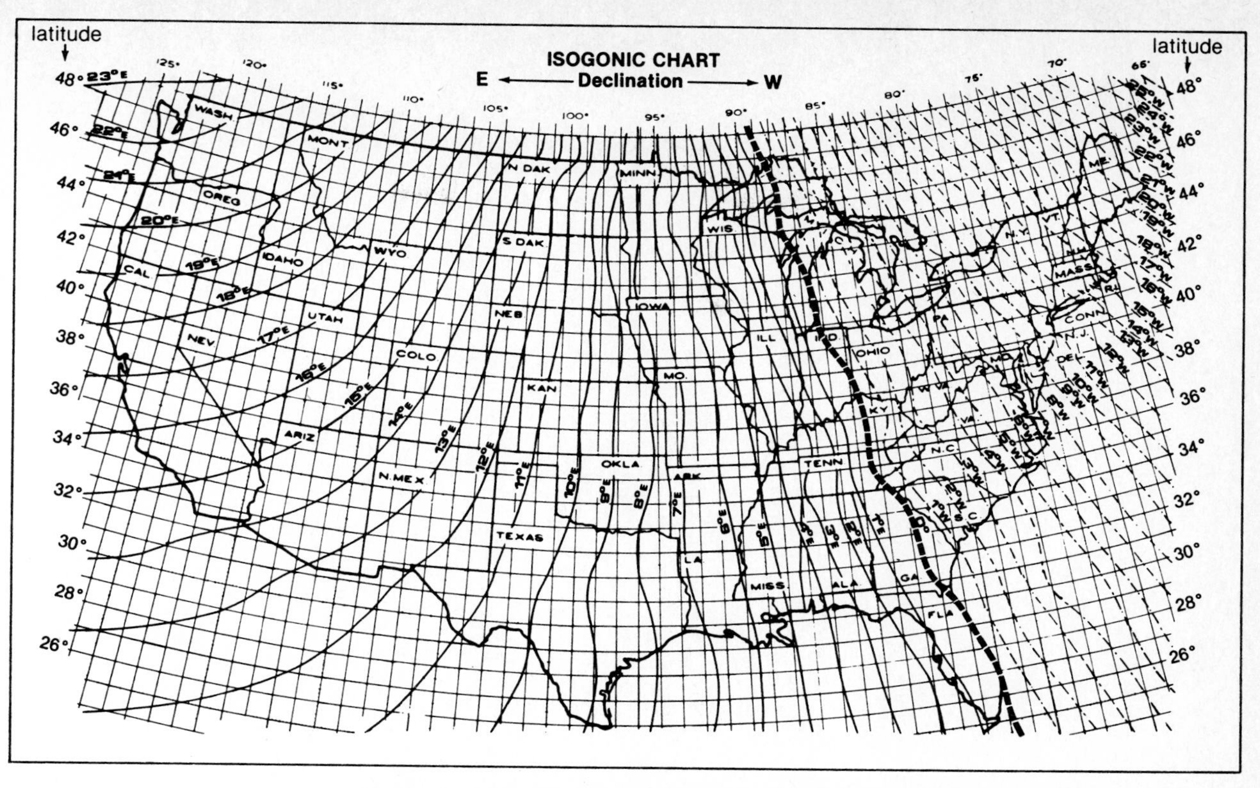

Figure 3-8 Isogonic chart. (Courtesy of Lewis and Associates, 105 Rockwood Drive, Grass Valley, CA 95945.)

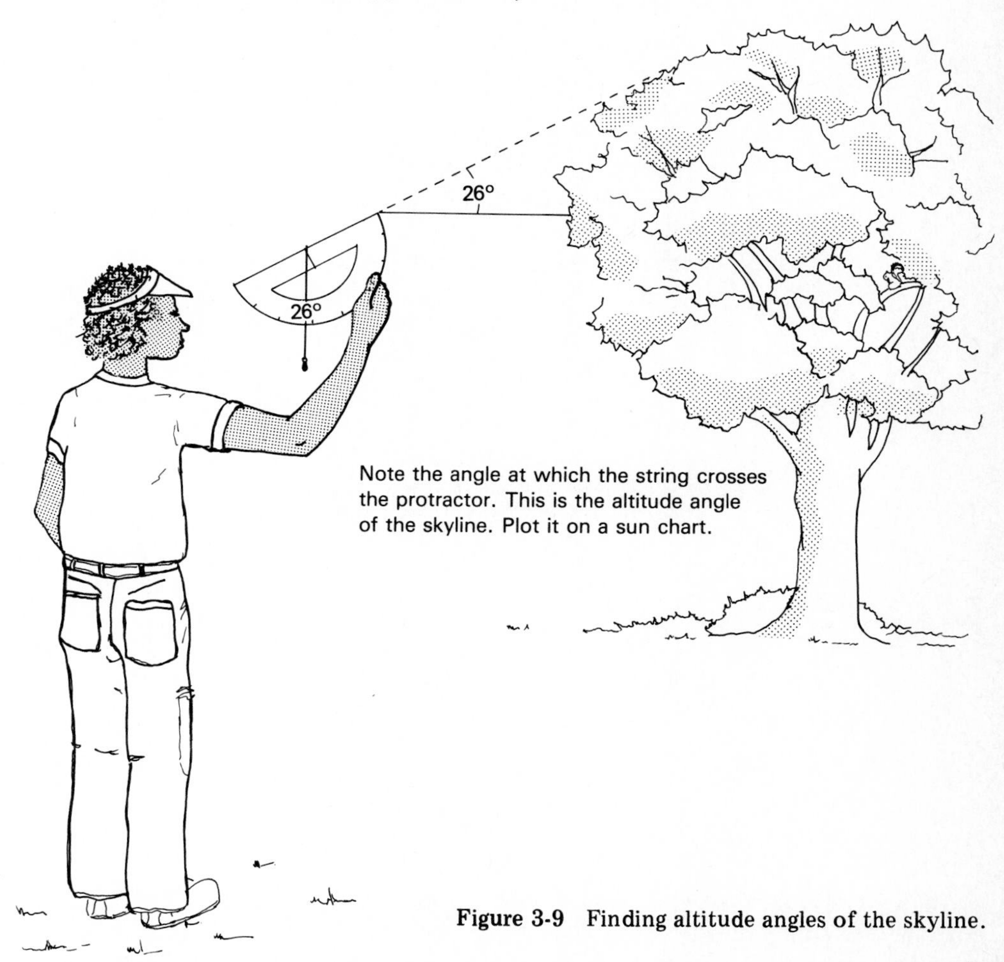

Figure 3-9 Finding altitude angles of the skyline.

connect the dots. Figure 3-7 depicts such a plot. The figure also includes sketches of the terrain, including obstructions. Similar sketches may be constructed on the reader's plots if desired.

Next, make a plot of the sun's path in the sky on December 21. (If the solar energy system will only be used for part of the year, plot the sun's path for the month of the solar system's season during which the sun is lowest in the sky.) The sun's altitude and azimuth angles at different latitudes and times of the year are listed in Table 3-2. For instance, on December 21, at a latitude of 32° north, the sun's altitudes and azimuths are as follows:

Solar Time		*Sun's Position*	
A.M.	*P.M.*	*Altitude*	*Azimuth*
8	4	10.3	53.8
9	3	19.8	43.6
10	2	27.6	31.2
11	1	32.7	16.4
12		34.6	0.0

These are plotted in Figure 3-7. Notice that at 8 A.M., the sun is at an azimuth of 53.8 *east.* At 4 P.M., its azimuth is 53.8 *west.*

Now compare the two plots. Whenever the skyline plot is *above* the sun path, the building site is in shade. This occurs until almost 11 A.M. and after 2:10 P.M. Whenever the sun path is above the skyline, the building site receives direct solar energy.

A good solar site should receive direct insolation from at least 9 A.M. to 3 P.M. throughout the winter. The site in the foregoing example is thus far from ideal. Either severe pruning of the tree or moving to another location should be considered.

Obstructions such as deciduous trees will shade a building less in winter (when they shed their leaves) than will evergreen trees, but even the trunk and bare branches can cause problems, for they can block as much as 50 percent of the sunlight that the fully leafed tree would.

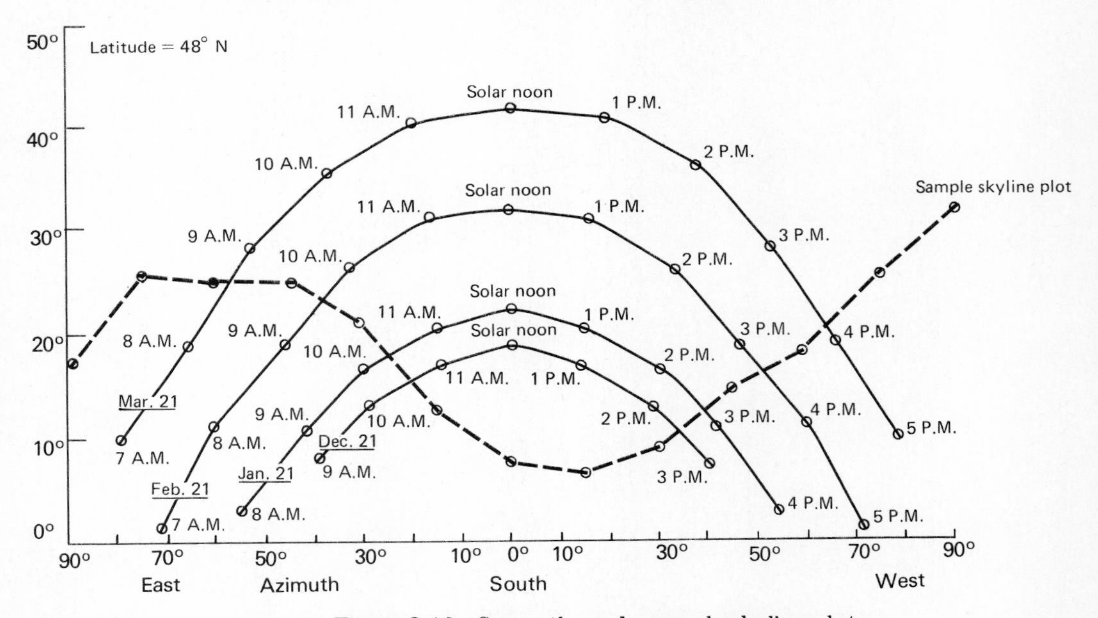

Figure 3-10 Sun paths and a sample skyline plot.

To fully evaluate a site for the coldest months of the year, sun paths for January 21, February 21, and March 21 (which are virtually the same as sun paths on November 21, October 21, and September 21) should also be plotted and compared with the skyline plot (see Figure 3-10). At the very least, construct an additional sun path graph for February 21. All the graphs can be plotted on the same piece of paper.

TABLE 3-2 Solar Altitude and Azimuth

Table I 24 Degrees North Latitude

Date	Solar Time A.M.	Solar Time P.M.	Solar Position Altitude	Solar Position Azimuth
Jan 21	7	5	4.8	65.6
	8	4	16.9	58.3
	9	3	27.9	48.8
	10	2	37.2	36.1
	11	1	43.6	19.6
	12		46.0	0.0
Feb 21	7	5	9.3	74.6
	8	4	22.3	67.2
	9	3	34.4	57.6
	10	2	45.1	44.2
	11	1	53.0	25.0
	12		56.0	0.0
Mar 21	7	5	13.7	83.8
	8	4	27.2	76.8
	9	3	40.2	67.9
	10	2	52.3	54.8
	11	1	61.9	33.4
	12		66.0	0.0
Apr 21	6	6	4.7	100.6
	7	5	18.3	94.9
	8	4	32.0	89.0
	9	3	45.6	81.9
	10	2	59.0	71.8
	11	1	71.1	51.6
	12		77.6	0.0
May 21	6	6	8.0	108.4
	7	5	21.2	103.2
	8	4	34.6	98.5
	9	3	48.3	93.6
	10	2	62.0	87.7
	11	1	75.5	76.9
	12		86.0	0.0
Jun 21	6	6	9.3	111.6
	7	5	22.3	106.8
	8	4	35.5	102.6
	9	3	49.0	98.7
	10	2	62.6	95.0
	11	1	76.3	90.8
	12		89.4	0.0

Date	Solar Time A.M.	Solar Time P.M.	Solar Position Altitude	Solar Position Azimuth
Jul 21	6	6	8.2	109.0
	7	5	21.4	103.8
	8	4	34.8	99.2
	9	3	48.4	94.5
	10	2	62.1	89.0
	11	1	75.7	79.2
	12		86.6	0.0
Aug 21	6	6	5.0	101.3
	7	5	18.5	95.6
	8	4	32.2	89.7
	9	3	45.9	82.9
	10	2	59.3	73.0
	11	1	71.6	53.2
	12		78.3	0.0
Sep 21	7	5	13.7	83.8
	8	4	27.2	76.8
	9	3	40.2	67.9
	10	2	52.3	54.8
	11	1	61.9	33.4
	12		66.0	0.0
Oct 21	7	5	9.1	74.1
	8	4	22.0	66.7
	9	3	34.1	57.1
	10	2	44.7	43.8
	11	1	52.5	24.7
	12		55.5	0.0
Nov 21	7	5	4.9	65.8
	8	4	17.0	58.4
	9	3	28.0	48.9
	10	2	37.3	36.3
	11	1	43.8	19.7
	12		46.2	0.0
Dec 21	7	5	3.2	62.6
	8	4	14.9	55.3
	9	3	25.5	46.0
	10	2	34.3	33.7
	11	1	40.4	18.2
	12		42.6	0.0

TABLE 3-2 (cont.)

Table II 32 Degrees North Latitude

	Solar Time		Solar Position			Solar Time		Solar Position	
Date	*A.M.*	*P.M.*	*Altitude*	*Azimuth*	*Date*	*A.M.*	*P.M.*	*Altitude*	*Azimuth*
Jan 21	7	5	1.4	65.2	Jul 21	6	6	10.7	107.7
	8	4	12.5	56.5		7	5	23.1	100.6
	9	3	22.5	46.0		8	4	35.7	93.6
	10	2	30.6	33.1		9	3	48.4	85.5
	11	1	36.1	17.5		10	2	60.9	74.3
	12		38.0	0.0		11	1	72.4	53.3
						12		78.6	0.0
Feb 21	7	5	7.1	73.5	Aug 21	6	6	6.5	100.5
	8	4	19.0	64.4		7	5	19.1	92.8
	9	3	29.9	53.4		8	4	31.8	84.7
	10	2	39.1	39.4		9	3	44.3	75.0
	11	1	45.6	21.4		10	2	56.1	61.3
	12		48.0	0.0		11	1	66.0	38.4
						12		70.3	0.0
Mar 21	7	5	12.7	81.9	Sep 21	7	5	12.7	81.9
	8	4	25.1	73.0		8	4	25.1	73.0
	9	3	36.8	62.1		9	3	36.8	62.1
	10	2	47.3	47.5		10	2	47.3	47.5
	11	1	55.0	26.8		11	1	55.0	26.8
	12		58.0	0.0		12		58.0	0.0
Apr 21	6	6	6.1	99.9	Oct 21	7	5	6.8	73.1
	7	5	18.8	92.2		8	4	18.7	64.0
	8	4	31.5	84.0		9	3	29.5	53.0
	9	3	43.9	74.2		10	2	38.7	39.1
	10	2	55.7	60.3		11	1	45.1	21.1
	11	1	65.4	37.5		12		47.5	0.0
	12		69.6	0.0					
May 21	6	6	10.4	107.2	Nov 21	7	5	1.5	65.4
	7	5	22.8	100.1		8	4	12.7	56.6
	8	4	35.4	92.9		9	3	22.6	46.1
	9	3	48.1	84.7		10	2	30.8	33.2
	10	2	60.6	73.3		11	1	36.2	17.6
	11	1	72.0	51.9		12		38.2	0.0
	12		78.0	0.0					
Jun 21	6	6	12.2	110.2	Dec 21	8	4	10.3	53.8
	7	5	24.3	103.4		9	3	19.8	43.6
	8	4	36.9	96.8		10	2	27.6	31.2
	9	3	49.6	89.4		11	1	32.7	16.4
	10	2	62.2	79.7		12		34.6	0.0
	11	1	74.2	60.9					
	12		81.5	0.0					

TABLE 3-2 (cont.)

Table III 40 Degrees North Latitude

Date	Solar Time A.M.	Solar Time P.M.	Solar Position Altitude	Solar Position Azimuth
Jan 21	8	4	8.1	55.3
	9	3	16.8	44.0
	10	2	23.8	30.9
	11	1	28.4	16.0
	12		30.0	0.0
Feb 21	7	5	4.8	72.7
	8	4	15.4	62.2
	9	3	25.0	50.2
	10	2	32.8	35.9
	11	1	38.1	18.9
	12		40.0	0.0
Mar 21	7	5	11.4	80.2
	8	4	22.5	69.6
	9	3	32.8	57.3
	10	2	41.6	41.9
	11	1	47.7	22.6
	12		50.0	0.0
Apr 21	6	6	7.4	98.9
	7	5	18.9	89.5
	8	4	30.3	79.3
	9	3	41.3	67.2
	10	2	51.2	51.4
	11	1	58.7	29.2
	12		61.6	0.0
May 21	5	7	1.9	114.7
	6	6	12.7	105.6
	7	5	24.0	96.6
	8	4	35.4	87.2
	9	3	46.8	76.0
	10	2	57.5	60.9
	11	1	66.2	37.1
	12		70.0	0.0
Jun 21	5	7	4.2	117.3
	6	6	14.8	108.4
	7	5	26.0	99.7
	8	4	37.4	90.7
	9	3	48.8	80.2
	10	2	59.8	65.8
	11	1	69.2	41.9
	12		73.5	0.0
Jul 21	5	7	2.3	115.2
	6	6	13.1	106.1
	7	5	24.3	97.2
	8	4	35.8	87.8
	9	3	47.2	76.7
	10	2	57.9	61.7
	11	1	66.7	37.9
	12		70.6	0.0
Aug 21	6	6	7.9	99.5
	7	5	19.3	90.0
	8	4	30.7	79.9
	9	3	41.8	67.9
	10	2	51.7	52.1
	11	1	59.3	29.7
	12		62.3	0.0
Sep 21	7	5	11.4	80.2
	8	4	22.5	69.6
	9	3	32.8	57.3
	10	2	41.6	41.9
	11	1	47.7	22.6
	12		50.0	0.0
Oct 21	7	5	4.5	72.3
	8	4	15.0	61.9
	9	3	24.5	49.8
	10	2	32.4	35.6
	11	1	37.6	18.7
	12		39.5	0.0
Nov 21	8	4	8.2	55.4
	9	3	17.0	44.1
	10	2	24.0	31.0
	11	1	28.6	16.1
	12		30.2	0.0
Dec 21	8	4	5.5	53.0
	9	3	14.0	41.9
	10	2	20.7	29.4
	11	1	25.0	15.2
	12		26.6	0.0

TABLE 3-2 (cont.)

Table IV 48 Degrees North Latitude

	Solar Time		*Solar Position*			*Solar Time*		*Solar Position*	
Date	*A.M.*	*P.M.*	*Altitude*	*Azimuth*	*Date*	*A.M.*	*P.M.*	*Altitude*	*Azimuth*
Jan 21	8	4	3.5	54.6	Jul 21	5	7	5.7	114.7
	9	3	11.0	42.6		6	6	15.2	104.1
	10	2	16.9	29.4		7	5	25.1	93.5
	11	1	20.7	15.1		8	4	35.1	82.1
	12		22.0	0.0		9	3	44.8	68.8
						10	2	53.5	51.9
						11	1	60.1	29.0
						12		62.6	0.0
Feb 21	7	5	2.4	72.2	Aug 21	6	6	9.1	98.3
	8	4	11.6	60.5		7	5	19.1	87.2
	9	3	19.7	47.7		8	4	29.0	75.4
	10	2	26.2	33.3		9	3	38.4	61.8
	11	1	30.5	17.2		10	2	46.4	45.1
	12		32.0	0.0		11	1	52.2	24.3
						12		54.3	0.0
Mar 21	7	5	10.0	78.7	Sep 21	7	5	10.0	78.7
	8	4	19.5	66.8		8	4	19.5	66.8
	9	3	28.2	53.4		9	3	28.2	53.4
	10	2	35.4	37.8		10	2	35.4	37.8
	11	1	40.3	19.8		11	1	40.3	19.8
	12		42.0	0.0		12		42.0	0.0
Apr 21	6	6	8.6	97.8	Oct 21	7	5	2.0	71.9
	7	5	18.6	86.7		8	4	11.2	60.2
	8	4	28.5	74.9		9	3	19.3	47.4
	9	3	37.8	61.2		10	2	25.7	33.1
	10	2	45.8	44.6		11	1	30.0	17.1
	11	1	51.5	24.0		12		31.5	0.0
	12		53.6	0.0					
May 21	5	7	5.2	114.3	Nov 21	8	4	3.6	54.7
	6	6	14.7	103.7		9	3	11.2	42.7
	7	5	24.6	93.0		10	2	17.1	29.5
	8	4	34.7	81.6		11	1	20.9	15.1
	9	3	44.3	68.3		12		22.2	0.0
	10	2	53.0	51.3					
	11	1	59.5	28.6					
	12		62.0	0.0					
Jun 21	5	7	7.9	116.5	Dec 21	9	3	8.0	40.9
	6	6	17.2	106.2		10	2	13.6	28.2
	7	5	27.0	95.8		11	1	17.3	14.4
	8	4	37.1	84.6		12		18.6	0.0
	9	3	46.9	71.6					
	10	2	55.8	54.8					
	11	1	62.7	31.2					
	12		65.5	0.0					

TABLE 3-2 (cont.)

Table V 56 Degrees North Latitude

Date	Solar Time		Solar Position	
	A.M.	P.M.	Altitude	Azimuth
Jan 21	9	3	5.0	41.8
	10	2	9.9	28.5
	11	1	12.9	14.5
	12		14.0	0.0
Feb 21	8	4	7.6	59.4
	9	3	14.2	45.9
	10	2	19.4	31.5
	11	1	22.8	16.1
	12		24.0	0.0
Mar 21	7	5	8.3	77.5
	8	4	16.2	64.4
	9	3	23.3	50.3
	10	2	29.0	34.9
	11	1	32.7	17.9
	12		34.0	0.0
Apr 21	5	7	1.4	108.8
	6	6	9.6	96.5
	7	5	18.0	84.1
	8	4	26.1	70.9
	9	3	33.6	56.3
	10	2	39.9	39.7
	11	1	44.1	20.7
	12		45.6	0.0
May 21	4	8	1.2	125.5
	5	7	8.5	113.4
	6	6	16.5	101.5
	7	5	24.8	89.3
	8	4	33.1	76.3
	9	3	40.9	61.6
	10	2	47.6	44.2
	11	1	52.3	23.4
	12		54.0	0.0
Jun 21	4	8	4.2	127.2
	5	7	11.4	115.3
	6	6	19.3	103.6
	7	5	27.6	91.7
	8	4	35.9	78.8
	9	3	43.8	64.1
	10	2	50.7	46.4
	11	1	55.6	24.9
	12		57.5	0.0
Jul 21	4	8	1.7	125.8
	5	7	9.0	113.7
	6	6	17.0	101.9
	7	5	25.3	89.7
	8	4	33.6	76.7
	9	3	41.4	62.0
	10	2	48.2	44.6
	11	1	52.9	23.7
	12		54.6	0.0
Aug 21	5	7	2.0	109.2
	6	6	10.2	97.0
	7	5	18.5	84.5
	8	4	26.7	71.3
	9	3	34.3	56.7
	10	2	40.5	40.0
	11	1	44.8	20.9
	12		46.3	0.0
Sep 21	7	5	8.3	77.5
	8	4	16.2	64.4
	9	3	23.3	50.3
	10	2	29.0	34.9
	11	1	32.7	17.9
	12		34.0	0.0
Oct 21	8	4	7.1	59.1
	9	3	13.8	45.7
	10	2	19.0	31.3
	11	1	22.3	16.0
	12		23.5	0.0
Nov 21	9	3	5.2	41.9
	10	2	10.0	28.5
	11	1	13.1	14.5
	12		14.2	0.0
Dec 21	9	3	1.9	40.5
	10	2	6.6	27.5
	11	1	9.5	13.9
	12		10.6	0.0

TABLE 3-2 (cont.)

Table VI 64 Degrees North Latitude

	Solar Time		*Solar Position*			*Solar Time*		*Solar Position*	
Date	A.M.	P.M.	*Altitude*	*Azimuth*	*Date*	A.M.	P.M.	*Altitude*	*Azimuth*
Jan 21	10	2	2.8	28.1	Jul 21	4	8	6.4	125.3
	11	1	5.2	14.1		5	7	12.1	112.4
	12		6.0	0.0		6	6	18.4	99.4
						7	5	25.0	86.0
						8	4	31.4	71.8
						9	3	37.3	56.3
						10	2	42.2	39.2
						11	1	45.4	20.2
						12		46.6	0.0
Feb 21	8	4	3.4	58.7	Aug 21	5	7	4.6	108.8
	9	3	8.6	44.8		6	6	11.0	95.5
	10	2	12.6	30.3		7	5	17.6	81.9
	11	1	15.1	15.3		8	4	23.9	67.8
	12		16.0	0.0		9	3	29.6	52.6
						10	2	34.2	36.2
						11	1	37.2	18.5
						12		38.3	0.0
Mar 21	7	5	6.5	76.5	Sep 21	7	5	6.5	76.5
	8	4	20.7	62.6		8	4	12.7	62.6
	9	3	18.1	48.1		9	3	18.1	48.1
	10	2	22.3	32.7		10	2	22.3	32.7
	11	1	25.1	16.6		11	1	25.1	16.6
	12		26.0	0.0		12		26.0	0.0
Apr 21	5	7	4.0	108.5	Oct 21	8	4	3.0	58.5
	6	6	10.4	95.1		9	3	8.1	44.6
	7	5	17.0	81.6		10	2	12.1	30.2
	8	4	23.3	67.5		11	1	14.6	15.2
	9	3	29.0	52.3		12		15.5	0.0
	10	2	33.5	36.0					
	11	1	36.5	18.4					
	12		97.6	0.0					
May 21	4	8	5.8	125.1	Nov 21	10	2	3.0	28.1
	5	7	11.6	112.1		11	1	5.4	14.2
	6	6	17.9	99.1		12		6.2	0.0
	7	5	24.5	85.7					
	8	4	30.9	71.5					
	9	3	36.8	56.1					
	10	2	41.6	38.9					
	11	1	44.9	20.1					
	12		46.0	0.0					
Jun 21	3	9	4.2	139.4	Dec 21	11	1	1.8	13.7
	4	8	9.0	126.4		12		2.6	0.0
	5	7	14.7	113.6					
	6	6	21.0	100.8					
	7	5	27.5	87.5					
	8	4	34.0	73.3					
	9	3	39.9	57.8					
	10	2	44.9	40.4					
	11	1	48.3	20.9					
	12		49.5	0.0					

Reprinted with permission from American Society of Heating, Refrigerating and Air-Conditioning Engineers, Inc., Atlanta, Georgia.

NOTES

1. J. Douglas Balcomb et al., *Passive Solar Design Handbook*, Vol. 2, Los Alamos National Laboratory, Los Alamos, NM, for U.S. Department of Energy, 1980, p. viii.

2. The figure is from *Solar Heating and Cooling of Residential Buildings, 1980 Edition*, U.S. Department of Commerce, Washington, D.C., pp. 3-6.

3. "Rating the Site Survey Tools," *Solar Age*, Vol. 8, No. 6, June 1983.

SUGGESTED READING

ANDERSON, BRUCE, *The Solar Home Book*. Andover, MA: Brick House Publishing Company, 1976.

ANDERSON, BRUCE, AND WELLS, MALCOLM, *Passive Solar Energy*. Andover, MA: Brick House Publishing Company, 1981.

REVIEW QUESTIONS

1. Define *active* and *passive* systems.
2. Compare characteristics of the two types of systems.
3. What are the three functions of a passive building?
4. Solar walls and sunspaces both employ glazing and massive materials to store thermal energy. The function is to heat on adjacent living space. So what is the difference between solar walls and sunspaces?

PROBLEMS

1. Using any of the shading assessment tools discussed, stand near your home and plot the skyline's altitude angles between 90° east and 90° west, taking a reading every 10°. Plot the readings on a sun chart.
2. During which hours is the location referred to in Figure 3-10 shaded on February 21? On April 21?

4 direct-gain systems I: solar apertures and movable insulation

Direct-gain systems allow ample amounts of sunlight to enter the living and working spaces of the building, where its energy is absorbed and stored in floors, walls, ceilings, fireplaces, and to some extent even furniture. To increase energy storage capacity, massive materials such as concrete, brick, and adobe are used in construction of the building's components.

This chapter examines methods of admitting sufficient solar energy into a building to satisfy its needs, while permitting only a minimum to escape. The next chapter is concerned with storing that energy, as well as preventing too much of it from entering during warmer times of the year when it is not wanted.

Adobe home with direct-gain solar apertures. (Courtesy of Ben Gilbert, Gilbert's Adobe Homes, Los Lunas, NM.)

LETTING THE SUNSHINE IN: THE DESIGN OF SOLAR APERTURES

A *solar aperture* is a glazed opening that is designed to admit more solar energy into a building than the amount of energy it allows to escape. Solar apertures are, in other words, net energy gainers.

To fulfill their function, solar apertures must be designed with care. They are usually located on the south side of the building in places that receive full sunlight during the heating season of the year. They are usually double-glazed and should always be tightly weather-stripped. Their performance is greatly improved if covered at night with some type of movable insulation to reduce heat losses.

Types of Apertures: Windows, Clerestories, and Skylights

There are three kinds of sunlight-admitting apertures commonly used in passively heated buildings: windows, clerestories, and skylights. Clerestories are vertical or almost vertical openings projecting upward from the plane of the roof and above the line of sight of the building's occupants. Skylights are openings that follow the plane of the roof (see Figure 4-1).

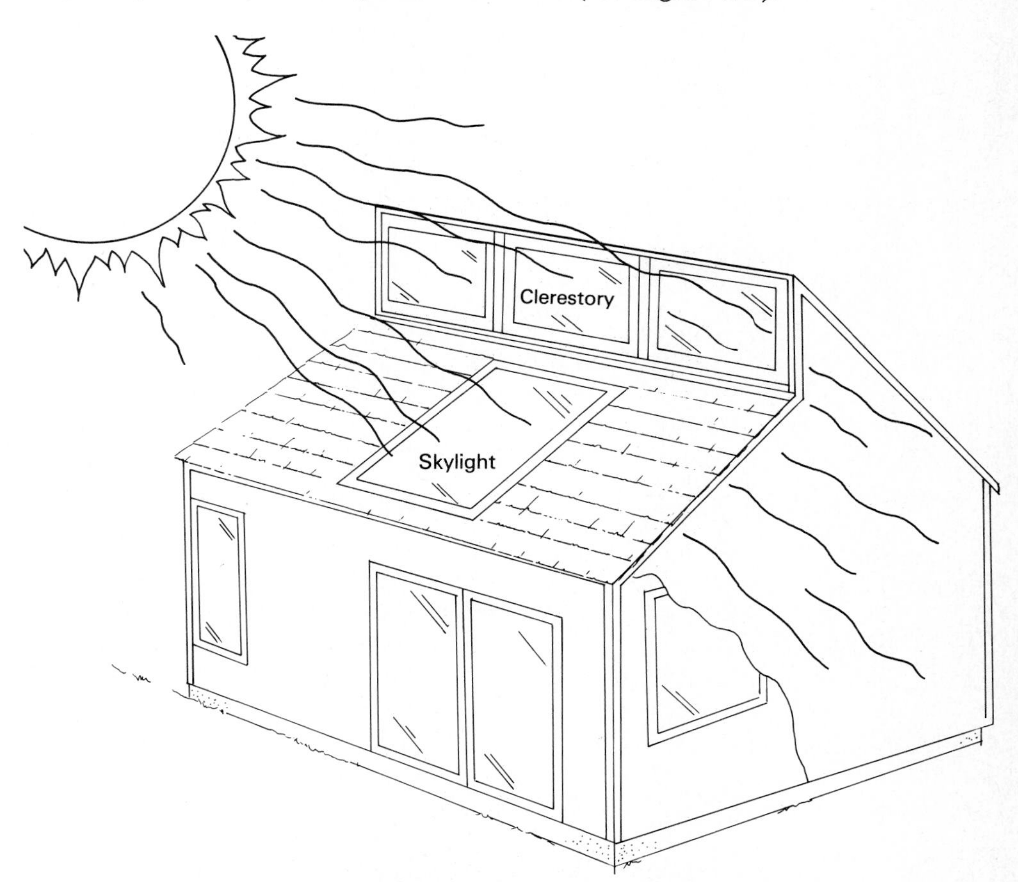

Figure 4-1 Clerestories and skylights make it possible for sunlight to reach areas deep in the interior of the building.

Windows are the most common type of solar aperture. Sometimes, however, low obstructions near the building's south wall severely limit the amount of sunlight that a window can collect, making roof-mounted apertures preferable. Clerestories and skylights also make it possible for surfaces deep in the interior of the building to receive sunlight (see Figure 4-1).

Skylights in shallow-pitched roofs are not at effective angles for wintertime solar energy collection when the sun is low in the sky. In summer, when

Figure 4-2 Disadvantages of skylights in shallow-pitched roofs. (a) In winter, they are not at favorable angles for solar collection. (b) In summer, they tend to collect too much energy.

the sun is high in the sky, shallow-angled skylights often collect too much energy and cause the building to overheat, as illustrated in Figure 4-2. If skylights are to be installed on the south slope of a roof, it is recommended that either insulated coverings be fabricated for them or that overhangs that provide summer shade be installed. Insulated coverings will not only reduce summer overheating; they can also minimize winter heat losses if used at night. Skylights are more prone to heat losses than are windows, since by being in the roof they come into contact with the warmest air in the building, which has been carried upward by convection currents.

Energy Transmittance

Solar apertures are designed to admit a maximum amount of solar energy into the building while allowing a minimum amount to escape. Tinted glass and glass with reflective surface coatings are not recommended for solar apertures because they reduce the transmittance of light. This, in fact, is what these special materials were originally designed for, to prevent summer overheating of a building.

Reflection and Absorption. No pane of glass transmits 100 percent of the sunlight falling on it. Some energy is reflected away by the inside and outside surfaces of the glazing, and some is absorbed by the glass itself. A ⅛-inch-thick pane of ordinary window "float" glass reflects about 8 percent of normally (perpendicularly) incident sunlight and absorbs 7 percent.[1] Absorbed energy is not totally lost, however. It is converted into thermal energy and then reradiated in both directions from the window in the form of infrared "heat waves." The portion radiated toward the interior of the house helps to warm it.

The amount of energy absorbed by a pane of glass is somewhat dependent on its iron oxide content. Float glass contains 0.1 to 0.13 percent iron oxide, while low-iron glasses made for solar energy applications contain as little as 0.01 percent. In a comparative study of solar apertures, it was found that a double glazing of low iron glass leads to a 4.7 percent greater energy gain than if float glass had been used.[2] Float glass is, however, far easier to obtain and usually less expensive than low-iron glass. Iron impurities occur naturally in glass, in varying amounts, and give the glass a greenish tinge that can be seen when it is held edge on. Additional iron oxide is added to float glass during the manufacturing process to bring its color up to a standard level of tint, rather than having different batches of glass be of slightly different shades.

Reflective losses in glass can be reduced by surface etching techniques or by the addition of antireflective coatings. These methods appear to offer a greater energy savings potential than the reduction of absorptive losses by removing iron oxide from the glass. Reflective losses for double glazing average about 16 percent for normally incident sunlight, but could be reduced to 4 percent.[3]

Translucent Glazing. Looking through translucent glazing is similar to looking through a wet shower door. Light is transmitted but in a diffused form so that images appear fuzzy. When translucent glazing is used in solar apertures, it diffuses sunlight in all directions, distributing energy more uniformly throughout the room than transparent glass does. This is of help in keeping living space temperatures even and the comfort level high. Because one cannot see clearly through translucent glazing, however, it is perhaps more suitable for skylights and clerestories than for windows.

When used in multiply-glazed solar apertures, translucent glass should constitute the inside pane only. Otherwise, light will be diffused before it reaches the inside pane and impinge upon it with a large range of angles of incidence. This can impair performance, for energy transmittance falls off rapidly for angles of incidence greater than 50 or 60°.[4]

Thermal Resistance

The higher the R-value, or thermal resistance, of a window, the slower heat will escape through it. A single-paned window has an R-value of only 0.9. If the walls surrounding such a window have a composite resistance of R-20, the rate of thermal energy escape through each square foot of glass is 20/0.9, or *22 times* greater than through each square foot of wall.

Double-glazed windows have R-values about double that of single glazing and are now standard equipment in solar-heated buildings. Air spaces between the panes are commonly 0.25 inch, although for optimum performance a 0.75-inch spacing should be used.[5]

Although double-paned windows allow less heat to escape than single-paned ones, they also transmit less sunlight. Typical transmission values are 75 percent for double panes and 85 percent for single panes for normally (perpendicularly) incident sunlight. But despite their lower transmissivity, the overall performance of double panes is superior to that of single panes in all but the mildest of climates.

In extremely cold climates such as inland Canada, the insulating value of double glazing is considered by some to be insufficient, and triple glazing is installed instead. Studies indicate, however, that double glazing with insulated window coverings installed each night provides greater energy savings at a lower cost than triple glazing without night insulation.[6]

The edges of the air spaces between multiple glazings are typically sealed with materials such as butyl rubber. Sometimes the seal fails and condensation forms between the panes, obscuring the transparency of the glass. An example of this problem is depicted in Figure 4-3. Possible causes

Figure 4-3 Condensation between panes of double glazing.

of seal failure are high thermal or wind stress or improper installation. If the glazing is mounted so that more of its weight is supported by one pane than the other, the seal will be subject to a shearing stress that can damage it. In addition, oil-based caulks can deteriorate butyl rubber seals if they come in contact with it.[7] It is strongly recommended that wherever it is available, *double-sealed* multiple glazing be used. The edges of this type of glass are sealed with layers of both silicone and butyl and are much less likely to leak.

Multiple glazing is also available that is made of one continuous piece of glass. The edges of the panes, in other words, are fused together. Although it is the ultimate in reliability, fused-edge glass is generally quite expensive. Windows made with it typically cost $15 to $20 per square foot, while those with butyl-sealed edges sell for about $5 to $7 per square foot.[8]

Heat-Retentive Glazing. The ideal window would be one that is totally transmissive for incident solar energy but completely opaque to the escape of thermal energy from the building's interior. There are a number of products now on the market which, in an attempt to come closer to the ideal than traditional windows, employ materials other than multiple panes of glass.

Coatings have been developed for plastic and glass that selectively reflect infrared radiation of approximately the wavelengths emitted by room-temperature objects. Normal glazings do not reflect this radiation but instead absorb it, heating up and then reradiating much of that energy to the outside, where it is lost.

One of these infrared reflecting materials is Heat Mirror,[9] made from a specially coated polyester film. The film is suspended between the two panes of a double-glazed window. The window has a thermal resistance of R-4.4,[10] considerably better than the resistance of normal double glazing (R-1.8) or triple glazing (R-3.3).

Projected costs of Heat Mirror windows are only slightly greater than for triple glazing. The product's disadvantage, however, is that its transmissivity to solar energy is only 53 percent, while for normal double and triple glazing it is 75 percent and 66 percent, respectively. Its low transmissivity makes a Heat Mirror window perhaps more suitable for north walls, where heat retention rather than solar energy collection is the primary function.

A product originally developed for use in one of MIT's solar testing buildings is the Airco ITO window.[11] ITO stands for indium tin oxide, a semiconductive coating deposited onto the glass. Tests at both Airco's facility and at MIT have indicated that the system has a very respectable transmissivity of 70 percent, with an R-value of 3.0. But indium is an expensive metal, and the vacuum-deposition process Airco used was rather involved. An alternative material is copper tin oxide (CTO), considerably cheaper and easier to apply. Testing by Guardian Industries[12] indicates that the product's performance and cost is close to that of triple glazing. Its usable lifetime, however, is not yet known.

A different type of heat-retentive window is Weathershield's Quad-Pane,[13] composed of two layers of polyester film sandwiched between two panes of glass. The window has a solar transmissivity close to that of triple glazing but a higher thermal resistance, R-3.8. Its insulating qualities are mainly due to the three dead-air cavities. In tests at Los Alamos National Laboratory, Quad-Pane windows performed significantly better than did Heat Mirror, probably because of Quad-Pane's increased transmissivity.

Heat-retentive glazing offers great energy-saving potential for the coming decade. As prices decrease and performance characteristics improve, products such as those described above should see widespread use in the solar energy field.

Movable insulation systems are already employed widely. If used consistently and correctly, they are an efficient way of preventing thermal energy escape through a window.

SAVING ENERGY WITH MOVABLE INSULATION

During the heating season, solar apertures become sources of energy loss whenever the sun is not shining. Insulating devices that are easily installed and removed can drastically reduce these losses. Movable insulation comes in the form of insulated shutters, curtains, or shades. Most devices are designed for use on the interior of the window, although some are built for external use and a few for application between the panes of double- or triple-glazed windows.

Window insulation devices can be constructed on site by a homeowner or contractor or purchased commercially. Of the many products on the market, a number were originally designed for other applications and are of questionable value as movable insulation. To choose among the various options, it is first necessary to understand the features that properly functioning window insulation should have.

All edges of the window covering need to be held tightly to the window frame or surrounding wall. Otherwise, the cold air next to the glass will leak into the living space, being replaced by warm room air that will in turn become chilled by contact with the windowpane. This type of air circulation driven by temperature differences is called natural convection and was discussed in detail in Chapter 1.

Studies at Los Alamos National Laboratory indicate that the minimum effective thermal resistance for movable insulation is about R-14.[14] System performance drops off rapidly for window coverings with R-values below this level.

Movable insulation should incorporate a layer of aluminized Mylar or aluminum foil in its construction, the purpose of which is to lower radiative losses by reflecting infrared waves back into the room. The reflective layer also serves as a vapor barrier to prevent moisture in room air from migrating to the cold window and condensing. Mildew and rot in the insulated covering as well as in the window frame could result from condensation.

Window insulations need to be strong and durable, since during the heating season they are installed and removed every day. They should also be made of fire-retardant material. Curtains and shades need to be disassembled and cleaned occasionally and so should be constructed accordingly. Finally, movable insulation must be convenient, easy, and quick to put in place; otherwise, it is unlikely to be used regularly.

Traditional window coverings that were designed mainly for decoration and privacy, such as drapes, shades, and venetian blinds, should not be considered for insulation applications unless a way of tightly sealing *all* their edges—top, bottom, and sides—to the window frame or surrounding wall is worked out. In the case of blinds, the interfaces between slats must also seal tightly.

Reflective Mylar sheets have been in use for a number of years; these are applied to the surface of a window and prevent summertime overheating of the building interior by limiting the amount of sunlight admitted. Because they also prevent some thermal radiation from escaping the building by reflecting it back inside, they are sometimes sold as window insulation. If installed in solar apertures, however, whose purpose is to admit maximum amounts of solar energy, they will probably do more harm than good. These films should not be confused with products such as Heat Mirror, discussed earlier in the chapter, that selectively transmit visible solar radiation but reflect the infrared back into the building's interior.

Interior Window Insulation

Movable insulation systems for the interior of a window, the type most commonly installed, come in a melange of different designs and prices. Figure 4-4 depicts a simple and inexpensive shutter that was built for the home of the author at a cost of only $2 per square foot, including both parts

Figure 4-4 Owner-built shutters made of rigid insulation, decorative fabrics, and a wood frame.

and labor. The shutter was made from a piece of rigid insulation sandwiched between two layers of fabric and surrounded by a wood frame. Foam rubber weather stripping on the edges of the shutter's frame prevents convective edge losses. The shutters are very light and easy to move around, since only the frame is made of wood and the rest is either insulation or fabric. The shutters in the photograph are placed in the windows every night during the heating season, removed in the morning and stored in the closet during the day. For greater convenience, they could also have been hinged to the window frame and simply swung out of the way each morning.

The interior fabric of the shutter can be of any material that is compatible with the room's decor. The outer fabric should be of a light-colored material, for if it is dark and the shutter is left in place during the day, the fabric will efficiently absorb solar energy and heat the air space between it and the window to the point where the glass might crack.

The foil covering of the rigid insulation provides both a vapor barrier and an infrared reflection surface. Because the insulation is isolated from the living space only by layers of fabric, polystyrene insulation is recommended over rigid foam, which sometimes outgases formaldehyde vapors.

Shutters of the type described above have been installed in the author's home since 1981 and have added greatly to the energy efficiency and interior comfort of the house. Their effect is especially noticeable when sitting near a window, for they eliminate cold convection currents and radiative losses from the body that can be chilling and uncomfortable.

A commercial window insulation for which the homeowner is able to choose the interior fabric is the Window Warmer roman shade.[15] It is sold in do-it-yourself kit form or can be custom sewn by the manufacturer. The product consists of four layers: the decorative interior fabric, a vapor barrier/thermal radiation reflector, a layer of Thinsulate insulation and a drapery lining fabric. Thinsulate is a nonwoven batting made by 3M that has been used for years in the manufacture of winter clothing. It has nearly twice the insulating value of goose down per inch of thickness. Two-thirds to three-fourths of an inch of Thinsulate is used in the shade, bringing its total thermal resistance up to a value of R-4.

All materials used in the Window Warmer except for the hardware are washable. Suggested 1984 list price for the sew-it-yourself kit was $2 per square foot, while a custom-sewn shade cost $4.50 per square foot. Hardware is included in these prices, but the decorative interior fabric is not. Typical prices of decorative coverings run from 30 to 60 cents per square foot.[16]

The Window Warmer roman shade is stored at the top of the window in horizontal folds when not in use (see Figure 4-5). When lowered, hinged wooden clamps on the sides and velcro strips on the head provide air seals to guard against convective losses. The bottom seal is furnished by a bar in the lower hem of the shade which rests on the window's sill if there is one. If not, a hinged clamp is used on the bottom as well as the sides. Lowering a Window Warmer shade each evening and sealing its edges is but slightly more time consuming than drawing a set of curtains across a window.

An alternative edge-sealing device for either flexible or rigid window coverings is a set of magnetic strips. Zomeworks Corporation[17] markets such strips, which they call Nightwall Clips. Each clip is made up of a matching flexible magnet and steel strips. Both parts have adhesive backs. The steel mounts on the window frame or the glass itself, and the magnet on the window covering. Nightwall slips are inexpensive (1984 prices for a 3-inch by ½-inch clip were 43 cents each) and easily installed.

(a)

(b)

Figure 4-5 Window Warmer roman shade. Insulation is provided by Thinsulite, a material used for lightweight winter clothing. (Courtesy of Creative Energy Products, Madison, WI.)

InsulShutter[18] is a durable hinged window covering that if cared for properly should have a lifetime equal to that of the house. The shutters are constructed of two layers of hardwood plywood enclosing a ¾-inch core of polyisocyanurate foam insulation (see Figure 4-6). Dead air spaces on both sides of the foam add to the system's insulation value, which is about R-7. Edge losses are controlled by an interlocking tongue-and-groove design at all vertical joints, lap joints at the sill and head, and polyurethane weather stripping on both vertical and horizontal joints.

InsulShutters fold up compactly for convenient daytime storage at the sides of the window. The shutters are suitable for either new construction or retrofits. Although more expensive than the Window Warmer (1984 suggested list prices for InsulShutters ran from $13.50 to $21.50 per square foot), they should outlast roman shade insulating devices by many years.

Figure 4-6 InsulShutter. InsulShutter's hardwood plywood construction contains 0.75 inch of isocyanurate, and two dead air spaces. (Courtesy of First Law Products, Keene, NH.)

Figure 4-7 The InsulLouver is also of hardwood construction, with an isocyanurate core. It is sold in prehung banks, and may be mounted on windows of any angle, from full vertical to full horizontal. (Courtesy of First Law Products, Keene, NH.)

Sunflake Corporation[19] sells a "pocket shutter" system in which the window covering slides into a wall recess when not in use and is thus totally hidden from sight. The system includes the window, a shutter envelope and a shutter built to slide in the envelope. The components are made to be mounted in an ordinary 2 × 4 wall but can be adapted to walls of thicknesses up to 12 inches. Although readily installed in new construction, pocket shutter systems are very difficult to *retrofit* into existing buildings.

The shutters of the Sunflake system are heavily insulated with 1½ inches of Thermax isocyanurate foam. Edge-loss protection is provided by half laps in the shutter's edges that correspond to the shape of the window jamb.

It must again be stressed that it is the combination of insulating value *and* edge protection that makes a shutter effective. Several inches of insulation are of little value in keeping a room warm if cold air can freely enter that room around the edges of the insulation. A shutter's edge-sealing mechanisms can function properly initially but wear out over time and require repair. They should thus be checked regularly at least several times during the heating season.

Exterior Window Insulation

External shutters and shades protect glazing from vandalism and storm damage as well as reducing street noise and heat loss. Most systems sold can be opened and closed from the *inside* of the building. This is an important feature, for one of the common problems with window insulation is that inconvenience of operation causes them to fall into disuse.

A popular type of external shutter is one made of interlocking horizontal slats that are stored on a roller above the window when not in use (see Figure 4-8). Insulation in this type of shutter is provided by dead air spaces and/or foam insulation inside the slats. The shutters can be operated either manually using pull straps or cranks, or automatically with electric motors. The shutters usually ride up and down in vertical tracks on either side of the window. Before purchasing, it is important to make sure that the tracks, the top and bottom of the shutter, and the joints between slats are tightly weather-stripped or sealed in some way to prevent cold air infiltration.

Between-Glazings Insulation

The best known between-glazings insulation is Beadwall, marketed by Zomeworks Corporation.[20] In this system, Styrofoam beads are blown into a 2½-inch air space between the window's panes when insulation is required and removed by means of a vacuum at other times. Zomeworks sells the vacuum/blower and bead storage units as well as the special window frame necessary. The consumer supplies the glazing and the electrical and bead transport connections between the windows. When filled, a Beadwall window has an insulating value of R-8.

When Should Movable Insulation Be Used?

A question often asked is should movable insulation be installed only at night or will there be significant additional savings if it is also put in place whenever the sky is heavily overcast? J. Douglas Balcomb[21] has provided strong evidence that the additional benefits of insulating windows whenever the sun is not shining are very minimal. This practice would also be quite inconvenient unless the system were totally automatic. The effects of insulation put in place at 4:30 P.M. and removed at 7:30 A.M. the next day were compared with a system run by an automatic controller that would cover the windows whenever it was "thermally advantageous." The controller-operated system's energy savings was only 1.9 percent greater on the average than that of the system run on "clock time."

ORIENTING SOLAR COLLECTION SURFACES

The direction in which solar collection surfaces face is a crucial matter in all types of solar systems. Improper orientation can drastically reduce the amount of sunlight received, severely impairing the system's usefulness.

On sites that receive full sunlight from at least 9 A.M. to 3 P.M., solar collection surfaces for direct-gain systems should be oriented at between 20° east and 32° west of true south.[22] Following this rule of thumb will lead to system performances that are almost always within 10 percent and usually within 6 percent of optimum. The rule is weighted toward westerly rather than easterly orientations because of the fact that solar energy collected in the afternoon is usually more effective than that collected in the morning. Afternoon energy does not have to be stored as long before it is used during the evening and nighttime hours, the times of peak residential heating requirements.

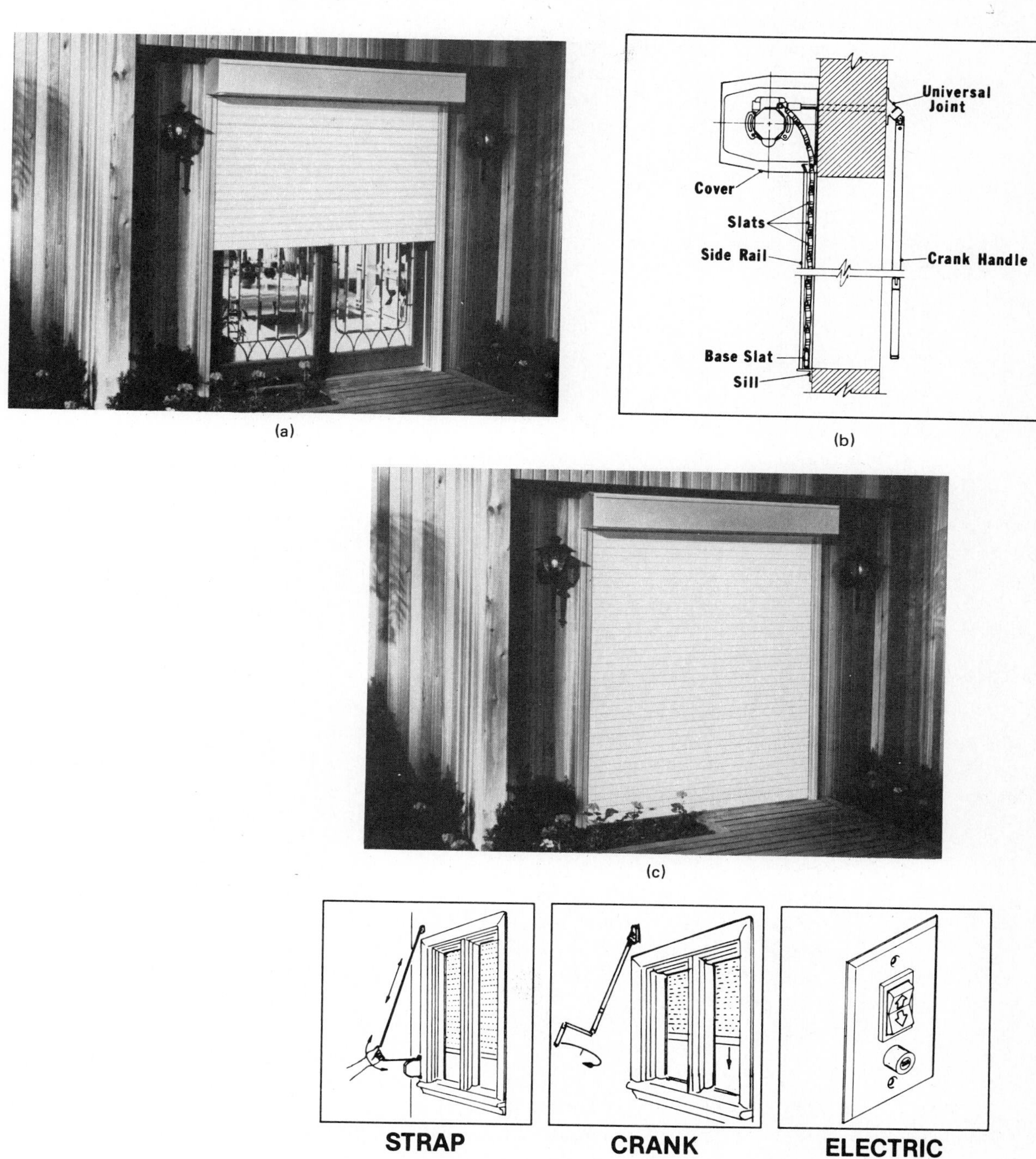

(a) (b) (c) (d)

Figure 4-8 Exterior rolling shutters. Note that the shutters can be operated from the interior of the building using either a pull strap, a crank, or an electric motor. (Courtesy of Pease Industries, Fairfield, OH.)

On sites that experience shading or heavy fog problems during certain parts of the day, solar collection surfaces should be oriented to take advantage of the sunny times. For instance, in a location prone to frequent afternoon fog it would be well to face the collection surfaces east of due south.

Orienting the long axis of a building in an east-west rather than north-south direction has several advantages. It exposes the long side of the structure to the midday sun, which allows maximum area on which to mount sunlight collecting glazing. Also, the energy that is admitted through the glazing has less distance to travel to reach the north side of the building. East-west-oriented buildings are thus able to rely more on natural convective processes rather than expensive-to-operate air blowers to circulate thermal energy throughout the living space (see Figure 4-9).

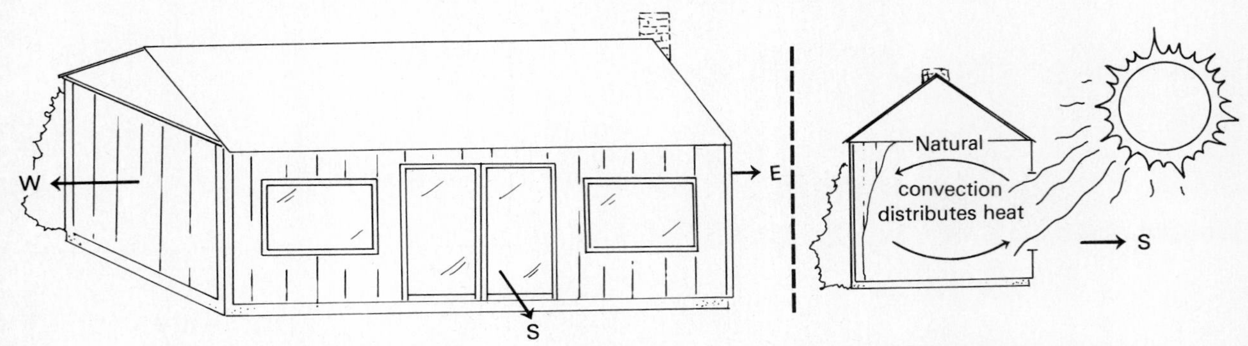

Figure 4-9 East-west building orientation. Buildings with their long axes running east-west allow maximum area in which to install glazing for solar collection. These buildings are also able to rely more on natural convection to distribute the heat than north-south-oriented buildings. This is because the heat has less distance to travel to reach the north wall of the building.

SIZING A SYSTEM

Calculating the Aperture Area

It is critical that a direct-gain system be sized correctly if it is to operate successfully. If the glazing area is too small, there will be a greater dependence on backup sources of heat than is necessary. If the solar apertures are too large, overheating problems and long payback periods result. To circumvent these pitfalls, the sizing procedures outlined below should be followed closely. The procedure involves three parts: determination of the heat-retention capabilities of the building under design, a decision as to what part solar energy will play in the total heating needs of the building, and calculation of the recommended solar aperture square footage using the data given in Table III-2. The three steps are carried out as follows:

1. Calculate the building load coefficient using the methods detailed in Chapter 2. The BLC is the structure's daily heat loss per degree of indoor-outdoor temperature difference, disregarding the contribution of the solar system.[23]
2. Decide whether the structure is to be "heavily solar heated" (80 percent of its annual heating needs supplied by the sun), "moderately solar heated" (60 percent of the heat supplied by the sun), or "solar tempered" (30 percent from the sun). These percentages are termed *solar savings fractions*, for they indicate the part of the building's space heating needs that are met by the solar system. The map in Figure 4-10 gives suggestions as to which percentage is most cost-effective for various areas of North America.

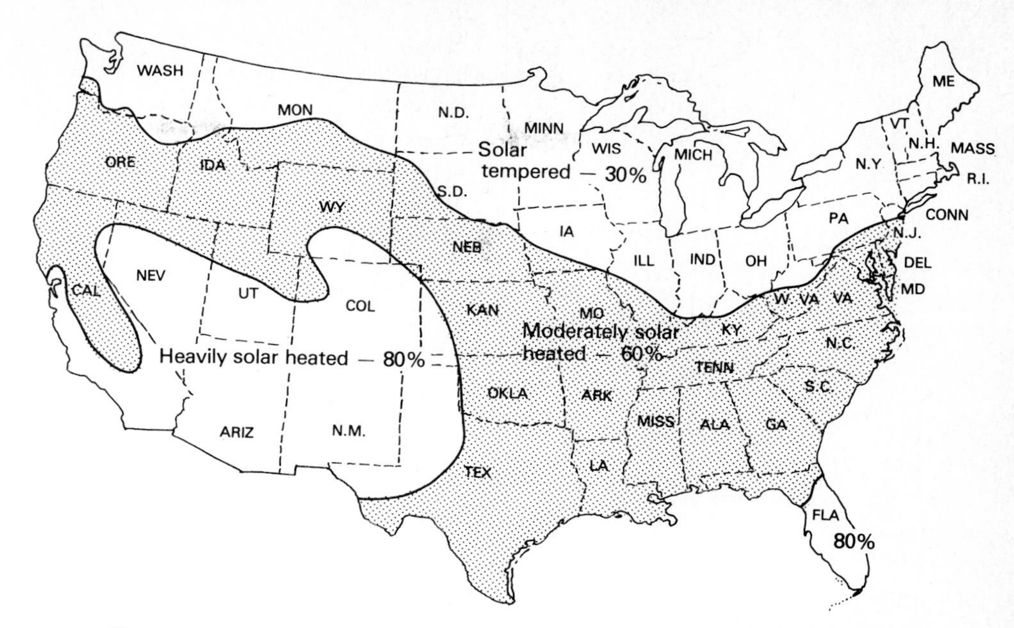

Figure 4-10 Recommended solar savings fractions for different areas of North America.

3. Table III-2 in Appendix III lists values for a number called the *load collector ratio*[24] for solar systems in 219 locations throughout the United States and Canada. To obtain the recommended solar aperture area for the building, choose the LCR appropriate to the system type and location and divide the building load coefficient by it:

$$\text{solar aperture area} = \frac{\text{BLC}}{\text{LCR}}$$

Example Calculation. Determine the recommended direct-gain aperture area for a house in Springfield, Missouri, that has a BLC of 6000 Btu/degree-day and R-4 night insulation over all apertures. Repeat the calculation for the case where no movable insulation is used.

The map in Figure 4-10 suggests that buildings in the Springfield, Missouri, area be "moderately solar heated" (60 percent solar savings fraction). From Table III-2, the LCR values pertaining to a direct-gain design in this location are 7 for a system without night insulation and 24 for one with night insulation. These LCR values yield recommended aperture areas as follows:

Without movable insulation:

$$\frac{6000}{7} = 857 \text{ ft}^2 \text{ of aperture}$$

With movable insulation:

$$\frac{6000}{24} = 250 \text{ ft}^2 \text{ of aperture}$$

The dramatic difference in the two square footages makes movable insulation almost mandatory for this system, for the expense of including

857 square feet of aperture in the south wall of a house would be very high. As we will see in Chapter 5, the amount of thermal storage mass necessary to balance that size of aperture would also be almost prohibitively large.

It sometimes happens that even with movable insulation, the required aperture size for the solar savings fraction desired is still too large. When this is the case, ways should be sought to reduce the building's heat losses by adding weather stripping and insulation. This will have the effect of lowering the BLC and thus reducing the required aperture size. If the house is already thermally efficient, it might be necessary to settle for a lower solar savings fraction by reducing the square footage of the apertures.

Sizing Procedures When the Site Is Not Listed in Appendix III

Many cities in the United States and Canada are not included in Appendix III. It is usually possible, however, to find a listed city with a climate similar to that of the building site. To do this, the designer needs to collect the following data about the site:

1. Annual degree-days
2. Average January temperature
3. Latitude
4. Solar energy data

This information can often be obtained from such places as local farm bureaus, airports, and building inspectors. The references included at the end of this chapter list detailed climatological data for many North American locations.

Solar energy data are the most difficult to obtain. The best types to locate are annual and monthly insolation statistics such as those in Table III-1. These statistics are frequently unavailable for small towns and rural areas, however. In lieu of them, *skycover* statistics can be used. Skycover measurements indicate the fraction of the sky that is obscured by clouds and are standard observations taken by even the smallest of weather stations. Data for the station nearest the building site can be obtained from the National Climatic Center.[25] An inexpensive method for securing enough data for a rough comparison with skycover information for the cities listed in Table III-1 is to order "surface observation" records of the local weather station for January 15 of at least 15 different years of observation (this will cost about $5). By averaging the skycover fractions at 12:00 noon for each, an approximate January skycover statistic is obtained. January data are generally the most important for solar design considerations, for it is at that time that the severest weather usually occurs.

After the necessary data are collected they should be compared with those of the cities included in Table III-1. The six to eight cities with data closest to those of the site should be selected and their climatic statistics carefully examined to pick the location most representative of the building site.

Example Calculation. Ukiah, California, experiences 2997 degree-days annually, has an average January temperature of 46°F, and is at a latitude of 39.1° north.[26] Data supplied by the National Climatic Center for the years 1950 through 1964 list skycover fractions at 12:00 noon on January 15 as follows:

1950	8	1958	10
1951	10	1959	10
1952	10	1960	2
1953	7	1961	8
1954	10	1962	2
1955	10	1963	0
1956	10	1964	0
1957	8		

A skycover figure of 10 indicates complete coverage by clouds; a figure of 0 indicates a clear sky. The mean skycover is arrived at by taking the sum of the measurements (105) and dividing it by the number of measurements taken (15). This yields an average January skycover of 7.0.

Six cities with climate characteristics close to that of Ukiah are:

City	*Degree-Days*	*January Temperature*	*Latitude*	*January Skycover*
Fresno, CA	2650	45	36.8	6.7
Sacramento, CA	2843	45	38.5	6.0
San Francisco, CA	3042	48	37.6	5.7
Las Vegas, NV	2601	44	36.1	5.2
Columbia, SC	2598	45	33.9	6.0
Atlanta, GA	3095	42	33.6	6.3

Degree-day figures for Sacramento and Atlanta deviate by 150 degree-days or less from that of Ukiah. Sacramento, Fresno, and Columbia have average January temperatures only 1°F different from Ukiah. The skycover fraction is quite close for Fresno and 0.7 different for Atlanta. Sacramento and Columbia are one point away from Ukiah's value. Sacramento's latitude is only 0.6° different from that of Ukiah. Since Sacramento's data are very close on three counts to those of Ukiah, and reasonably close on the fourth, it appears to be the best choice for a representative location.

The sizing methods detailed in this chapter yield good estimates of the solar savings fractions that result from certain-size apertures. For a more detailed analysis that includes month-by-month performance estimation, refer to Volumes 2 and 3 of the *Passive Solar Design Handbook.*[27]

NOTES

1. *Solar Applications Seminar*, Pacific Sun, Inc., Palo Alto, CA, 1977, p. B-8.
2. J. Douglas Balcomb et al., *Passive Solar Design Handbook*, Vol. 2, Los Alamos National Laboratory, Los Alamos, NM, for U.S. Department of Energy, 1980, p. 106.
3. Ibid., p. 108.
4. *Energy Primer*, Portola Institute, Menlo Park, CA, 1974, p. 22.
5. Balcomb et al., Vol. 2, p. 108.
6. Joe Kohler and Dan Lewis, "Seeing Through Window Options," *Solar Age*, Vol. 7, No. 1, January 1982, pp. 42–44.
7. Private communication with Builder's Glass and Supply, Inc., Ukiah, CA.

8. Private communications with Anderson Corp., Bayport, MN, and Builder's Glass and Supply, Inc., Ukiah, CA.

9. James A. Moore, "Super Windows: The Long Debut," *Solar Age*, Vol. 7, No. 6, June 1982, pp. 36–69.

10. Ibid.

11. Ibid.

12. Ibid.

13. Ibid.

14. Balcomb et al., Vol. 2, p. 55.

15. *Solar Products Specifications Guide*, SolarVision, Inc., Harrisville, NH, 1983, pp. 62–63.

16. Based on 45-inch-wide material selling for $3.50 to $5 per running yard.

17. *Solar Products Specifications Guide*, p. 72.

18. Ibid., p. 64.

19. Ibid., p. 67.

20. Ibid., p. 59.

21. Balcomb et al., Vol. 2, p. 55.

22. Ibid., p. 28.

23. Ibid., p. 32.

24. Ibid., p. 38.

25. Call (704) 258-2850, extension 682, for surface observations.

26. Paul Bardahl et al., *California Solar Data Manual*, California Energy Commission and Department of Energy, March 1978, pp. 242, 250, 257.

27. Balcomb et al., Vols. 1 and 2. No specific pages are given because much of these books is devoted to system design.

SUGGESTED READING

BALCOMB, J. DOUGLAS, ET AL., *Passive Solar Design Handbook*, Vols. 2 and 3. Los Alamos, NM: Los Alamos National Laboratory for U.S. Department of Energy, DOE/CS-0127/3, 1982. Voluminous design data. An absolutely essential part of a solar designer's reference library.

MAZRIA, EDWARD, *The Passive Solar Energy Book*. Emmaus, PA: Rodale Press, Inc., 1979. Graphical methods of designing overhangs.

OLGYAY, ALADAR, AND OLGYAY, VICTOR, *Solar Control and Shading Devices*. Princeton, NJ: Princeton University Press, 1957.

REVIEW QUESTIONS

1. Define *solar aperture*.
2. Should tinted glass be used in a solar aperture? Why?
3. What is the advantage of installing translucent glazing in solar apertures?

4. List three important features that movable insulation should have to function properly.
5. Within what range of directions should solar apertures face?
6. Define *solar savings fraction* (SSF).
7. Define *load collector ratio* (LCR).

PROBLEMS

1. Determine the recommended direct-gain aperture for a house in Detroit, Michigan, that employs night insulated apertures and has a BLC of 7500 Btu/degree-day.
2. Do the same for a house in Louisville, Kentucky, with a BLC of 5500 Btu/degree-day. Perform the calculation both for night insulated and uninsulated apertures.
3. Do the same for a house in Phoenix, Arizona, with a BLC of 6000 Btu/degree-day. Again, consider both night insulated and uninsulated apertures.
4. Solarville is an imaginary city that experiences 5000 degree-days annually, has an average January temperature of 32° F, is at a latitude of 43° north, and has a mean January skycover of 6.3. It is not one of the cities listed in the tables of Appendix III. Find a city from that appendix with as similar a climate as possible to that of Solarville.
5. Lunarville is another imaginary city. It has 7200 degree-days annually, an average January temperature of 24°F, is at a latitude of 49° north, and has a mean January skycover of 4.8. Find a city from Appendix III with as similar a climate as possible to that of Lunarville.

5 direct-gain systems II: energy storage and overheating protection

ENERGY STORAGE

Characteristics of Thermal Storage Materials

Thermal storage mass is material included in the building for the purpose of storing heat. In direct-gain systems, it is typically installed in the walls, floors, and ceilings of the building and can have load-bearing and aesthetic roles as well as functioning as thermal storage.

Solar apertures that are designed using the procedures discussed in Chapter 4 are meant to admit enough sunlight into a building during clear days to provide for some or all of the nighttime heating needs as well as daytime needs. This implies that more energy must be collected during daytime hours than the building requires at that time. The function of the thermal mass is to absorb and store the excess energy for use after sunset, when the lower outside temperature and absence of solar energy increases the building's heat loss over that during the daytime. If insufficient mass is installed for the size of solar aperture, the building will overheat during the day (or the heat must be vented and thus lost) and become too cold at night.

Thermal storage mass also helps to cool a house during the warm summer months. If exposed to outside air at night, the temperature of the mass usually drops considerably, helping to moderate living-space temperatures the next day by absorbing excess heat.

There are many materials that can be used as thermal storage, but no matter which are chosen, they should have the following characteristics:

1. *An ability to store a large amount of heat per cubic foot of volume.* This ability is measured by the *heat capacity* of the material, which in British units is defined as the number of Btu of thermal

energy 1 cubic foot of a substance stores as its temperature is raised 1 degree Fahrenheit.

2. *An ability to transmit thermal energy rapidly between its surface and its interior.* This allows the deeper regions of the mass to participate in the daily cycle of storing and releasing energy. In solids, this ability is measured by the material's thermal conductivity.

Common Storage Media

The heat capacities and thermal conductivities of a number of common substances are listed in Table 5-1. Insulating materials such as glass wool and polyurethane foam have low heat capacities as well as low conductivities and are thus poor choices for storage mass. Wood has a medium heat capacity but a fairly low conductivity. Water has the highest heat capacity on the list. Because it is a liquid, it can transport thermal energy very efficiently through the process of convection, which was described in detail in Chapter 1. Thus, provided that the storage tank does not inhibit convective flow, it is not important that water has a lower thermal conductivity than many solid materials. Figure 5-1 depicts direct-gain systems in which water housed in translucent plastic columns functions as thermal storage mass, absorbing solar energy during the day and releasing warmth at night. Water is also a popular storage medium for solar wall systems, which are studied in Chapter 6.

The most commonly used thermal storage media for direct-gain systems are dense solids such as concrete, brick, and adobe. Both their heat capacities and thermal conductivities are fairly high and unlike water, methods of building with them are well known. Walls, foundations, and floors made from them serve as load-bearing members and integral parts of the building envelope as well as repositories of thermal energy.

Concrete is a mixture of two basic elements: an aggregate and a paste. The aggregate, typically composed of sand and gravel or crushed rock, is held together by the paste, which is made from portland cement and water.[1] Concrete can either be poured on site or purchased in the form of blocks, bricks, and block paving units. If hollow concrete blocks are used, they should be filled with mortar to increase their thermal storage capacity.

TABLE 5-1 Heat Capacities and Thermal Conductivities of Common Substances

Substance	*Heat Capacity* ($Btu/°F\text{-}ft^3$)	*Thermal Conductivity* ($Btu/ft\text{-}°F\text{-}h$)	*Source*[a]
Concrete	30.0	1.0	(1)
Brick			
Common	22.4	0.42	(2)
Facing	25.0	0.77	(2)
Adobe	24.0	0.33	(1)
Dry sand	18.0	0.19	(1)
Water	62.4	0.35	(2)
Pine wood	20.8	0.097	(1)
Stone			
Limestone	22.4	1.04	(2)
Granite	31.8	1.28	(2)

[a](1) J. Douglas Balcomb et al., *Passive Solar Design Handbook*, Vol. 2, Los Alamos National Laboratory, Los Alamos, NM, for U.S. Department of Energy, January 1980, p. 182. (2) *Energy Conservation Design Manual for New Residential Buildings*, State of California, February 1978, Table 3-5.

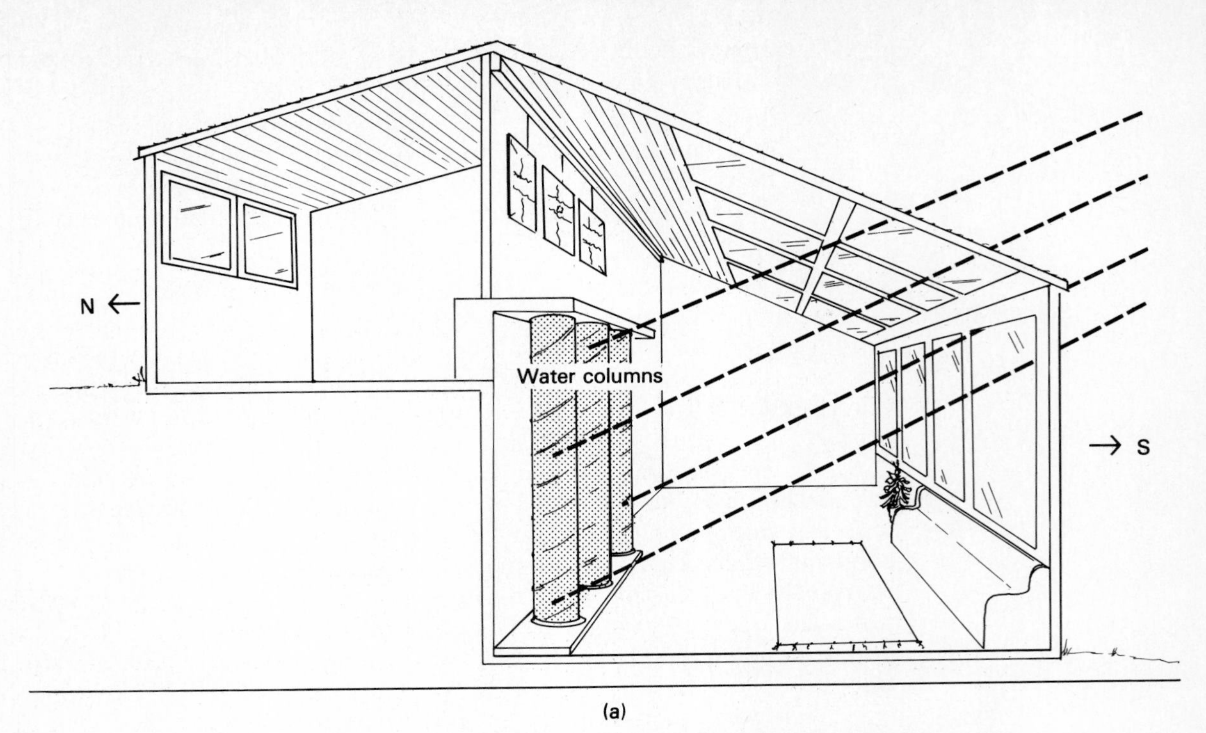

(a)

(b)

Figure 5-1 (a) Thermal energy storage with water. (b) Water wall thermal storage tubes. [(b), Courtesy of Kalwall Corp., Manchester, NH.]

Concrete is available in a broad range of densities, determined by the amount and density of the aggregate used. Normal-weight concrete, 135 to 160 pounds per cubic foot, is the type most commonly used for passive solar energy applications. Structural lightweight concrete, 85 to 115 pounds per cubic foot, is used as thermal storage mass for a floor that is supported by a wooden subfloor. Other types of concrete, not usually appropriate for passive applications, are insulating lightweight concrete (15 to 90 pounds per cubic foot) and heavyweight concrete (160 to 400 pounds per cubic foot), often used for radiation shielding.[2]

Brick is made from clay or shale hardened by heat. The three most common types are building brick, facing brick, and brick paving. The latter two types have slightly higher heat capacities and thermal conductivities than those of building brick. All three varieties are suitable for storage mass applications, although the high density of some brick pavers leads to such a low capacity to absorb water that the bricks tend to "float" on the mortar during wall construction, and are thus best used for floors only.

Far less common than the above are refractory bricks—those designed to line kilns, boilers, and furnaces. Although the heat capacity and thermal conductivity of refractory brick such as magnesite are superior to those of ordinary brick, their high price prohibits their use in most solar systems. The price for a magnesite brick is about $3.50. The cost of a building or face brick only slightly smaller in size is about 25 cents.[3]

Bricks often have cavities built into them that improve their firing characteristics. As with concrete blocks, these cavities should be grouted with mortar during construction of the thermal mass. Since the heat storage properties of mortar are close to those of brick, there will be little difference in performance between completely solid bricks and grouted ones.

Although black is the color with the highest solar absorptivity, the darker natural brick colors—browns, blues, and reds—perform almost as well and are often preferred for aesthetic reasons. The glazed ceramic coatings that are deposited on some bricks, however, are frequently highly reflective and can thus significantly reduce the brick's solar absorptivity.

Adobe is sun-dried brick made from clay and is popular in the arid southwestern United States. The word is derived from the Spanish *adobar*, meaning "to plaster." In a dry climate, adobe thermal storage walls last indefinitely if protected from ground moisture and to some extent from rain. The main cause of crumbling adobe walls is absorption of water either from the ground or through leaky roofs.

Phase Change Materials

The materials discussed above store thermal energy through excitations of their molecules and subatomic particles as their temperatures rise. As discussed in Chapter 1, the higher a substance's temperature, the more energy of motion its molecules have. There is another way that materials can store heat, however, which is quite different from this. Many materials absorb substantial amounts of thermal energy when they melt, reemitting it when they solidify. (This energy is called *latent heat.*) Materials that undergo a change of phase at temperatures that are both easy to attain in solar systems, and yet high enough to be useful for heating purposes, offer great potential for applications in the solar energy field.

Phase-change materials (PCMs) being employed in solar energy systems include Glauber's salt (sodium sulfate decahydrate), calcium chloride hexahydrate, and paraffin wax. Glauber's salt melts at 89°F, absorbing 104 Btu per pound as it does so. Calcium chloride hexahydrate melts at 81°F and absorbs 75 Btu per pound; and paraffin wax melts at 116°F, absorbing 65

Btu per pound. These three materials also have fairly high heat capacities, which enable them to absorb considerable amounts of energy in addition to their latent heat.

The advantage that phase-change materials offer is large thermal energy storage capacity in a small volume and mass. A cubic foot of Glauber's salt, for instance, that is melted and then heated to 120°F absorbs about 13,000 Btu, as much as 13 cubic feet of concrete heated from 89°F to 120°F.

Because smaller volumes of PCMs are needed, they can be installed in spaces not convenient for other types of mass. Dow Chemical Company manufactures thin calcium chloride hexahydrate Enerphase panels made to fit between the studs of a conventionally framed building.[4] Energy Materials, Inc., fills black polyethylene tubes with calcium chloride decahydrate. Each Thermal rod is 6 feet long and 3½ inches in diameter and may be placed behind windows, along walls, or in any other location where it will receive direct sunlight throughout the day.[5]

Phase-change media offer much promise as compact thermal storage mass for solar energy systems, but their successful use depends to a great extent on proper packaging of the materials. A continuing problem with them in past years has been that once melted, they did not resolidify well. This was due to separation and stratification of the various chemicals while the PCMs were in their liquid state. When temperatures dropped, total solidification did not take place.

The stratification and separation problems have been greatly reduced by storing the PCMs in thin tubes or shallow pans, such as the Thermal rods and Enerphase panels mentioned above. Adding thickening and nucleating agents has also helped.

Computer analyses have indicated that the optimum packaging for PCMs may be in containers even thinner than most of those sold. In comparisons of containers from ½ inch to 4 inches in thickness, the highest performances for direct-gain systems were obtained from the 1-inch-thick packages. It was also found that containers situated so that room air could circulate past both sides instead of just one side performed slightly better.

Test results from Los Alamos National Laboratory of thermal storage walls made from stacked PCM-filled cans were somewhat varied. One wall of commercially manufactured calcium chloride-filled containers performed far worse than was hoped for, which led to uncomfortable temperature extremes in the test room. In addition to this problem, leaks and corrosion were observed in many of the cans.[6]

Boardman Energy System's sodium sulfate wall performed considerably better. Its test results in the areas of "useful efficiency" and "useful solar fraction" were, in fact, superior to those of all the other systems that were examined, which included those using concrete and water as thermal storage. Temperature in the room with the sodium sulfate wall was also comparatively stable.[7]

The *productivity* of a building's thermal storage mass can be defined as the amount of energy saved per cubic foot of mass compared to a similar building have no mass except for that normally used in construction. A method of calculating the cost-effectiveness of storage mass is to divide its initial cost, in dollars per cubic foot, by its productivity. In one study of this type,[8] the cost-effectiveness of a variety of calcium chloride PCM systems was between $170 and $700 for every 1 million Btu saved annually. The same calculation on a 6-inch-thick concrete floor slab yielded a value of $230. This is an indication that after many years of research, the cost of a PCM system finally appears to be dropping into the range of other thermal mass systems. It remains to be seen, though, whether problems with eventual separation and stratification of the PCMs' chemicals that have plagued these materials in the past have indeed been solved.

A Comparison of Solid Thermal Storage Materials

The graphs in Figure 5-2 compare the amount of thermal energy that varying thicknesses of different substances are able to collect, store, and then return to the living space during a 24-hour period. The heat capacity and conductivity of each substance was taken into consideration when constructing the graphs. The materials are part of a vertical wall in a direct-gain system. The wall is insulated on its back side.

Notice from the graph that concrete outperforms brick, adobe, sand, and wood as a thermal storage medium for all thicknesses studied, but especially for thicknesses of 4 inches or greater. Notice also that each material has a thickness at which it performs best. For concrete it is about 7 inches; for brick and adobe it is 4 to 5 inches. When thicknesses greater than optimum are used, the time delay associated with thermal energy penetrating the materials and then reappearing at its surface can be 24 hours or more. In this situation a portion of thermal energy collected one day reappears at the surface a day later and can interfere with the efficiency of the mass's charging cycle.[9]

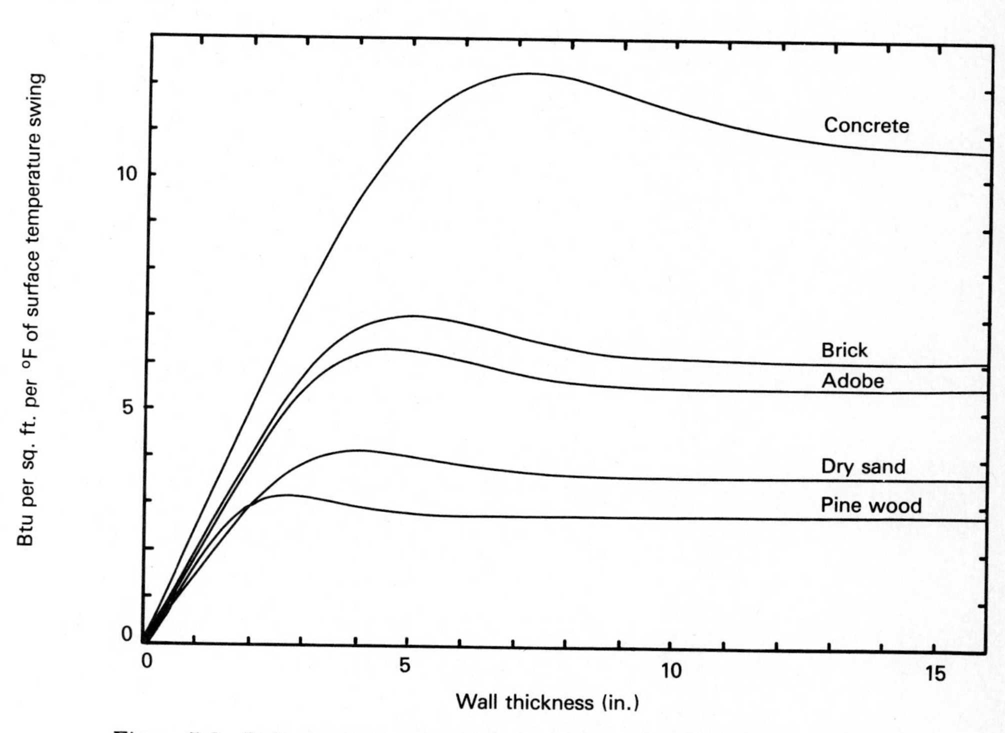

Figure 5-2 Daily heat stored and returned. The diurnal (daily) heat storage capacity of a directly heated wall for different values of the wall thickness. The wall is assumed to be insulated on the back side (that is, the side away from the room). (From J. Douglas Balcomb et al., *Passive Solar Design Handbook*, Vol. 2, Los Alamos National Laboratory, Los Alamos, NM, for U.S. Department of Energy, 1980, p. 181.)

Locating Thermal Storage Mass

The interior wall, floor, and ceiling surfaces of a building can be divided into the following three zones:

1. *Primary radiation zones* are those areas that receive direct sunlight.
2. *Secondary radiation zones* are those irradiated either by sunlight

reflected off, or by infrared radiation emitted from, other solar-heated surfaces in the area.

3. *Remote zones* are those heated mainly by convection currents carrying warm room air from other parts of the building.

These zones are depicted in Figure 5-3. The most advantageous location for mass is in zone 1, in direct sunlight. Surfaces in this area attain temperatures higher than those of the air. The next choice for mass is the secondary radiation zone. Most of the energy admitted through a solar aperture is eventually absorbed, sometimes after multiple reflections, by a surface in either a primary or a secondary radiation zone, and thus as much of the storage mass as possible should be located in these two areas. Not only is mass located in radiation zones more effective as storage than if in a remote zone, but it also aids more in reducing or eliminating clear-day overheating problems that would otherwise necessitate venting of excess energy to the outside or installation of electrically powered blowers to move heated air to other parts of the building.

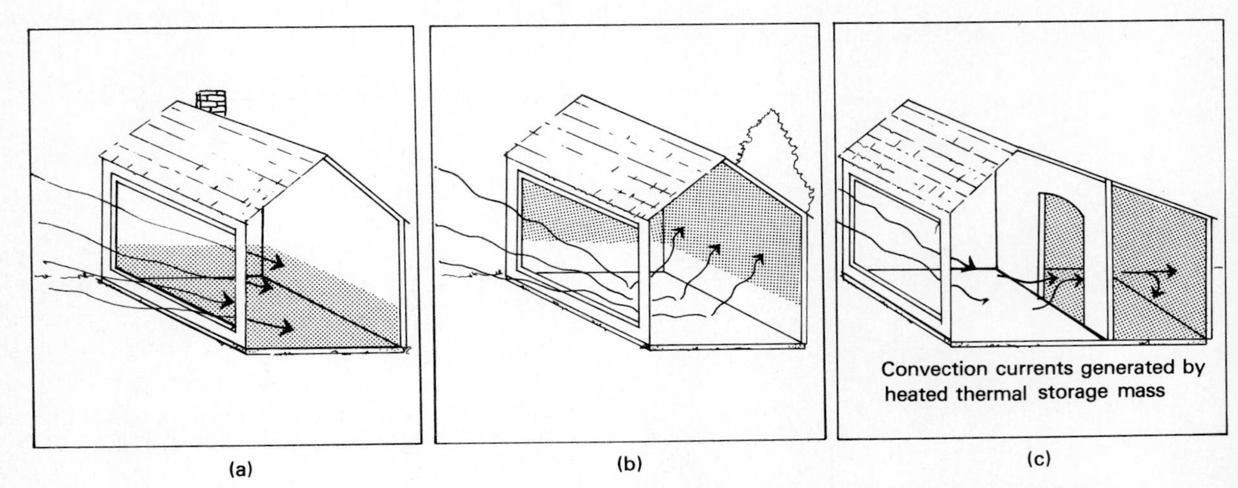

Figure 5-3 Solar heating zones of a building. (a) Primary radiation zone (shaded area). (b) Secondary radiation zone. (c) Remote zone.

Boundaries between primary and secondary radiation zones are not fixed throughout the day but vary depending on the position of the sun in the sky. For the purpose of energy calculations, a full "day" is considered to be the hours between 9 A.M. and 3 P.M. (solar time), for it is during this period that the most intense energy is received. A section of floor, for example, might receive primary radiation from 9 A.M. until 1 P.M. each day but be shaded and receive only secondary radiation for the remainder of the time. When designing a building, the hours of primary and secondary radiation should be determined for *wintertime* positions of the sun.

Because translucent glazing diffuses incoming sunlight throughout the room, it has the ability to *expand* the boundaries of a primary radiation zone (see Figure 5-4). Since thermal storage mass performs best when located in a primary zone, the use of translucent glazing can increase the effectiveness of mass in the primary zone.

Remote zones in a passive building depend on natural convection processes to deliver their thermal energy, and these processes do not occur until there is a temperature difference between the two areas. Thus remote zones often heat up later in the day and stay cooler than other zones. Also, tem-

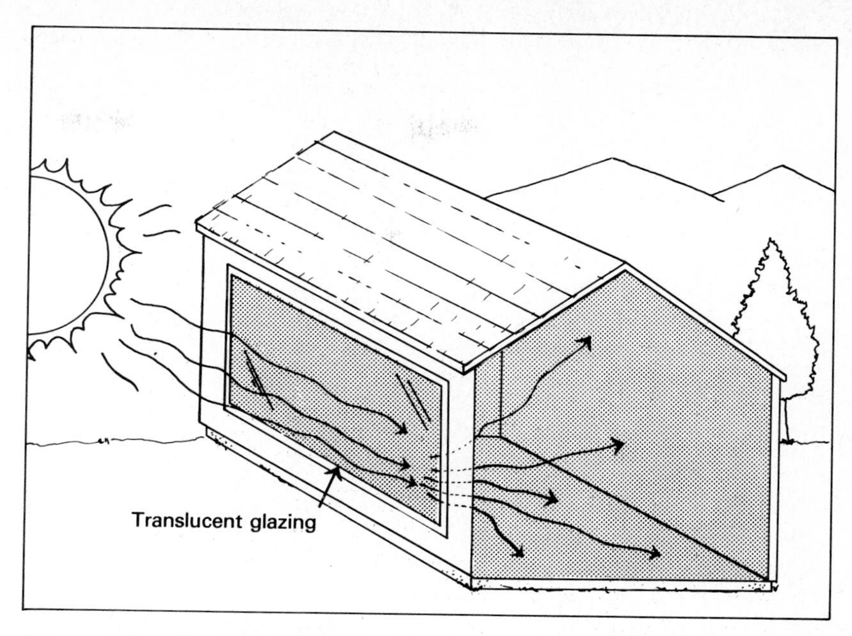

Figure 5-4 Effect of translucent glazing. By diffusing light, translucent glazing expands the boundaries of a primary radiation zone.

perature stratification in the room air leads to storage mass located near the ceiling absorbing thermal energy at a different rate and thus being more effective than mass near or in the floor. Mass in other zones is heated mainly through radiative processes, so temperature stratification is less of a problem.

Thermal storage mass performs better when spread thinly over a large area in the zone than thickly over a small area. A 100-square-foot layer of 3-inch-thick concrete, for instance, stores about 740 Btu of thermal energy each day for each degree of temperature swing. A layer twice as thick but of only half the area stores but 600 Btu per degree.[10] In addition, widely distributed mass leads to more uniform temperatures throughout the zone, an important factor for comfort.

Storage Mass Sizing Procedures

After the building's required solar aperture area has been calculated and the types and locations of thermal storage mass to be used have been selected, the *amount* of mass necessary can be determined with the aid of Table 5-2. This table lists the area of mass required per square foot of aperture as a function of the mass thickness, the material used, and the location of the mass. As an example of the table's use, assume that the aperture area is 150 square feet and concrete located in a primary radiation zone is to be used as mass. Table 5-2 tells us that 5 square feet of concrete per square foot of aperture is needed if the concrete is 3 inches thick, 4 square feet per square foot of aperture if the layer is 4 inches thick, 3 square feet for a 6-inch layer, and so on. If a 6-inch-thick-layer is decided on, the total area necessary is equal to the aperture area of 150 square feet times 3, or 450 square feet.

When the mass is in a primary radiation zone some of the day but due to the movement of the sun through the sky is in a secondary radiation zone the rest of the day, the designer must interpolate between Table 5-2a and b. As an example, assume that a designer is planning to install a 4-inch-thick layer of concrete in an area that receives direct sun (primary radiation) from 9 to 11 each morning but secondary radiation from 11 to 3 P.M. How many square feet of mass is required per square foot of solar aperture?

TABLE 5-2 Storage Mass-to-Aperture Area Ratios

	Thickness (in.)							
Material	*½*	*1*	*1½*	*2*	*3*	*4*	*6*	*8*
(a) Primary Radiation Zones								
Concrete		14		7	5	4	3	3
Brick		17		8	6	5	5	5
Adobe		17		8	6	5	6	6
Gypsum	76	38	26	20				
Oak		17		10	10	11	11	11
Pine		21		12	12	12	13	13
(b) Secondary Radiation Zones								
Concrete		25		12	8	7	5	5
Brick		30		15	11	9	9	10
Adobe		31		15	12	10	10	11
Gypsum	114	57	39	31				
Oak		28		17	17	19	19	19
Pine		36		21	20	21	22	22
(c) Remote Zones								
Concrete		27		17	15	14	14	15
Brick		32		20	17	17	18	19
Adobe		33		21	18	18	19	20
Gypsum	114	57	42	35				
Oak		32		24	26	24	28	28
Pine		39		27	28	30	31	31

Sources: Data drawn mainly from *The Thermal Mass Pattern Book*, Total Environmental Action, Inc., Harrisville, NH, 1980, and used by permission of T.E.A., Inc. Data for adobe interpolated using diurnal heat capacity comparisons from J. Douglas Balcomb et al., *Passive Solar Design Handbook*, Vol. 2, Los Alamos National Laboratory, Los Alamos, NM, for U.S. Department of Energy, 1980, p. 182.

The value in Table 5-2a for 4 inches of concrete is 4 square feet per square foot of aperture. The value in the secondary radiation table is 7. Since the mass receives only secondary radiation for four of the six hours between 9 A.M. and 3 P.M., the square footage of mass should be four-sixths of the way from 4 to 7. The calculation is performed as follows:

$$\text{difference between the two values from the table} = 7 - 4 = 3$$

$$\frac{4}{6} \times 3 = 2$$

This is then added to the primary radiation zone value:

$$4 + 2 = 6 \text{ ft}^2 \text{ of mass per ft}^2 \text{ of aperture}$$

The sizing computations become more involved if different types of storage mass are used. Consider the following example.

Example Calculation. An architect is designing a direct-gain passive home that will have 300 square feet of solar apertures in its south wall. The rooms on the south side of the building have a 6-inch-thick concrete slab floor with an area of 700 square feet. The floor is in direct sunlight from 9 A.M. until 2 P.M. during winter days and receives secondary radiation after that. The architect also wants to put 3-inch-thick brick on some of the wall area of the rooms. The walls receive primary radiation for an average of three of the six hours between 9 A.M. and 3 P.M. and secondary radiation for the other three hours. How many square feet of brick should be used?

The first step in solving this problem is to determine what part of the mass requirement is met by the concrete slab. The values from the primary and secondary radiation tables for 6 inches of concrete are 3 square feet and 5 square feet, respectively, per square foot of solar aperture. Since the slab receives secondary radiation for only one of the six hours between 9 A.M. and 3 P.M. (and primary radiation the rest of the time), the mass square footage should be one-sixth of the way from 3 square feet to 5 square feet, or in other words, about 3.3 square feet of concrete is the right amount to use per square foot of aperture. The area of solar aperture "balanced" by the 700-square-foot concrete slab storage mass is equal to the number of 3.3-square-foot sections that are contained in the floor, or stated a different way, to 700 square feet divided by 3.3:

$$700 \text{ ft}^2 \div 3.3 = 212 \text{ ft}^2$$

Thus 212 square feet of the 300 square feet of aperture is balanced by the concrete slab. The remaining 88 square feet of aperture must be balanced by the brick on the walls.

Since the 3-inch wall brick is exposed to direct sun for half of the day and secondary radiation for the other half, a square footage midway between the values in the primary and secondary radiation tables should be used. These values are 6 and 11, respectively. Thus a value of 8.5 should be used. The total square footage of wall mass is obtained by multiplying the remaining aperture area of 88 square feet by 8.5, yielding a brick area of 748 square feet.

Insulating Thermal Storage Media

Walls. In traditional construction practices, massive materials such as concrete or brick that are located in the exterior walls of a building are either thermally insulated from the living space or are not insulated at all (see Figure 5-5). Both practices destroy the material's effectiveness as thermal storage. Massive elements should be exposed to the building's interior environment and insulated from the outside climate. Outside insulation or *outsulation*[11] systems such as the one depicted in Figure 5-6 can be applied over most types of buildings and are suitable for retrofit situations as well as for new construction.

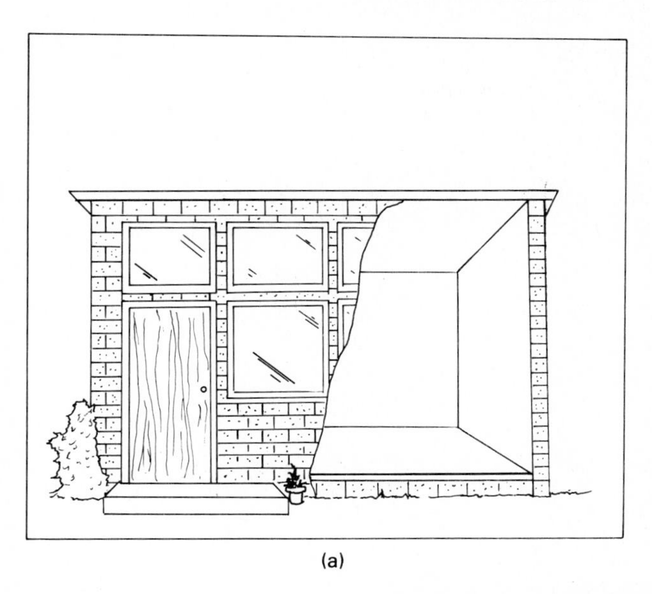

(a)

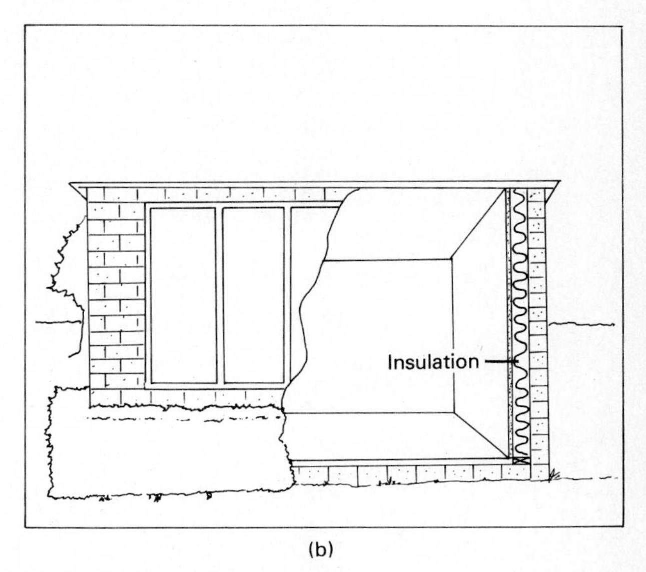

(b)

Figure 5-5 Traditional masonry buildings. Masonry walls have in the past either been (a) left uninsulated, or (b) insulated on their interior surfaces.

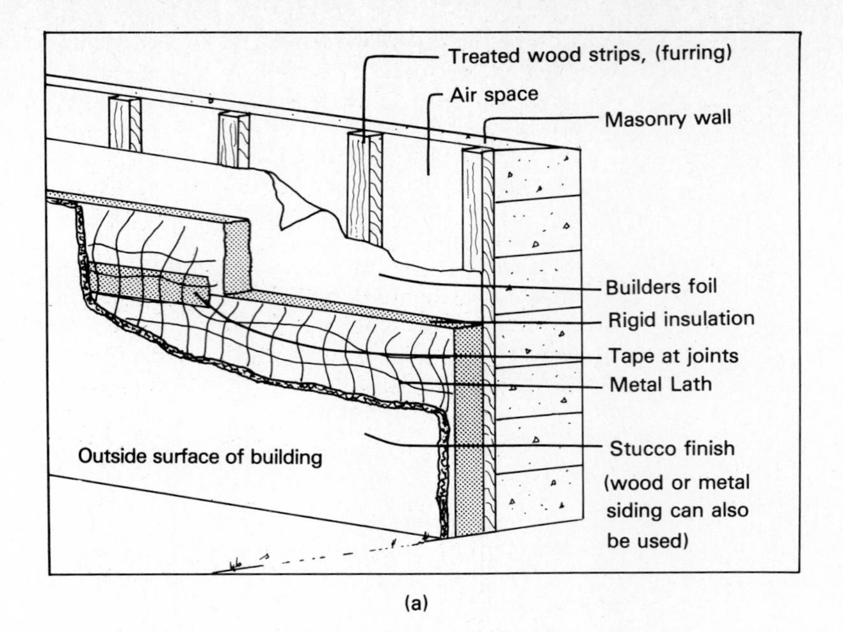

(a)

(b)

Figure 5-6 Outsulation. (a) Outsulation systems are applied onto the exterior surface of masonry walls. (b) Outsulation system being installed on a building in Wayne and Susan Nichols's La Vereda compound, Santa Fe, NM.

Outsulation systems consist of several layers. The first is a grid of treated wood strips called *furring* that are attached directly onto the storage mass surface with masonry nails. A layer of builder's foil is applied over the furring, with the reflective side facing inward. Rigid insulation, typically 2 inches thick, is applied next and nailed to the furring. Joints between foam batts are sealed with fiberglass tape. If the foam is covered by a reflective surface, the builder's foil is not necessary.

Outsulation systems are finished in many ways. The rigid insulation is often covered by a cement mixture, after which a finish coat of stucco is applied. Sometimes metal lath is used as the foundation for the cement, while at other times bonding agents mixed with the cement help it to adhere directly to the foam. Wood siding or shingles may also be installed over the insulation.

The thermal resistance of an outsulation system using two inches of extruded polystyrene board and including an air space formed by the furring strips is about R-14 for a stucco finish and R-15 for a wood siding finish.

It is critical that water be kept out of the outsulation system. If allowed to seep in and freeze, it can lead to separation of a stucco layer from the rigid insulation next to it, with resultant deterioration of the stucco. Careful installation of flashing over the top edges of the outsulation helps to solve this problem. A location sometimes overlooked is the top edge of outsulation near window sills.

Although rigid foam insulations such as urethane and isocyanurate have high R-values per inch of thickness, they can be damaged by continued exposure to water leaking into the system, especially if the water freezes. Because of its closed-cell structure, extruded polystyrene is considerably more resistant to moisture damage.

Outsulation is especially convenient for retrofit situations. Since it is applied to the outside of the building, it does not reduce floor space, nor do the occupants need to leave during installation. Over half of the systems in Germany—a country that relies heavily on outsulation to conserve energy—were retrofitted.

Thermal mass that is located in the interior walls of the building needs no insulation, since all heat that it discharged is contained in the living space. Even if only one side of an interior massive wall is directly exposed to the sun, both sides of the wall will radiate energy into the living space.

Floors. Massive floors should be thermally protected from the environment outside the building. Dry sand is sometimes used as insulation under slabs-on-grade, but if the winter moisture content of the surrounding ground is high, or if its average temperature is below 50°F, the use of rigid foam batts is recommended.[12] Two inches of smooth-skin, extruded polystyrene has a thermal resistance about equal to 2 *feet* of dry sand.[13]

When thermally massive materials such as concrete and/or brick pavers are installed on top of a wooden subfloor, insulation needs to be installed between the joists unless the space under the floor is already thermally protected from the outside climate. Methods of insulating under massive floors of various types are illustrated in Figure 5-7.

Finishing a Massive Surface

A critical part of constructing direct-gain systems is the finishing off of the thermal storage mass surfaces in ways that do not impair the mass's function but that are attractive and fit into the decor of the house. Covering a slab floor with a carpet impairs its performance by insulating it to some extent from the living space, as can covering massive walls with wood paneling or gypsum board. It is especially harmful if there are air spaces between the covering and the mass.

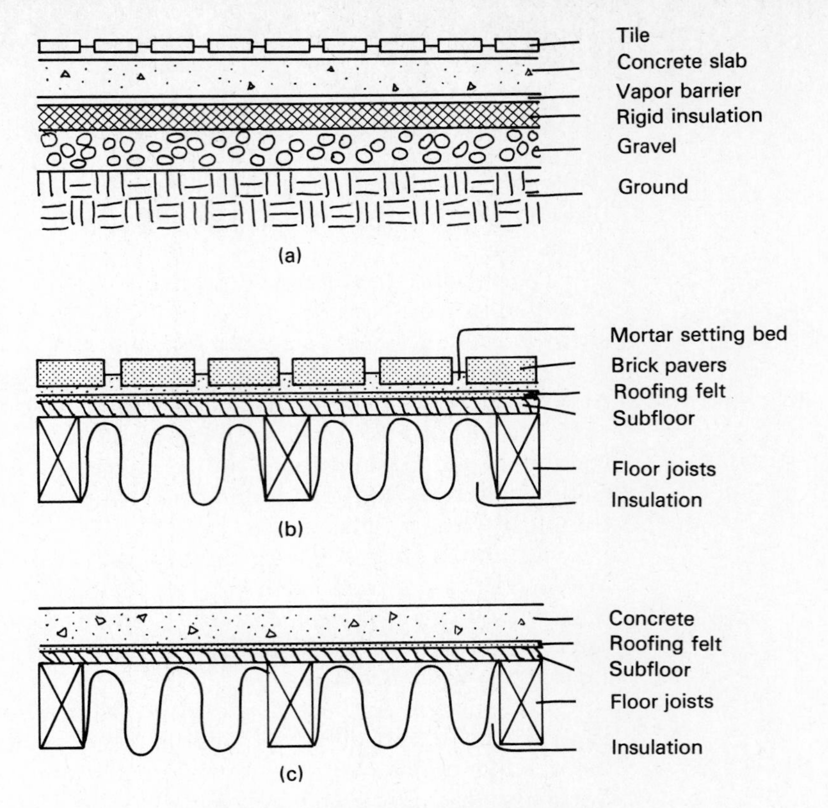

Figure 5-7 Insulating under massive floors.

Finishes that do not diminish storage mass's efficiency include plaster, stucco, and tile or flagstone facing. The surfaces may also be painted with appropriate colors, stained, waxed, and/or polished. In addition, if poured concrete is used as mass, it may while still wet be stamped with various patterns, textured in a number of ways, or embedded with multicolored rock aggregate.

Thermal storage mass can and should be attractive. Figure 5-8 illustrates how mass that is imaginatively finished can add to the charm of a living or work space rather than detract from it.

The Effect of Surface Coloring. The surface color of thermal storage mass affects the percentage of incident sunlight that it absorbs and thus has some effect on the total system performance. In direct-gain systems, however, it has been found that the surface color is much less critical than it is in other types of collection systems, such as solar walls or flat-plate collectors. The reason for this is that most of the energy entering a direct-gain aperture is eventually absorbed by some surface in the room, even if it is first reflected several times off other surfaces. Only a minority of the energy manages to find its way out of the solar aperture (see Figure 5-9).

The solar absorptance of a surface is a measure of the percentage of normally incident sunlight that it captures and converts into thermal energy. The energy that is not absorbed is reflected from the surface. As long as the surfaces in a direct-gain space have absorptances of 0.80 or higher, it has been found that the system's performance will be within a few percent of what it would be if the surfaces were perfect absorbers.[14] This is fortunate, for the best absorbing surfaces are black in color and thus not very aesthetic

(a)

(b)

(c)

(d)

(e)

Figure 5-8 Decorating a house with thermal storage mass. (a) Adobe brick is an attractive building material as well as a functional thermal storage mass. Tile and brick floors also add to a home's beauty in addition to storing heat. (b) A masonry fireplace can store heat from both the sun and the fire. (c) Besides being an enjoyable place to sit, the tiled concrete bench in the foreground is a storehouse of heat. (d) Embedded pebbles dress up a concrete wall or floor. (e) The thermally massive columns, floor, and plant enclosure in this photo are useful heat storage constructions for either residential or commercial situations. Both the concrete plant enclosure and the earth contained within can store thermal energy. The floor of the building is not made of discrete bricks, but is a continuously poured concrete slab that has been colored and stamped with a pattern while still wet. These types of floors generally cost considerably less than either brick floors or tile installed over a concrete subfloor. [(a), (b), Courtesy of Ben Gilbert, Gilbert's Adobe Homes, Los Lunas, NM; (e), courtesy of Bomanite Corporation, 81 Encina Avenue, Palo Alto, CA 94301.]

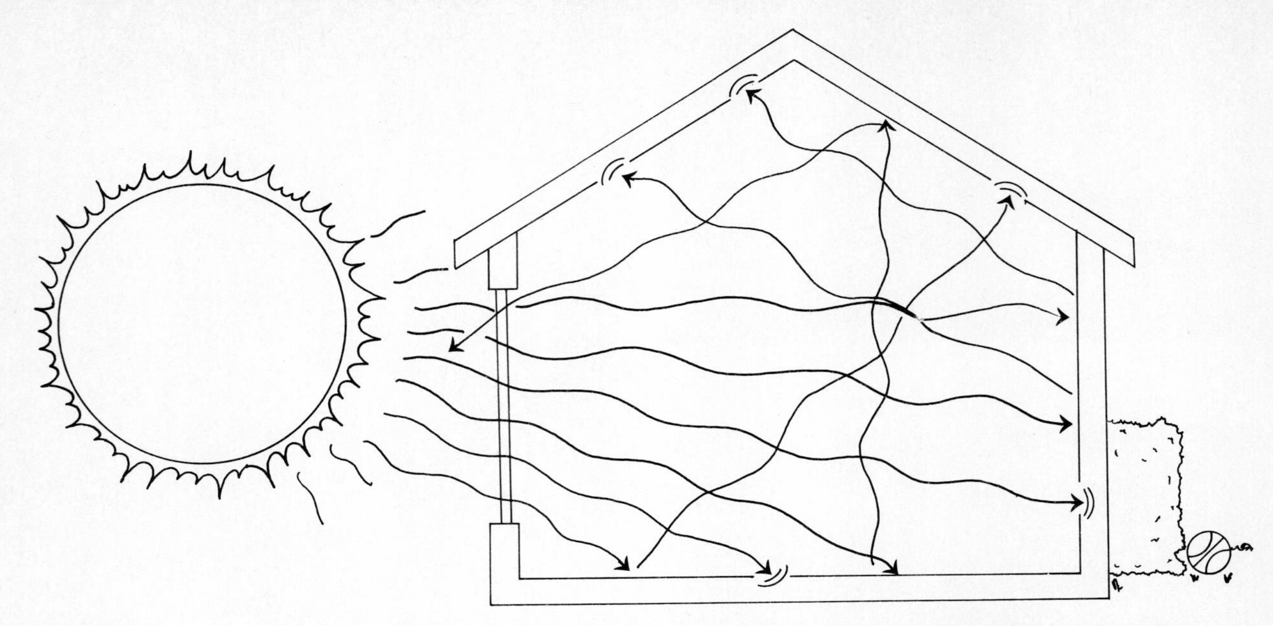

Figure 5-9 Most of the energy that enters a solar aperture in a well-designed direct-gain system is eventually absorbed by some surface in the room.

as room decor. Note from Table 5-3, which lists absorptances of various colors, that those with values of 0.80 or higher include shades of blue, brown, green, and gray.

If colorings with absorptances as low as 0.50 are used, the system will still perform about 80 percent as well as if perfect absorbers had been used. Most of the colorings listed in Table 5-3 have absorptances of 0.50 or higher and include bright hues such as reds, yellows, and oranges.

The Effect of Furniture, Carpets, and Wall Decorations

Furniture, carpets, and wall decorations are low-mass objects unable to store a significant amount of heat. When sunlight strikes such objects, their surfaces heat up quickly, especially if they are dark in color. Much of this heat is then transferred to the room air shortly thereafter. In extreme cases where a large part of the room's surfaces are covered by such objects, daytime overheating can occur, requiring venting of thermal energy and a loss in overall system performance.

The question is, how to determine when carpeting, furniture, and wall hangings in a building significantly detract from the solar system's efficiency. In computer modeling calculations at Los Alamos National Laboratory, it was found that if no more than 20 percent of incoming solar radiation is absorbed by low-mass surfaces, the overall drop in performance is 5 percent or less.[15] Some guidelines for preventing problems are as follows:

1. Primary radiation zones should be kept as free of low-mass objects as possible.
2. Excessive clutter should be avoided in secondary radiation zones.
3. Low-mass objects in primary or secondary radiation zones should be relatively light in color.

TABLE 5-3 Absorptivities of Various Colored Surfaces[a]

Surface	Absorptivity
Optical flat black paint	0.98
Flat black paint	0.95
Black lacquer	0.92
Dark gray paint	0.91
Black concrete	0.91
Dark blue lacquer	0.91
Black oil paint	0.90
Stafford blue bricks	0.89
Dark olive drab paint	0.89
Dark brown paint	0.88
Dark blue-gray paint	0.88
Azure blue or dark green lacquer	0.88
Brown concrete	0.85
Medium brown paint	0.84
Medium light brown paint	0.80
Brown or green lacquer	0.79
Medium rust paint	0.78
Light gray oil paint	0.75
Red oil paint	0.74
Red bricks	0.70
Uncolored concrete	0.65
Moderately light buff bricks	0.60
Medium dull green paint	0.59
Medium orange paint	0.58
Medium yellow paint	0.57
Medium blue paint	0.51
Medium Kelly green paint	0.51
Light green paint	0.47
White semigloss paint	0.30
White gloss paint	0.25
Silver paint	0.25
White lacquer	0.21
Polished aluminum reflector sheet	0.12
Aluminized Mylar film	0.10
Laboratory-vapor-deposited coatings	0.02

[a] This table is meant to serve only as a guide. Variations in texture, tone, overcoats, pigments, binders, and so on, can vary these values.

Source: J. Douglas Balcomb et al., *Passive Solar Design Handbook,* Vol. 2, Los Alamos National Laboratory, Los Alamos, NM, for U.S. Department of Energy, January 1980, p. 75.

OVERHEATING PROTECTION

The same insulation that keeps heat in the building in winter helps keep excessive warmth out in summer. Solar apertures, however, can allow an overabundance of thermal energy into the building during the warmer months, leading to severe overheating problems. Movable insulation devices such as those discussed in Chapter 4 can be of help but will cause problems if applied to too many of the apertures, for although they prevent excessive interior heat gain, they can also make interior spaces dim and dreary by blocking natural daylight.

A technique that should be used whenever possible is to install overhangs above south-facing apertures to provide seasonal shading but still allow natural daylight to enter the building.

Figure 5-10 Overhangs. (a) Summer conditions. (b) Winter conditions.

Figure 5-10 illustrates how an overhang is able to block direct solar energy from impinging on the aperture during the summer months, when the sun is high in the sky, but allows the apertures to receive full sunlight during the winter, when the sun is considerably lower.

As illustrated in Figure 5-11, overhangs can be designed in many different ways. Some are slanted, while others are horizontal. In many situations, roof eaves or balconies serve as overhangs; at other times simple extensions from the building are used that fulfill no other function than that of providing seasonal shading.

The graphs in Figure 3-10 depict the path of the sun through the sky at various times of the year, for observers at a number of different latitudes. Notice that the sun follows the same path on May 21 as it does on July 21. This is also true for April 21 and August 21, for March 21 and September 21, and so on. Thus, assuming that the sky is clear, a window overhang

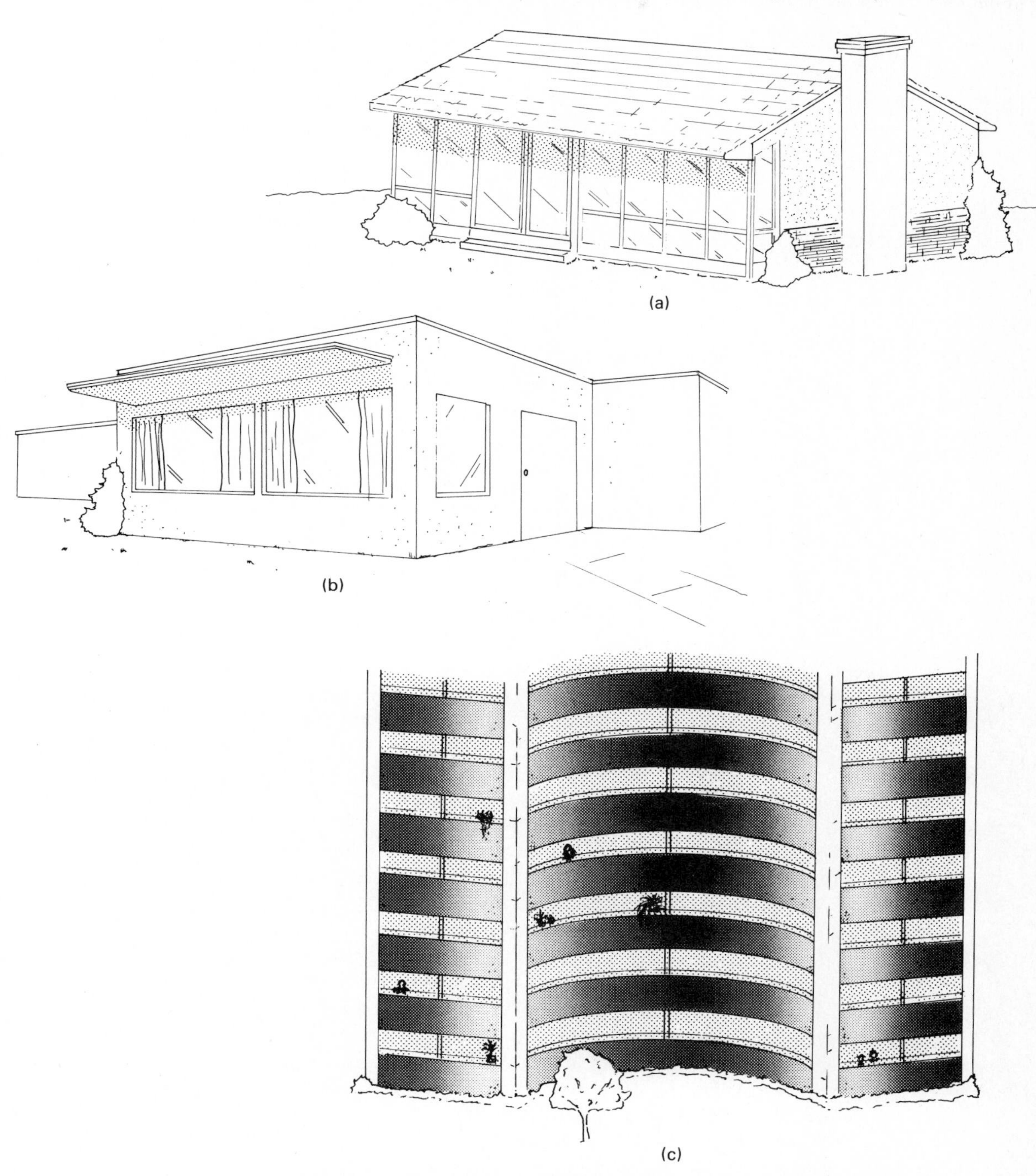

Figure 5-11 Overhang designs. (a) Roof eaves in this home serve as overhangs. (b) Horizontal extensions from the building provide shade. (c) Balconies shade the windows below.

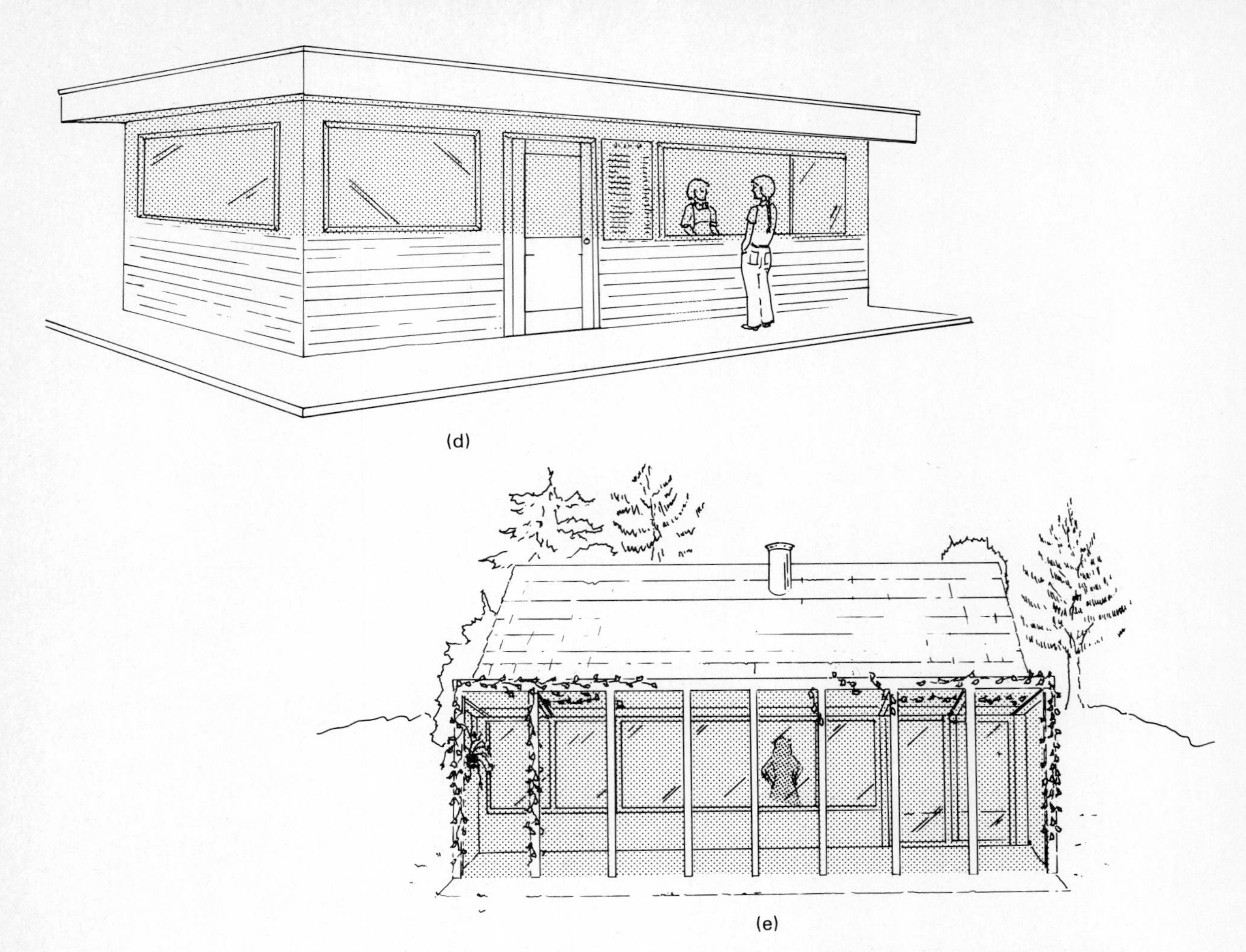

(d)

(e)

Figure 5-11 (cont.) (d) Overhangs can be part of the building's architectural styling. (e) Trellisses with deciduous vegetation growing on them provide seasonal shading. Even leafless vegetation, however, can lead to unwanted light reduction in winter.

casts an identical shadow pattern on, for instance, March 21 as on September 21. Unfortunately, the weather in March is usually quite a bit colder than in September. An overhang that would completely shade a window might be welcome on September 21 but most unwelcome on March 21, when solar energy might be sorely needed.

One way to circumvent such a problem is to construct adjustable overhangs that can be extended to various lengths according to the time of year. Fabric awnings lend themselves to adjustment but usually do not stand up for more than a few years to intense rain, wind, and sunshine. More durable are slats of wood or metal that are attached to permanently fixed frames as the weather warms up and taken down during the heating season (see Figure 5-12). But if these installation and removal operations are at all tedious, many homeowners will not bother to do them. What is more commonly done is to design fixed overhangs that are compromises between allowing full winter sun to enter the building and completely blocking summer sun.

It is recommended that a fixed overhang be small enough to allow full sun to strike the aperture between 9 A.M. and 3 P.M. during the month of January, when the weather is generally the most severe. To meet this re-

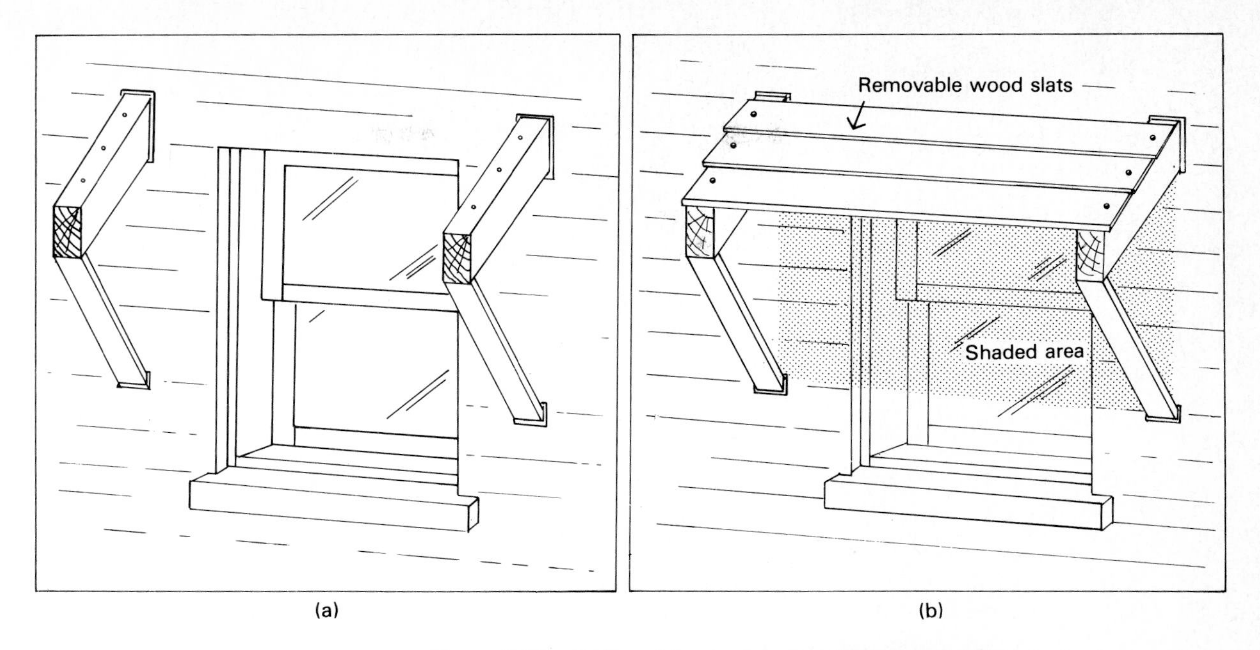

Figure 5-12 Removable overhangs. (a) Winter. (b) Summer.

quirement, the overhang must be designed within certain limits, which can be expressed by a limit on angle A depicted in Figure 5-13. Angle A is that formed by a line running from the top of the window to the end of the overhang and a horizontal line starting at the top of the window. It must be greater than some value, depending on the latitude of the site and the orientation of the house, in order for the window to receive total sunshine for a full day in January.

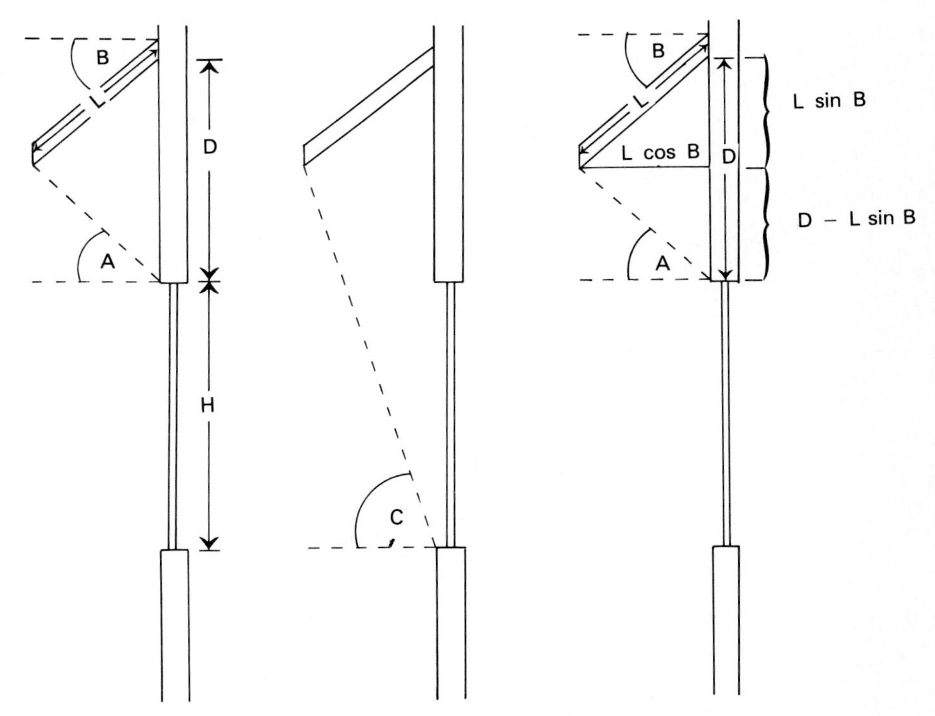

Figure 5-13 Fixed-overhang dimensions and angles.

TABLE 5-4 Limiting Values for Angles A and C (Degrees)

	Angle A_{min} Aperture Orientation			Angle C_{max} Aperture Orientation		
Latitude	Due South	15° SE or SW	30° SE or SW	Due South	15° SE or SW	30° SE or SW
28°N	41	45	62	77	78	79
32°N	37	40	56	73	74	75
36°N	33	36	49	70	71	72
40°N	29	32	44	66	66	68
44°N	26	28	38	62	63	65
48°N	22	23	30	58	59	61
52°N	18	19	24	54	56	58
56°N	14	15	18	50	51	54

Table 5-4 lists limiting values for angle A for different latitudes and building orientations. For a building located at a latitude of 36° north and facing 15° southeast, for instance, the table requires that angle A be greater than 36°. The versatility of Table 5-4 is that it can be applied to any type of overhang—slanted or horizontal, and with a base that is located at any distance above the top of the window. The table can also be used for clerestories and skylights.

Once angle A_{min} is determined, the overhang can be designed. The first steps in doing this are to decide on the window height (H), the distance from the top of the window to the base of the overhang (D), and the overhang's slope (angle B). Figure 5-13 depicts these quantities, which are chosen largely on the basis of aesthetic considerations and local code requirements. The next design step is to use the formula below to determine the maximum overhang length for full sun on January 21 (L_{winter}).[16]

$$L_{winter} = \frac{D}{\sin B + (\cos B)(\tan A_{min})}$$

No previous knowledge of trigonometry is required to use this formula. The values for sin B, cos B, and tan A_{min} are simply read from Table 5-5 and inserted into it.

Example Calculation. A building located at a latitude of 32° north has apertures that face 30° southwest. An overhang whose base is 1.5 feet above the top of the windows and whose slope is 15° is to be installed. What should its limiting length be?

From Table 5-4, angle $A = 56°$. The information given in the problem tells us that angle $B = 15°$ and $D = 1.5$ feet. From Table 5-5,

$$\sin 15° = 0.259$$
$$\cos 15° = 0.966$$
$$\tan 56° = 1.48$$

$$L_{winter} = \frac{1.5}{0.259 + (0.966)(1.48)} = 0.89 \text{ foot}$$

Note that when overhangs are horizontal, both B and sin B equal zero.

Overhangs designed according to the method described above provide partial shading of the solar aperture during the summer months. The closer the length is to its limiting value, the more shading is provided. It is recommended that in locations that experience hot summers, the maximum allowable overhang length be used. The overhang should provide total shading of the aperture at noon of each day between May 15 and August 1.[17] To accomplish this, angle C, formed by a line from the bottom of the window to

TABLE 5-5 Values of Sin, Cos, and Tan

When Angle *B* Equals	Sin *B* Equals	Cos *B* Equals	When *A* or *B* Equals	Tan *A* or *B* Equals
0°	0	1	0°	0
5°	0.0872	0.9962	5°	0.08749
10°	0.1736	0.9848	10°	0.1763
15°	0.2588	0.9659	15°	0.2679
20°	0.3420	0.9397	20°	0.3640
25°	0.4226	0.9063	25°	0.4663
30°	0.5	0.8660	30°	0.5774
35°	0.5736	0.8192	35°	0.7002
40°	0.6428	0.7660	40°	0.8391
45°	0.7071	0.7071	45°	1
50°	0.7660	0.6428	50°	1.1918
55°	0.8192	0.5736	55°	1.4281
60°	0.8660	0.5	60°	1.7321
65°	0.9063	0.4226	65°	2.1445
70°	0.9397	0.3420	70°	2.7474
75°	0.9659	0.2588	75°	3.7321
80°	0.9848	0.1736	80°	5.6713
85°	0.9962	0.0872	85°	11.4301
90°	1	0	90°	—

the tip of the overhang and a horizontal line intersecting the window bottom, must be less than a certain value that is a function of the building's latitude and orientation. These values are listed in Table 5-4 (angle *C* is depicted in Figure 5-13).

If it is found that angle *C* is greater than the value from Table 5-4, the system should, if possible, be redesigned. Decreasing the window height (*H*) while keeping the distance (*D*) from the top of the window to the base of the overhang the same will decrease angle *C* (see Figure 5-14). If this is done,

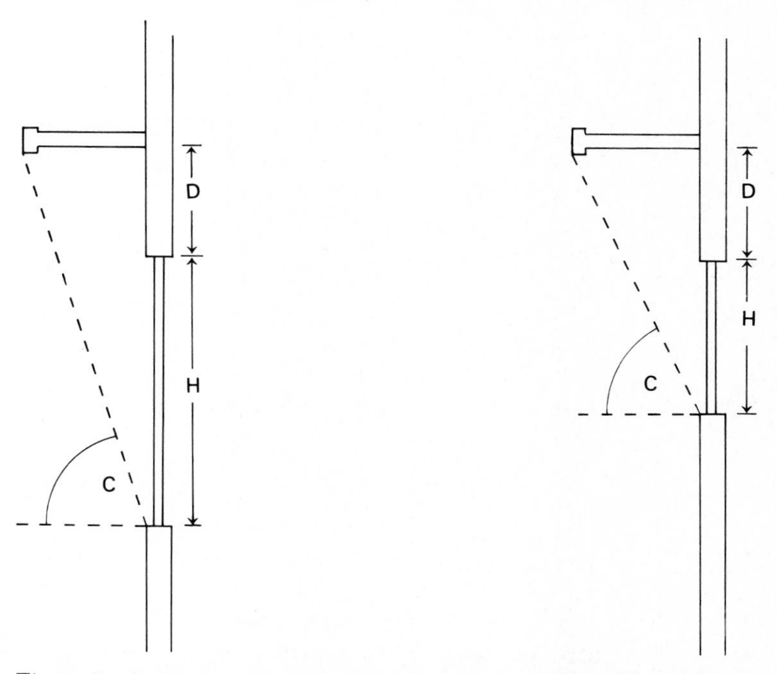

Figure 5-14 Decreasing the window height while keeping *D* the same will decrease angle *C*.

it might be necessary to increase the width of the apertures or add new apertures so that the total solar aperture square footage is not changed.

Example. The aperture and shading combination depicted in Figure 5-15 has angle C equal to 70°. The aperture orientation is 15° southwest and the latitude is 40° north. Table 5-4 tells us that the maximum value for angle C should be 66°. Redesigning is thus advisable. A small decrease in the window height, while keeping D constant, will bring angle C into the acceptable range.

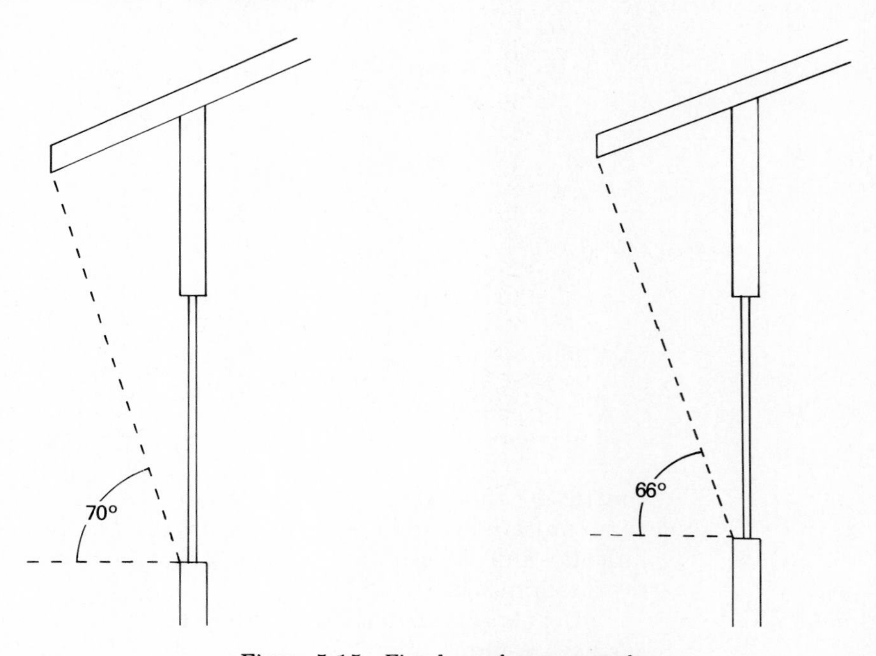

Figure 5-15 Fixed-overhang example.

If additional protection against summer overheating is needed (which often is not known until the building has been occupied for a time), several measures may be taken.

The more that the overhang *width* extends past the sides of the window, the more useful it is in preventing morning or afternoon sun from shining under its edge and hitting the window (see Figure 5-15). It is recommended that the overhang extend a distance of $D + H$ or more past the sides of the window.

Two additional overheating control measures are to install vertical fins on the sides of the window (these are illustrated in Figure 5-16) or design the windows to be recessed into the wall. Olgyay and Olgyay's *Solar Control and Shading Devices*[18] and Mazria's *The Passive Solar Energy Book*[19] are two references that discuss the design of side fins.

Overhangs and fins are useful for shading apertures that face in a direction close to south. Windows that face east or west catch the full brunt of near-perpendicular insolation when the sun is low in the sky and require different treatment. They can be blocked off by movable insulation or light-colored curtains and shades or, since they are generally not used as solar apertures, can be fitted with tinted or reflective glazing. Deciduous vegetation is also an aid in shading the house during summer, but as mentioned in Chapter 3, even its bare branches can lead to unwanted light reduction during winter. Natural ventilation processes coupled with shading will often provide sufficient comfort for the interior of the house during the warm months. Figure 5-17 depicts several building configurations that enhance air movement through the house.

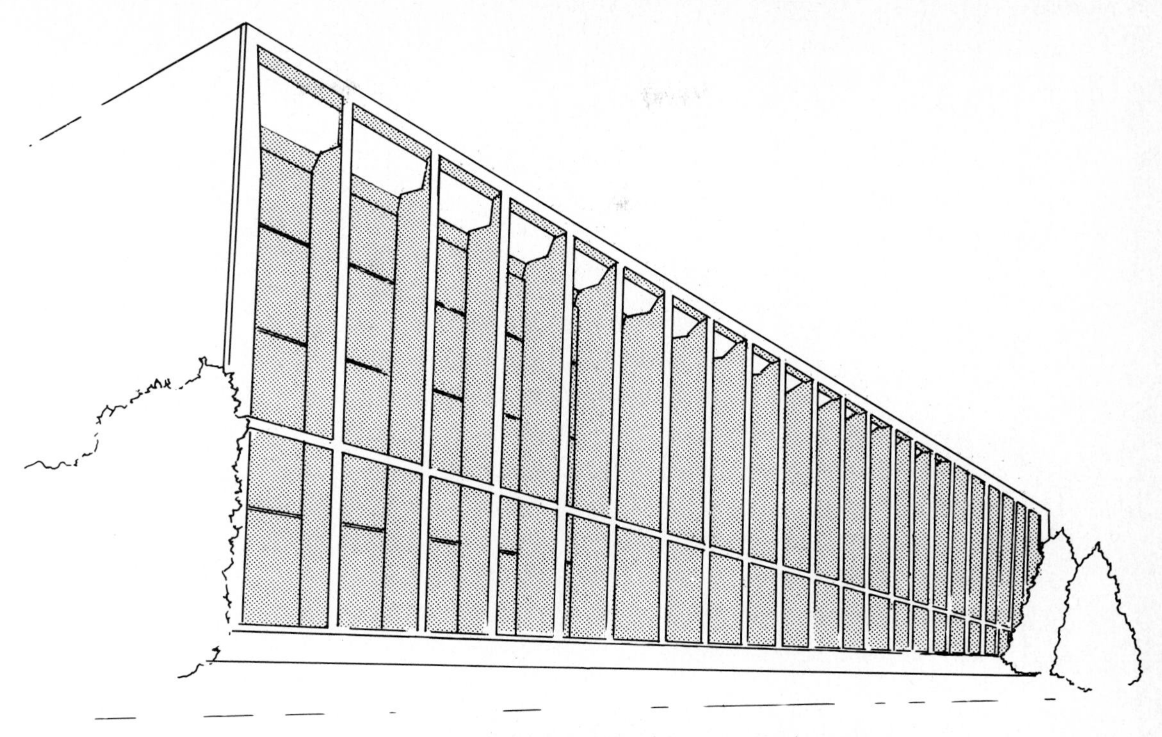

Figure 5-16 Example of vertical side fins.

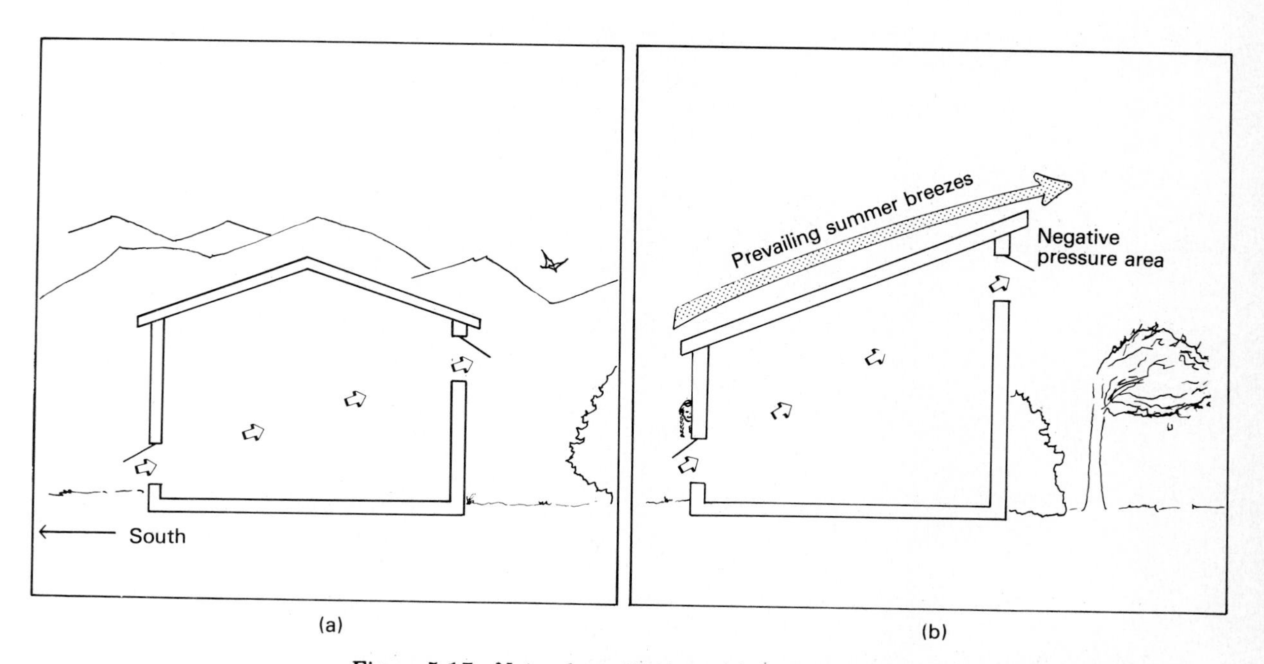

Figure 5-17 Natural ventilation. (a) Low south windows and high north windows such as clerestories create a "thermal chimney" that helps draw hot air out of the house and cool air in. (b) Wind sweeping up a roof sloped toward the direction of prevailing summer breezes also helps draw hot air out of the house by creating a negative pressure area at the downstream end of the roof. This is called a Venturi effect.

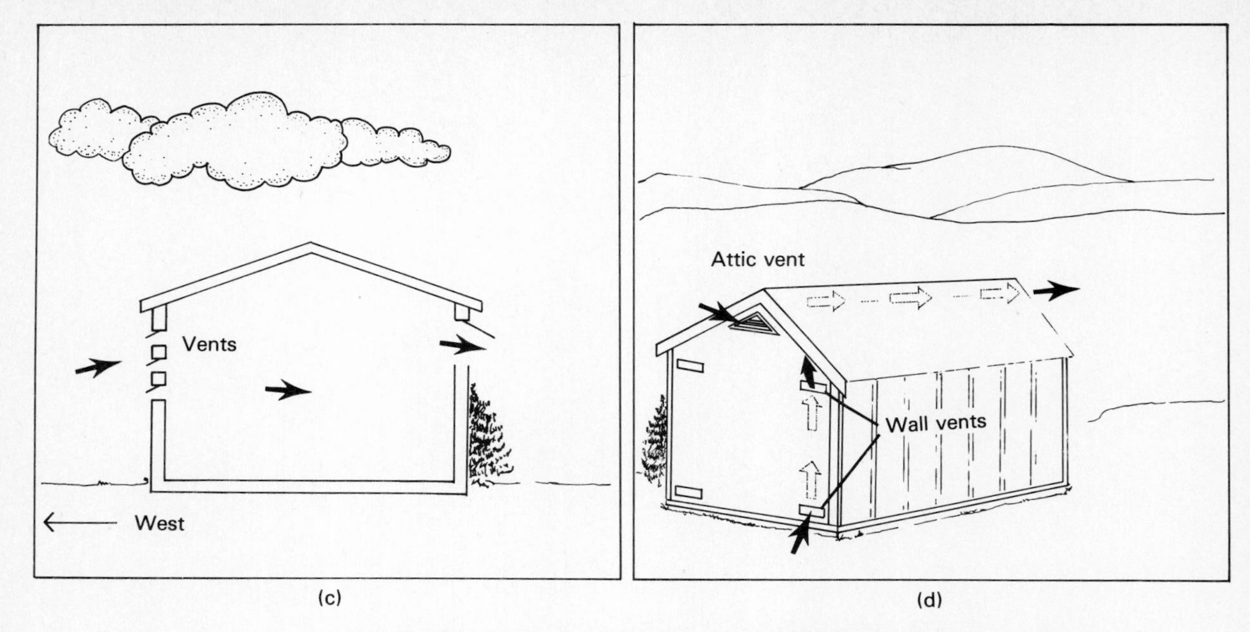

Figure 5-17 (cont.) (c) Good cross-ventilation is important. Vents instead of windows are sometimes added to the west wall to allow ventilation while limiting overheating due to intense late afternoon sun. (d) In areas with warm summers, attics and sometimes west walls should be extra-well vented (1 square foot of vent for every 100 square feet of wall or attic floor area) to prevent heat build-up within them.

NOTES

1. *Passive Solar Construction Handbook*, Steven Winter Associates, New York, for Southern Solar Energy Center and U.S. Department of Energy, 1981, p. 4.2200.

2. Ibid., Sec. 4.2.

3. Stephen S. Szoke, "Brick: Properties and Performance in Thermal Storage," *Solar Age*, Vol. 6, No. 7, July 1981 plus price updates.

4. Joe Kohler and Dan Lewis, "Phase-Change Products for Passive Homes," *Solar Age*, Vol. 8, No. 5, p. 66; Dow Chemical Co., FP&S Dept., 2020 Dow Center, Midland, MI 48640.

5. Kohler and Lewis; Energy Materials, Inc., 2622 South Zuni, Englewood, CO 80110.

6. R. D. McFarland and J. D. Balcomb, "Los Alamos Test Room Results," *Proceedings of the 7th National Passive Solar Conference*, American Solar Energy Society, Inc., 1982, p. 155.

7. Ibid.

8. Kohler and Lewis, p. 65.

9. J. Douglas Balcomb et al., *Passive Solar Design Handbook*, Vol. 2, Los Alamos National Laboratory, Los Alamos, NM, for U.S. Department of Energy, 1980, p. 180.

10. Ibid., p. 181.

11. This is the term used by Dryvit Systems, Inc., of West Warwick, RI, but it is becoming a generic term.

12. *Passive Solar Construction Handbook*, p. 3.1009.

13. *Energy Conservation Design Manual for New Residential Buildings,* State of California Energy Resources Conservation and Development Commission, 1978, Appendix A. *Extruded* rather than expanded polystyrene should be used in all applications where it might contact ground moisture. The extruded variety will last far longer.

14. Balcomb et al., Vol. 2, pp. 73-74.

15. Ibid., pp. 78-81.

16. The derivation of this formula is as follows: It can be shown, using the information in Figure 5-13, that

$$\tan A = \frac{D - L \sin B}{L \cos B}$$

Thus,

$$L \cos B \tan A = D - L \sin B$$

$$L(\sin B + \cos B \tan A) = D$$

$$L = \frac{D}{\sin B + \cos B \tan A}$$

17. These dates span a time frame from five weeks before until five weeks after the summer solstice. If the aperture is in total shade at midday during these dates, overheating of the living space will be greatly reduced.

Adjusting the overhang length so that full shade of the aperture is provided during more of the year is not advisable in many locations, for it will cut down on solar heating at times when it might be needed.

18. Aladar Olgyay and Victor Olgyay, *Solar Control and Shading Devices,* Princeton University Press, Princeton, NJ, 1957.

19. Edward Mazria, *The Passive Solar Energy Book,* Rodale Press, Emmaus, PA, 1979.

SUGGESTED READING

HOLLAND, ELIZABETH. "Insulation Moves Outside," *Solar Age,* Vol. 6, No. 11, November 1981. A discussion of outsulation methodology.

Passive Solar Construction Handbook. New York: Steven Winters Associates, Inc., for Southern Solar Energy Center and U.S. Department of Energy, 1981. Useful information on constructing thermal storage walls and floors.

SZOKE, STEPHEN S., "Brick: Properties and Performance in Thermal Storage," *Solar Age,* Vol. 6, No. 7, July 1981.

REVIEW QUESTIONS

1. Define *thermal storage mass.* What are two characteristics that it should have?
2. Compare the advantages of using water as a thermal storage mass with those of using masonry materials.
3. What are phase-change materials (PCMs)? What advantages do they offer as thermal storage materials? What problems have they had in the past?
4. Of the commonly used masonry materials, which has the highest heat capacity? The highest conductivity?

5. Define *primary* and *secondary* radiation zones.
6. Which thermal mass configuration will store more energy each day: 200 square feet of 4-inch-thick brick located in a primary radiation zone or 100 square feet of 8-inch-thick brick located in a primary radiation zone?

PROBLEMS

1. A building with 250 square feet of direct-gain aperture employs adobe walls as thermal storage mass. If the adobe is 8 inches thick and located in primary radiation zones, how many square feet of it should be used?
2. Repeat the calculation of Problem 1 for concrete and brick thermal storage mass.
3. A 6-inch-thick layer of concrete receives direct sun from 1 P.M. to 3 P.M. each day, but is in a secondary radiation zone from 9 A.M. to 1 P.M. How many square feet of mass is required per square foot of aperture? If 225 square feet of aperture are installed, how many square feet of mass are needed?
4. A 3-inch-thick layer of brick is in a primary radiation zone from 9 A.M. to 2 P.M. and a secondary radiation zone after that. If 175 square feet of aperture is installed, how many square feet of mass is needed?
5. A direct-gain passive home is to have 375 square feet of aperture in its south wall and roof. A 4-inch-thick concrete slab floor with an area of 900 square feet is in a secondary radiation zone until 10 A.M. and a primary zone after that. The walls of the room are to be of 8-inch-thick adobe. They will receive direct sunlight for about 4 hours every day and secondary radiation the rest of the time. What is the recommended square footage of the walls?
6. A direct-gain passive building is to have 200 square feet of solar aperture. A 3-inch-thick concrete slab floor with an area of 500 square feet is in a primary zone for the entire day. A 300-square-foot wall faced with 3 inches of brick is also in a primary zone all day. There is an additional wall in the building that is in a primary zone for half the day and in a secondary zone the remainder of the time. How many square feet of it should be faced with 3-inch-thick brick in order to save enough total thermal storage mass to just balance the size of aperture? (It is assumed that the remaining wall area that is left unfaced with brick has a low enough thermal storage capacity that it can be neglected.)

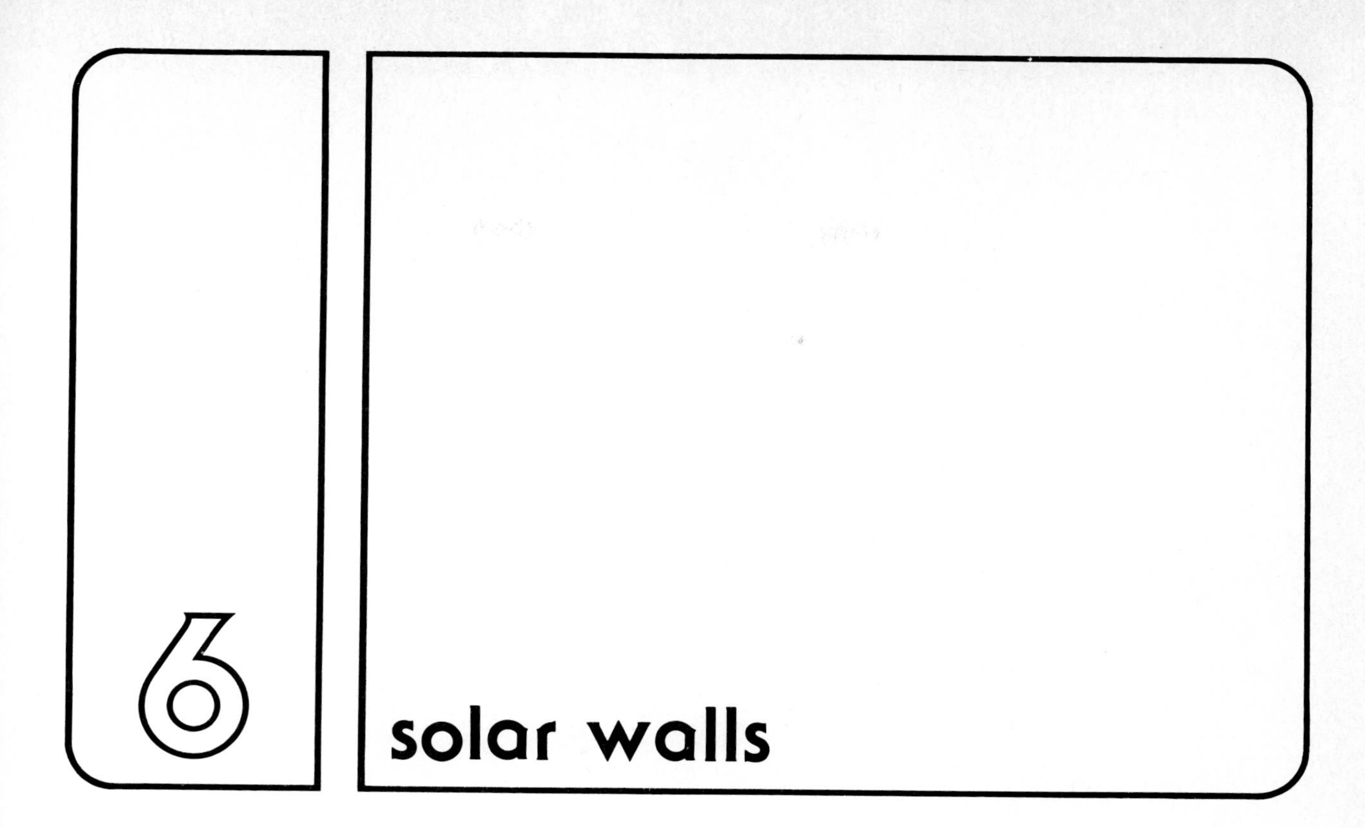

6 solar walls

BASIC CONFIGURATIONS OF SOLAR WALL SYSTEMS

Solar wall systems differ from direct-gain systems in that the energy collecting surface is located *between* the sun and the interior of the building rather than within the interior.[1] Solar walls are generally not insulated from the living space and are constructed of thermally massive material.

The most common solar wall designs are variations of the systems depicted in Figure 6-1. Sunlight passing through glazing in the south side of a building impinges upon a wall located several inches away. The wall is of a dark color in order to improve its absorptivity. Absorbed sunlight is converted into thermal energy that eventually travels through the wall and is radiated into the living space. In addition, the heated wall raises the temperature of the air between it and the glazing. Some solar walls have vents in them that allow this air to flow into the living space through the process of natural convection.

Solar walls are generally constructed of either a masonry material such as concrete, brick, and adobe or are filled with water. Most walls are designed to release collected energy *gradually* to the building's interior. Massive walls continue giving off heat 8 or more hours after the sun has set, thus helping to keep the house at a comfortable temperature throughout the night. As with direct-gain systems, solar wall collecting surfaces should face in a direction between 20° east and 32° west of true south.

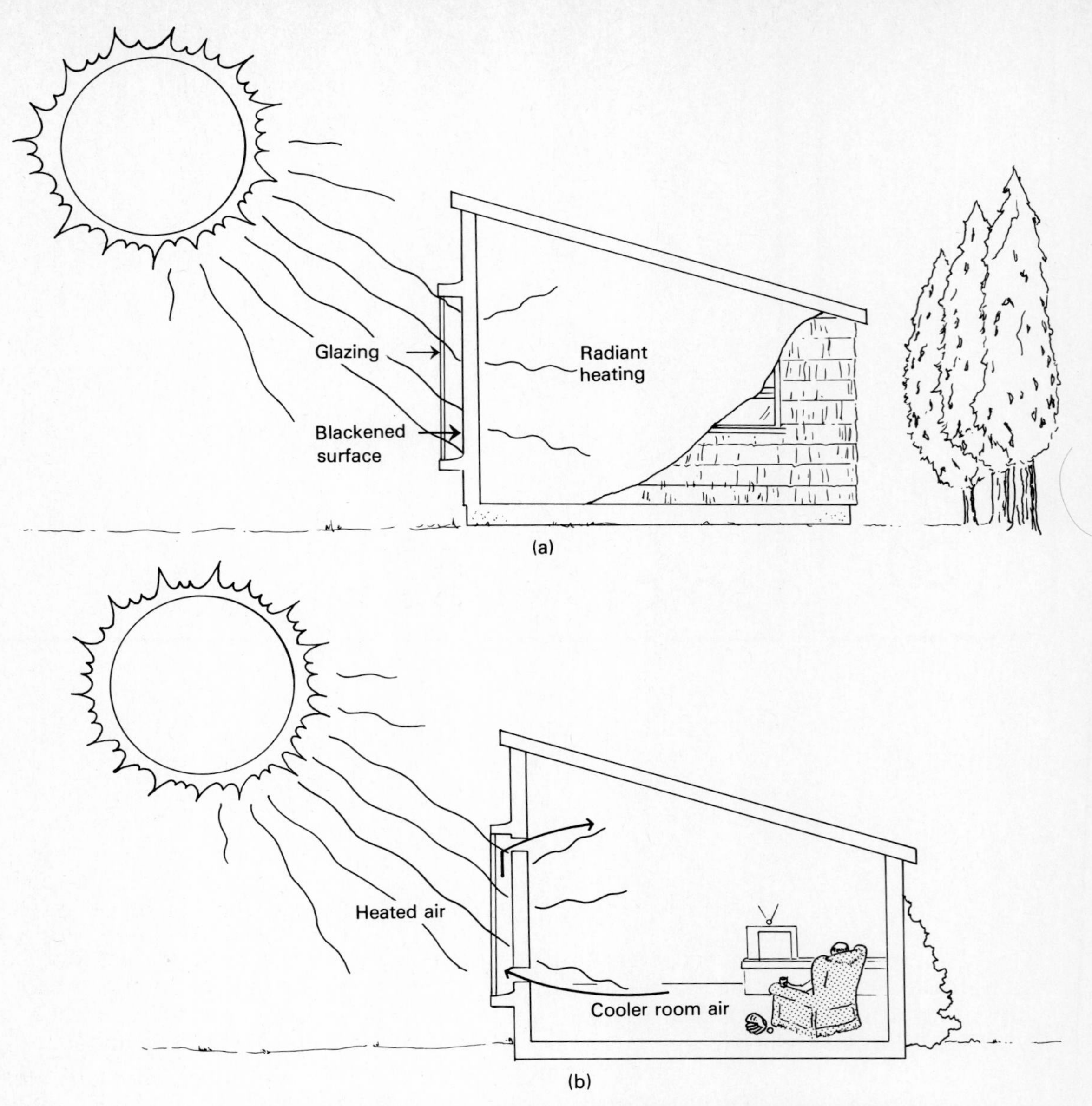

Figure 6-1 Solar wall. (a) Unvented design. The wall is heated by the sun. Heat is conducted through the wall and radiated into the living area. (b) Vented design. The solar-heated wall surface warms the air between it and the glass. Natural convection carries this air into the living space. In addition, the living space is warmed by radiant energy from the wall, as in the unvented design.

TROMBE WALLS

Masonry solar walls are often called *Trombe walls* due to the extensive work of Felix Trombe and Jacques Michel at Odeillo, France, in the design of passively heated and cooled solar structures. Their first dwelling that made use of south-facing concrete collecting surfaces mounted behind panes of glass was built in 1967.

Trombe walls are often designed to serve a load-bearing function as well as to collect and store the sun's energy and to help enclose the building's interior spaces. Multiple uses of solar energy components help greatly to reduce the overall labor and material cost of constructing a passively heated building.

A critical part of solar wall design is choosing the proper thickness. Excessively thick walls take too long to transmit the thermal energy they collect. A typical result of this might be a living space that does not receive enough heat during the evening hours when it is needed most, for the energy is still working its way through the wall. Walls that are too thin, on the other hand, transmit energy too rapidly, resulting in overheating of the living space during the day and little energy left for the evening. The solar savings fraction, the percentage of the building's heating needs that are met by the sun's energy, is also decreased in poorly designed wall systems.

The optimum thickness of a Trombe wall is to a great extent dependent on properties of the material from which it is made such as thermal conductivity and heat capacity. (These quantities are defined in the beginning of Chapter 5.) The higher the conductivity, for instance, the faster heat flows through the wall and thus the thicker it can be. Table 6-1 lists recommended thicknesses for walls made of a variety of commonly used materials.

TABLE 6-1 Recommended Trombe Wall Thicknesses

Building Material	*Recommended Thickness (in.)*
Concrete	14–18
Brick	
Facing	12–16
Common	10–14
Adobe	8–12
Stone: limestone, sandstone, granite	14–18

Solar walls made of different materials lead to different levels of performance, whether the walls are optimally thick or all of equal thickness. Table 6-2 lists the solar savings fractions that would result from a solar wall being constructed of three different materials. Notice that the higher the

TABLE 6-2 Effect of Wall Material on the Solar Savings Fraction[a]

			Solar Savings Fraction (%) for Four Sample Systems (load collector ratio = 18 for each)			
Wall Material	*Conductivity (Btu/ft-°F-hr)*	*Heat Capacity (Btu/°F-ft³)*	*Raleigh, North Carolina, System*	*Des Moines, Iowa, System*	*Reno, Nevada, System*	*Toronto, Ontario, System*
Concrete	1.0	30.0	60	34	67	30
Brick (common)	0.42	22.4	53	31	60	28
Adobe	0.33	24.0	51	30	58	28

[a]Wall thickness for all cases = 12 inches. These results are for four houses in four different cities. The load collector ratio (LCR) for all systems is 18 (LCRs were introduced in Chapter 4). The absolute value of solar savings fractions is of course different for systems of other designs or in different locations. The usefulness of this table is to evaluate the *relative* dependence of SSF on wall material.

Source: Information for the calculation of values in this table was obtained from the LCR Tables in J. Douglas Balcomb et al., *Passive Solar Design Handbook*, Vol. 3, Los Alamos National Laboratory, Los Alamos, NM, for U.S. Department of Energy, DOE/CS-0127/3, 1982.

conductivity of the substance, the greater the solar savings fraction (SSF) that results. (Can the reader explain why? See the problems at the end of this chapter.) Thus, from the point of view of energy savings, concrete appears to be the best of the three masonry materials for the solar wall, with brick second best.

Vented and Unvented Trombe Walls

The original Trombe walls built at Odeillo, France, used top and bottom venting that allowed heated air to flow by means of natural convection from the space between glass and masonry into the living area. To reduce reverse convection currents at night that would allow air cooled from exposure with the glass to flow through the lower vents into the house, the Odeillo dwellings had these vents located above the bottom of the collecting wall. This created a space between glass and wall into which cold air could settle (see Figure 6-2).

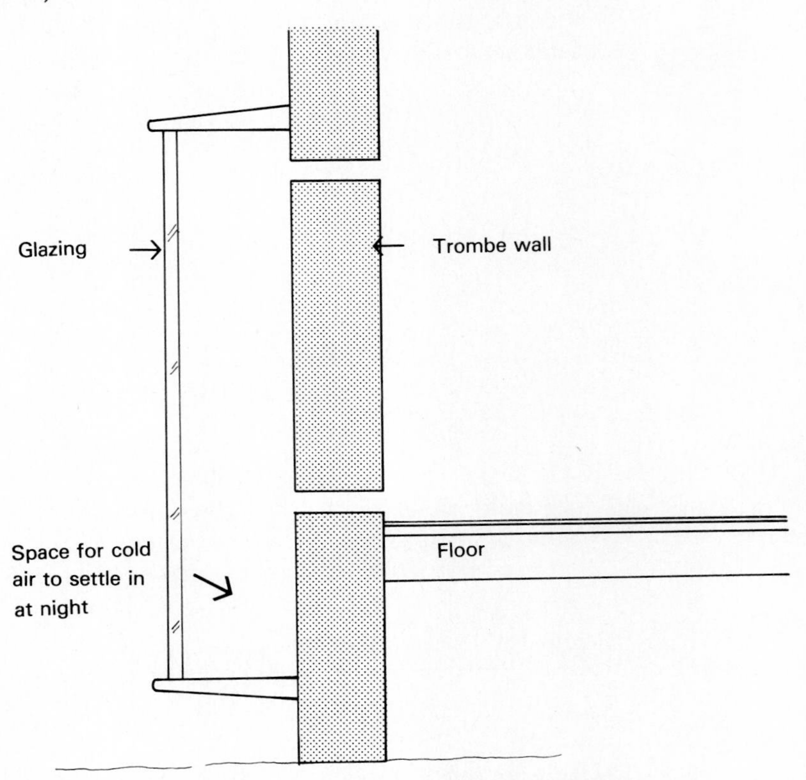

Figure 6-2 Early Trombe wall configuration.

A later Odeillo building employed a cold air trap much like the dirt trap under a kitchen or bathroom sink. The lower wall vent was placed below floor level, as in Figure 6-3, retarding the flow of dense, cold air into the living space during the night by forcing it to flow upward after passing through the vent.

In addition to the designs described above, today's Trombe walls make use of backdraft dampers that allow airflow through the vents in one direction only. The dampers are constructed so as to permit thermosyphoning of warm air into the house but prohibit reverse flows (see Figure 6-4). The damper flaps must be extremely light if they are to be opened and closed by gentle convection currents in the air. Thin plastic films or sheer fabrics such as silk are used as material for the flaps.

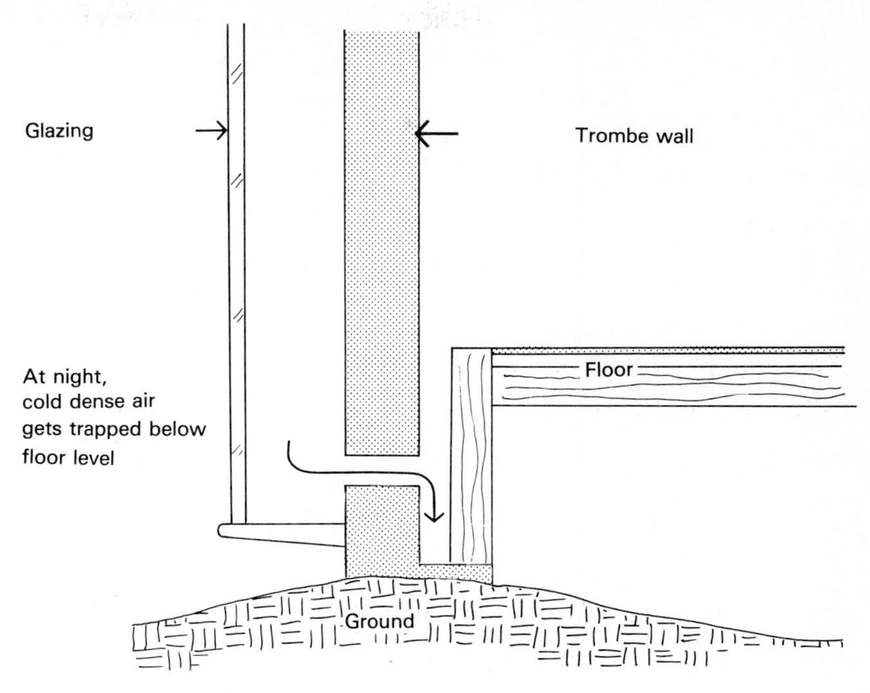

Figure 6-3 Cold air trap.

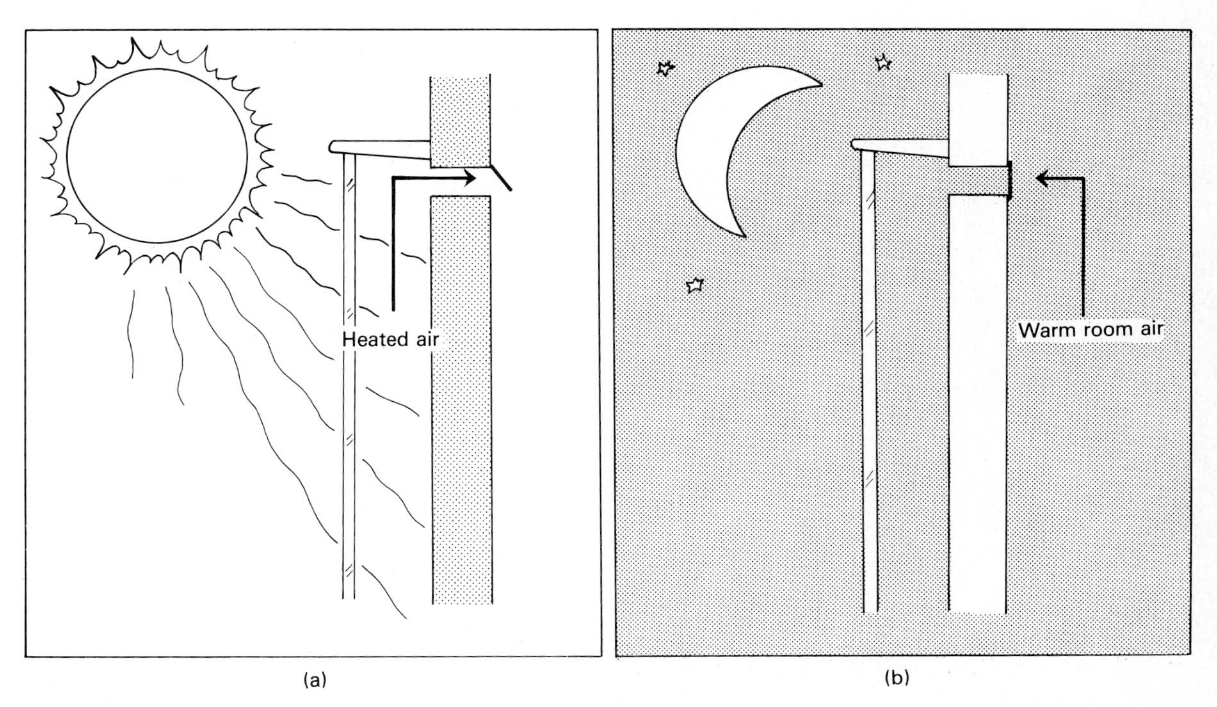

Figure 6-4 Backdraft dampers. (a) During the day, solar-heated air in the space between glass and wall can rise and push past the damper, entering the building's interior. (b) At night, warm room air cannot escape through the vent.

In recent years an increasing number of buildings are being designed with *unvented* Trombe walls. The main advantage of eliminating the vents is to reduce the interior temperature swings typical of vented walls. A vented wall allows heated air between it and the glazing to readily flow into the living space. This occurs soon after the sun first strikes the solar wall and often results in living-space temperatures that rise to an uncomfortably warm level after a number of hours. Unvented walls, on the other hand, transmit thermal energy through the process of conduction only. Heat is gradually, rather than suddenly, released into the living space, resulting in far more stable temperatures. One series of comparisons carried out in the Los Alamos National Laboratory solar energy test structures,[2] for instance, yielded room temperature swings of 27°F with vented Trombe walls but swings of only 10°F with unvented walls.

Although they often tend to produce large temperature swings, vented walls have some advantages over unvented designs. Vented walls heat a living space more rapidly on cold mornings, which can be a very welcome feature. In addition, the use of vented walls results in a higher solar savings fraction (see Table 6-3). This is because it takes collected energy less time to reach the living space through a vented wall and thus there is less time for heat to escape through the glazing.

TABLE 6-3 Solar Savings Fractions of Vented and Unvented Trombe Walls

	Thickness and Material of the Wall	*Load Collector Ratio of the System*	*SSF: Vented Wall (%)*	*SSF: Unvented Wall (%)*
Atlanta, GA	18-in. concrete	20	60	53
Boston, MA	12-in. concrete	13	40	37
Moline, IL	12-in. brick	30	21	16
Phoenix, AZ	11-in. adobe	25	81	75

Note that in each of the four cities, venting a Trombe wall results in a higher SSF than leaving it unvented.

A method of obtaining advantages of *both* vented and unvented designs is to build a solar wall with vents equal to about 6 percent of the wall area (3 percent for the upper vents and 3 percent for the lower), and if there are large interior temperature swings, gradually reduce the size of the vents by partially blocking them with wood or fabric until temperatures remain within an acceptable range. By doing this, the system's solar savings fraction is lowered no more than is necessary to achieve temperature stability. The vents should be blocked off from the top downward so that backdraft damper operation is not impaired.

Other methods of reducing living-space temperature swings include adding mass to the solar walls or designing the building so that thermal energy is efficiently and quickly transferred from the rooms directly behind solar walls to other parts of the building, either through natural or forced convection. The construction of large openings between rooms, vents near the tops of interior walls, and northern rooms several feet higher than those on the south side are practices that facilitate heat transport throughout the building. If necessary, forced-air blowers can also be installed.

Overheated air can be vented to the outside through the building's windows, although the thermal energy the air contains would then be lost.

Daytime venting is a common practice in spring and fall, when there is more solar energy available than can be used. The house should be designed, however, so that it is not necessary to vent in winter, for heat allowed to escape during the day could probably have been all used during the evening, and would have reduced the need for auxiliary heating. Adding thermal mass to rooms that are heated by solar walls but that receive no direct sunlight has a limited effect on temperature swings, due to the inefficiency of transferring thermal energy from air to a mass surface.

Estimating Temperature Swings

J. Douglas Balcomb and others of New Mexico's Los Alamos National Laboratory have developed ways of estimating the clear weather daily temperature swing in a building once the load collector ratio for its solar system has been calculated. (LCRs for solar walls are, as will be discussed later in the chapter, computed in the same way as for direct-gain systems.) The following formulas are used for these estimations:

Direct-gain systems:

$$\text{temperature swing, } Sw = 0.74 \times \Delta T$$

Vented Trombe wall (3% of wall area in vents):

$$Sw = 0.65 \times \Delta T$$

Unvented Trombe wall:

$$Sw = 0.13 \times \Delta T$$

Water wall:

$$Sw = 0.39 \times \Delta T$$

ΔT is the expected temperature difference on a clear January day between the average inside temperature and the average outside temperature. ΔT can be read from the graphs in Figure 6-5, which were calculated for systems without night insulation. Systems with night insulation would have somewhat reduced swings.

Example Calculation. The temperature swing for a building with a vented Trombe wall system, an LCR of 50, and located at a latitude of 40° is as follows:

$$Sw = 0.65 \times 18 = 11.7°\text{F}$$

With an unvented Trombe wall, it is noticeably less:

$$Sw = 0.13 \times 12 = 1.6°\text{F}$$

For lower LCRs, the temperature swings increase dramatically. Buildings that are strongly "solar driven"—in other words, have large aperture areas compared to the heat load—tend to have lower LCRs, and thus are more in need of night insulation to stabilize temperatures. Increasing the surface area of the thermal storage mass exposed to direct sunlight also has a stabilizing influence.

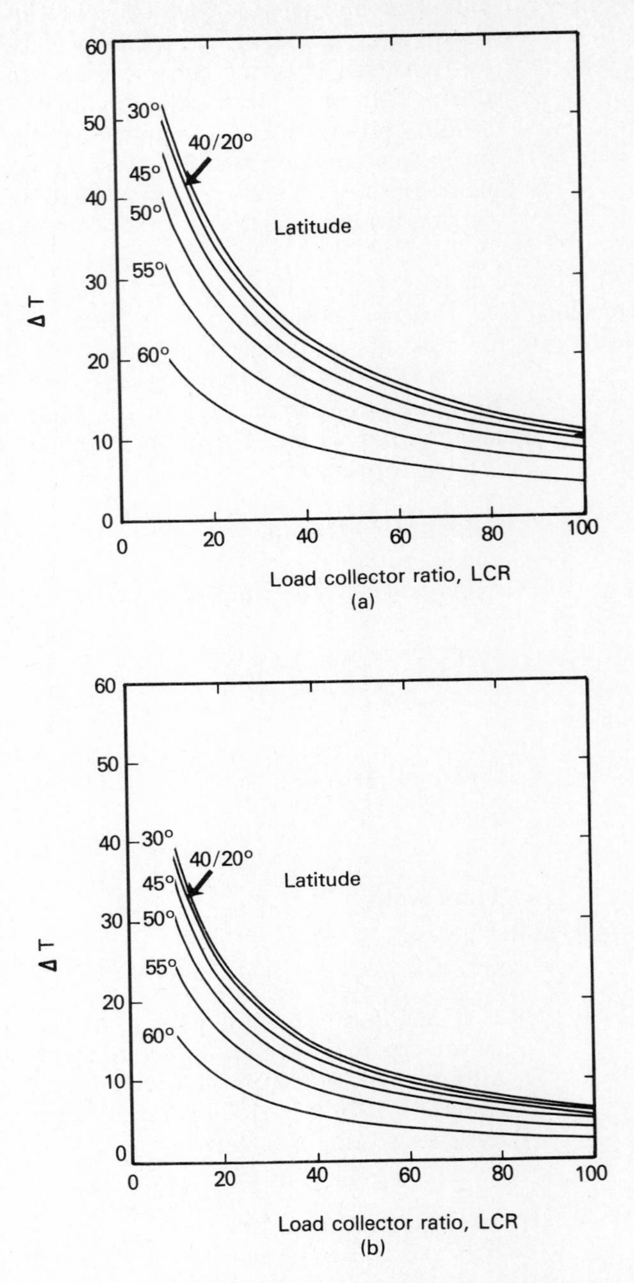

Figure 6-5 (a) Temperature difference to be expected on a clear January day between the average inside temperature and the average outside temperature (ΔT). This curve applies to the situation of a direct-gain, water wall, or vented Trombe wall passive design. The curve marked 40/20° applies at both a 40° latitude and a 20° latitude. (b) ΔT for a clear January day for the case of an unvented Trombe wall. (From J. Douglas Balcomb et al., *Passive Solar Design Handbook*, Vol. 2, Los Alamos National Laboratory, Los Alamos, NM, for U.S. Department of Energy, 1980, p. 45.)

Summer Cooling

A vented wall system will help circulate fresh air through a building in summer if an additional vent that opens to the outside is installed at the top of the glazing, the upper vent in the Trombe wall itself is sealed, and windows or vents in the north side of the building are opened (see Figure 6-6). Heated air flowing upward and out through the Trombe wall's external vent draws air in the lower vent, which in turn causes air to be drawn through the open windows or vents in the building's north side.

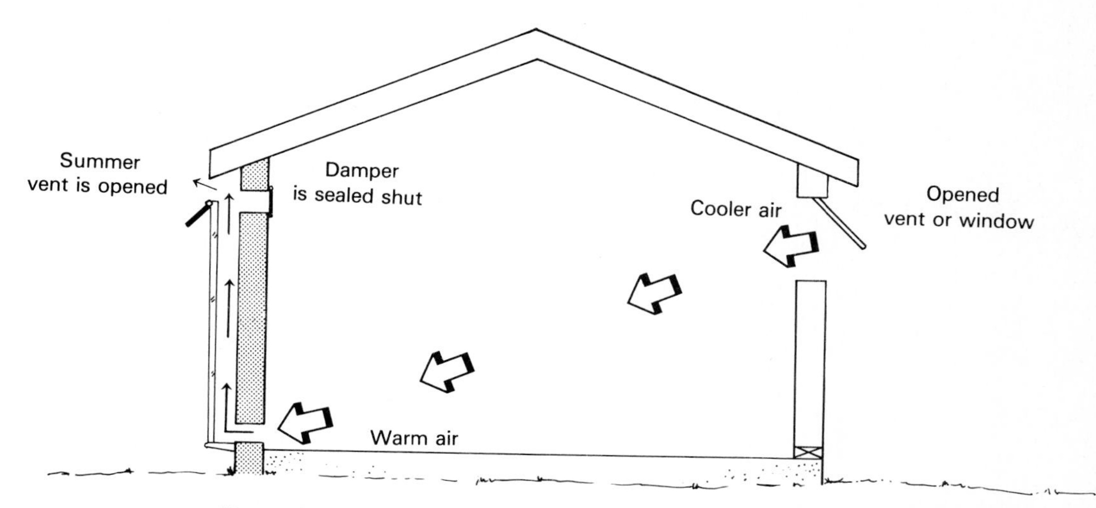

Figure 6-6 Summer ventilation. Heated air between the glazing and Trombe wall rises and escapes out of the summer vent, helping to draw warm air out of the house and cooler air in.

Construction Details of Trombe Wall Systems

Upper and lower vents in a solar wall should each have areas equal to about 3 percent of the total wall area. A vented 250-square-foot wall, for instance, should have upper vents totaling $250 \times 0.03 = 7.5$ square feet, and lower vents of the same size.

Because Trombe walls are generally made of masonry, they are quite heavy and must be firmly supported by the building's foundation. Details of how the wall is tied into the foundation are given in Figure 6-7. As indicated by the figure, a foundation wall under a solar wall should be insulated on both sides (except in the case of heated basements or crawl spaces) to prevent energy losses to the air and ground. The insulation should extend below ground level as far as the footing, or on deep foundations, 2 feet into the ground. Exposed above-grade insulation must be protected in some way, typically by a coating of cement plaster applied over wire lath.[3]

The glazing frame of a Trombe wall system may be constructed of wood or metal. If metal is used, it should be separated from the wall by either air or wood blocking to reduce conductive heat losses from the wall through the metal. If wood is used for the aperture framing, it should be kiln dried and of high quality. Paint applied to the frame should be able to withstand temperatures of 180°F. The aperture frame can either be supported by the wall or by the building foundation (see Figure 6-8). Both site built and commercially available frames are used.

Tempered glass or plastic materials able to withstand high temperatures are used for the glazing. Table 7-2 lists the maximum temperatures that various glazings can be subjected to. To allow for expansion, ½ inch of space should be left between the edges of the glass pane and the frame

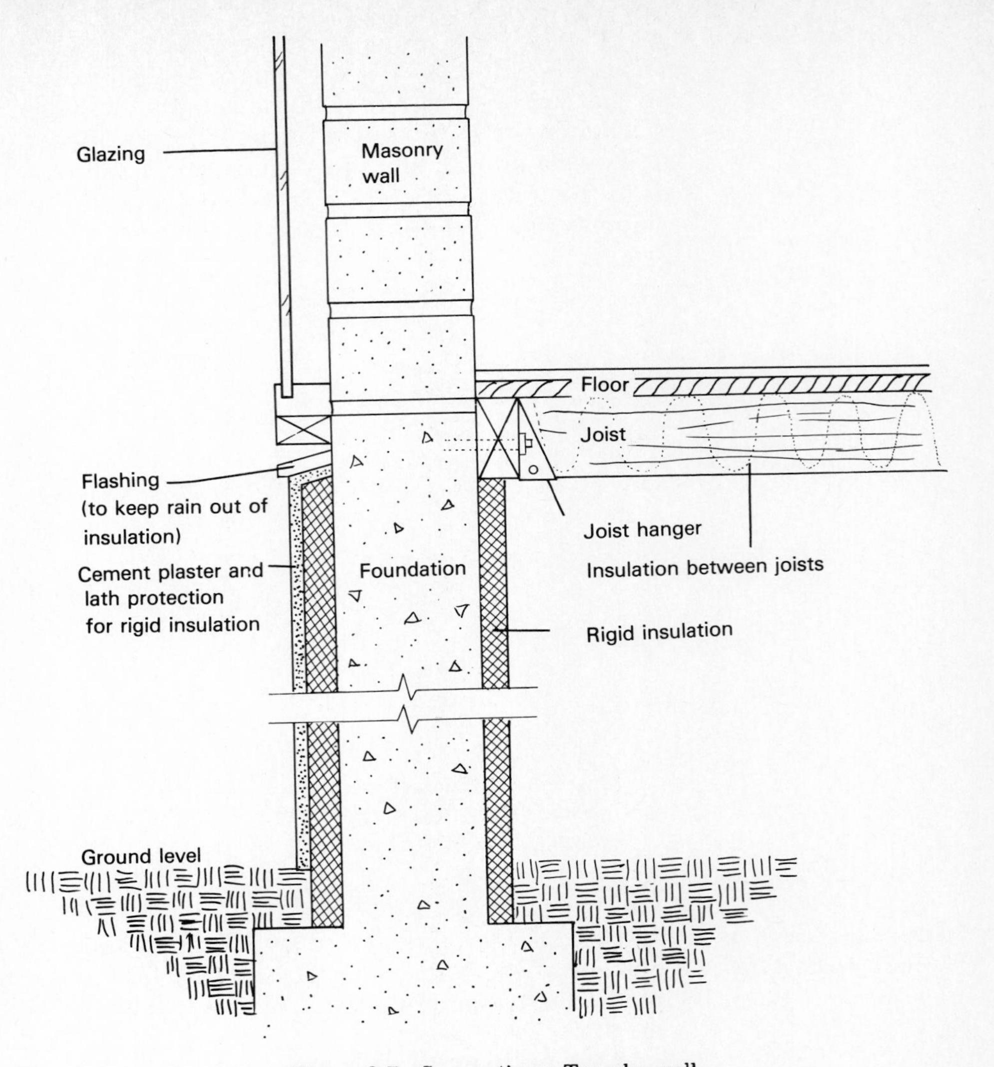

Figure 6-7 Supporting a Trombe wall.

mullion. Double glazing is best for most locations, although in severe climates triple glazing is sometimes used.

Preframed glazing units such as patio doors are often employed in solar wall systems. Whichever glazing is installed, it must be weathertight to prevent infiltration losses but must also be removable so that it and the solar wall behind it can be cleaned.

Of particular importance in weatherizing an installation is the caulking applied to joints around glazing and frame. Although the better caulks such as silicone are more expensive than, for example, butyl rubber sealant, they are well worth the extra expense. Their superior performances when exposed to intense heat and ultraviolet radiation result in longer life expectancies. Table 2-1 lists pertinent characteristics of a number of popular construction sealants.

Even with high-quality products, some separation between caulking and the surfaces to which it is applied might occur during the first few months after installation, especially if a solar wall is unvented. For this reason, framing joints should be examined regularly to keep them and the rest of the system in peak operating condition.

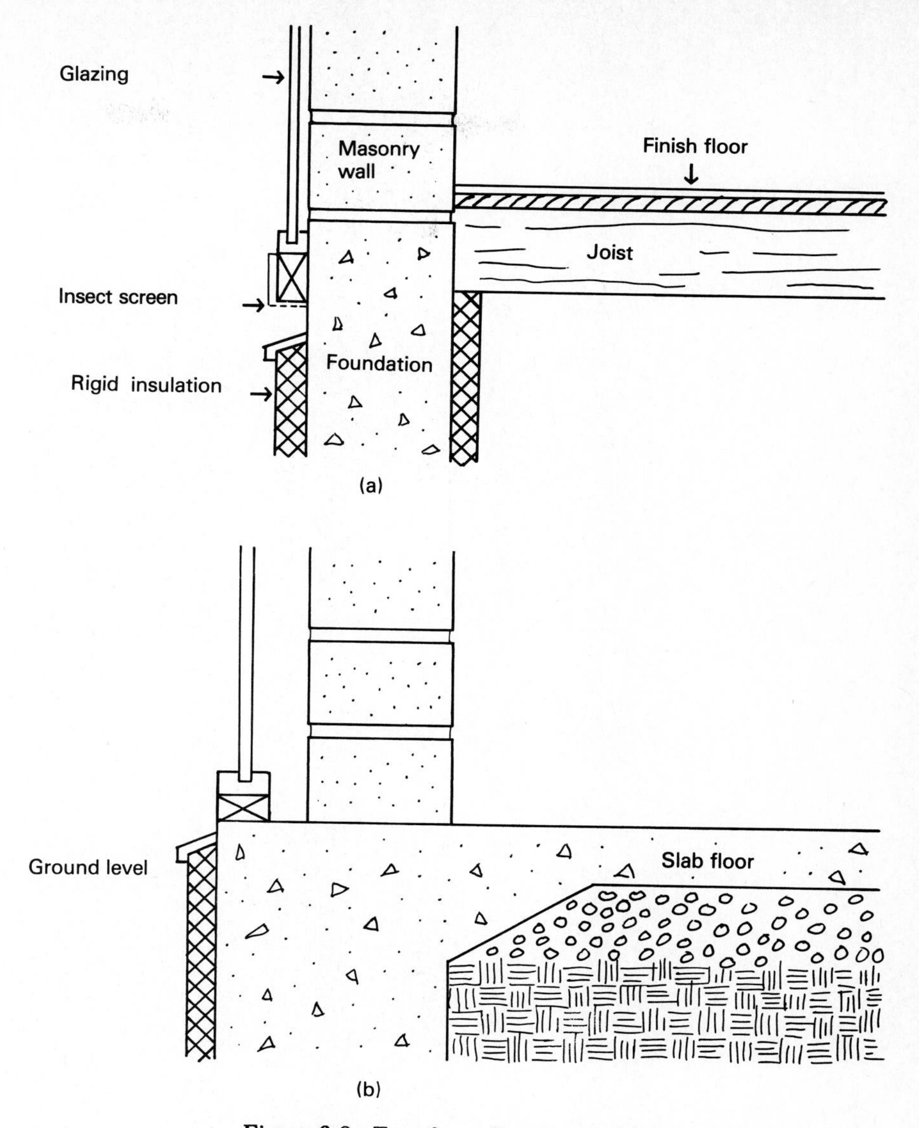

Figure 6-8 Trombe wall cross sections.

While the exterior surface of a solar wall should be of a dark color for good absorptivity, the interior surface can either be left untreated, painted, plastered, or wallpapered. Gypsum board is not recommended as a finish unless it can be installed in such a way as to eliminate air spaces between it and the masonry. Even small air gaps decrease performance dramatically by causing an insulating barrier to heat flow. A thick coat of construction adhesive applied before the gypsum board is installed helps minimize this problem. Wood finishes nailed to furring strips on the wall should be avoided because of air gaps as well as the insulating qualities of wood.

WATER WALLS

The advantage of using water as a thermal storage mass is that it stores considerably more heat per cubic foot than masonry. As listed in Table 5-1, the heat capacity of water is 62.4 Btu per cubic foot per degree of temperature rise, while for concrete it is 30.0 Btu and for common brick it is only 22.4 Btu.

Water-filled walls must be engineered to collect energy efficiently and transfer it to the living space. A common practice is to build a wall out of vertical Kalwall fiberglass tubes placed in a line parallel to the glazing, as illustrated in Figure 6-9. If spaces are left between the tubes, the system becomes a combination of direct gain and solar wall. Although tubes are frequently painted with dark colors to improve their absorptivity, it is also

(a)

(b)

Figure 6-9 Water walls. (a) Water wall tubes used in Merchants Bank, Winooski, Vermont, and made of Sun-Lite Premium II, by Kalwall. (b) Water wall tubes used in a residential application. (Courtesy of Kalwall Corp., Manchester, NH.)

common to leave transparent or translucent tubes unpainted. Light that passes through them and the water they contain help to illuminate the living space. Solar energy that is not absorbed by the water in the tubes is, for the most part, absorbed by surfaces in the room and so is not wasted. If necessary, the absorptivity of a translucent water wall may be adjusted by dyeing the water different colors. If it is found, for instance, that too much light enters a room and causes glare problems, the water color can be darkened.

Water walls transmit thermal energy mainly through convection currents, while masonry walls rely on conduction. Convection is a more rapid method of heat transport, which means that energy travels through a water wall sooner than through a masonry wall of equal thickness. Although rapid heating of a building's interior is welcome on cold mornings, it can also lead to overheating problems. In tests performed at Los Alamos, it was found that unvented masonry walls were much more effective in reducing large temperature swings within the living space.[4]

As with masonry walls, the greater the thickness of a water wall, the smaller room temperature swings will be. Water walls with thicknesses of 16 to 20 inches have been shown to result in stable living-space temperatures.

For aesthetic reasons, a conventional wall is sometimes placed between the water containers of a system and the building's living space. This is usually a mistake, for radiant heat from the water is blocked by the wall. It is better either to design the water containers to be attractive and leave them exposed or to construct a Trombe wall instead.

Another alternative is to build a combination Trombe–water wall. This type of system, designed by Harold Hays,[5] employs water-filled bags placed in cavities of hollow cement blocks and can bring together the best of both types of solar walls. The design will work well as long as there is good thermal contact between concrete and water. Such a system has the advantages of a Trombe wall, such as the strength to support a building load and an interior surface that can be finished aesthetically, while at the same time enjoying the added heat storage capacity of a water wall.

SYSTEM DESIGN

Thermal Protection for Solar Walls

As with all passive systems, the aim of a solar wall is to collect the maximum amount of solar energy while losing a minimum amount of heat to the outside. Three heat-retention options to be considered are: increasing the number of glazings beyond the two typically used, installing night insulation, and adding a *selective surface* to the wall.

Additional Glazing. Each layer of glazing added to the system increases the thermal resistance of the south wall by about R-0.9, but it also decreases the amount of transmitted sunlight. The dependence of performance on the number of glazings is listed in Table 6-4. Notice that the difference in solar savings fraction between one glazing and two is very large, which is why double glazing is standard in the solar industry. But as successive glazing layers are added, the increment in performance becomes steadily less. Notice also that the addition of a third layer of glazing has far more effect in a cold climate like Madison, Wisconsin, than it does in a gentle one like Santa Maria, California.

Adding multiple layers of glazing to a system can be an expensive undertaking unless thin films such as Mylar or Teflon are used for at least some of the layers. The advantages of installing multiple glazing layers for thermal

TABLE 6-4 Effect of Number of Glazings on System Performance

System Location and Description	*LCR*	*SSF (%) for the Following Number of Glazings:*		
		Single	*Double*	*Triple*
Madison, WI: vented 12-in. concrete Trombe wall, no night insulation	10	—	35	48
Santa Maria, CA: unvented 12-in. concrete Trombe wall, no night insulation	42	50	62	67
Sioux City, IA: vented 12-in. concrete Trombe wall, no night insulation	48	10	19	21

Source: J. Douglas Balcomb et al., *Passive Solar Design Handbook*, Vol. 3, Los Alamos National Laboratory, Los Alamos, NM, for U.S. Department of Energy, DOE/CS-0127/3, 1982, LCR tables.

protection, though, is that once in place they require no user participation. Night insulation, on the other hand, requires daily installation and removal (unless automatically operated), but it also provides higher insulating values.

Night Insulation. Only several of the movable insulation designs discussed in Chapter 4 are suitable for solar wall systems. Since there is but a narrow space between glass and wall, the user must be able to operate the insulation system without need of entering the space. Of the systems made for the *interior* surface of the window, pocket shutter systems that slide out from a wall recess are perhaps the most adaptable. Between-glazings insulation such as Beadwall or exterior insulation such as rolling shutters are also suitable.

Selective Surfaces. A less expensive and, according to recent tests, excellent alternative to movable insulation is a selective surface applied to the solar wall. The flat black paint that is commonly used as a wall finish is a good absorber of sunlight, but it is unfortunately a good *emitter* of thermal energy as well. As a flat black wall's surface temperature rises, much of its energy is given off in the form of infrared "heat" radiation that is absorbed by the glazing and eventually lost to the outside. A selective surface is able to reduce these emissions radically.

Selective surfaces for solar applications are made of materials that absorb sunlight nearly as well as flat black paint while, for reasons related to their ability to carry electric current, have a low emissivity of energy in the infrared wavelengths. Since infrared energy flow makes up a large part of the total energy loss of the system through the glazing,[6] selective surfaces can be quite effective in improving performance.

Experiments with oxides of various metals such as chromium, nickel, and copper have shown them to be selective surfaces with great potential for use in solar applications. Berry Solar Products[7] manufactures a material of particular value for solar wall systems: a copper foil on which chromium oxide, commonly called "black chrome," has been deposited (Table 6-5). When glued onto the surface of a Trombe or water wall,[8] the foil provides an energy-retaining selective surface whose performance compares admirably well with that of movable insulation. Numerous experiments at the Los

TABLE 6-5 Absorptivity and Emissivity of Solar Collecting Surfaces

	Solar Absorptivity	*Thermal Emissivity*
Black chrome (Berry Solar Products)[a]	0.94	0.10
Flat black paint	0.95	0.90

[a] Berry Solar Products, 2850 Woodbridge Avenue, Edison, NJ 08820.

Alamos solar test structures, as well as those done by other researchers, have confirmed the value of selective surfaces as a means of dramatically reducing heat losses.[9] R. D. McFarland, for instance, notes in a paper on testing at Los Alamos that the use of a selective surface on a double-glazed Trombe wall resulted in a 40 percent increase in efficiency.[10]

Unlike night insulation, selective surfaces do not require daily installation and removal. They are also significantly less expensive to purchase than most night insulations. Prices for Berry Solar Products' black chrome foil in 1984 were only $1.65 to $2.20 per square foot,[11] while commercial movable insulation systems cost from $3.50 to over $20 per square foot.

When applying a selective foil to rough surfaces such as brick walls, it is advisable to first use a *conformal coating*[12] on the wall to make it smoother, thus minimizing insulating air pockets between the wall and foil. Even with such a coating, there will often be small bubbles and creases in the foil after it is installed. For this reason, as well as because of a selective coating's dull black color, it is often more aesthetically pleasing to use translucent rather than transparent glass in front of the wall.

Black chrome is not a new material, but has been employed for decades in the metallurgical industry in nonsolar applications. Based on this field experience and on manufacturer's tests, it is expected to have a lifetime of over 20 years when applied to solar collecting surfaces. The adhesives that hold the selective foil to the wall have had less extensive testing. R. Judkoff has reported that after a year of use, foil glued with silicone building sealant[13] has shown no evidence of delamination or bubbling except for that noted at the time of installation. J. Hyde notes that foil attached to masonry with Devcon epoxy cement and to Kalwall fiberglass water tubes with National Adhesive's Bondmaster No. 701 has held in place for 2 years.[14] And after 5 years of field testing, Berry Solar Products reports excellent durability of its adhesives for temperatures up to 250°F.[15]

The Effect of Wall Color

For aesthetic reasons, solar walls are sometimes coated with colors other than black. While this can be done in direct-gain systems with little overall impairment of performance, solar walls are somewhat more sensitive, due to the fact that reflected sunlight is not reabsorbed by other surfaces but is for the most part lost through the glass (see Figure 6-10). Changing the color from flat black to brown, for instance, corresponds to an absorptivity change from 0.95 to 0.85. In testing performed at Los Alamos, this produced a 15 percent drop in the solar savings fraction.[16] The results for a naturally colored brick wall were even more dramatic. Red brick has an absorptivity of about 0.70, which led to a solar savings reduction compared to a flat black surface of 38 percent. Performances for other absorptivities are illustrated in Figure 6-11.

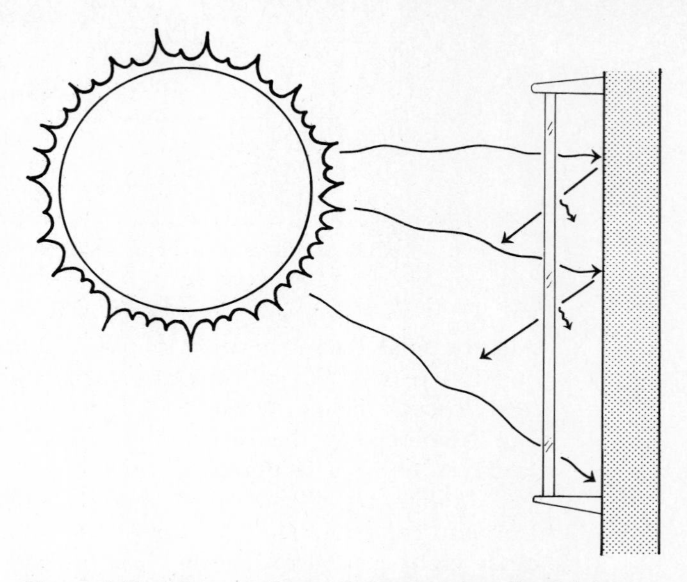

Figure 6-10 Most of the sunlight that is reflected from a solar wall is lost out through the glazing. Dark coloring should thus be used on a solar wall to reduce reflection and improve absorption as much as possible.

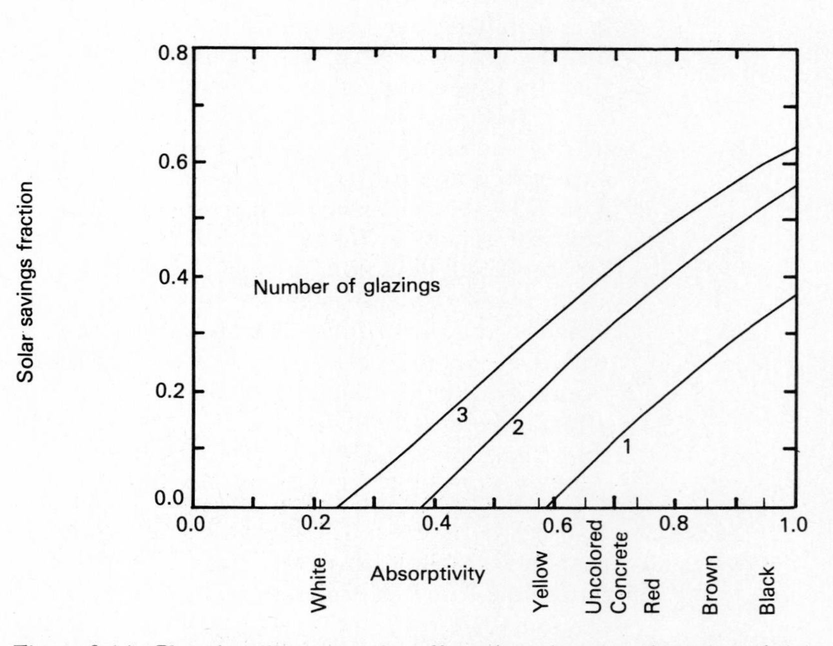

Figure 6-11 Plot that illustrates the effect that changing the value of the surface absorptance has on a solar wall's performance. The study was done on a water wall in Los Alamos, New Mexico. (From J. Douglas Balcomb et al., *Passive Solar Design Handbook*, Vol. 2, Los Alamos National Laboratory, Los Alamos, NM, for U.S. Department of Energy, 1980.) A number of colors corresponding to various absorptivities are listed on the horizontal axis of the graph.

The foregoing results suggest that if the flat black color of a painted or selectively surfaced wall is not aesthetically pleasing to the designer, it would be advisable to use translucent glass to hide the wall rather than change its color.

Retrofitting

A south-facing exterior wall with a good solar exposure can be converted into a solar wall by painting it black and covering it with glazing. An air space must be left between glass and wall and edges must be tightly sealed so that cold outside air cannot enter. If the wall is insulated or if rapid living-space heating and maximum solar savings fraction is desired, upper and lower vents should be cut in the wall. Concrete, concrete block, brick, stone, and adobe walls are best for retrofits, since they are able to store as well as collect heat. Wood and stucco walls are also usable, however. They simply transfer collected energy more rapidly into the living area, storing little of it. Construction details of site built retrofitted solar walls are depicted in Figure 6-12.

Solar wall collectors such as the Heliopass air heater can also be retrofitted onto existing south walls. The Heliopass is a 3- by 6½-foot panel that

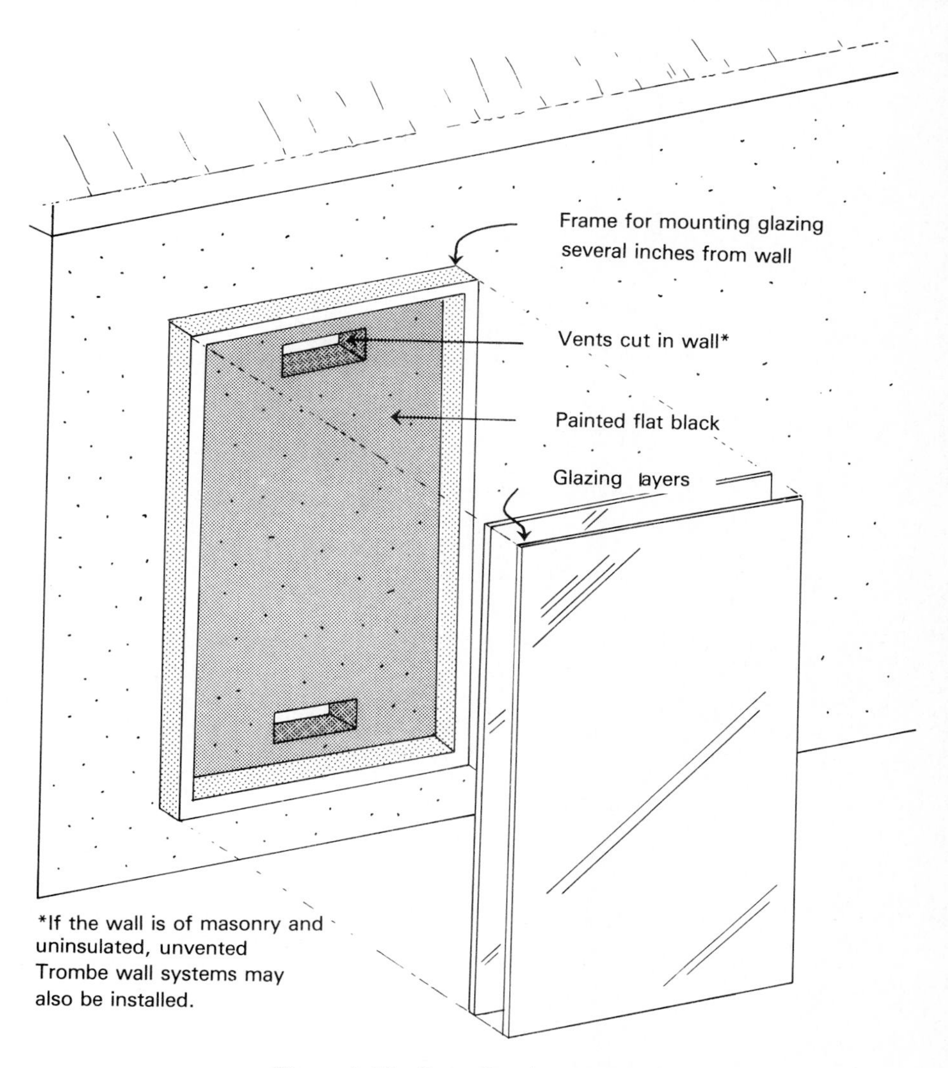

Figure 6-12 Retrofitted solar wall.

is attached to the building by means of lag screws through a mounting flange. Upper and lower vents must be cut in the building envelope for air circulation. There are several inches of air space between the heater's tempered, textured glazing and its semiselective collecting surface.[17] Lightweight passive dampers that prevent reverse thermosyphoning at night are included with the heaters.

Backup Heater System

Backup heater systems are commonly used to provide thermal energy during times when a solar system is not putting out sufficient heat for the building's needs, such as during periods of overcast skies. The backup heaters are often operated by thermostat and designed to come on whenever living-space temperatures drop below a certain point. As mentioned in Chapter 2, a surprisingly large amount of power (and money) can be saved by setting the thermostat at no more than 68°F during the day and lowering it to 60°F for 9 hours every night. This practice boosts considerably the fraction of the building's total energy needs that are met by its solar system.

Table 6-6 illustrates the effect of the backup heater's thermostat setting on the solar savings fraction. Notice that for the Madison, Wisconsin, site a change in the setting of only 5 degrees (from 70°F to 65°F) results in an SSF change of 20 percent in one system and 23 percent in the other.

TABLE 6-6 Effect of Backup Heater Thermostat Setting on the Solar Savings Fraction

Location of System	*Load Collector Ratio of System*	*Percent of SSF Change[a] for the Following Thermostat Settings (°F):*					
		55	*60*	*65*	*70*	*75*	*80*
Albuquerque, NM	13	28	19	12	Reference temperature	−12	−25
	28	51	32	15		−15	−29
	62	67	43	19		−19	−33
Madison, WI	3	68	46	23		−20	−41
	20	70	40	20		−15	−30

[a] SSF changes are calculated in comparison with the SSF for a 70°F thermostat setting.

$$\% \text{ of SSF change} = \frac{\text{SSF at actual thermostat setting}}{\text{SSF at } 70^\circ\text{F setting}} - 1$$

Source: The SSF values are taken from J. Douglas Balcomb et al., *Passive Solar Design Handbook*, Vol. 2, Los Alamos National Laboratory, Los Alamos, NM, for U.S. Department of Energy, January 1980, p. 127.

SIZING CALCULATIONS

Solar wall systems are sized using the same steps listed in Chapter 4 for direct-gain systems. A summary of those steps are as follows:

1. Calculate the building load coefficient.
2. Determine the solar savings fraction appropriate for the building site location.
3. Obtain the load collector ratio from Appendix III and divide the BLC by it to obtain a recommended solar aperture area.

A point to remember is that in a solar wall system the aperture is of the same area as the thermal storage wall that lies directly behind it. Thus, by sizing the aperture, the storage mass is also sized.

Table III-2 lists LCR values for six different solar wall systems: a concrete, a brick, an adobe, and a water wall (all of which are flat black in color), a concrete selective surface wall, and a concrete wall with night insulation. All systems are double glazed.[18]

Example Calculation. Consider a house in Oakland, California, whose BLC is 9000 Btu/°F-day. What aperture area should a vented solar wall system in this house have if the wall is made of (a) concrete with a flat black surface, (b) concrete with a selective surface, or (c) brick with a flat black surface? None of the systems use night insulation.

The map in Figure 4-10 recommends that a house in Oakland, California, be designed with a solar savings fraction of 80 percent. Table III-2 yields LCRs for cases (a), (b), and (c) above of 24, 36, and 20, respectively. Dividing the building load coefficient of 9000 by these values gives us the recommended aperture areas of

(a) 375 square feet for the flat black concrete wall.
(b) 250 square feet for the selectively surfaced concrete wall.
(c) 450 square feet for the flat black brick wall.

DIRECT-GAIN AND SOLAR WALL SYSTEMS COMPARED

Table 6-7 compares the solar aperture square footage that various direct-gain and solar wall systems need in order to provide 60 percent of the building's annual heating load. The study was carried out for a structure with a building load coefficient of 10,000 Btu/°F-day, a value that is typical of a well-insulated three-bedroom house. Systems both with and without night

TABLE 6-7 Solar Aperture Area (Square Feet) Comparison for Direct-Gain and Solar Wall Systems[a]

	Aperture Area Required to Provide a Solar Savings Fraction of 60% in the Following Locations:			
System Description	*San Diego, CA*	*Houston, TX*	*St. Louis, MO*	*Toronto, Ontario*
Systems without thermal protection (night insulation or selective surfaces)				
Direct gain	92	222	1250[c]	556[d]
Trombe wall[b]	115	256	909	2000
Water wall	85	189	667	1429
Systems with night insulation				
Direct gain	84	172	455	667
Trombe wall[b]	86	179	500	714
Water wall	74	152	400	625
Systems with selective surfaces				
Trombe wall[b]	83	175	500	769
Water wall	75	156	435	667

[a] This table was computed for a house with a building load coefficient of 10,000 Btu/°F-day.

[b] Vented concrete Trombe wall.

[c] This is for an SSF of only 50%; aperture areas are not listed for higher SSFs.

[d] This is for an SSF of only 20%; aperture areas are not listed for higher SSFs.

Source: J. Douglas Balcomb et al., *Passive Solar Design Handbook*, Vol. 3.

insulation are considered. The results were tabulated for four cities representing four different climatological zones.

Notice from Table 6-7 that when no thermal protection is used, all of the systems perform considerably poorer than with protection. Notice also that water walls need slightly less collecting area than that required for masonry. The rapid convective heat transfer that occurs in water walls allows their exterior surface temperatures to remain somewhat lower than those of masonry walls, and thus less thermal energy is radiated toward the glazing and lost. The use of a selective surface rather than flat black paint would probably lessen the performance gap between these two systems, for it would reduce dramatically the radiative losses from the walls.

In spite of the slightly lower performance, many people still prefer Trombe walls over those of water. Among their reasons are that the building methods for masonry walls are well known, they serve a structural as well as a solar function, fluid leakage is not a problem and their interior surfaces can easily be finished in an aesthetic manner.

Direct-gain systems *with* night insulation show a dramatic improvement in performance over those without, especially in areas with cold climates. Required aperture areas of the insulated systems are comparable to those of Trombe and water walls with some form of thermal protection (night insulation or selective surface).

Finally, the data in Table 6-7 indicate that the thermal protection offered by selective surfaces is, as mentioned previously in the chapter, very similar to that provided by night insulation.

Mixed Systems

There are advantages and disadvantages to both direct-gain and solar wall systems. Direct-gain systems can be built very economically, for a large part of the design is simply locating windows and massive wall and floor elements in the right places. Rooms heated by direct gain warm up in the morning more rapidly than those behind solar walls, which block light from entering the living space. There is often a sharp temperature drop, though, in direct-gain spaces at sunset.

Rooms behind direct-gain apertures tend to be bright and cheery due to the large amount of natural daylight that they receive. Some direct-gain spaces suffer from glare and fabric fading problems, however, because of too much light. Reducing excessive daylight when this occurs by drawing the drapes or putting movable insulation in place is not a good solution, for it also cuts down on the warming effect of the sun.

Rooms behind solar walls do not generally experience problems due to excessive light. In fact, it is necessary to make sure that these spaces contain enough window, clerestory, and/or skylight area to prevent them from being dim and dreary. Summertime overheating is, however, more easily controlled in solar wall systems.

By adjusting the vent size in a solar wall, a degree of living-space temperature control is attainable that is usually not possible in direct-gain systems. Temperatures are able to be kept far more uniform through the daytime, evening, and night hours.

Advantages of both direct-gain and solar wall designs can be realized by building mixed systems that use elements of each design. For instance, a percentage of a building's south wall can have direct-gain apertures installed in it, while the remainder is filled by a solar wall system. Since all buildings need windows,[19] it makes sense to place as many of them as possible in the south wall, where they will be net energy gainers rather than energy losers. Windows can even be placed *in* a solar wall, as in Figures 3-3 and 6-13a

(a)

(b)

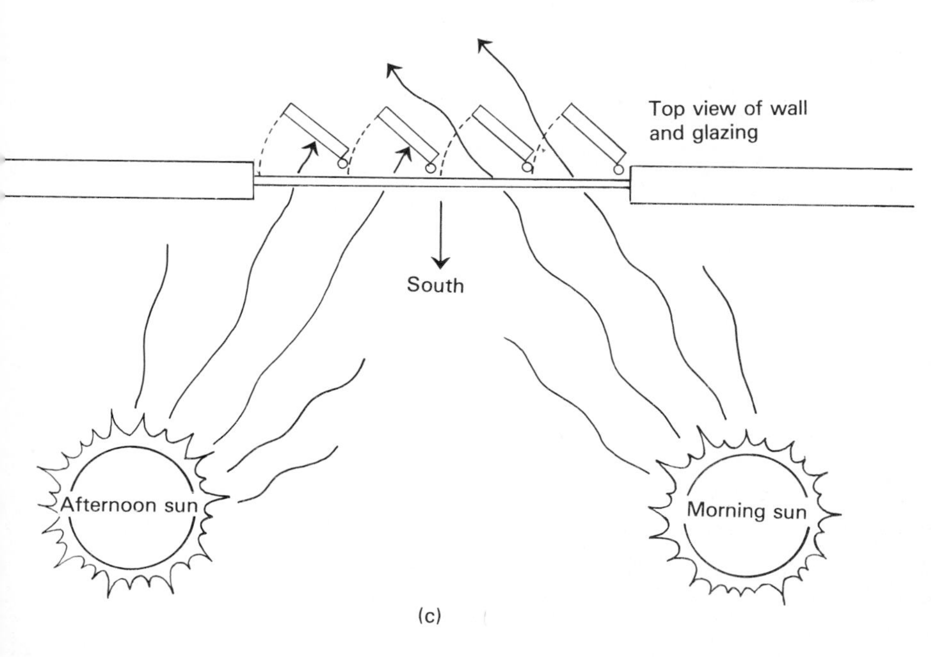

(c)

Figure 6-13 (a) Large Trombe wall system in Mark Jones's Alta Mira subdivision, Santa Fe, New Mexico. Notice the many direct-gain apertures embedded in the Trombe wall. (b) Interior view of an unvented Trombe wall with direct-gain apertures, also in the Alta Mira subdivision. Apertures in Trombe walls allow much needed daylight into the living spaces behind those walls. (c) Segmented and rotated solar wall.

and b. This practice is especially effective in preventing dark, shadowy living areas immediately behind the solar walls.

If a solar wall is segmented and the pieces rotated as in Figure 6-13c, morning sun will enter and warm the living space, while afternoon solar energy is absorbed by the wall segments, to be gradually released throughout the evening. This configuration helps to reduce overheating problems. A water wall design is easily made into a mixed system by leaving spaces between and above the water columns, as illustrated in Figure 6-9.

Sizing Mixed Systems. To size a system that involves passive designs of several different types, it is necessary to find a single load collector ratio representative of the particular combination chosen. To do this, the LCRs for each configuration are multiplied by the percentage of the total aperture area that that design fills. The sum of the results yields the composite LCR of the whole system. An example of this method of combining LCRs, called *weighted averaging*, is worked out below.

Example Calculation. An architect has designed a house in which 50 percent of the collection area is to be an adobe Trombe wall system, 25 percent is to be night-insulated direct gain, and 25 percent is to be direct-gain clerestories that, for reasons of convenience and expense, are not to be night insulated. The LCRs for each of the three collection systems have been found to be 9, 24, and 11, respectively. What is the composite LCR of the system?

$$\text{composite LCR} = \underset{\substack{\text{Trombe}\\ \text{wall}}}{(0.50 \times 9)} + \underset{\substack{\text{night-}\\ \text{insulated}\\ \text{direct}\\ \text{gain}}}{(0.25 \times 24)} + \underset{\substack{\text{uninsulated}\\ \text{direct}\\ \text{gain}}}{(0.25 \times 11)} = 13.25$$

As an illustration of how the aperture areas of each of the three systems are calculated, assume that the building's load coefficient is 3500 Btu/°F-day (see "The Building Load Coefficient" in Chapter 2):

$$\text{total aperture area for the house} = \frac{\text{BLC}}{\text{LCR}} = \frac{3500}{13.25} = 264 \text{ ft}^2$$

The aperture of each of the three systems is calculated as follows:

$$\text{Trombe wall aperture area} = 264 \times 0.50 = 132 \text{ ft}^2$$

$$\text{night-insulated direct gain} = 264 \times 0.25 = 66 \text{ ft}^2$$

$$\text{uninsulated direct gain} = 264 \times 0.25 = 66 \text{ ft}^2$$

NOTES

1. In this text, the term *solar wall* is taken to mean a wall that collects and stores solar energy, and that is located between the sun and building interior. Bruce Anderson and Malcolm Wells (*Passive Solar Energy*, Brick House Publishing Company, Andover, MA, 1981, p. 60) define this term simply as a wall built primarily to store heat. The definition of solar wall used in my text is close in meaning to the term *thermal storage wall* used in J. D. Balcomb et al., *Passive Solar Design Handbook*, Vol. 2, Los Alamos National Labo-

ratory, Los Alamos, NM, for U.S. Department of Energy, 1980, p. 4, although my definition of solar wall can include nonmassive walls not designed to store heat, as well as massive walls.

2. Balcomb et al., Vol. 2, p. 95.

3. *Passive Solar Construction Handbook*, Steven Winter Associates, New York, for Southern Solar Energy Center and U.S. Department of Energy, 1981, p. 3.2227.

4. Balcomb et al., Vol. 2, p. 95.

5. Bruce Anderson, *The Solar Home Book*, Brick House Publishing Company, Andover, MA, 1976, p. 133.

6. Although glass is not *transparent* to most of the infrared spectrum, it will *absorb* incident infrared waves. Some of this energy will be emitted back toward the building interior, while some will be emitted outward and lost.

7. Berry Solar Products, 2850 Woodbridge Avenue, Edison, NJ 08820.

8. Berry Solar Products (see note 7) sells selective foils with high-temperature acrylic or silicone adhesives already applied to the reverse side of the foils. Alternatively, adhesives such as Dow Corning Silicone Sealant or National Adhesives Bondmaster resins can be used with selective foils that do not have adhesives precoated onto them.

9. Jack C. Hyde, "Performance of Night Insulation and Selective Absorber Coatings in LASL Test Cells," *Proceedings of the 5th National Passive Solar Conference*, 1980; J. Douglas Balcomb, "Dynamic Measurement of Nighttime Heat Loss Coefficients through Trombe Wall Glazing Systems," *Proceedings of the 6th National Passive Solar Conference*, 1981; R. D. McFarland and J. Douglas Balcomb, "Los Alamos Test Room Results," *Proceedings of the 7th National Passive Solar Conference*, 1982; R. Judkoff, "Performance of a Selective-Surface Trombe Wall in a Small Commercial Building," *Proceedings of the Annual Meeting of the American Section of the International Solar Energy Society*, 1981.

10. McFarland and Balcomb, p. 155.

11. Private communication with Berry Solar Products.

12. The purpose of a conformal coating is to provide a uniformly textured wall surface for application of the selective foil. Berry Solar Products has tested various conformal coatings. Some that they recommend are Bondmaster 414-22 and 414-32 (National Adhesives, Bridgewater, NJ); Laticrete 3701 (Laticrete International, Woodbridge, CT) and Hysold Epoxi-Patch (Rudolph Bros. and Co., Groveport, OH).

13. R. Judkoff and F. Sokol, *Performance of a Selective Surface Trombe Wall in a Small Commercial Building*, Solar Energy Research Institute Technical Paper 721-1158. Presented at the AS/ISES Annual Meeting, 1981.

14. Jack C. Hyde, "Performance of Night Insulation and Selective Absorber Coatings in LA SL Test Cells," *Proceedings of the 5th National Passive Solar Conference*, 1980.

15. Private communication with Berry Solar Products, April 1983.

16. Calculated from information in Balcomb et al., Vol. 2, p. 107.

17. From information supplied by Photic Corporation, 2668 South Memorial Highway, Traverse City, MI 49684.

18. Heat capacity and thermal conductivity values for concrete, brick, and adobe were matched with those for the Trombe walls considered in Appendix F of Balcomb et al., Vol. 3. Those Trombe walls from Appendix F whose characteristics were closest were used in the design tables in Appendix III of this book.

19. It is the view of the author that this is so, in spite of the fact that buildings—even schools—are now being constructed that are totally windowless. The reasons for building with windows include the continued sanity of the inhabitants of the building, as well as for daylighting, ventilating, and solar-heating advantages.

SUGGESTED READING

BALCOMB, J. DOUGLAS, ET AL., *Passive Solar Design Handbook*, Vol. 3. Los Alamos National Laboratory, Los Alamos, NM: for U.S. Department of Energy, DOE/CS-0127/3, 1982. LCR tables on 42 different types of Trombe walls and 15 types of water walls.

REVIEW QUESTIONS

1. What is a solar wall? How does it differ from a Trombe wall?
2. Compare the advantages and disadvantages of vented versus unvented Trombe walls.
3. How many glazing layers are recommended for most solar wall systems?
4. Compare the characteristics of water walls with those of Trombe walls.
5. Compare the characteristics of the different types of thermal protection used for preventing heat loss from solar walls.

PROBLEMS

1. In Table 6-2 it can be seen that the highest solar savings fractions are obtained from Trombe walls made of materials with the highest conductivities. Explain why.
2. What is the expected temperature swing for a building with a direct-gain system, an LCR of 40, and located at a latitude of 45°?
3. What is the expected temperature swing for a home with an unvented Trombe wall system, an LCR of 50, and located at a latitude of 55°?
4. How many square feet of upper vents should a 300-square-foot vented Trombe wall system have? How many square feet of lower vents?
5. How many square feet of solar aperture should a 325-square-foot Trombe wall have? (No calculations are needed for this. It's very easy.)
6. Consider a house in New York City whose BLC is 14,000 Btu/°F-day. What aperture area should an unvented solar wall system in this house have if the wall is:
   (a) Concrete with a flat black surface?
   (b) Concrete with a selective surface?
   (c) A water wall with a selective surface?
7. An architect has designed a house in which 40 percent of the collection area is to be a selective-surfaced, concrete Trombe wall, 40 percent is to be night-insulated direct gain, and 30 percent is to be uninsulated direct gain. The LCRs for the three collection systems are 13, 26, and 18, respectively. What is the composite LCR of the system?

   If the building's load coefficient is 5000 Btu/°F-day, what should the total solar aperture area be? What should the aperture area of each of the three systems be?

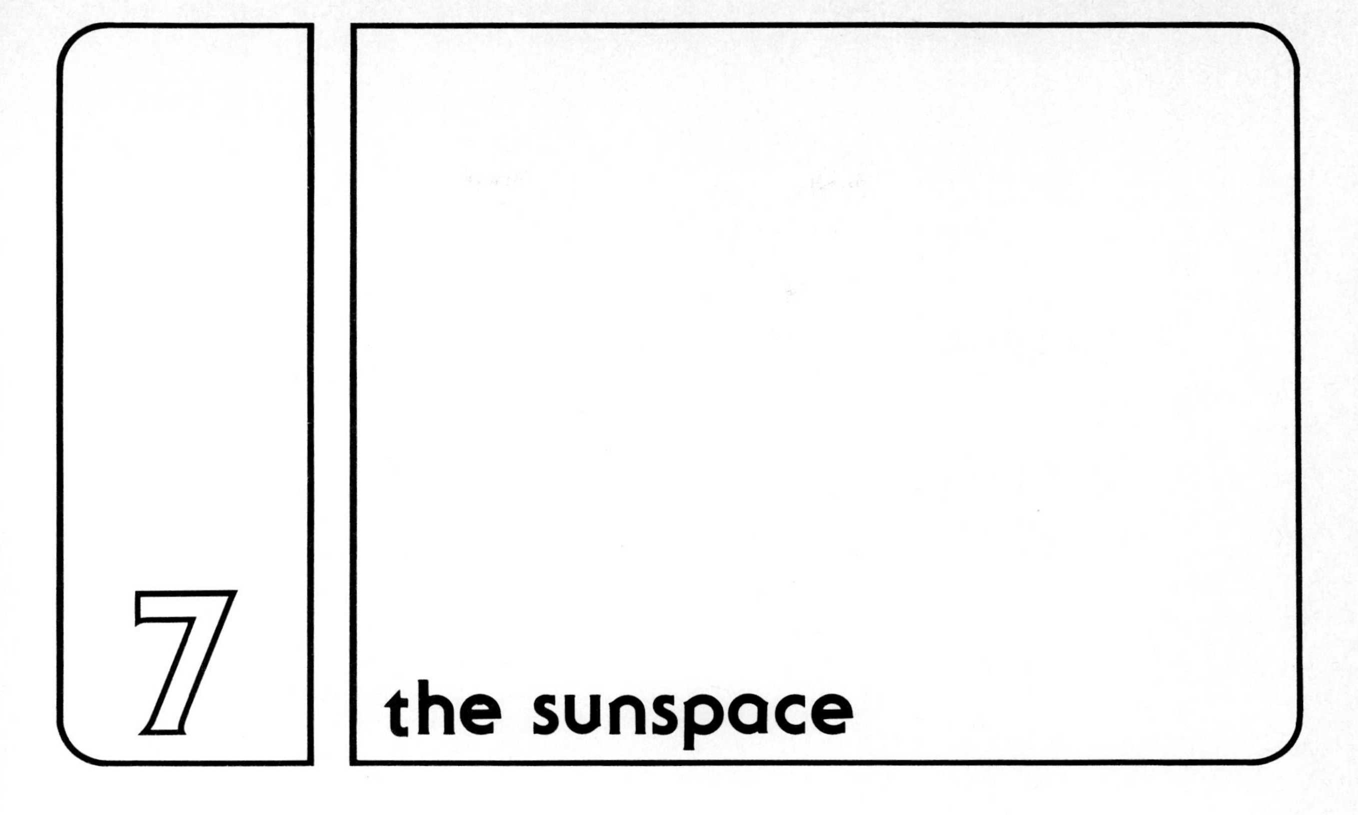

7 the sunspace

The *sunspace* is an enclosed area whose primary function is to collect solar energy for the purpose of heating an adjacent building. Like other passive solar systems, sunspaces are usually located on the south side of a building. The space's south wall and very often part of its roof are constructed mainly of glazing. Sunspaces can be made large enough to provide an addition to the building's living area or they can be quite small, enclosing only enough area for solar collecting surfaces and thermal storage mass. The space's floor and walls function as solar collecting surfaces and are typically dark in color. Thermal mass is provided by masonry in the building envelope, by the earth in dirt floor sunspaces, and/or by containers of water or phase-change salts. The outsides of these containers also serve as solar collecting surfaces.

The heat collected by a sunspace is transmitted to living areas in the adjacent building by vents and windows in the walls and/or by conduction through those walls. In this respect a sunspace's mode of operation resembles that of a solar wall system.

Sunspaces are often called *solar* or *attached greenhouses* because they are frequently used for growing plants. The term "sunspace" is used in this text, however, to represent the structure whose *main function* is to provide space heating for an adjacent building, and one of whose secondary functions might be to serve as a greenhouse. Other names for sunspaces include *solariums* and *solar rooms.*

Sunspaces may provide additional living space for the building as well as act as a buffer zone between interior and exterior climates. The sunspace makes an especially good buffer region when the building must be entered through it.

Sunspaces differ from direct-gain spaces in that the latter are part of the main living area of a building, while the former exist chiefly to heat that main living area. Large temperature swings are allowable in sunspaces but need to be suppressed in direct-gain spaces for comfort reasons. Sunspace temperatures are often considerably warmer than the rest of the building during daylight hours but drop to rather chilly levels at night.

SUNSPACE CONFIGURATION AND ORIENTATION

In designing a sunspace, several facts should be taken into account. Among them are the following:

1. The greater the area of south-facing glazing, the more solar energy will be collected.
2. The more shared wall area between living space and sunspace, the faster the energy transfer into the building will be and the better buffer the sunspace will be against cold outside temperatures.
3. Energy transfer will also be enhanced if the shared wall area is directly irradiated by the sun.
4. The less the distance from the shared wall to the south wall of the sunspace, the better will be the convective energy flow of heated air from the sunspace through vents or windows and into the living space.[1]

Designs that are in harmony with the considerations listed above are the long and narrow configurations illustrated in Figures 7-1 and 7-2. In these figures, the sunspace is connected to the building's south wall and its long axis is parallel to the wall. The designs in Figure 7-1 are termed *attached sunspaces.*

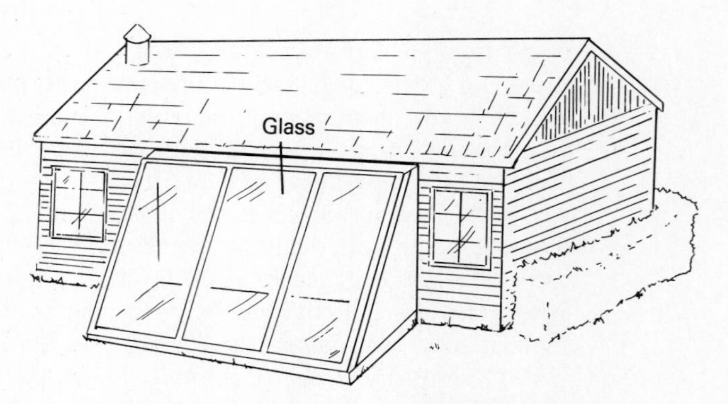

Figure 7-1 Attached single-plane sunspace.

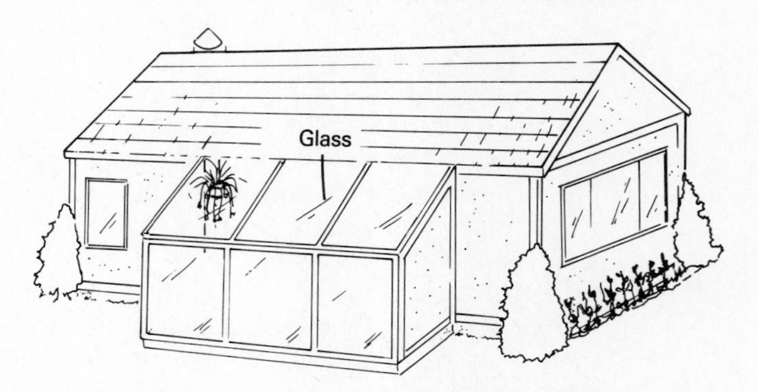

Figure 7-2 Attached shed roof sunspace.

Attached sunspaces of the geometry depicted in Figure 7-1 are the least expensive to construct, requiring only one pane of glazing. Although effective as solar collection systems, much of their floor area is unusable as living space due to the lack of headroom.

The shed roof configuration in Figure 7-2 offers more headroom and is a very popular design in situations where the sunspace is also meant to serve as an additional room of the building. The vertical or near-vertical glazing plane is at a good angle for wintertime collection as well as for receiving radiation reflected off the ground immediately in front of it.

Configurations in which there is even more shared wall area than in the designs above are illustrated in Figure 7-3 and are termed *semienclosed sunspaces.* A very desirable arrangement is depicted in Figure 7-3a, in which not only the sunspace's north wall but its end walls as well are shared with the building. Thermal energy transmitted through the end walls in this design is not lost. Also, the building itself shields the sunspace from winter winds that can carry away heat.

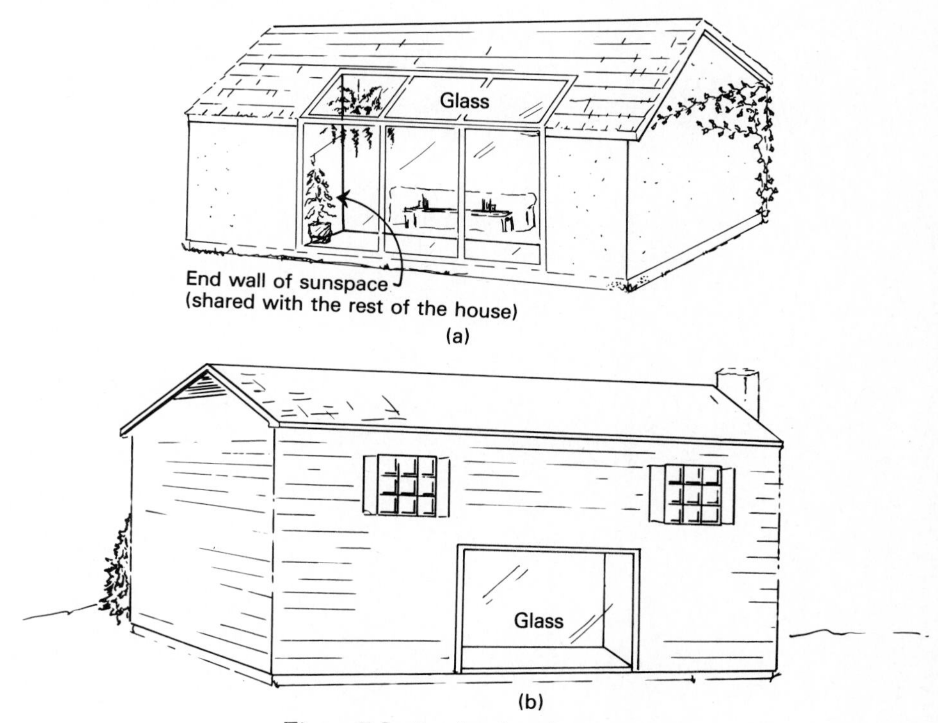

Figure 7-3 Semienclosed sunspaces.

One step beyond the indented configuration just described is the design depicted in Figure 7-3b, in which the sunspace roof is a shared area with a portion of the building's second floor. In this design all energy leaving the sunspace, except for that which exits through the south wall, goes into the living space.

Orientation of Apertures

The solar collecting apertures of sunspaces in most North American locations should face within 30° east or west of true south.[2] In extremely cold areas, they should face within 20° of south. Following this rule of thumb will result in performances that are typically within 5 percent of that for a sunspace facing due south.

There is at present some debate as to whether sunspaces are far better off containing only vertical glass and doing away with sloped roof glass completely. Some distinct disadvantages of sloped glazing are as follows:

1. It often leads to severe sunspace overheating. This is especially true in summer, but frequently occurs in winter as well. This disadvantage can be greatly ameliorated by adding sufficient external venting to the sunspace (equal to at least 20 percent of the total aperture area) as well as providing summertime shading of the sloped glazing.
2. Sloped glazing is far more prone to leaking than vertical glass. The impact of rain on a nonvertical surface is greater, and protruding components of the window frames can act as dams that trap water, preventing it from running off the glass. This standing water is likely eventually to work its way through small cracks. Also, if not mounted carefully, the top layer of sloped double glazing tends to slide downward after a while, breaking the edge seals around it.
3. It is very hard to install movable insulation on sloped glass. It tends to sag or fall off. Edge seals are especially difficult to maintain.
4. It generally costs considerably more to install sloped glazing than it does vertical glazing.

The points listed above are elaborated upon in several articles by John W. Spears, a strong critic of sloped glazing (see "Suggested Reading" at the end of this chapter).

In contrast, some *advantages* of sloped glass are as follows:

1. Unless the sunspace is quite narrow, the shared wall in a vertical-glazing-only configuration will be largely in shade. This cuts down its effectiveness as a thermal storage wall and slows down heat transfer into the rest of the house.
2. Because it faces the sun more directly, sloped glazing admits more solar energy per square foot than vertical glazing (60° glazing admits roughly 10 percent more energy).
3. Many sunspaces serve in addition as greenhouses. Unless plants are placed directly in front of windows, they will receive considerably less light in vertical glazed spaces than in those with roof glazing. On the other hand, spaces with sloped glazing must not be allowed to overheat severely if the plants they contain are to survive.

THE COMPONENTS OF A SUNSPACE

The Shared Wall

Shared walls between sunspaces and building interiors may be divided into two types: those that are thermally conductive and those that are insulated. Both usually contain vents for convective transfer of heat to the living spaces. Thermally conductive walls are generally made of masonry, which has the advantage of storing heat for release to the living space after sunset. In addition to convective heat transport through the vents, thermal energy is conducted through these walls and then radiated into the living space. Table 7-1 lists recommended thicknesses of uninsulated shared walls made from various masonry materials.

Sunspaces with insulated shared walls offer more temperature control

TABLE 7-1 Recommended Thickness of Uninsulated Shared Walls[a]

Material	*Thickness (in.)*
Adobe	10–12
Brick	10–14
Concrete	12–16
Hollow-core block filled with concrete	10–12
Rock	10–14
Water	8 or more

[a] These thicknesses take into consideration the time lag between when the sunlight is absorbed and when the thermal energy reaches the living space after being transmitted through the wall. Typical wall temperatures on the sunspace side are 100 to 120°F and 68 to 78°F on the living-space side. Transmission time through the wall averages 10 to 14 hours.

Source: Reprinted from Bill Yanda and Rick Fisher, *The Food and Heat Producing Solar Greenhouse,* John Muir Publications, Inc., Santa Fe, NM, 1980.

of the living space than do those with uninsulated ones. Overheating of the building interior can be avoided both in winter or summer by simply closing the vents. The price for this temperature control, though, is impaired performance. Tests show that solar savings fractions for sunspaces with insulated shared walls are less than for those with uninsulated masonry walls.[3] A probable reason for this is that in the latter case thermal energy is transmitted into the living area both by convection through the vents and by conduction through the masonry. Thermal energy transfer to the living space is more rapid, which means that sunspace air temperatures remain lower, resulting in reduced heat loss through the glass.

End Walls Unlike greenhouses, attached sunspaces generally operate more effectively when their end walls are insulated rather than clear, for the thermal energy losses prevented by such walls usually outweigh the energy gains that would occur if the walls were clear. In addition to being insulated, it is advantageous for end walls to be constructed of massive materials, for this increases the thermal storage capacity of the system. End walls of semienclosed sunspaces do not have to be insulated since they are shared walls with the living areas. The walls should be designed similarly to the shared wall on the north side of the sunspace.

The Glazing There are a variety of different glazing materials used in sunspaces. They can be broken down into four groups: glass, fiberglass, acrylic and polycarbonate panels, and thin plastic films. The characteristics of each group are discussed below and summarized in Table 7-2.

Glass. Glass has an excellent aesthetic appearance and a long lifetime if treated with care. It does not yellow or embrittle with exposure to sunlight, and it has a high solar transmissivity. It does not sag or ripple as some plastics do and is available in both transparent and translucent varieties. It is suitable for both vertical and nonvertical surfaces, although in the latter case safety or extra-strength glass should be used.

TABLE 7-2 Characteristics of Various Glazing Materials*

Material	*Brand Names*	*Typical Thickness (in.)*	*Approximate Solar Transmissivity at Normal Incidence (%)*	*Losses due to Reflection (%)*	*Losses due to Absorption (%)*	*Maximum Continuous Operating Temperature (°F)*	*Infrared Transmissivity (%)*	*Expected Lifetime (yr)*	*Comments*
Glass									
Float glass (0.10–0.13% iron oxide)	Many	1/8	Single: 85 (Double: 75)	8	7	400	2	50+	Long-lasting and easily obtainable. Glasses can span greater distances than many plastics, which are prone to sagging.
Low-iron glass (0.01% iron oxide)	Solatex, Heliolite, Sunadex	1/8	90	8	2	400	2	50+	Its extra transmissivity is often not worth its considerably higher expense.
Fiberglass-reinforced polyester (FRP)	Filon, Sunlite, Lascolite	0.027–0.060	85	15	0	180–200	2–8	7–20	Impact resistant. Needs UV-resistant coating in order to be long-lasting. Translucent rather than transparent.
Acrylic	Plexiglass, Lucite	1/8	90	7	3	180	2	10–15	Impact resistant, but surface is easily scratched. Both transparent and translucent types are available.
Double-skinned acrylic	Exolite		86			160			
Polycarbonate	Lexan Poly-Glaz	1/8	81			250	2	5–15	Very high impact resistance. Surface is easily scratched. Embrittles and discolors with time. Both transparent and translucent types are available.

Double-skinned polycarbonate	Exolite		74	10	9	240			
Thin film									Inexpensive. Optical clarity of thin films is often poor, due to wrinkling and buckling. Because of tearing due to high winds or mistreatment, thin films are more suitable for use as inner layer(s) of multiple glazing. High transmissivity of infrared wavelengths allows considerable thermal energy to escape through the film.
Polyvinyl fluoride	Tedlar	0.004 (4 mils)	90	10	0	150	30		Resistant to ultraviolet degradation.
Fluorinated hydrocarbon	Teflon	0.001 (1 mil)	96			400	58	20	Excellent solar transmissivity. No ultraviolet degradation.
Polyester	Mylar, Flexiguard	0.003-0.014	85				10-30	4-10	
UV-resistant polyethylene	Monsanto 602	0.004	85			140	70	1	Temporary glazing, due to short lifetime.

*Note: Values in table are for single glazing, unless otherwise indicated.

Sources: Ametek, Inc., *Solar Energy Handbook*, Chilton Book Company, Radnor, PA, 1979, p. 165; E. Levy, D. Evans, and C. Gardstein, *Passive Solar Construction Handbook*, Steven Winter Associates, Inc., New York, 1981, pp. 4.1101-4.1105; *Solar Products Specifications Guide*, SolarVision, Inc., Harrisville, NH, 1983, Section 8.1; *Solar Applications Seminar*, Pacific Sun, Inc., Palo Alto, CA, 1977, p. B-8.

Figure 7-4 Single-plane sunspace near Mendocino, California. The Pacific Ocean is in the background.

A drawback of glass is that it is subject to breakage from vandalism, falling tree limbs, or careless handling during transportation and installation. Safety or extra-strength glasses are the only types that should be installed at nonvertical angles. Glass tends to be a rather expensive material to buy, although prices vary tremendously. Typical costs for tempered glass range from about $2.50 to $5 per square foot for single pane and $4 to $8 for double pane (1984 prices). Bargain prices can often be obtained, however, by buying surplus stock, seconds, or used materials scavenged from remodeled and demolished buildings. In a 1984 survey by the author, for instance, it was found that tempered patio-door-pane seconds, perfect for sunspaces, could be purchased for as low as 82 cents per square foot.

When multiple-layer glass is used, as it usually should be, it is strongly recommended that it be of a type whose edges are "double-sealed" with both silicon and a butyl rubber. Otherwise, condensation and fogging are very likely to occur between the panes, reducing solar transmissivity as well as visibility, taking away from the aesthetic beauty of the sunspace and possibly leading to rot in the window frame.

Fiberglass. Fiberglass is the common name for materials made from grids of glass fibers embedded in a polyester resin. Fiberglass is easy material to work with, lighter and less expensive than glass, and has high impact strength, a distinct advantage in situations where vandalism is a problem. It is very popular for sunspace and greenhouse applications and can be cut at the building site to any shape desired.[4] It is a translucent material, which is advantageous in that it diffuses light evenly over the interior surfaces in a sunspace. Translucency can be a drawback, however, if the sunspace is to serve as a living space as well and a view of the outside is desired.

Although fiberglass initially possesses a high transmissivity (85 percent is typical), this tends to deteriorate with age, heat, and exposure to ultraviolet light. The longest-lasting fiberglasses are those coated with thin protective films. Lascolite Crystalite-T[5] is surfaced with a polyvinyl fluoride film called Tedlar. Filon is another popular material whose surface is protected in various ways. Filon Solar Plate/20 Greenhouse Panels, for instance, are coated with ultraviolet and heat-resistant Filoplate resin that polymerizes to the core resin of the panels.[6] Suggested prices for the Filon panels in 1984 were $0.87 to $1.11 per square foot, while Lascolite ran from $0.90 to $1.30 per square foot.[7] Filon guarantees their products for up to 20 years, depending on the location where they will be used and the type of use they will be subjected to.[8] Vertically mounted glazings last longer than horizontal ones, for they receive less wear from rain, hail, falling objects, and intense summer sun. Panels in northern locations usually have longer useful lives than do those in warm, sunny locations.

When the surfaces of fiberglass panels have been worn down to the point where the glass fibers begin to show, the surfaces glisten as if they had a coat of frost on them. Refinishing compounds such as those manufactured by DuPont or Filon[9] can be brushed or sprayed onto the panels at this time to give them years' more useful life. If the refinishing is not done promptly, however, the panels will quickly deteriorate, for their protective coatings have been worn away.

Only fiberglass *without* pigmentation should be used in solar applications, in order that the highest transmissivity possible is achieved. Fiberglass is sold in both corrugated and flat form. Corrugated panels are especially suitable for roofs due to their strength and impact resistance. Because of their undulations, however, they have about 20 percent more surface area than those of flat panels. This cuts down on their efficiency, for the more glazing area with which warmed sunspace air can come into contact, the greater the rate of heat loss.

Solar Components Corporation[10] manufactures preassembled, multipli-glazed panels designed specifically for passive applications. They are constructed of Sun-Lite fiberglass sheets sandwiched with aluminum grids and have thermal resistances as high as R-3.6. For easy mounting, they can be ordered preattached to aluminum frames.

Acrylic and Polycarbonate Panels. Acrylic and polycarbonate panels are attractive, lighter than glass and easier to handle, and are available in either transparent or translucent styles. Brand names include Plexiglas and Lucite (acrylics), and Lexan and Poly-Glaz (polycarbonates). Because of their high impact resistance, acrylics and polycarbonates are often installed in place of glass where vandalism is likely to occur. A main drawback to their use is that their surfaces are easily scratched and abraded. Their price and transmissivity are comparable to that of glass.

Acrylics have excellent weather-resistance characteristics, while polycarbonates haze or yellow to some extent with time. Tests performed on Qualex,[11] a double-walled extra-high ultraviolet-resistant grade of Lexan polycarbonate, showed that the loss of light transmissivity due to ultraviolet deterioration was less than 10 percent in a period of 10 years.

Exolite[12] is another type of double-walled panel made either of acrylic or polycarbonate. The panels have insulating values comparable to double-paned glass with a ⅝-inch air gap and are sold in 4-foot-wide strips of varying lengths.

Thin Plastic Films. Thin films have the advantage of extreme lightweight, low cost, ease of installation, and high solar transmissivity. Their main drawback is their fragility. A strong wind can tear them, and they can easily be destroyed by vandalism. They also transmit far more infrared radiation than the materials discussed above, which means that they lose thermal energy at a more rapid rate. In addition, they are prone to sagging problems on hot days.

Thin films are most useful either as temporary glazing or as the inner layer(s) of a multiply-glazed solar aperture, where they will be protected from damage by the more sturdy outer glazing. Many thin films are translucent, though, and so do not provide a view of the outside. Even the transparent films such as Tedlar and Teflon provide a somewhat distorted view due to ripples and sagging.

Polyethylene is the least expensive thin film, coasting only a few cents per square foot. It is also the material with the shortest lifetime, deteriorating rapidly with exposure to ultraviolet radiation and heat. Monsanto 602 UV-resistant polyethylene and Mylar polyester are somewhat more durable but also more expensive (see Table 7-2).

Tedlar and Teflon films have excellent weatherability characteristics, being able to withstand UV bombardment and severe temperatures. Their solar transmissivity is among the highest of any glazing material (unfortunately, so is their infrared transmissivity). They are both highly recommended as inner glazing. Prices in 1984 were $0.80 per square foot for Tedlar and roughly $1.00 per square foot for Teflon.

Layers of Glazing. Double glazing is standard in the solar industry and is usually the best choice for sunspace applications. If inexpensive thin films are used for the inner glazing, however, multiple layers should be considered, especially in severe northern climates, where triple glazing is often more cost-effective than double.

The Roof Sunspace roofs differ from those of conventional greenhouses in that the section nearest the north wall is opaque in order to provide summer shading that will prevent overheating of the wall and the main living space. The opaque roof section must be small enough, though, to allow winter sun to

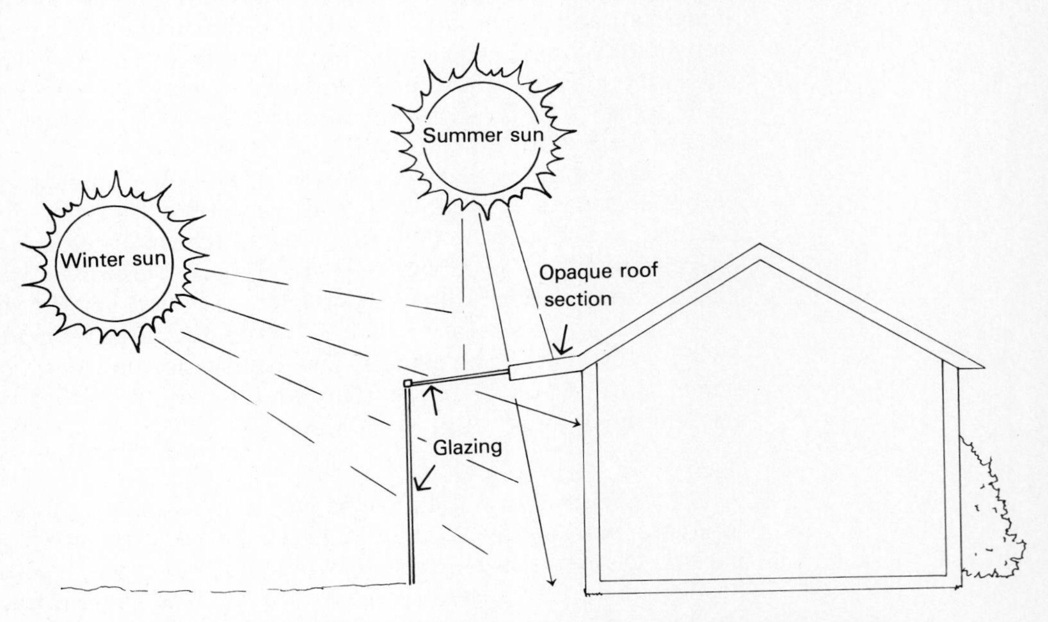

Figure 7-5 The opaque roof section must be long enough to provide summer shading, but short enough to allow winter sun to strike well up on the sunspace's north wall.

strike well up on the sunspace's north wall (see Figure 7-5). The information in Table 7-3 is useful in determining the right opaque roof section dimension.

TABLE 7-3 Guide for Calculating a Sunspace's Opaque Roof Section Length

Step 1: Obtain angle A from Table 5-4 (you need to know latitude and orientation of sunspace).

Step 2: Calculate maximum recommended opaque roof section length, using the formula

$$\text{maximum opaque roof section length} = \frac{(\text{height of sunspace at shared wall}) - 6\text{ ft}}{\sin B + (\cos B)(\tan A)}$$

where B is the roof slope.

Example Calculation. Calculate the maximum recommended opaque roof section length for a sunspace located at 48° north latitude, facing 15° southeast, with a roof slope of 25° and a height at the shared wall of 8 feet.

From Table 5-4, $A = 23°$.

$$\text{Maximum length} = \frac{8\text{ ft} - 6\text{ ft}}{\sin 25° + (\cos 25°)(\tan 23°)}$$

$$= \frac{2}{0.4226 + (0.9063)(0.4245)}$$

$$= 2.5\text{ ft}$$

Note: This is a guide for single-story sunspaces only.

The opaque roof section should be insulated, for it is at the highest part of the sunspace and thus comes in contact with the warmest air.

A major factor in determining the roof's slope is the aesthetic appearance of the sunspace—for instance, how it fits in with the roof slope of the building to which it is attached. The steeper the roof, the better its angle for collecting wintertime solar energy and the less vulnerable it is to snow-loading problems.

Movable Insulation

Movable insulation markedly improves sunspace performance, as it does for direct-gain and solar wall systems. The insulated roman shade and external rolling shutters described in Chapter 4 may be employed in sunspaces as well, for they require little storage room when not in use. Special provisions might have to be made, though, if the roman shades are to insulate the shallow-angled glazing of sunspace roofs. Figure 7-6 depicts a design for using folding shutters to insulate the roof glazing. By day they are stored under the opaque section of the roof.

Beadwalls may also be installed in sunspaces. Their bead storage containers can be placed outside the space if lack of room is a problem.

Pocket shutters that slide into wall recesses are hard to integrate into the design of an attached sunspace south wall, for the wall is usually made mostly of glass and thus there is no place for the shutters to be stored. In semienclosed greenhouses, however, the shutter can slide into a pocket in the adjacent south wall of the building.

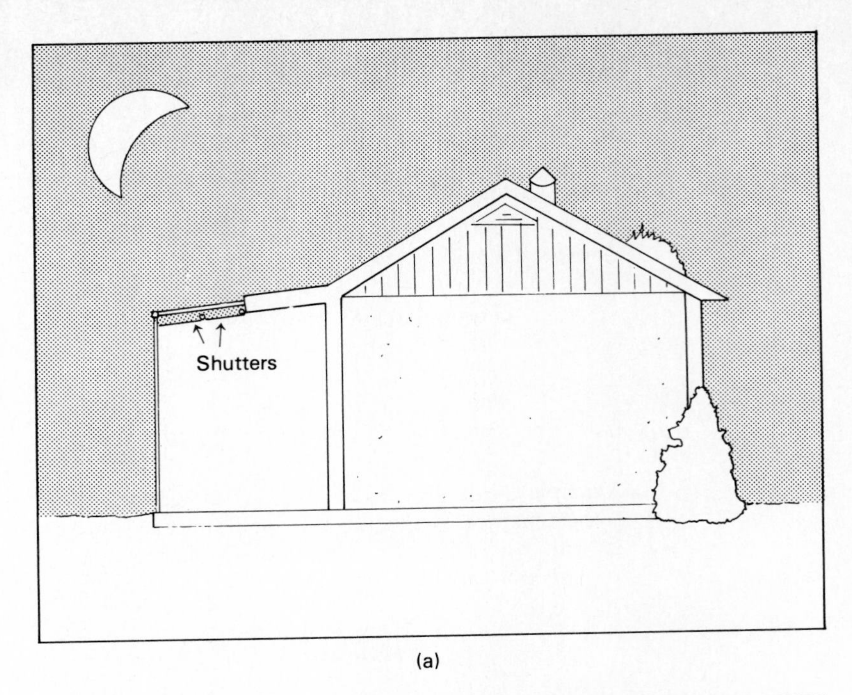

(a)

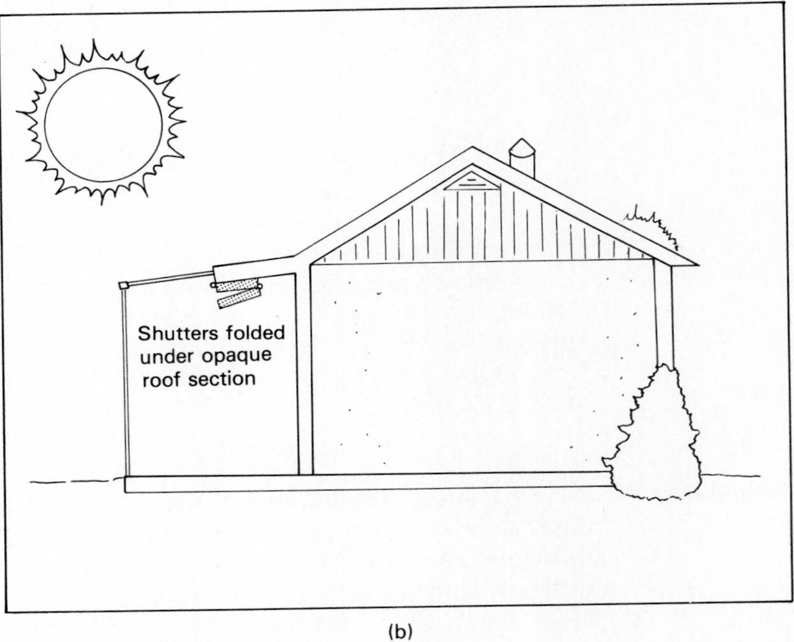

(b)

Figure 7-6 Insulating the glazed roof section with folding shutters. (a) Nighttime. (b) Daytime.

The Floor Floors in sunspaces are usually either of dirt or masonry. In addition to providing a place for ornamental and food-producing plants, dirt floors act as fairly good thermal storage masses if they are insulated on all sides to a depth of about 2 feet, as shown in Figure 7-7. The insulation must be of a type that will not deteriorate after years of contact with the earth. An extruded polystyrene board insulation often referred to as *blueboard* is made for purposes such as this. If there is a perimeter foundation around the sunspace, the insulation should be placed on the outside so that the concrete is included in the thermal mass. If there is no such foundation, the insulation panels can be placed directly in trenches. Insulation with an R-value of 10 is recommended, which can be provided by 2 inches of smooth-surface extruded polystyrene.[13] A secondary advantage of insulating the perimeter of a dirt floor is that the raised temperature of the earth promotes root growth and extends the seasons of most plants.

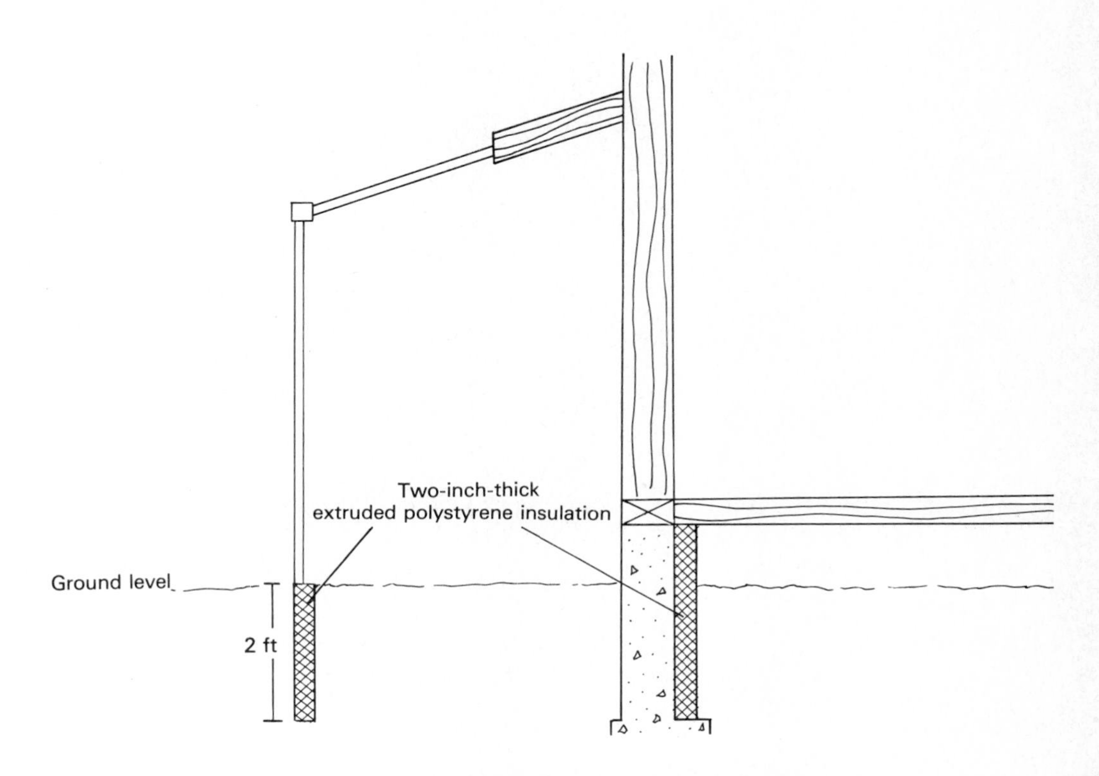

Figure 7-7 Perimeter insulation for dirt floor greenhouses.

Masonry floors should be treated like those in direct-gain systems. If the winter moisture content of the ground beneath them is high or if the average winter temperature several inches into the ground is below 50°F, the floor should be insulated underneath and on the sides, as shown in Figure 5-7.

Thermal Water Storage In addition to masonry and earth thermal storage in the walls and floor of a sunspace, heat is often accumulated in water containers such as metal drums, and sometimes in open troughs (see Figure 7-8). Some sunspaces even contain swimming pools or hot tubs. Large amounts of water mass are especially useful in greenhouse sunspaces in order to minimize temperature extremes that can be detrimental to plants. Mass containers are usually of a dark color so that they are able efficiently to absorb radiative energy as well as that carried by air convection.

Figure 7-8 Heat is stored in the open water trough of this sunspace as well as in its masonry floor. Evaporating moisture from the open pool can be beneficial to plant growth. (Courtesy of General Glass International Corp., manufacturers of SolaKleer low-iron glass, New Rochelle, NY; and Mekler/Ansell Associates, New York.)

Summer Overheating Control

A sunspace must be adequately vented to the outside in summer if overheating is to be avoided. This is especially important when sloped glazing is employed, or when plants are grown in the space. Operable windows and skylights, roof vents, and doors are devices that let in cooling breezes. Movable insulation provides shading and helps keep the heat out, although if it is on the inside of the glazing it should be light in color to reflect sunlight. Otherwise, the heat collected might be enough to damage the glazing. Glazing panels can also be designed for removal in summer, converting the sunspace into a patio or covered breezeway as is done in the Wilbur Springs dwellings depicted in Figure 7-9.

Whatever method of summer venting is used, the total vent area should be equal to 20 percent or more of the sunspace floor area. The venting devices should not be placed on the end walls only but should be installed in the south wall and roof as well. Staggered high and low vents such as those in Figure 7-10 are especially effective, for they will initiate a "chimney effect" that draws hot air out and cooler air in.

Figure 7-9 The lower row of windows are designed to be removed in summer, converting the sunspace behind them into a covered breezeway. This is the Solar Lodge at Wilbur Springs Hotel and Health Sanctuary, near Williams, California.

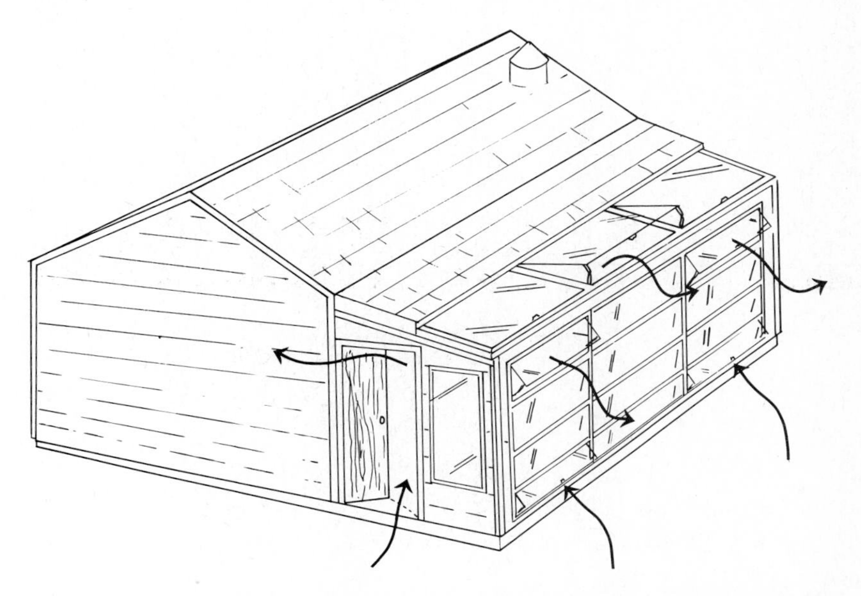

Figure 7-10 Sunspace venting. Venting devices should be placed in end walls, south wall, and roof. A combination of high and low vents creates convection currents that draw cool air in and push hot air out.

Another effective method of overheating prevention is the use of "shade cloth" over the apertures during times of the year when they are likely to collect too much insolation. Often made of polypropylene, shade cloth is sold in a variety of solar transmissivities, allowing the sunspace to be naturally daylighted while maintained at a comfortable temperature. The cost of shade cloth is typically between 12 and 60 cents per square foot (1984 prices).

A situation that sometimes occurs is one in which a sunspace overheats while other rooms of the house remain chilly and in need of warmth. Poor circulation is most likely the cause of this problem. Larger vents between house and sunspace, and/or the use of circulating fans and possibly ducting to the cold areas, will help ameliorate the problem.

SUNSPACE KITS

Sunspace kits differ from site-built units in that they include prefabricated components such as framing members and wall and roof sections. The kits are engineered for quick assembly, which cuts on-site labor costs. Because they are predesigned, however, it is sometimes difficult to integrate their style aesthetically into that of the rest of the building, especially if they are being retrofitted onto that building. To help lessen this problem, many kits have provisions for the owner to supply the exterior and interior finishing materials (see Figure 7-11).

Figure 7-11 Sunspace kit that has been well integrated into the design of the house. The kit is sold by Northern Sun of Lynwood, Washington. The system was designed by Integrated Solar Systems of Anacortes, Washington. Note that there is vertical glass above as well as below the tilted glass section. (Courtesy of Integrated Solar Systems, Inc., Anacortes, WA.)

The precut framing members of a sunspace kit are generally either of wood or aluminum. Wall and roof sections, often preinsulated, are cut in modular sizes for ease of construction. The preinsulated sections typically have a cellular foam core sandwiched by plywood, hardboard, or metal panels.

The kits are attached to exterior building walls and are supported either by concrete foundations or sit directly on the ground. Typical kit prices range from $10 per square foot for a lightweight aluminum structure with polyethylene glazing[14] to $30 to $60 per square foot for spaces with complete walls, roof, tempered glass windows, and some finishing materials. Added to this is the cost of the foundation, floor (if it has one), and thermal mass. Movable insulation for the glazing is also usually extra. Some sample sunspace kits, both of the simple and the more complex varieties, are described in Table 7-4.

TABLE 7-4 Examples of Sunspace Kits

Name of Kit	*Manufacturer*	*Description of Kit*	*Typical Sizes*	*Price per Square Foot (1984 Prices)*
Radiant Room	Abundant Energy, Inc. 116 Newport Bridge Road Warwick, NY 10990 (914) 258-4022	Foundation, roofing, and siding is provided by the builder so that they can match those of the existing house. The kit includes a choice of three south wall glazing angles (57°, 75°, and 90°). R-22 insulation is provided for the roof; wall insulation is furnished by the builder. The windows can be either glass or Exolite. The glass windows are double-sealed with silicon and butyl to prevent condensation between the panes. Open-beam construction, with a choice of cedar or pine interior paneling. Optional accessories include water thermal storage containers and night insulation for the windows.	Width: 10–13 ft; length: 12 ft, 16 ft, 20 ft, and longer	\$30–\$42
Solar Shed	Green Mountain Homes Royalton, VT 05068 (802) 763-8384	Fairly low-cost sunspace kits. The foundation, roofing, insulation, and interior finishing materials must be supplied by the builder.	Width: 8½ ft; length: 14 ft, 17 ft, 20 ft	\$18–\$24
Sun Space	Northern Sun 21705 Highway 99 Lynwood, WA 98036 (206) 771-3334	The kit includes detailed instructions for the do-it-yourselfer. Prefinished cedar construction. Double-sealed, insulating glass panels (Heat Mirror glazing can be provided at an extra charge). A variety of roof pitches are available. Other options include circulating fans, and summer shade screens.	Width: 8 ft; length: 12 ft, 15 ft, 18 ft, 21 ft	With end walls: \$41–\$53 Without end walls: (semienclosed design): \$28–\$29
Sun Room	Garden Way 2540 Ferry Road Charlotte, VT 05445 (800) 343-1908	Foundation, roofing, and siding must be supplied by the builder to match the existing home. Walls consist of precut three layer panels (exterior particleboard, polyurethane insulation, and interior MDO plywood). Insulation: R-27 in roof, R-20 in walls. The design features attractive, arching laminated beams. Night insulation for glazing is supplied at additional cost, as are Kalwall tubes for heat storage.	Width: 10 ft; length: 13 ft, 17 ft, 21 ft	\$42–\$54

THE ATTIC SUNSPACE

Although sunspaces are not commonly located in attics, in some situations it is advantageous to do so, for it utilizes an area of the house that is normally not used for anything but storage and as a buffer zone between internal and external temperatures. Because light is admitted into the space through skylights, clerestories, or dormer windows in the roof, attic systems sometimes have better solar exposures than sunspaces built on the ground and shaded by neighboring buildings and trees. Attic sunspaces are especially suitable for high-density housing developments in which solar access is a problem.

The glazing in attic sunspaces is located in the south-facing sections of the roof. The rest of the roof, as well as the end walls, are heavily insulated. The attic floor must be insulated as well to prevent summer overheating of the living space. Interior surfaces of the attic are painted black and water containers are used to add to the thermal storage mass. As with other sunspaces, movable insulation dramatically reduces nighttime heat losses. A blower is generally necessary in order to move the heated air down into the living areas. Backdraft dampers are installed on air ducts to prevent warm house air from rising into the attic at night. The layout of an attic sunspace system is depicted in Figure 7-12.

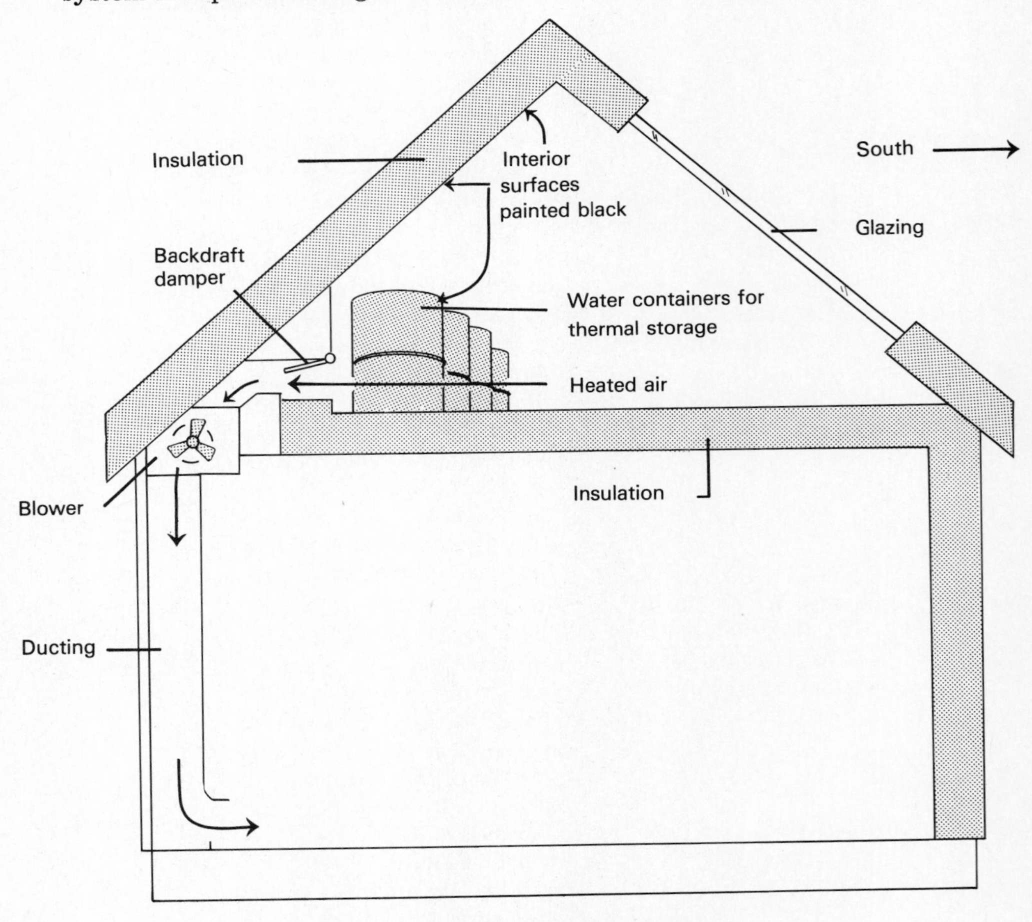

Figure 7-12 Attic sunspace.

RETROFITS

Sunspace retrofits should be compatible with the existing architecture and must conform to local building codes. A retrofit problem sometimes encountered when the south side of a house faces the front yard or the street is that city ordinances require a minimum yard size, and adding a structure to the front of the house will violate this rule. If there is a porch on the south wall of the house, however, it can be enclosed and converted into a sunspace. Sunporches are often the most cost-effective type of retrofit for older homes, next to basic energy conservation measures such as weather stripping to reduce infiltrative losses and adding insulation to cut down on conductive losses. Figure 7-13 depicts retrofitted sunporches. Note that skylights can be added to the porch roof to increase the amount of light. Air exchange with the house is provided by the windows and door fronting on the porch.

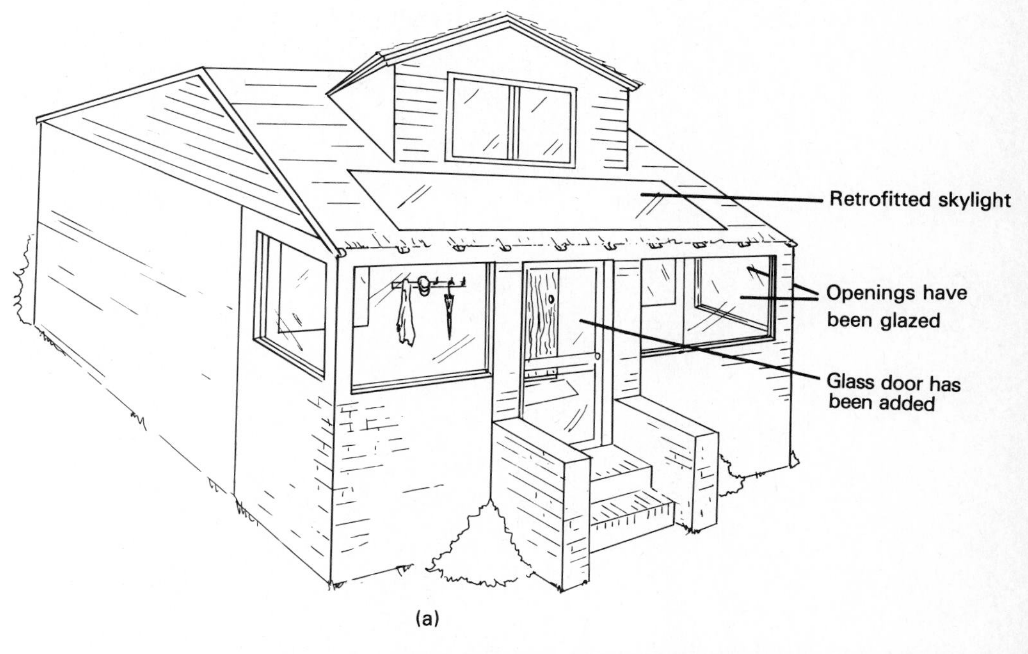

(a)

(b)

Figure 7-13 Retrofitted sun porches.

Another problem encountered with retrofits is what to do when the south face of the building is very close to the property line. One solution might be to add a Trombe wall system to the south face. Another possibility is to extend a sunspace from the east or west side of the building. The sunspace depicted in Figure 7-14a still receives good solar exposure, although the shared wall area between it and the house is not as large as if it were on the building's south side.

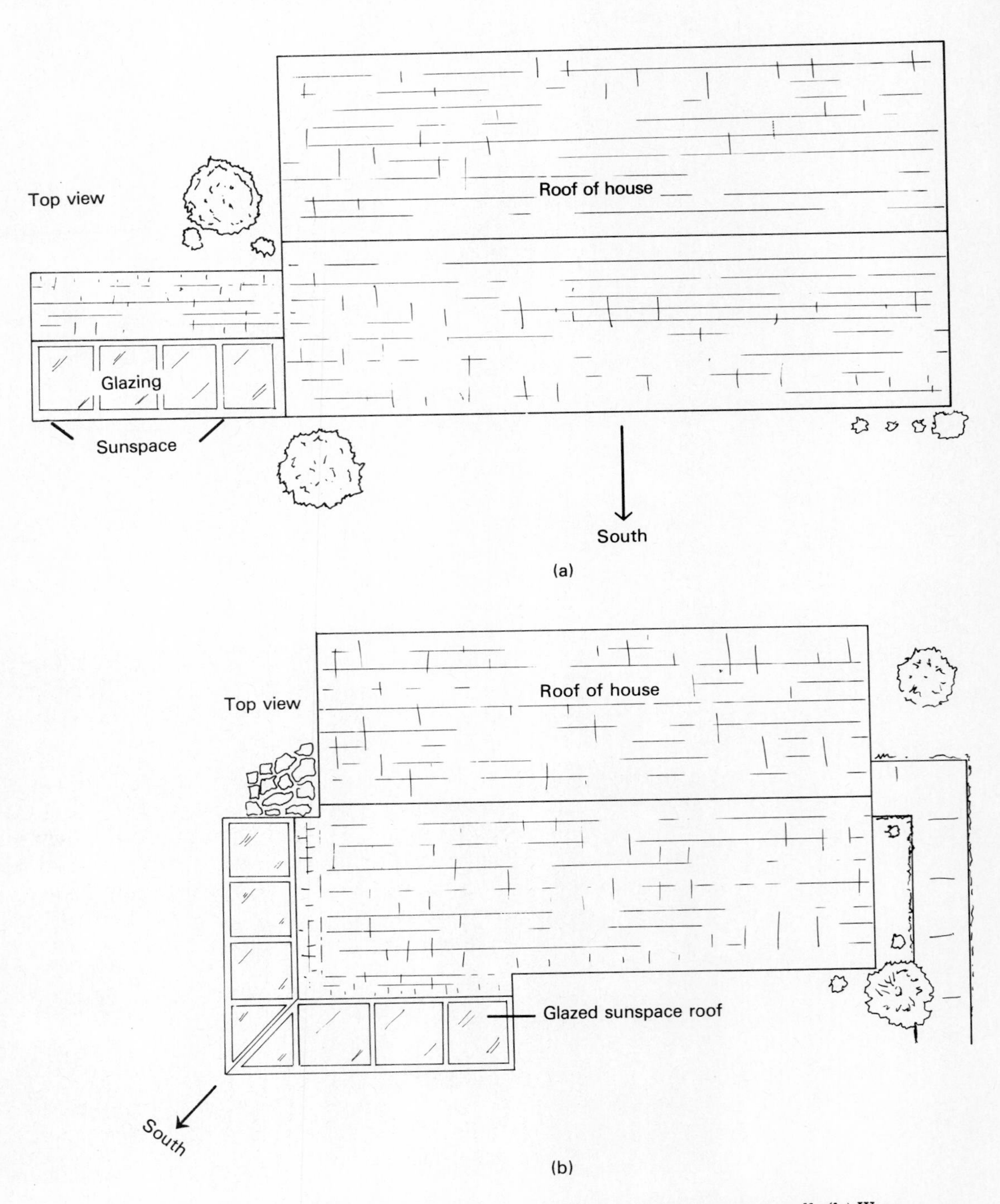

Figure 7-14 (a) Extending a sunspace from the east or west wall. (b) Wrapping a sunspace around a corner.

If a house happens to be situated so that one corner is facing due south, a sunspace located on either of the adjacent sides will not have the optimum exposure to the sun. In such a case, the sunspace can be built *around* the corner, as in Figure 7-14b. This configuration is especially advantageous for plant growth, because sunlight is admitted for more hours each day than if the glazing were facing due south.[15] In early morning and late afternoon, though, one of the two faces of the space receives no direct sunlight and ideally should be covered with movable insulation to reduce heat losses.

SECONDARY SUNSPACE FUNCTIONS

Use as a Greenhouse

Greenhouse gardening, according to Bill Yanda and Rick Fisher,[16] takes the natural patterns and rhythms of growing and harvesting and intensifies them in space and time. Space in greenhouses is limited and must be used efficiently. Growing time is extended, for a solar-heated greenhouse can produce summer vegetables throughout the year.

The success of greenhouse crops depends, among other things, on temperature stability. Temperatures that are too high or too low will kill plants. Thermal storage mass is as important in greenhouses as it is in sunspaces for keeping nighttime temperatures up to an acceptable level. The soil beds of the greenhouse garden make fairly good thermal masses if the sunspace perimeter is insulated to a depth of about 2 feet (see "The Floor" section of this chapter). The stored warmth in the soil adds root growth and helps the plants withstand cold periods.

Greenhouse and sunspace engineering differ somewhat from each other in the area of maximum acceptable temperatures. It would be fine for a sunspace used for no other purposes to attain winter temperatures well over 100°F, for this energy could be employed in space heating the adjacent house. But temperatures that high would be detrimental to many plants, and so if the sunspace is to serve as a greenhouse as well, usable heat must sometimes be vented, even in winter. Adding additional thermal storage mass minimizes the amounts of heat that must be wasted in this way.

Enclosing Swimming Pools and Hot Tubs

When swimming pools, hot tubs, and spas are housed in sunspaces, they become part of its thermal storage mass. The solar collecting and heat-retaining properties of the sunspace in turn provide a gentle environment, even in winter, for swimming and soaking and help keep the water at a temperature high enough for comfort. In addition, much of the auxiliary heat employed for keeping the water temperature up is eventually used in space heating the adjacent building. Sunspace pools and tubs are usually covered when not in use, not only for heat retention but also to prevent excessive humidity from entering the living area.

SIZING THE SUNSPACE

Table III-2 is employed for determining glazing areas of sunspaces in much the same way as it is used for direct-gain and solar wall systems. It is important to note, however, that the table does not yield the actual, or "net" glazing area of the sunspace but rather its *projection* onto a vertical plane (see Figure 7-15). This makes the sizing procedure versatile and adaptable to many sunspace geometries. It is explained later in the section how to design a sunspace given the projected glazing area.

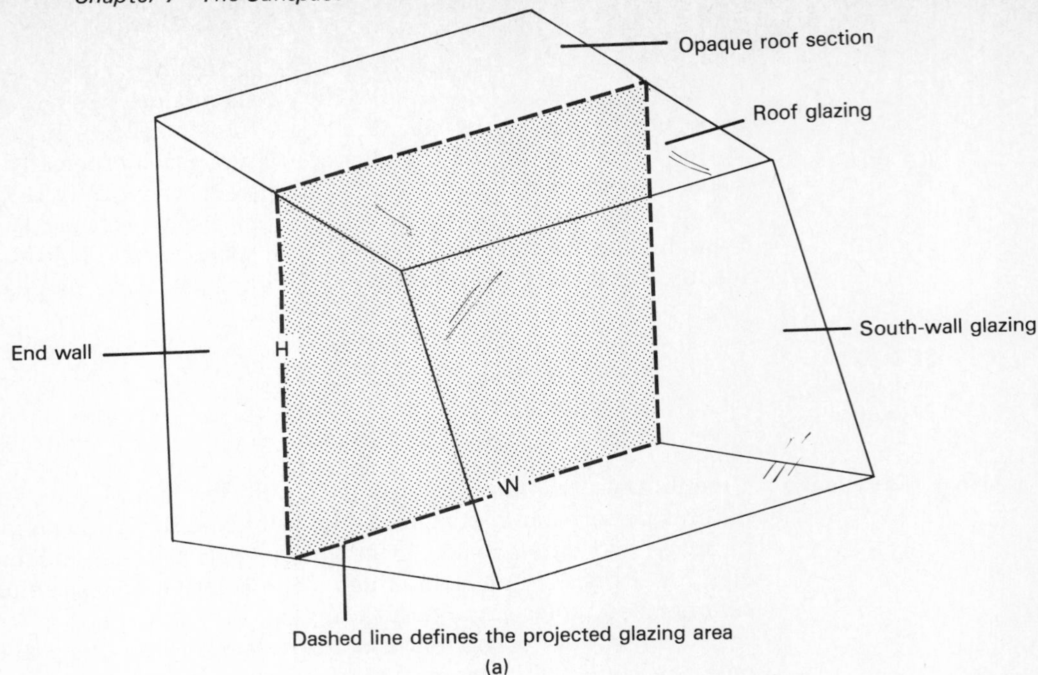

Figure 7-15 Projection of glazing area onto a vertical plane. (a) Three-dimensional view. (b) Side and front views. H_R, height of glazed roof section; H_S, height of glazed south wall; H, total glazing height $= H_S + H_R$.

Data for the following sunspace configurations are listed in Table III-2:

1. The *attached shed roof sunspace*, composed of a vertical or nearly vertical glazed south wall and a partially glazed roof.
2. The *attached single plane sunspace*, in which the glazing plane is oriented at a nearly optimum angle to the winter sun.
3. The *semienclosed shed roof sunspace*, in which the east and west as well as the north walls are shared with the living area.

The configurations listed above are sketched in Figures 7-1, 7-2, and 7-3a. One of the configurations should be chosen for the sunspace under design and the sizing procedure described below applied to it.

Table III-2 lists data for sunspace designs in which movable insulation is used over the glazing at night, for this practice greatly improves the system's efficiency. In addition, data are included for an attached shed roof configuration that does not use movable insulation, as a means of comparing performances.

Double glazing was used in all of the reference designs from which Table III-2 was constructed. End walls of attached sunspace configurations were assumed to be opaque and insulated. All walls that are shared with the living space were constructed of 12-inch-thick concrete.[17] Inside surfaces of the sunspace have absorptances of 0.80, which corresponds to the absorptance of paint colors such as medium light brown or medium rust.[18] This absorptance is used on the reference models because it was thought that since the sunspace usually serves as some sort of living space as well, cosmetic considerations would weigh against the use of more absorptive colors such as flat black (0.95) or dark gray (0.91).

The recommended projected glazing area for the chosen design is determined in the same way that aperture areas were arrived at in previous chapters. First, the building load coefficient is calculated; then it is decided whether the house is to be heavily solar heated, moderately solar heated, or solar tempered; and finally, the building load coefficient is divided by the appropriate load collector ratio from Table III-2 to yield the glazing area (see Chapter 4 for a more detailed explanation of this procedure).

The dimensions and tilt angles of the reference sunspaces for which the values in Table III-2 were calculated are detailed in Figure 7-16. Because of aesthetic considerations or a desire for more or less floor area, the designer

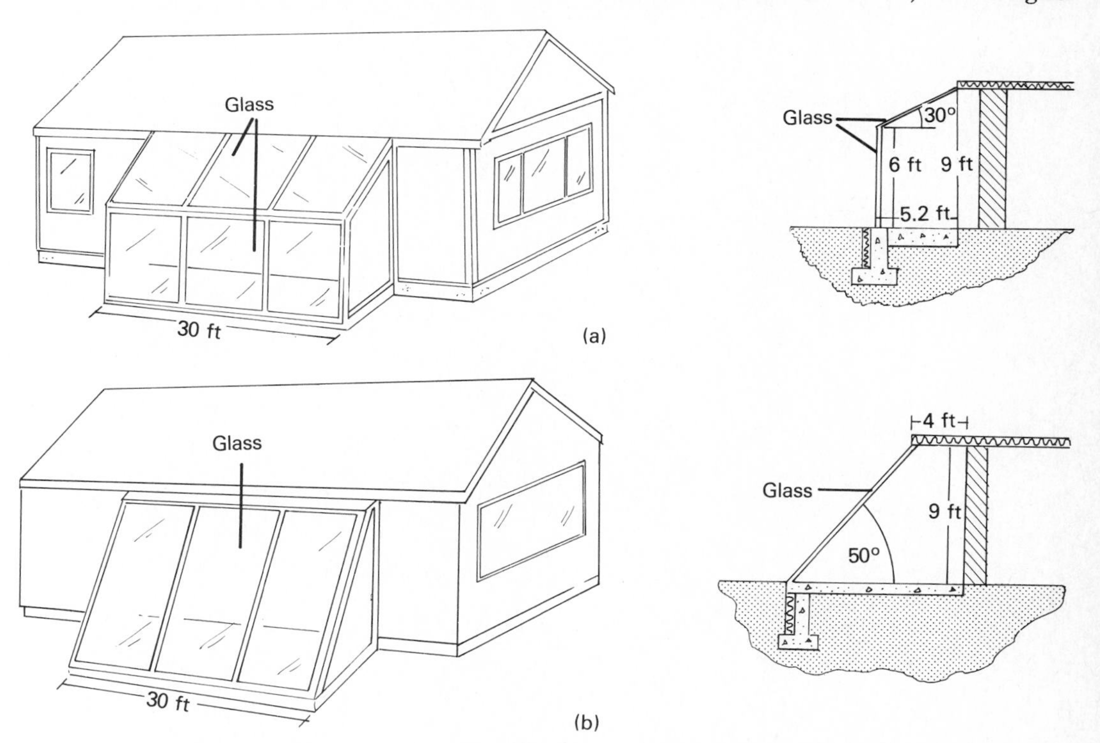

Figure 7-16 Reference sunspaces. Included in this figure are the dimensions and tilt angles of the sunspace designs for which the values in Table III-2 were calculated. (a) Attached shed roof sunspace. (b) Attached single-plane sunspace. (From J. Douglas Balcomb et al., *Passive Solar Design Handbook*, Vol. 3, Los Alamos National Laboratory, Los Alamos, NM, for U.S. Department of Energy, DOE/CS-1027/3, 1982, p. 88.)

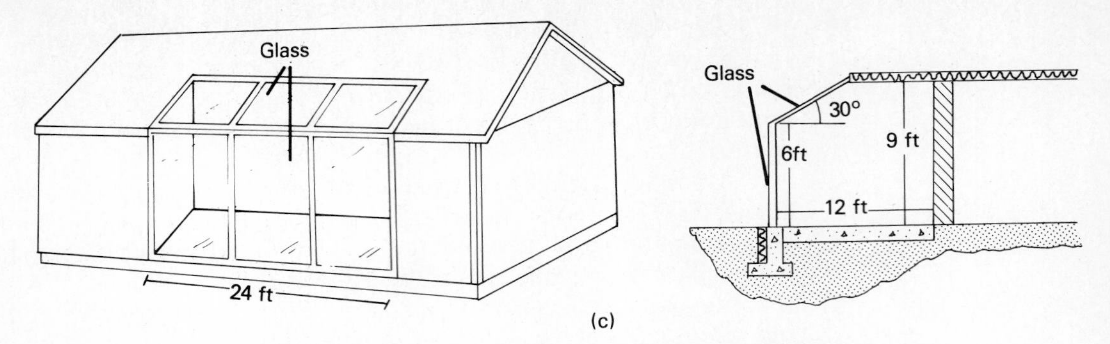

Figure 7-16 (cont.) (c) Semienclosed shed roof sunspace.

might wish to build a structure of different size and shape from the reference configuration.

Changing the reference dimensions will of course change the total amount of energy collected. If the guidelines listed below are observed, the sunspace should have high solar savings fractions that are also fairly close to those of the reference designs.

1. *Single-plane sunspace* glazing should be installed at a 50 to 65° tilt angle. This will result in good water and snow drainage, adequate headroom and reduction of summer overheating, as well as nearly optimum solar savings fractions.
2. The total glazing height of a *shed roof sunspace* is the sum of the heights of the south wall and roof glazing (see Figure 7-17). As long as the total height remains constant, the height of the south wall can be increased or decreased as much as 2 feet from the reference design height of 6 feet with a solar savings fraction change of usually not more than 5 percent.[19] The less the height of south wall glazing, the better the performance, for more of the incident solar energy will strike the roof glazing, which is at a better solar collection angle. Decreasing the vertical glazing height, however, reduces headroom and usable floor space.
3. Changing the depth of a *shed roof sunspace* while keeping the vertical glazing height constant, as shown in Figure 7-18, can be done with little change in SSF for most locations in the United States and Canada.
4. Variations in the east-west width of a sunspace changes the total glazing area and so of course also affects the solar savings fraction. As long as the sunspace is at least 10 feet wide, though, the energy collected and utilized *per square foot* of glazing varies little.

With the foregoing guidelines in mind, the sunspace can be designed. Notice from Figure 7-15 that the total glazing height (H) times the width (W) is simply equal to the actual glazing area projected onto a vertical plane. H and W should be selected so that their product is equal to the recommended projected area calculated above from the information in Table III-2. After this, the roof glazing and south wall glazing heights (H_R and H_S) and tilt angles (R and S) should be decided upon. (These quantities are also depicted in Figure 7-17.) Aesthetics as well as the guidelines listed above play a large part in the choice of glazing heights and tilt angles, for a sunspace's design should be harmonious with that of the rest of the house.

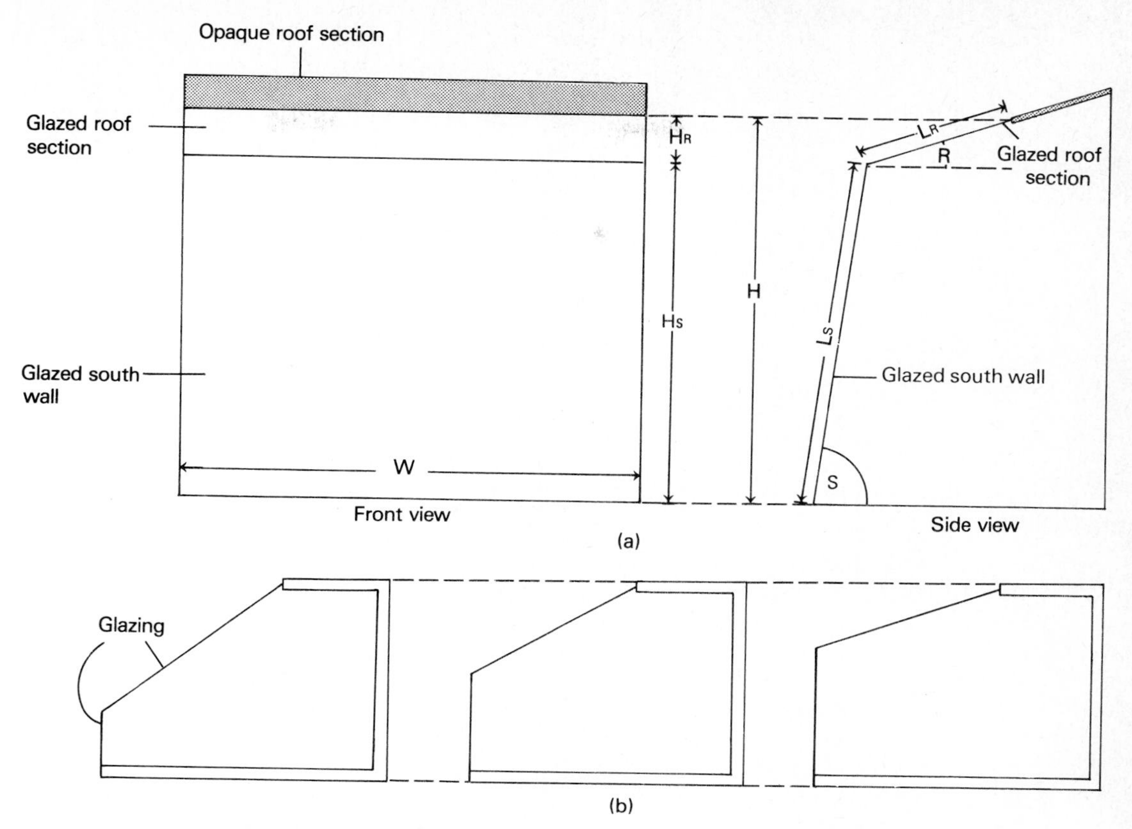

Figure 7-17 (a) Dimensions of a sunspace: Front and side views. (b) Three shed roof glazing configurations in which the total glazing height (H) remains constant, even though the south-wall height (H_S) and the roof glazing height (H_R) vary.

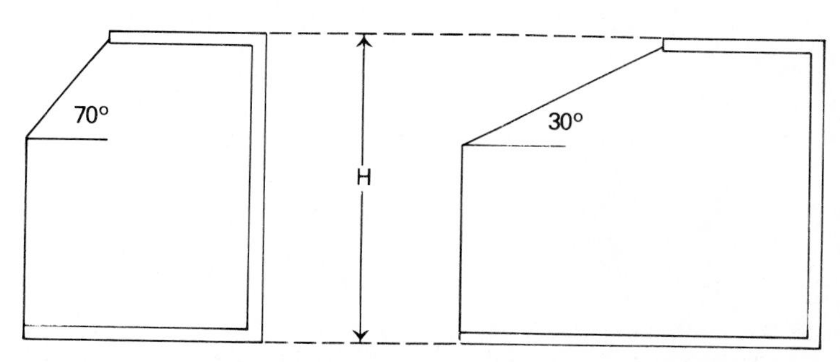

Figure 7-18 Acceptable shed roof sunspace depth variations. The tilt angle of the upper glazing can be varied between 70 and 30° with little change in sunspace performance for most locations in the United States and Canada.

The actual glazing area of the south wall or roof is its width (W) times its length (L_S or L_R). Its width is the same as that of the projected glazing area. Its length can be calculated from its height and tilt angle as follows:

$$L_R = \frac{H_R}{\sin R}$$

$$L_S = \frac{H_S}{\sin S}$$

If a calculator with a built-in table of sines is not available, Table 5-5 may be used.

Example Calculation. The recommended projected glazing area of a sunspace is 300 square feet. The glazing width is 30 feet and the total glazing height is 10 feet. This height is divided between the roof glazing height (H_R) of 3 feet and the south wall height (H_S) of 7 feet. The tilt angles of roof and south wall are 25° and 75°. (The sunspace is depicted in Figure 7-19.) What are the lengths and areas of the roof and south wall glazing?

Lengths:

$$L_R = \frac{H_R}{\sin R} = \frac{3 \text{ ft}}{\sin 25°} = \frac{3}{0.42} = 7.1 \text{ ft}$$

$$L_S = \frac{H_S}{\sin S} = \frac{7 \text{ ft}}{\sin 75°} = \frac{7}{0.97} = 7.2 \text{ ft}$$

Areas:

$$\text{Roof:} \quad L_R \times W = 7.1 \times 30 = 213 \text{ ft}^2 \text{ of glazing}$$

$$\text{South wall:} \quad L_S \times W = 7.2 \times 30 = 216 \text{ ft}^2 \text{ of glazing}$$

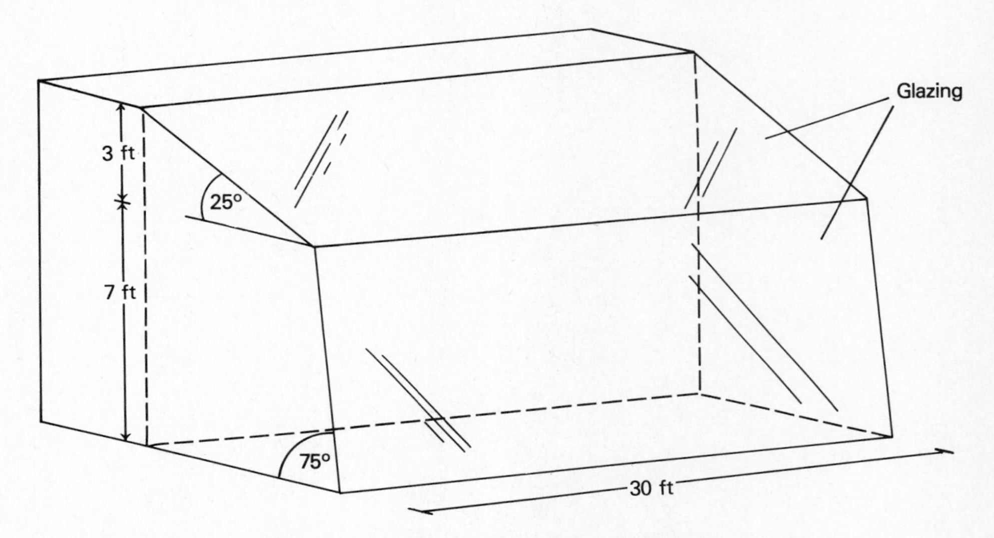

Figure 7-19 Shed roof sunspace example.

SUNSPACES COMPARED WITH OTHER PASSIVE SYSTEMS

In a comparative study of night-insulated sunspaces, direct-gain, and Trombe wall systems in three cities, it was found that the *productivities* of each system (the total energy saving due to the system divided by the area of solar aperture) in any one location were surprisingly close to one another. In Nashville and Denver, for instance, the differences were less than 10 percent. In Boston, the variation was about 20 percent.[20] This, and the similar study on solar walls in Chapter 6, suggest that other factors are most important in the choice of a system, such as the comfort level each will provide, the aesthetic value, and of course, the cost.

Any of the types of passive systems studied will provide a comfortable environment if constructed correctly and meticulously. Direct-gain systems, though, are the most likely to result in large interior temperature swings, especially if night insulation is not used.

The admission of thermal energy into the house or the prevention of its loss is more easily controlled in sunspace and solar wall systems since sunlight does not directly enter the living space but first strikes an intermediate mass. Sunspaces also have the advantage of providing additional living space for the building in the form of a recreational area, a greenhouse, or an enclosure for a hot tub or swimming pool. However, a sunspace generally undergoes larger temperature swings than the rest of the house and thus is frequently unoccupied during certain times of the year.

Some guidelines that might help a building designer in planning a solar system are as follows:

1. Direct-gain systems are extremely economical choices for aperture areas up to about 8 percent of the structure's floor area. This is because no additional thermal mass other than the materials normally used in building construction need to be installed.
2. If *additional* aperture area is needed, it can be installed as part of a direct-gain, solar wall, or sunspace system. Trombe walls are frequently good choices, due to the multiple functions that they serve, such as providing structural support and thermal mass as well as a collection surface. Remember, also, that the use of movable insulation in a Trombe wall system is not as important as it is in a direct-gain configuration, especially if the Trombe wall is covered with a selective surface.
3. Before deciding to construct a sunspace, its advantages over other types of systems of providing an extra recreational space or greenhouse must be weighed against its additional cost per square foot of aperture provided (its additional cost is due to the extra building materials and time that is usually required for construction). Sunspaces offer benefits on which it is difficult to put a monetary value: They can add a beautiful, naturally daylighted, plant-filled enclosure to the house that can be a joy to enter on dreary winter days (see Figure 7-20). The choice to build such a structure is a matter of taste and the homeowner's life-style, as well as simply an economic decision.

(a)

Figure 7-20 In addition to providing heat, sunspaces can be attractive, naturally daylighted, plant-filled enclosures that are a joy to enter. (a) Notice the skylights and windows on the sides and at the bottom of the sunspace that can be opened to reduce overheating. [(a), Courtesy of General Glass International Corp., manufacturers of "SolaKleer" low-iron glass, New Rochelle, NY; and Mekler/Ansell Associates, New York.]

(b)

(c)

(d)

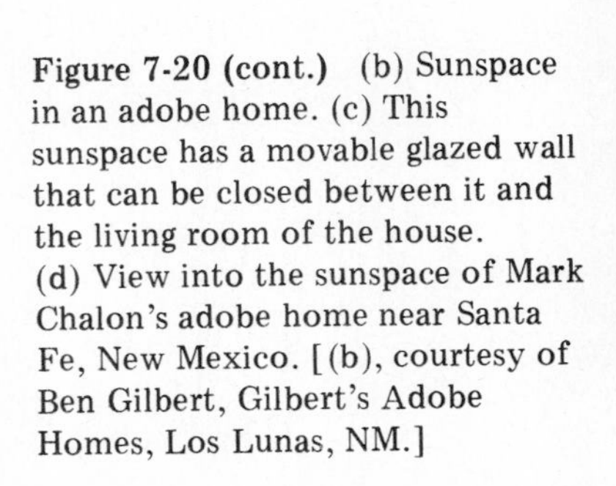

Figure 7-20 (cont.) (b) Sunspace in an adobe home. (c) This sunspace has a movable glazed wall that can be closed between it and the living room of the house. (d) View into the sunspace of Mark Chalon's adobe home near Santa Fe, New Mexico. [(b), courtesy of Ben Gilbert, Gilbert's Adobe Homes, Los Lunas, NM.]

EXAMPLES OF SOLAR SYSTEM COSTS

When deciding on a solar system configuration, many variables are of interest to the designer. Besides wanting to know how big the system must be and what solar savings fraction it will produce, the architect, builder, or future homeowner is also concerned with its *incremental cost*, that is, the extra amount of money that the house will cost compared to one without solar heating. Of additional interest is the system's cost per square foot of aperture, for this datum can then be compared with that of other solar designs collecting roughly the same amount of energy.

Table 7-5 lists the foregoing parameters for a number of typical passive systems of various types. The examples should give the designer a rough idea of how the costs of various systems compare with one another.

TABLE 7-5 Examples of Incremental Costs of Solar Systems

System Location	*Description*	*Year of Study*	*Projected Solar Aperture Area (ft^2)*	*Floor Area of Residence (ft^2)*	*Incremental Cost of System (dollars)*	*Incremental Cost per Square Foot of Aperture (dollars)*
	Direct-Gain Systems					
Seattle, WA	Thermal energy is stored in a concrete floor slab.	1981	238	1400	1,080	4.54
Massachusetts	Windows, both solar and nonsolar, in these multifamily dwellings are triple glazed. No thermal storage mass other than normal building materials have been added to the structures. The reason for the zero incremental cost is that the amount of window area was no more than that used in normal construction. It was simply oriented for maximum solar gain.	1981	46	520	0	0
Denver, CO	A 310-ft^2 room addition to an existing house. The $500 incremental cost covers additional glazing and thermal storage mass in the floor, consisting of brown brick pavers.	1982	70	310	500	7.14
Denver, CO	A storage shed in the same house as above that was converted into a bathroom and direct-gain office space. Incremental cost is largely due to mass added to floor, in the form of brick pavers on a sand leveling bed overlying an existing slab.	1982	25	200	200	8.00
	Solar Wall Systems					
Phoenix, AZ	The 1-foot-thick concrete Trombe wall is one of a variety of passive features used in this home. Its cost is comparable to that of an insulated wood frame wall.	1982	58.9	1517	242	4.11

TABLE 7-5 (cont.)

System Location	*Description*	*Year of Study*	*Projected Solar Aperture Area (ft^2)*	*Floor Area of Residence (ft^2)*	*Incremental Cost of System (dollars)*	*Incremental Cost per Square Foot of Aperture (dollars)*
Denver, CO	In this *retrofitted* system, 360 ft^2 of single glazing was added to an uninsulated brick veneer south wall. Labor was provided by the homeowner, including applying black stain to the brick. Recycled materials were used. State and federal tax credits reduced the incremental cost from $641 to $192, which was recovered in fuel savings during the first year of operation. Note that this solar wall collector is not a Trombe wall, for it has very little thermal storage mass in it. Heated air is circulated with the aid of two 20-in. fans into the house. The system's incremental cost includes the price of the fans.	1982	360	1270	641	1.78
Massachusetts	Compare this Trombe wall system to those above, in order to see the tremendous range of prices possible, depending on what materials are used and who does the labor. The glazing is triple (and cost a whopping $17.50/$ft^2$!), and the total price includes canvas awnings for shading. When the "alternative wall tax credit" is applied, the cost per square foot is reduced somewhat, to $23.48.	1982	104	520	2,931	28.18
Knoxville, TN	The *water wall* in this system is composed of 20 1-foot-diameter tubes 6 feet tall, behind 144 ft^2 of double-pane south-facing glass. Total amount of water is 676 gallons. Window Quilt insulated curtains help retain the heat during nighttime hours.	1982	144	1200	2,120	14.72
		Sunspace Systems				
Denver, CO	A sunspace retrofit on a 60-year-old house with uninsulated double brick walls. The single-glazed sunspace encloses a large amount of thermal mass: 525 ft^2 of double-brick shared wall, 280 ft^2 of 6-inch-thick concrete floor slab, tinted brown, and 2900 gallons of water in 55-gallon drums. The	1982	309	1700	12,000	38.83

TABLE 7-5 (cont.)

System Location	*Description*	*Year of Study*	*Projected Solar Aperture Area (ft^2)*	*Floor Area of Residence (ft^2)*	*Incremental Cost of System (dollars)*	*Incremental Cost per Square Foot of Aperture (dollars)*
Denver, CO (cont.)	$12,000 construction price includes the cost of several conservation measures in the house itself (such as additional attic insulation).					
Sedro-Woolley, WA	A Northern Sun sunspace kit integrated into the construction design of a new house. Incremental cost includes the brick shared wall and side walls, which function as thermal storage mass. Heat Mirror glazing was used in the solar apertures.	1984	225	1800	5,175	23.00

Sources: Information for all designs except the Sedro-Woolley, Washington, house came from the *6th* and *7th Proceedings of the National Passive Solar Conference*, 1981 and 1982, published by the American Solar Energy Society.

Direct gain systems: Seattle, WA: McDonald and Tsongas, *6th Proceedings*, pp. 237–241; Massachusetts: Noble and Lofchie, *7th Proceedings*, pp. 761–766; Denver, CO: Andrews and Snyder, *7th Proceedings*, pp. 231–236, and private communication with Steve Andrews.

Solar wall systems: Phoenix, AZ: Andrejko, Duffy, and Thomas, *7th Proceedings*, pp. 795–800; Denver, CO: Andrews and Snyder, *7th Proceedings*, pp. 231–236; Massachusetts: Noble and Lofchie, *7th Proceedings*, pp. 761–766; Knoxville, TN: Reid, McGraw, and Bedinger, *7th Proceedings*, pp. 747–750.

Sunspace systems: Denver, CO: Andrews and Snyder, *7th Proceedings*, pp. 231–236; Sedro-Woolley, WA: information furnished by Greg Snelson of Integrated Solar Systems, Anacortes, WA.

LIFE-CYCLE-COST ANALYSIS

Of interest to a solar designer is not only the solar system's incremental cost, but also the amount of money the system will save in heating bills throughout its lifetime. The *life-cycle cost* of an installation is the sum of its initial cost, maintenance, repair, and energy costs over the predicted lifetime of the system, minus any salvage value it has at the end of its useful life. When the life-cycle cost of a solar heating system is compared to that for a conventional heating system for the same building, an estimate of monetary saving due to reduced heating bills is obtained. A complicating factor in arriving at the life-cycle cost of a system, however, is that a dollar spent at some point in the future is not worth the same amount as a dollar spent today. Due to inflation, a dollar is worth less and less as time goes on. Also, a dollar invested today can earn interest. For instance, if it were possible to earn 7 percent per annum interest on an investment, an investor might consider a dollar today to be equivalent to $1.07 a year from now. If in addition there were a 6 percent inflation rate, an investor would consider a dollar today equivalent to $1.13 a year from now. Other factors sometimes considered by investors are credits for tax deductions.

To compare an energy system's initial expenditures with those incurred throughout the lifetime of the system, all cash amounts are converted to their "present value." In other words, if it is estimated that 10 years from now a system will need $500 worth of repairs, this $500 is converted into what its value would be in today's dollars. The conversion factor used to do this is the "rate of interest at which an investor feels adequately compensated for trading money now for money in the future."[21] This rate of interest is dependent on the above-mentioned factors, and is termed the *discount rate.*

The expenditures incurred during construction of a solar energy system can be divided into initial expenditures, one-time-only expenditures made at some point in time after the system goes into operation; and repeated expenses, such as maintenance or the yearly amount of energy used by the backup heating system. These annual costs can stay constant or can change from year to year. All expenses incurred after the initial ones must be "discounted" to their present value. The factors tabulated in Appendix IV make these discounting operations a fairly straightforward process.

To understand better how life-cycle-cost comparisons are set up, it is instructive to work through two examples.

Example Calculation 1. The initial cost of a solar system with a conventional backup heating system is $7000, and it is estimated that they have a useful life of 25 years. Once the solar system is in, heating costs for the house are expected to be $250 per year to start with and are expected to escalate at a rate of 5 percent per year, due to fuel cost increases. It is also estimated that without the solar system, a conventional system costing $3000 would be needed, and would use $750 worth of fuel to heat the building the first year, increasing at a rate of 5 percent per year. Assuming an annual discount rate of 8 percent, compare the life-cycle costs of the solar and the conventional systems.

First, we need to convert all cash amounts to their present value. If the $7000 incremental cost of the solar system is paid at the beginning of the system's lifetime, it is already at its present value. To convert the $250 total fuel cost over the system's 25-year lifetime to its present value, refer to Table IV-2 with a discount rate of 8 percent. Use the column in this table corresponding to a 5 percent energy price escalation, and the last line in the table, corresponding to a 25-year lifetime. The value obtained, 17.694, is multiplied by the $250 base-year energy price to yield the *total* energy cost over the 25-year lifetime, converted to its present value:

$$\$250 \times 17.694 = \$4423.50$$

The total life-cycle cost of the solar system is equal to the initial cost plus energy costs. (If there were expected to be maintenance or repair costs, they would also have to be converted to present values and added on.)

Initial cost of solar system	$ 7,000.00
Energy costs	4,423.50
	$11,423.50

Let's compare this to the life-cycle cost of heating the building using only conventional means. The $750 annual heating cost is converted to present value in the same way as the backup heating cost was for the solar system:

$$\$750 \times 17.694 = \$13,270.50$$

Adding this to the $3000 initial cost of the conventional system yields a total life-cycle cost for the conventional heating system of $16,270.50.

The net present value savings obtained over the 25-year lifetime of the solar system is significant:

$$\$16,270.50 - 11,423.50 = \$4847.00 \text{ savings}$$

Example Calculation 2. Let us now perform a more complicated comparison. Assume that the initial cost of a solar system is $20,000, the base-year (first-year) costs for backup heating are $400, the expected rate of energy price escalation is 4 percent per year, the discount rate is 12 percent, and the system lifetime is 20 years. Maintenance is expected to run $40 per year.

The conventional heating system's initial cost is $14,000 and its base-year energy costs are $1600. Maintenance is expected to cost $70 per year. It is estimated that repairs on the furnace will be necessary in the fifteenth year, costing about $2500. The salvage value of the conventional system after a 20-year period is expected to be $1800.

Table 7-6 outlines the life-cycle-cost comparison calculations. Let us investigate how the values in the table were obtained.

TABLE 7-6 Life-Cycle-Cost Analysis

	Solar System with Small Conventional Backup System		*Conventional Heating System*	
Initial cost		$20,000.00		$14,000.00
Maintenance	$40/yr × 7.469 =	298.76	$70/yr × 7.469 =	522.83
Energy costs	$400 × 10.047 =	4,018.80	$1,600 × 10.047 =	16,075.20
Repairs		—	$2,500 (during 15th year) × 0.183 =	457.50
Salvage value		—	$1,800 (during 20th year) × 0.104 =	−187.20
Total		$24,317.56		$30,868.33

Present-value cost savings using solar system rather than conventional system = $6,550.77

The present value of 20 years of fixed maintenance costs was calculated with the aid of Table IV-1. The factor for a 12 percent discount rate and 20 discount periods is 7.469. Multiplying this by the $40 annual maintenance cost for the solar system yields a total present value of maintenance of $298.76.

Since energy costs escalate at a rate of 4 percent in this example, the 4 percent column in Table IV-2c is used to compute present value of energy costs (Table IV-2c is selected because the discount rate in this problem is 12 percent). Maintenance and energy costs for the conventional heating system are calculated similarly to those for the conventional system.

The conventional system has a repair cost of $2500 during its fifteenth year. Since this is a once-only expense rather than an annual one, the repair cost is converted to its present value using Table IV-3. The 12 percent discount rate column is selected. Since the repair occurs in the fifteenth year, the value on the fifteenth line, 0.183, is the factor needed.

The conventional system's salvage value after 20 years is also converted to its present value using Table IV-3. The salvage value is not an expense, but just the opposite, so it is *subtracted* from the other values.

In order to find out more about life-cycle costing, refer to the reference in the "Suggested Reading" at the end of this chapter.

HOUR-BY-HOUR PERFORMANCE OF SOLAR SYSTEMS

In our solar system sizing calculations, we have been concerned with a building's daily energy losses as well as its annual solar savings fraction. It is also interesting and instructive to examine the hour-by-hour performance of systems, for there can sometimes be important information revealed by doing so.

The Public Service Company of New Mexico has carefully monitored a number of solar-heated homes on a continual basis throughout the day. In one study, six homes in Santa Fe's La Vereda subdivision were examined.[22] The buildings used combinations of direct-gain, Trombe wall, and sunspace passive systems to provide heating. Electric baseboard heaters were employed as backup systems. One parameter continually recorded was the amount of energy used by these backup heaters, for this is a good indication of how well the solar systems are meeting the space-heating needs of the house. (The less backup heat needed during the heating season, the higher the solar savings fraction.)

Backup heater data were recorded in watts used per square foot of floor space per degree-day (see Figure 7-21). The data can thus be compared to those taken in other houses of different size and to data recorded on other days in which different numbers of degree-days have accrued.

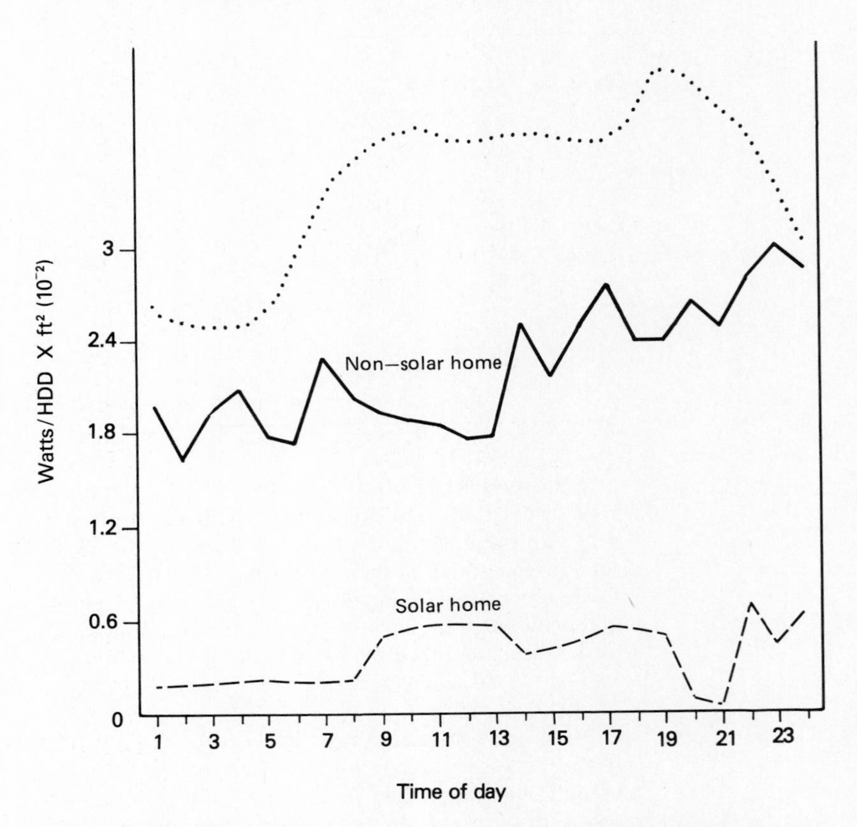

Figure 7-21 Power usage comparison between La Vereda solar homes (dashed line) in Santa Fe, New Mexico, and nonsolar, superinsulated totally electric (SMART) homes (solid line) in the same area. Public Service Company of New Mexico system profile (dotted line) represents the shape of PNM's winter peak-day load (not drawn to scale). HDD, heating degree-day. La Vereda is a joint project between the Public Service Company of New Mexico; Communico, Inc. (builders, Wayne and Susan Nichols); Los Alamos National Laboratory; and U.S. Department of Energy. (Courtesy of Stephen Pyde, P.E., La Vereda Project Manager, Public Service Company of New Mexico, Albuquerque, NM.)

Also included in the study were data from nonsolar, superinsulated, totally electric homes (the "SMART" homes in Figure 7-21). The difference between the two graphs in the figure is dramatic. As can be seen, the solar homes use far less space-heating energy, regardless of the time of day, than do the superinsulated but unsolarized dwellings. This is an indication that the thermal storage mass of the La Vereda dwellings continues to emit heat long after the sun has set.

Another interesting fact can be obtained by studying the dotted line in Figure 7-21, which represents the utility company's energy load at different times of day. The graph indicates that the heaviest usage occurs between 8 A.M. and 10 P.M., with a noticeable peak energy draw at 7 P.M. The times of peak demand are, in many areas, the times when electricity costs the consumer the most money per kilowatt-hour. During the times of peak draw in February 1981, the La Vereda homes used on the average 73 percent less energy per square foot than did the SMART nonsolar homes.

NOTES

1. The less distance the heated air has to travel, the less opportunity for some of its energy to be lost, through leaks in or contact with the building envelope, or through mixing with cooler air masses inside the sunspace.

2. J. Douglas Balcomb et al., *Passive Solar Design Handbook*, Vol. 3, Los Alamos National Laboratory, Los Alamos, NM, for U.S. Department of Energy, DOE/CS-0127/3, 1982, pp. 93–98.

3. This statement can be verified by studying the LCR tables in Balcomb et al., Vol. 3. Insulated shared walls, although allowing more control over temperature in the living space, also inhibit heat flow somewhat from the sunspace into the rest of the house, thus providing more opportunity for this heat to escape to the outside.

4. Glass can be scored and broken into many different shapes, but it takes a much higher skill level to do this than to cut fiberglass.

5. Available from Lasco, a division of Philips Ind. Inc., 3255 East Miraloma Avenue, Anaheim, CA 92806. (714) 992-1220. Call or write for distributors in your locale.

6. Available from Filon division of Vistron Corp., 12333 Van Ness Avenue, Hawthorne, CA 90250. (213) 757-5141. Call or write for distributors in your locale.

7. Private communications with the manufacturers.

8. Private communications with the manufacturers.

9. Write or call the manufacturers for details. Filon's address is given in note 6. DuPont Company is located in Wilmington, DE 19898.

10. Solar Components Corp., P.O. Box 237, Manchester, NH 03105. (603) 668-8186.

11. Available from Structural Sheets, Inc., 196 East Camp Avenue, Merrick, NY 11566. (516) 546-4868.

12. CYRO Industries, 155 Tice Boulevard, Woodcliff, NJ 07675. (800) 631-5384. [In New Jersey, call (800) 922-8032.]

13. Extruded polystyrene board insulation is not that easy to find. Do not substitute expanded polystyrene! It will not stand up to the same conditions. Dow makes an extruded polystyrene insulation sometimes referred to as "Dow blue." Marine supply companies sometimes carry blueboard, as it is used to insulate boats.

14. "Directory of Greenhouse Kit Manufacturers," *Solar Age*, Vol. 7, no. 5, pp. 35–40.

15. The "around-the-corner" configuration is depicted and discussed in Bill Yanda and Rick Fisher's *The Food and Heat Producing Solar Greenhouse*, John Muir Publications, Inc., Santa Fe, NM, 1980, pp. 112–114.

16. Ibid., p. 78.

17. The data in Appendix III were taken from the LCR tables in Balcomb et al., Vol. 3. Those LCR tables do not list sunspace shared walls, or solar walls, by the materials they are constructed of, but rather by the thermal properties of the material (specifically, ρct and ρck, where ρ is density, c is specific heat, t is thickness, and k is thermal conductivity). These thermal properties closely correspond to those of concrete in many of the cases listed in Balcomb's LCR tables. It is from a number of those cases with concrete-like properties that the sunspace LCR values in Table III-2 of this book were selected.

18. Balcomb et al., Vol. 2, p. 75.

19. This can be confirmed by studying the sensitivity data on pp. 568–569 of Balcomb et al., Vol. 3.

20. Joe Kohler and Dan Lewis, "Passive Principles: Choosing Your System," *Solar Age*, Vol. 6, No. 12.

21. Harold E. Marshall and Rosalie T. Ruegg, *Simplified Energy Design Economics*, U.S. Department of Commerce, Washington, DC, 1980, p. 16.

22. Stephen E. Pyde, "Load Management and the La Vereda Passive Solar Community," *Proceedings of the 6th National Passive Solar Conference, American Section of the International Solar Energy Society*, 1981, p. 20. In a private communication Mr. Pyde affirmed that recent data have continued to support the findings of the paper.

SUGGESTED READING

BALCOMB, J. DOUGLAS, ET AL., *Passive Solar Design Handbook*, Vol. 3. Los Alamos, NM: Los Alamos National Laboratory for U.S. Department of Energy, DOE/CS-0127/3, 1982.

LELEN, KENNETH, "A Buyer's Guide to Greenhouse Kits," *Solar Age*, Vol. 7, No. 5, May 1982.

MARSHALL, HAROLD E., AND RUEGG, ROSALIE T., *Simplified Energy Design Economics*. Washington, DC: U.S. Department of Commerce (National Bureau of Standards Special Publication 544), 1980. Available from U.S. Government Printing Office, Stock No. 003-003-02156-3. A good beginning book on life-cycle-costing methodology and other types of economic analysis.

SPEARS, JOHN W., "Goodbye, Sloped Glass," *New Shelter* (Rodale Press), Vol. 3, No. 4, April 1982.

SPEARS, JOHN W., "A New Slant in Sunspace Design," *Solar Age*, Vol. 8, No. 9, September 1983.

YANDA, BILL, AND FISHER, RICK, *The Food and Heat Producing Solar Greenhouse*. Santa Fe, NM: John Muir Publications, Inc., 1980.

REVIEW QUESTIONS

1. Define *sunspace*. How does it differ from a solar greenhouse?
2. Within what range of angles should the glazed south wall of a sunspace face?

3. Compare the characteristics of glass glazing with those of fiberglass.
4. Why do sunspaces have opaque roof sections? Why are opaque end walls popular in them?
5. How can a dirt floor in a sunspace be insulated?
6. Describe three ways that thermal energy is stored in sunspaces.
7. What other functions besides heating of the adjacent building do sunspaces often have?
8. Compare characteristics of sunspaces with those of other passive systems.
9. Define *incremental cost, life-cycle cost,* and *discount rate.*

PROBLEMS

1. The recommended projected glazing area of a sunspace is 440 square feet. Glazing width is 40 feet and total glazing height is 11 feet. This height is divided between roof and south wall heights of 2 feet and 9 feet. Roof tilt is 17° and that of the south wall is 68°. What are the lengths and areas of the roof and south-wall glazing?
2. The building load coefficient of a home in Dallas, Texas, is 7300 Btu/°F-day. What projected glazing area should a night-insulated attached shed roof sunspace have in order to provide an SSF of 80 percent? If the total height of the sunspace is 7.5 feet, what should its width be? If the height and slope of the south wall glazing is 6 feet and 70°, and that of the roof is 1.5 feet and 20°, what should their actual glazing areas be?
3. The building load coefficient for a house in Madison, Wisconsin, is 8000 Btu/°F-day. How many square feet of projected glazing area should a night-insulated, attached single-plane sunspace have to provide a solar savings fraction of 30 percent? If the sunspace's glazing slope is 50°, how many actual square feet of glazing should it have?
4. Do Problem 3 for an attached single-plane sunspace *without* movable insulation.
5. The BLC of a house in Sacramento, California, is 10,000 Btu/°F-day. What projected glazing area should a night-insulated, semienclosed shed roof sunspace have in order to provide an SSF of 60 percent? If the height and slope of the south wall glazing is 5 feet and 60°, and that of the roof is 2.5 feet and 30°, what should their actual glazing areas be?
6. The initial cost of a solar system with a backup heating system is $6000, and it is estimated that they have a useful life of 20 years. Once the solar system is installed, heating costs for the house are expected to be $350 per year to start with, and are expected to escalate at a rate of 6 percent per year, due to electricity cost increases.

   It is also estimated that without the solar system, a conventional system costing $3000 would be needed, and would use $950 worth of fuel the first year to heat the building, with costs increasing at the rate of 6 percent per year.

   Assuming an annual discount rate of 8 percent, compare the life-cycle costs of the solar and conventional systems.

7. Assume that the initial cost of a solar system is $18,000, the base-year backup heating costs are $450, the expected rate of energy price escalation is 5 percent per year, the discount rate is 10 percent, and the system lifetime is 23 years. Maintenance is expected to run $300 per year.

   The conventional heating system's initial cost is $13,000 and its base-year energy costs are $1300. Maintenance costs $105 per year. It is expected that $3000 worth of repairs on the furnace will be necessary in the twelfth year. The salvage value of the system after 23 years is $1000.

   Perform life-cycle-cost comparisons on the two systems.

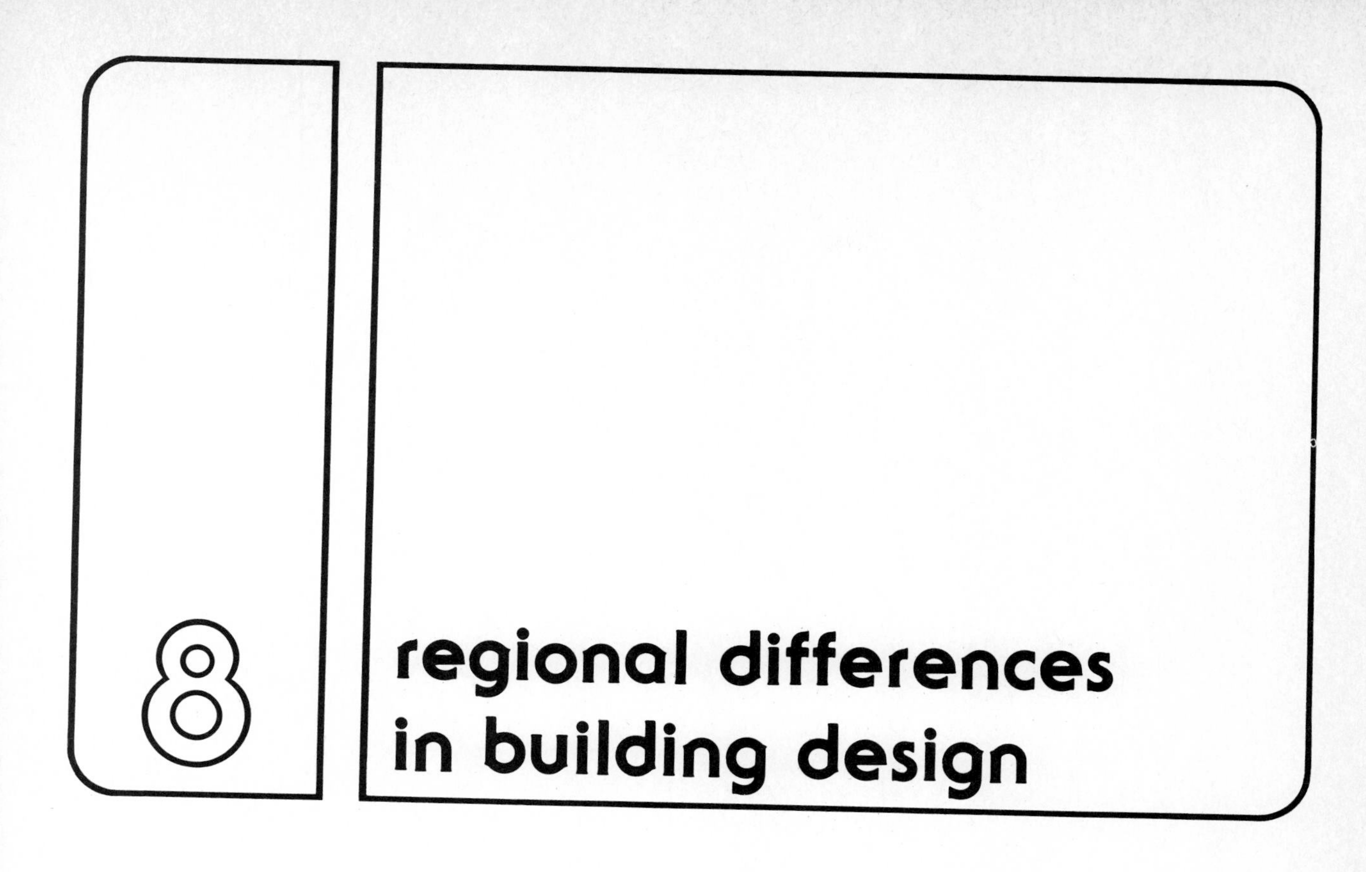

8 regional differences in building design

The preceding chapters provided guidelines for the design of solar-heated structures. Although many aspects of the design process are the same no matter where the construction site is, the radical climatic variations in different parts of North America must also be taken into account when planning a building. Climatic variations affect not only the areas of solar apertures and collectors, which are taken into account in preceding parts of this book, but also the construction methods used, the building materials chosen, and the orientation of the building.

NORTHERN INTERIOR CLIMATES

Climates in the interior regions of Canada and the northern United States are typified by extremely cold, long winters with short days. It is thus not surprising that "superhouses" are becoming increasingly popular in these areas (see Chapter 2 for a discussion of them). These buildings feature stringent anti-infiltration measures such as continuous polyethylene barriers, as well as massive amounts of insulation, multiple glazing layers in the windows, and "airlock"-type entry vestibules that inhibit energy losses each time the front door is opened.

One of the pioneering cities in the field of superhomes is Saskatoon, Saskatchewan, where both the city and provincial governments, with help from the Canadian government, have been pushing energy-efficient technology for years. Under their programs, builders in Saskatoon are constructing homes that use *75 percent less* heating fuel than standard houses in the same area.[1]

The farther north a site is, the lower the sun is in the sky and thus the more useful *vertical apertures* such as windows or clerestories are for solar

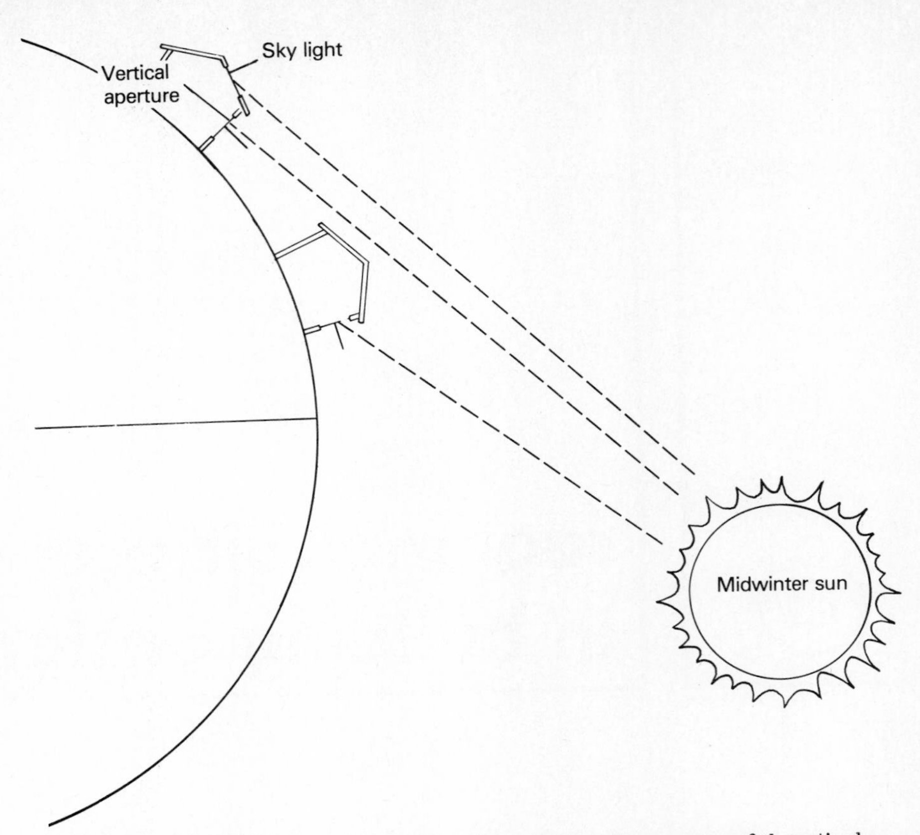

Figure 8-1 The farther north a building site is, the more useful vertical solar apertures such as windows and clerestories are as compared to skylights. This is because in northern locations, light from a midwinter sun strikes vertical apertures at angles of incidence close to perpendicular, while it strikes skylights at rather oblique angles.

collection purposes, as compared to skylights. This is because light from a low-altitude winter sun strikes vertical collection surfaces at closer to normal angles of incidence, resulting in more energy transmittance (see Figure 8-1).

PACIFIC NORTHWEST COASTAL REGIONS

Although winter temperatures in coastal Pacific Northwest areas are considerably warmer than those in northern interior regions, the short winter days and the frequent overcast and foggy conditions make heavily insulated and tightly weatherized superhouses attractive choices for this locale, too. As illustrated in Figure 8-2, western Oregon and Washington have fewer hours of sunshine in January than just about any other part of the 48 contiguous states. Because there is such a minimal amount of midwinter solar energy in this region, a building's energy conservation measures become increasingly important.

In a study by McDonald and Tsongas on Pacific Northwest residences,[2] it was found that added insulation and infiltration control was able to reduce average space-heating costs by 75 percent compared to those of conventional structures. "Conventional" structures were those following the Seattle Energy Code, requiring R-19 ceilings and floors and R-11 walls. Infiltration levels in these buildings averaged about 0.8 air change per hour. The more heavily weatherized buildings to which these structures were compared boasted R-38 ceilings, R-19 walls and floors, and 0.2 air change per hour due to natural infiltration (air-to-air heat exchangers were used to keep air quality high).

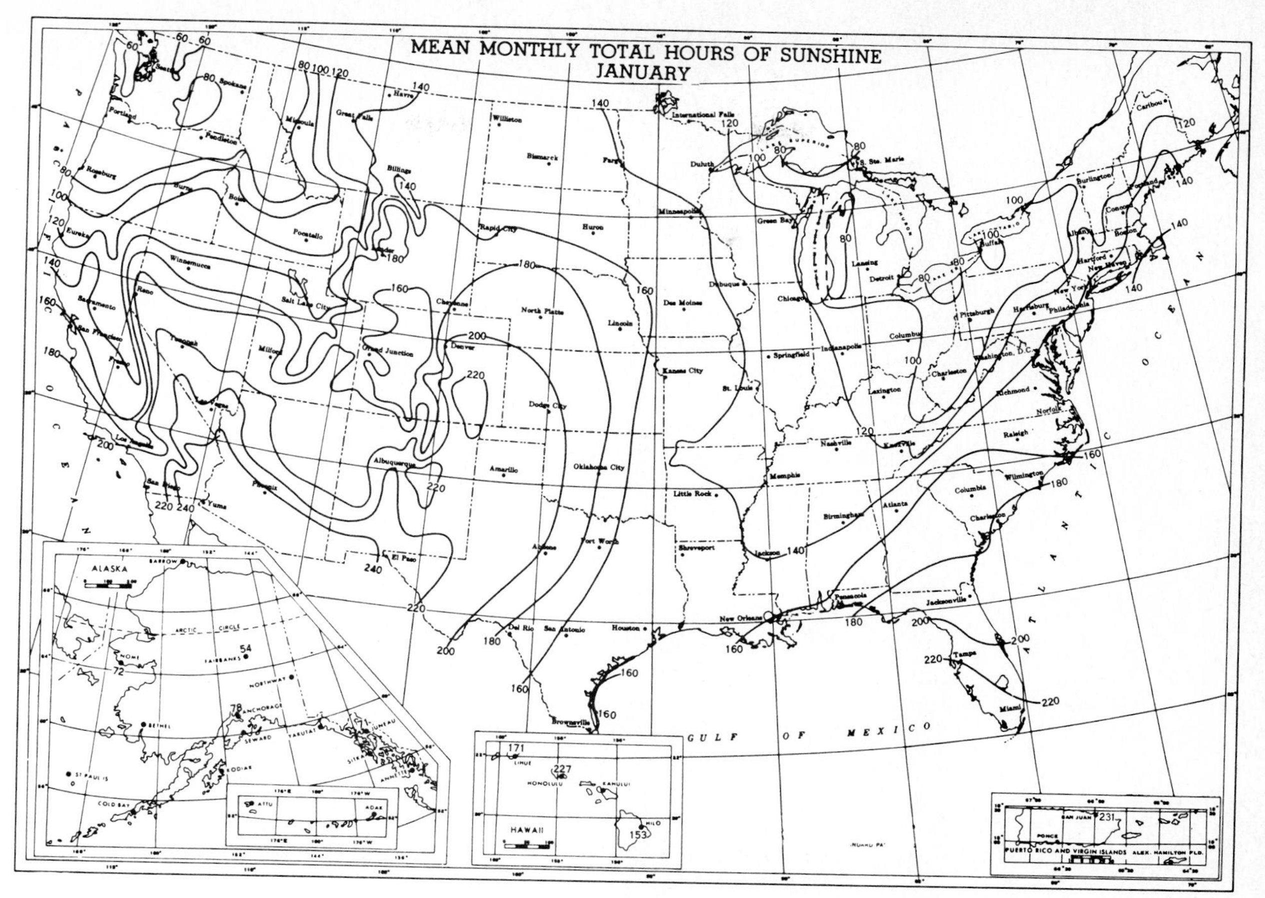

Figure 8-2 Mean monthly hours of sunshine for January. (From *Climatic Atlas of the United States*, U.S. Department of Commerce, Washington, DC, 1979, p. 67.)

Lowering the building's thermostat setpoint at night also had a dramatic impact on heating bills. In fact, of all the conservation measures examined in the McDonald study, it was the single most effective one, cutting the building's annual heat load from 32 million Btu per year to 23 million Btu.

THE SOUTH

In much of the American South, heat is needed for only about four months of the year, while for the remaining eight months, shielding solar apertures to prevent too much thermal energy from entering a building is a major concern. Thus, in southern regions proper design of window overhangs and other shading devices is of critical importance (see Chapter 5 for overhang design techniques).

Insulation and thermal mass are very beneficial in southern homes in summer as well as winter, for during warm months they help keep "coolth" inside and heat outside. Allowing cool nighttime breezes to circulate through the house lowers the mass's temperature, which helps keep the house comfortable the following day.

The humid, sticky conditions of summer can be somewhat alleviated if windows, doors, and the orientation of the building itself are engineered to make use of the prevailing breezes, which frequently blow from the south. One design that utilizes this idea employs low windows on the building's south face, high clerestories on the north side, and a roof that slopes upward from south to north. Air blowing over the roof creates a low-pressure Venturi effect near the clerestories that helps draw hot, stale air out of the house.

In colder climates, the reflective foil on building insulation is generally faced inward in order to contain heat within the structure. But in areas where cooling is of greater concern, the foil is sometimes faced outward to reflect away incoming thermal energy. Since the foil is not functional unless it faces an air space, the insulation must in this case be installed so as to leave an air space between it and the exterior siding or roofing. If this is done, a vapor barrier should also be installed between the insulation and the interior finish surface of the building to prevent moisture buildup inside the building envelope.

THE SOUTHWEST

The southwestern United States is largely a dry desert region typified by extreme temperatures, especially in summer. Buildings in this area are traditionally made of indigenous masonry materials such as adobe. In Tucson, Arizona, over 90 percent of the single-family homes are of masonry construction.[3]

Although masonry walls help greatly to stabilize interior temperatures, they are far more effective and energy conserving if they are constructed with an insulation layer. Masonry walls with insulation on their interior surfaces are popular in cities such as Phoenix, Tucson, and Yuma, Arizona. These walls perform well in summer but are not able to help heat the house in winter by collecting and storing solar energy admitted through the building's apertures. Far more preferable are walls insulated on the outside that function well at both times of the year (see Chapter 5 for a discussion of "outsulation"). The masonry in these walls gradually heats up through the course of a day, reaching its maximum temperature in late afternoon or evening, hours after the hottest outside air temperatures. In summer, this time lag is beneficial because interior air temperatures have fallen somewhat and evaporative coolers are better able to handle the heat given off by the wall (electric rates are lower at this hour of day, too). In winter, the wall gives off its maximum heat at the time when it is most needed.

Although they are energy efficient, high-mass walls tend to be more expensive than the 2 × 4 or 2 × 6 stud-framed walls that are common throughout North America. A more economical alternative to a house in which all parts of the building envelope contain heavy masonry construction is a high mass/low mass combination. In this design, high-mass walls enclose spaces such as living rooms and kitchens that are in continuous use, while less expensive walls containing little thermal storage mass enclose bedrooms and other occasionally occupied spaces that can quickly be cooled down or heated up when necessary.

Unlike in colder climates, it is not usually cost-effective in the Southwest to insulate under a concrete floor slab. This is because the average ground temperature 4 feet below grade in hot desert regions is about 70°F, which is an ideal interior temperature. The ground thus provides a source of thermal stability for the house above it.

The Arizona Solar Energy Commission recommends a minimum of R-30 insulation in the ceilings or roofs of desert buildings to minimize summer heat gain as well as winter heat loss. Any air space above the insulation blanket (such as an attic above an insulated ceiling) should be well vented to prevent heat buildup. A good rule of thumb to follow is to provide 1 square foot of vent area for every 100 square feet of attic floor space.[4]

In desert cultures in many parts of the world, thick masonry roofs are constructed on buildings that help keep the living spaces below comfortable. Water is often sprayed onto these roofs during hot weather. As it evaporates, the water carries away much excess thermal energy from the house.

A modern variation of the centuries-old high-mass roof designs that is especially suitable to the dry conditions of the Southwest is the solar roof

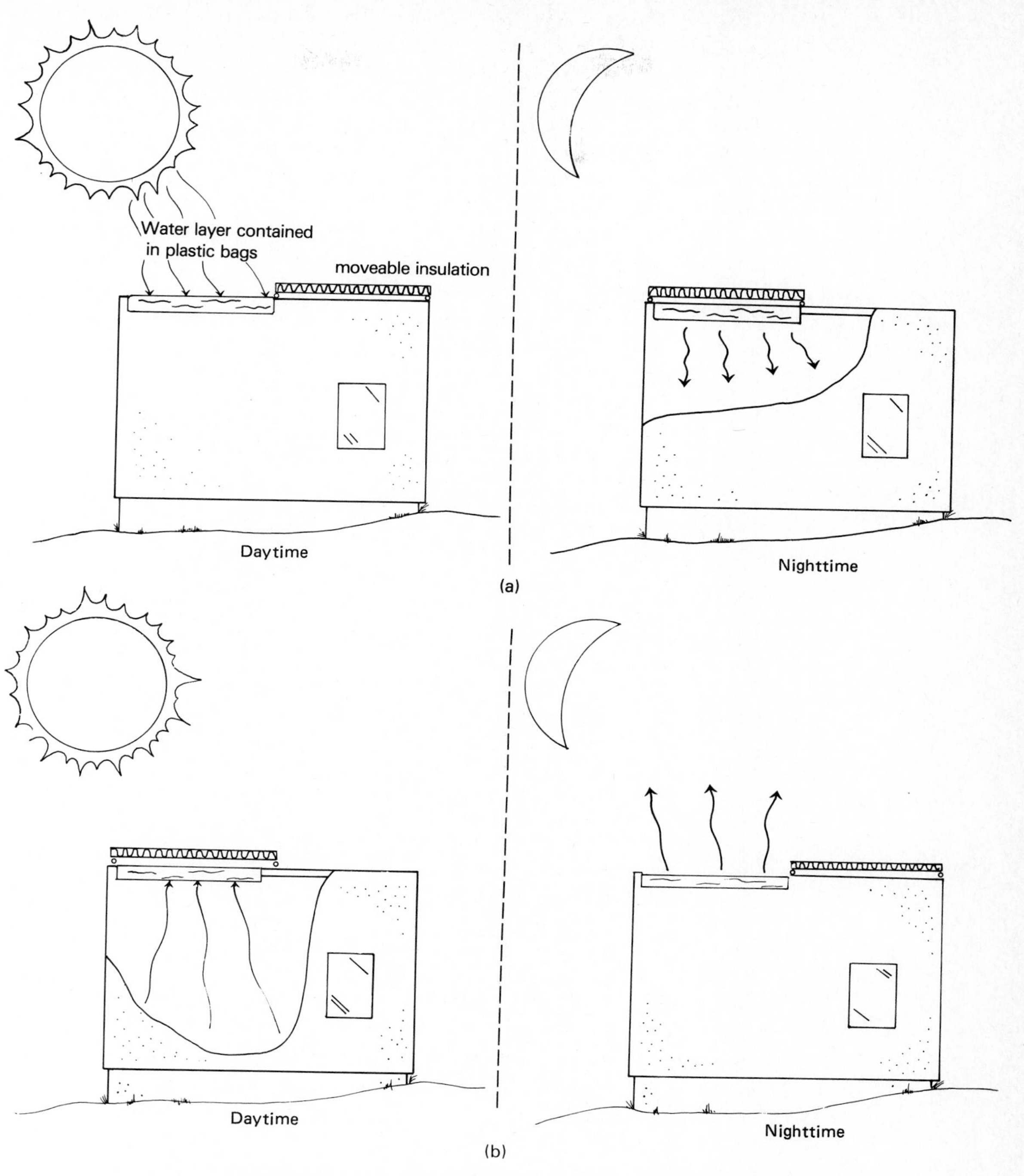

Figure 8-3 Solar roofs heat a house in winter and cool it in summer. (a) Winter conditions. Daytime: Water layer thermal storage mass is exposed to the sun. Nighttime: Movable insulation is rolled in place above the sun-warmed water. Thermal energy from the water is conducted through the roof and radiated into the house. A ceiling fan can also be employed to destratify the air. (b) Summer conditions. Daytime: Insulation protects the water layer from the sun's warmth. Meanwhile, the water is absorbing heat from the building, helping to keep it cool. Nighttime: The insulation is rolled back, allowing the water to radiate its collected heat away. Clear nights are essential for this process to occur efficiently.

depicted in Figure 8-3. The roof has an 8-inch- to 1-foot-thick water-filled layer in it resembling a water bed that can be covered when necessary by movable insulation.

During summer nights, the water bags are left exposed to the sky. The water's temperature drops considerably, not only because nights in the Southwest are typically cool but also because of the region's dependably clear skies. During clear nights, the water layer continuously radiates thermal energy away into space while receiving little in return. Its temperature can thus drop to well below that of the air. In overcast areas, the sky cover, which is of a considerably higher temperature than deep space, radiates energy back toward the solar roof and thus its temperature does not drop as much (see Figure 8-4).

Figure 8-4 During overcast summer nights, solar roofs do not lose their heat as quickly as on clear nights, for clouds radiate some thermal energy back toward the roofs.

During summer days, the cooled water bags are covered by movable insulation. Heat from the house below passes through the ceiling and is absorbed by the water.

In winter, the roof is left uninsulated during the day so that the sun can warm the water layer. In the evening, the layer is covered and the water's warmth is conducted through the ceiling and then radiated down into the living space below. Occupants of houses with solar roofs report that the comfort level, both winter and summer, is high compared to conventional heating systems.[5] Temperatures in homes of this type in the Southwest have been found to remain quite stable throughout the year.

NOTES

1. Jane Meyer and Craig Sieben, "Super Saskatoon," *Solar Age*, Vol. 7, No. 1, January 1982, pp. 26–32.

2. C. L. McDonald and G. A. Tsongas, "A Comparative Analysis of Conservation and Passive Solar Strategies in Multi-family Residences," *Proceedings of the 6th National Passive Solar Conference, American Section of the International Solar Energy Society*, 1981, pp. 237–241.

3. Jeffrey Cook, "Passive Design for Desert Houses," *Solar Age*, Vol. 6, No. 2, p. 30.

4. Ibid., pp. 32–33. To find out more about recommended desert building techniques, write the Arizona Solar Energy Commission, 1700 West Washington Street, Suite 502, Phoenix, AZ 85007.

5. Bruce Anderson and Malcolm Wells, *Passive Solar Energy*, Brick House Publishing Company, Andover, MA, 1981, pp. 70-72.

REVIEW QUESTIONS

1. What measures are recommended for solar buildings in northern interior climates?
2. Why is the reflective foil on building insulation sometimes faced toward the wall's exterior in climates with warm, long summers?
3. Describe some of the measures recommended for buildings in the dry deserts of the American Southwest.
4. Describe the operation of a solar roof system.

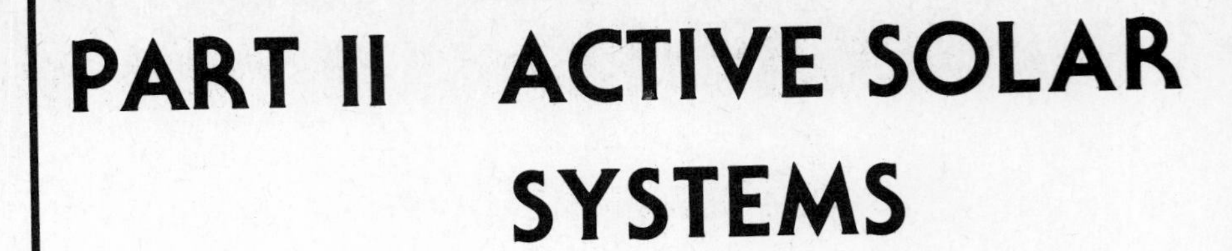

PART II ACTIVE SOLAR SYSTEMS

9 the flat-plate collector

Common to all solar systems is a means of collecting the sun's energy. This is done in a variety of ways, although in active systems the most common method is through the use of *flat-plate collectors*, so named because of their planar absorbing surface.

The first recorded use of a flat-plate collector was in the nineteenth century by the Swiss scientist Nicolas Saussure, who built a wooden box

Flat-plate collector. (Courtesy of American Solar King Corp., Waco, TX.)

with a black bottom, covered it with glass, and obtained near-boiling temperatures. The glass let in the visible part of the solar spectrum, where it was absorbed by the box's bottom and transformed into thermal energy. As the black surface heated up, it emitted infrared "heat radiation," but since glass is opaque to this part of the spectrum the energy was trapped inside the box, raising its temperature. Modern solar panels are built along much the same lines as Saussure's prototype, although improved materials and methods of construction allow higher temperatures and considerably longer collector lifetimes to be achieved.

Today's flat-plate collectors are either of a *hydronic* type, in which the medium being heated is either water or some other liquid, or of a type designed to heat air. Thermal energy absorbed by the collector is generally used to heat a building water supply, a living space, or a swimming pool or hot tub.

A collector has five basic parts: its absorber plate, the glazing, insulation, flow channels, and the case.

1. The collector's absorber plate is of a dark color so as to enhance its ability to capture solar radiation.
2. Glazing is placed in front of the absorber of all but low-temperature collectors, while insulation is installed on the sides and behind the absorber.
3. Embedded in the absorbing plates of hydronic collectors are flow channels to carry fluid from the collector's inlet to its outlet.
4. Flow channels in air collectors are formed by the spaces between the absorber plate and the insulation behind it, and/or between plate and glazing.
5. Supporting the glazing and enclosing insulation and absorber is a weatherproof case.

HYDRONIC COLLECTORS

Hydronic solar collectors are the most common types in the United States today, finding their main application in swimming pool and domestic water heating systems. There are a myriad of different designs on the market, so in order to shop intelligently it is first necessary to understand how each part of a collector works and the different methods used in their construction.

Absorber Plates

The heart of a hydronic collector is its absorber plate, whose function is to trap as much of the sun's radiation that impinges on it as possible. The plate's temperature rises as it absorbs solar energy, and this heat is then conducted to the liquid traveling through the plate's embedded flow channels.

The longer it takes for the transfer of thermal energy into the moving fluid, the more energy will be radiated away from the plate or convected away by currents in the air. These losses lower a collector's overall efficiency. To combat this, absorber plates of all but low-temperature collectors are generally made out of highly conductive metals.

The metals most commonly used are copper and aluminum. Aluminum absorber plates are less expensive but are also subject to certain corrosion problems between them and the flow channels unless precautions (discussed below) are taken during the collector's construction. Copper is not as prone to these problems and also has a higher thermal conductivity. Its extra cost is usually not that significant when integrated into the expense of an entire solar system.

Flow Channels

Copper flow channels are usually used in solar collector arrays designed for carrying potable water. If the array carries a medium such as propylene glycol or silicone oil, however, other metals such as aluminum can be employed.

Flow channels are connected to absorber plates in a variety of ways, one of which is by being pressed tightly into form-fitting grooves in the absorber. If small air gaps between plate and flow channel develop, though, heat flow will be greatly impaired and collector efficiency will plummet. To hold the flow channels more firmly in place, riveted clamps are sometimes used at intervals of several inches. In many owner-built collectors, thermally conductive cements such as Chemax Corporation's Tracit 1100[1] is applied between flow channels and absorber to fill any gaps and thus improve heat flow. A stronger and longer-lasting method of bonding is to solder, braze, or weld the flow channels into place.

If flow channels and absorber plate are of dissimilar metals, such as copper and aluminum, the danger of galvanic corrosion exists. Keeping the absorber plate totally dry by enclosing it in a tight, weatherproof case will help prevent this. Also used are thin "dielectric" layers between the two metals, such as certain paints or deposited coatings. The thermal cement mentioned above can be used as a dielectric layer.

Another problem that arises when dissimilar metals are used is severe strain on the flow channel-absorber plate bond due to different expansion rates as the collector heats up. For these reasons, totally copper absorber plate–flow channel combinations are very popular in today's collectors. One of the most reliable methods of bonding copper flow passages to copper absorbers is through the use of a "high-frequency resistance welding" technique. In this method, 400,000-hertz (cycles per second) electric current is passed through the surfaces to be joined, melting a 4-mil-thick layer on each surface. During the welding, pressure is also applied in order to improve the bonding. Any oxides that have formed on the surfaces are squeezed out from between the flow passages and absorber, together with a small amount of melted copper. Thermatool Corporation,[2] a large manufacturer of absorber plates, is confident enough of their manufacturing process to offer a 30-year guarantee on the integrity of their high-frequency resistance welds.

Flow Channel Patterns

Flow channels are usually laid out in one of two patterns: the parallel riser tube configuration and the serpentine configuration (see Figure 9-1). Parallel risers have the advantage of offering less resistance to flow than that of serpentines and are most commonly used. Circulating pumps do not have to work as hard or expend as much energy when pumping liquid through them. Serpentine channels are sometimes used for higher-temperature applications, for fluid travels a greater distance passing through such a configuration and thus comes in contact with more heated metal.

Flow channels are typically ½ inch in inner diameter, except for the headers in parallel riser designs, which are usually larger since they have to carry more water. (Headers are the passages on each end of the collector that transport fluid to and from the riser tubes.) Headers on collectors made for space heating and domestic hot water applications average about 1 to 1½ inches in diameter. Swimming pool heating collectors employ larger headers to handle the higher flow rates. Fafco solar panels, for instance, use 2⅜-inch-diameter headers.

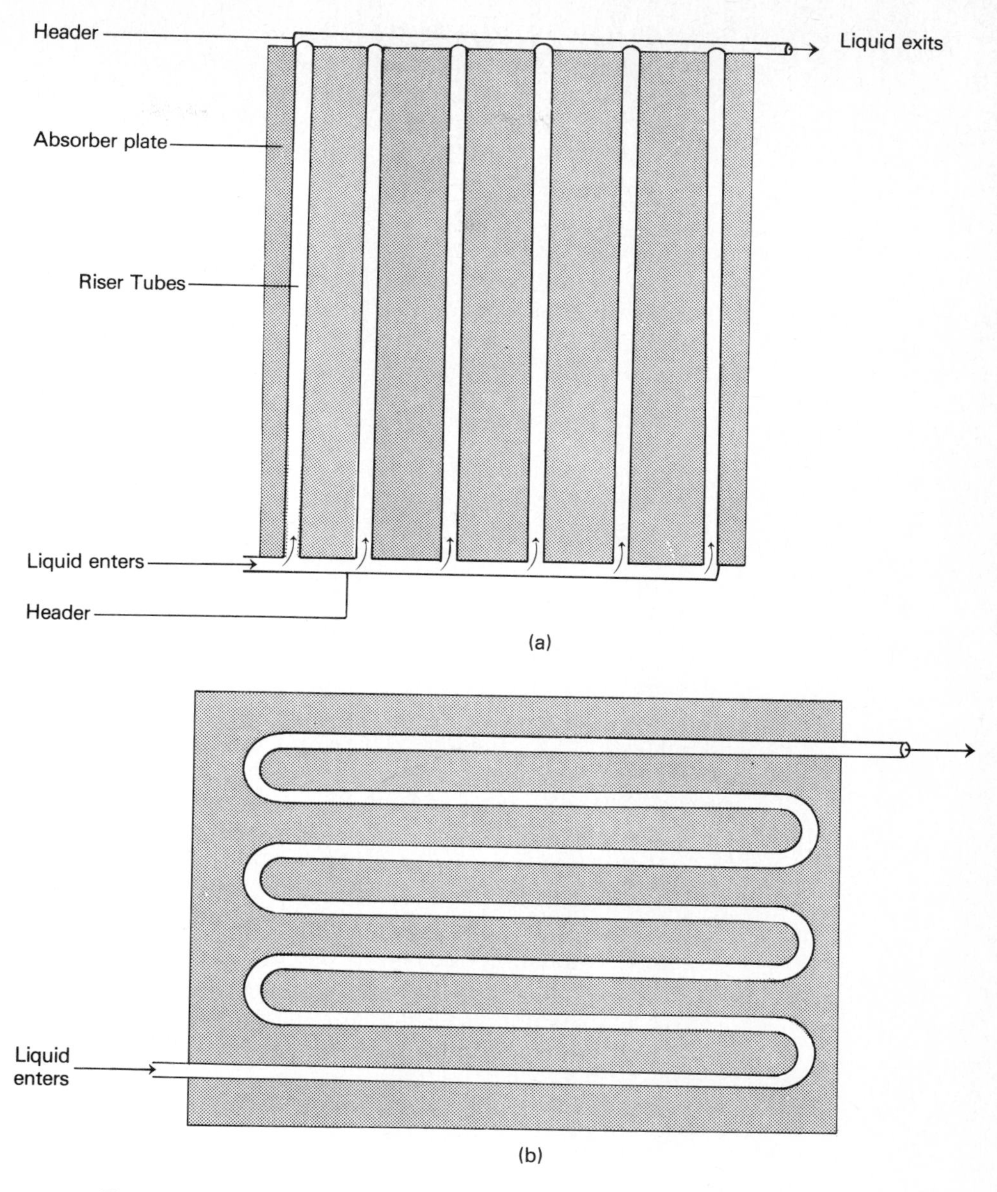

Figure 9-1 Flow channel patterns. (a) Parallel riser configuration. (b) Serpentine configuration.

Absorber Coating It was mentioned in the "Science of Heat" section of Chapter 1 that black materials absorb more incident solar radiation than do lighter materials. Since the function of an absorber plate is to collect as much energy as possible, it is usually of a black or nearly black color. As incident sunlight is captured, it is converted from radiant to thermal energy, which is then carried away by the heated water or other fluid to a storage tank.

The most common and least expensive absorber coating is flat black paint. It is better than glossy black, for shininess indicates higher reflectivity and thus less absorption of incident radiation. If the absorber surface is properly cleaned before applying, a baked-on layer of paint will last 20 years or more.

A flat black surface absorbs 95 percent of normally (perpendicularly) incident solar radiation. Its disadvantage, though, is that it *emits* large amounts of energy as the absorber warms up. The ideal coating would be one that absorbs a large amount of energy but in some way holds onto it as temperatures rise. *Selective surfaces* are coatings that do exactly that. As you will recall from our discussion in Chapter 6, they absorb solar energy almost as well as flat black paint but emit much less infrared energy than that emitted by flat black surfaces.

The difference between the efficiencies of selectively coated absorber plates and those with conventional coatings gets greater and greater as absorber temperatures rise. The added expense of selective coatings is generally cost-effective for applications such as domestic water heating, but they are not needed for low-temperature heating systems such as those for swimming pools.

Glazing

The function of the one or more layers of glazing in a solar collector is to let energy in but allow as little of it as possible to escape. As mentioned in Chapter 1, glass is transparent to the visible portion of the solar spectrum—the most intense part—but opaque to most of the infrared. The sunlight that reaches the absorber plate raises its temperature and causes it to emit infrared radiation that is contained within the collector by the glazing. This leads to high interior collector temperatures that could not be attained in unglazed solar panels. Without glazing, energy from the absorber plate would not simply be radiated away; it would also be convected away by currents in the air around it. In unglazed panels for swimming pool systems, in fact, this is a major source of heat loss.

Glazing is very useful in reducing thermal energy losses, but it is far from perfect. Infrared radiation emitted by the absorber plate cannot pass through glass but it can be absorbed by it, raising the glass's temperature. Air warmed from contact with the absorber eventually passes some thermal energy on to the glazing as well. Conduction carries this heat to the glass's outer surface, where convective and radiative processes carry it away.

The higher the collector temperature, the faster heat loss occurs. To reduce it somewhat, multiple layers of glazing with dead air spaces between them are sometimes used. Table 9-1 lists the thermal resistance values for single, double, and triple layers of glazing. Notice that a second layer increases the thermal resistance by about a factor of 2. The third layer increases it by a factor of 3 to 3.5, compared to single glazing.

TABLE 9-1 Thermal Resistance Values for Single, Double, and Triple Glazing

	R-Value
Single glass	0.91
Double glass	
With ¼-in. air space	1.69
With ½-in. air space	2.04
Triple glass	
With ¼-in. air spaces	2.56
With ½-in. air spaces	3.23

Source: ASHRAE Handbook—1981 Fundamentals, American Society of Heating, Refrigerating and Air-Conditioning Engineers, Inc., Atlanta, GA, 1981. Calculated from Table 8, pg. 23.28.

Although multiple glazing does reduce heat loss, it also cuts down on the amount of energy admitted into the collector, as well as significantly increasing the cost of the unit. It is the opinion of many flat-plate-collector manufacturers that the energy savings due to multiple layers of glass rarely counterbalances its added expense. Selective surface coatings on the absorbers are a cheaper yet quite effective way of reducing energy losses. Black chrome selective surfaces are standard, for instance, in Grumman Energy Systems[3] solar panels, but of the many thousands of collectors Grumman sells each year, very few are double-glazed and must usually be specially ordered. American Solar King[4] is another large manufacturer that has all but abandoned double glass layers for most space and domestic water heating applications.

Glass solar collector covers, whether single- or double-layered, must be tempered to prevent fracturing due to large temperature variations between outside and inside surfaces, or between different parts of the collector. The glass must also have high light transmissivity. Ordinary window glass is not the best for solar collectors due to its high iron oxide content of 0.10 to 0.13 percent. Iron oxide is added to glass for cosmetic reasons, but unfortunately it has the side effect of reducing transmissivity. Today's collectors generally use specially manufactured low-iron glass. Also beneficial, when available, is glass that is surface etched to reduce reflectivity.

Table 7-2 listed properties of a number of commonly used glazing materials. Notice that the iron oxide content and thickness of a layer of glass affect the percentage of solar energy lost through absorption but have little effect on reflective losses, which are much more dependent on surface preparation of the glazing.

In addition to glass, plastic glazings of various types are also used on solar collectors. Each glazing has its strong points. Glass can span greater distances than plastics and withstand far hotter temperatures. A further advantage of glass is that it presents an aesthetically neater appearance than that of many plastics, which tend to sag and ripple.

A big advantage of plastic glazing is that it is resistant to shattering, either through rough handling of the collector during shipping and installation or because of acts of vandalism such as thrown rocks. Plastic solar collector glazings are also generally lighter in weight than glass and cheaper in price, although due to the rising cost of petroleum products, price savings are less than they used to be.

Plastic glazings are manufactured in a wide range of thicknesses. *Thin films* such as Dupont's Tedlar are typically only a few mils thick (1 mil equals one thousandth of an inch), whereas polycarbonate glazings such as Lexan average ⅛ to 3/16 inch. Thin films are not generally used by themselves in commercial collectors, for although they have high transmissivity of sunlight, they are fairly fragile and can be torn by strong winds. They also allow far more infrared radiation to escape from the collector than do glass or other plastic glazings. Notice from Table 7-2, for instance, that a 1-mil layer of Teflon transmits *58 percent* of the infrared spectrum between 3 and 50 micrometers wavelength, which includes most of the infrared energy emitted by a solar collector absorber plate. Glass passes only 2 percent of this energy. In spite of their limitations, thin films are popular in owner-built collectors, for they are easy to handle and cut to size, are less expensive than many other glazings, and can quickly be repaired or replaced if damaged.

Tedlar is a thin polyvinyl fluoride film that stands up quite well to continuous exposure to ultraviolet radiation and is employed in a variety of solar energy applications. When coated onto a thicker, more durable glazing, many of the problems that thin films are prone to when used alone can be circumvented. Lascolite Crystalite-T, for instance, is a fiberglass-reinforced polyester material (abbreviated FRP) that is coated with a Tedlar layer on the side facing the sun. The Tedlar layer shields the FRP from ultraviolet radiation, which would quickly cause it to deteriorate, while the FRP provides strength and reduces infrared energy escape. Lascolite can withstand continuous collector operating temperatures up to 180°F, while it is recommended that if Tedlar is used by itself, it not be exposed to absorber plate temperatures above 150°F.

Filon, another manufacturer of FRP material, uses Tedlar on some of its greenhouse and sunspace glazing panels but a proprietary gel-coated surface called Filoplate for high-temperature flat-plate-collector applications. The Filoplate resin is designed to polymerize to the resin in the core of the panels, providing extra durability and preventing delamination of the surface film. Filoplated Solar-E panels have withstood rigorous desert testing in which the daytime temperature of the FRP was held at an average of 189°F for a full month. The only apparent change in the panels was a transmissivity loss of 1.29 percent.[5]

Table 7-2 includes energy transmission figures for a variety of glazings. The figures given are for sunlight that is normally (perpendicularly) incident on the glazing. But remember that as the sun moves through the sky in the course of a day, its angle of incidence with respect to the collector varies and thus so does the percentage of energy transmitted (see Figure 9-2). This percentage changes little between angles of incidence of 0 and 50°, but at greater angles it drops sharply as reflection increases.

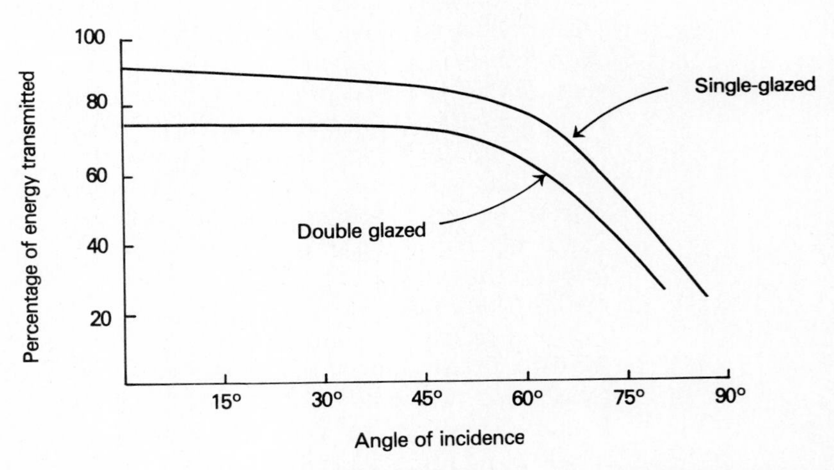

Figure 9-2 Energy transmitted through glazing versus angle of incidence of insolation.

Insulation

Both fiberglass and foam insulations are commonly used for flat-plate solar collectors. Foam insulations have higher R-values per inch of thickness, which means that less is required. But foams have a tendency to outgas vapors that can cloud the collector glazing and reduce its transmissivity. Edges of the absorber plate as well as inside walls of the collector case must thus be tightly sealed to prevent gases from the insulation behind them from reaching the glazing.

Certain foams, such as polystyrene, break down at temperatures within the range of many glazed flat-plate collectors. One of the few readily available foams that is able to withstand high temperatures is polyisocyanurate, a modified polyurethane insulation. It is sold under a number of different brand names,[6] and can withstand temperatures up to 400°F, which is above the range of collectors with flat-black-coated absorbing plates but *within* the range of some selectively surfaced collectors (temperatures as high as 450°F have been measured under stagnation conditions in these types of collectors).[7]

Fiberglass insulation is stable to temperatures well above the flat-plate-collector range. A gluelike substance called *binder* is often added, however, to insulations manufactured for use in walls. The binder helps to give the insulation shape but it also can outgas at high temperatures, clouding and reducing the transmissivity of the glazing. Any fiberglass insulation used in a solar collector should be of low or zero binder content.

The main disadvantages of fiberglass are that it takes up more space than does a layer of foam with an equivalent insulating value and that it is readily compressed, which means that more structural support must be added for its use than that required for foams. Fiberglass-insulated collectors must thus have somewhat larger, heavier, and more expensive cases than those using foam. Some solar panels, such as the Optimal series of the Sunbox,[8] use a combination of fiberglass and foam in different parts of the collector.

No matter what type of insulation is employed, a sufficient amount should be installed in the space behind the collector absorber plate to bring the thermal resistance up to a value of R-8 to R-11. This is especially important if the collector has a selective surface or is double glazed. This means that 1 to 1½ inches of polyisocyanurate foam or 2½ to 3½ inches of fiberglass are required. The sides of the collector should have about half this amount of insulation.

The Case

The collector case provides a weathertight covering for absorber plate and insulation and acts as a support system for both these elements as well as the glazing. Together with insulation and glazing, the case reduces convective and radiative losses from the collector. In commercially manufactured solar panels, cases are usually made of aluminum, and sometimes of steel or plastic. Owner-built collectors generally have wooden cases (see Lab Project A in Appendix I).

It is important that the collector case provide room for expansion of the glazing and that the gasket around the glass be made of materials that will not degrade with high temperatures or exposure to UV radiation. EPDM rubber has been used for years for this purpose and with good results.

Efficiency Studies

The efficiency of a solar collector can be defined to be the percentage of the energy striking it that is carried away to a storage tank by the fluid in the flow channels. For design purposes, efficiencies of various collectors are determined from graphs prepared by the manufacturer, or better, by an independent testing organization such as NASA (National Aeronautics and Space Administration) or DSET (Desert Sunshine Exposure Tests, Inc.). Figure 9-3 is an example of an efficiency graph for a single-glazed collector made by American Solar King. On the graph's horizontal axis is a quantity called the "fluid parameter," which is the ratio of the temperature difference between collector inlet and outside air to the insolation received (in units of Btu/hr-ft^2):

$$\text{fluid parameter} = \frac{\begin{array}{c}\text{collector inlet}\\ \text{temperature}\end{array} - \begin{array}{c}\text{outside (ambient)}\\ \text{air temperature}\end{array}}{\text{insolation}}$$

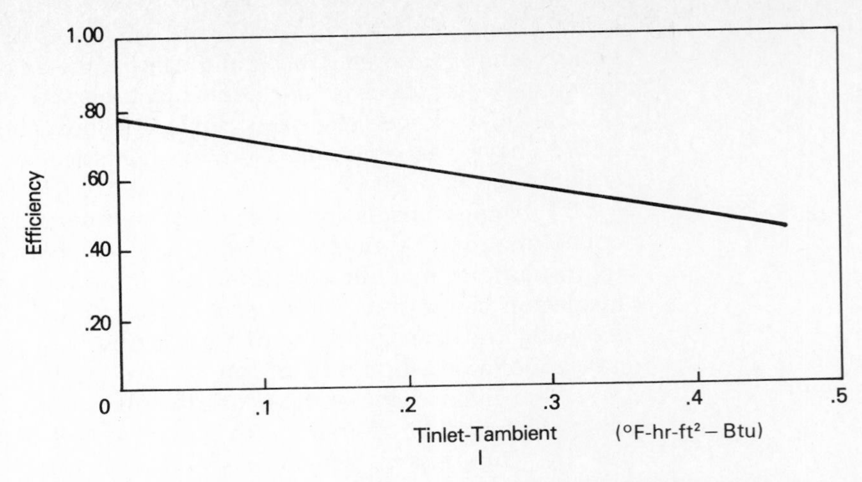

Figure 9-3 Efficiency graph. (Courtesy of Grumman Energy Systems Company, Melville, NY.)

As an example of how efficiency graphs are used, suppose that the fluid parameter for a certain system is calculated to be 0.2. The graph in Figure 9-3 then tells us that the collector's efficiency is 62 percent. This means that 62 percent of the energy striking the solar collector is actually carried away by the fluid in its flow channels. The other 38 percent of the energy either never gets through the glazing or, if it does, is lost in another of the ways discussed earlier in the chapter.

Efficiency curves are useful for calculating the necessary square footage of a solar collector array, as well as for comparing the performances of different collectors. But before the comparison is made, it is important to make sure that all graphs use the same parameters. Tests carried out in accordance with ASHRAE (American Society of Heating, Refrigerating and Air-Conditioning Engineers) guidelines use the collector inlet temperature and the collector's gross area in their calculations, while NBS (National Bureau of Standards) testing employs the *average* collector temperature and the *effective* area.

The efficiency graphs in Figure 9-4 are for two collectors manufactured by Grumman Energy Systems.[9] The Sunstream 332A is single-glazed, while the Sunstream 432A has double glazing. Both absorber plates are coated with black chrome selective surfaces. The efficiency graphs were made up in accordance with ASHRAE guidelines. In order to interpret the graphs, it is helpful to understand how the fluid parameter on the horizontal axis is related to the climatological conditions of the area. Notice that the colder the temperature, the higher is the fluid parameter. This parameter also grows larger as the latitude of the site increases (and clear-day insolation decreases). Notice also that the fluid parameter varies throughout the day as conditions change. For instance, if the ambient temperature is 55°F, collector inlet temperature is 110°F, and the collector receives 300 Btu/hr-ft^2 insolation, the fluid parameter is 0.18, a value typical of mild climate areas in the middle of a winter day. If the ambient temperature is 25°F, collector inlet temperature 110°F, and insolation on the collector 265 Btu/hr-ft^2, the fluid parameter is 0.32, more typical of a cold climate area during the middle of a clear winter day.

To analyze various collectors' performances for a given area, determine typical clear-sky midday winter fluid parameters for that area and compare efficiencies of the collectors (to be more thorough, the fluid parameters can be determined for several times throughout the day, such as 9 A.M., 12 noon, and 3 P.M.). Table 9-2 can be of help in determining clear-sky insolation values.

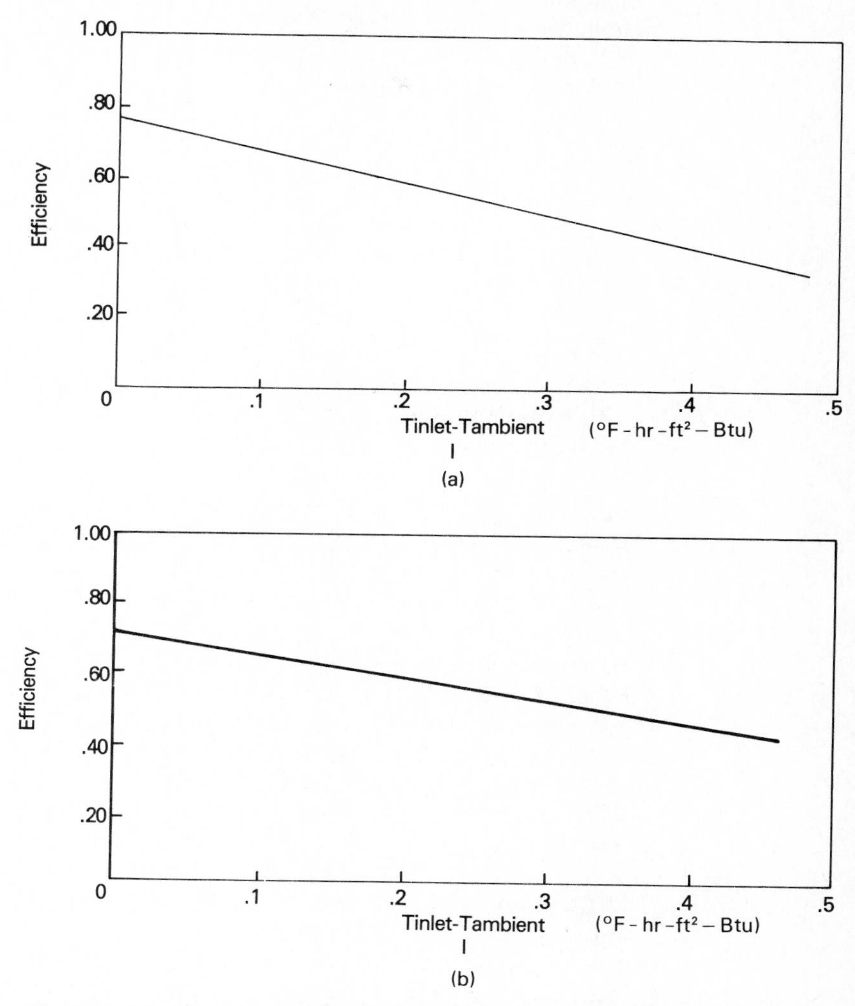

Figure 9-4 Efficiency graphs. (a) Grumman Sunstream 332A. (b) Grumman Sunstream 432A. (Courtesy of Grumman Energy Systems Company, Melville, NY.)

TABLE 9-2 Clear-Sky January Insolation Values on a Collector Surface (Btu/hr-ft²)

Date: January 21
Time: 12 noon

Latitude of Site	*Collectors Facing South and Oriented at a Tilt Equal to:*		
	Latitude − 10°	*Latitude*	*Latitude + 10°*
24°N	296	319	332
32°N	285	308	321
40°N	270	291	303
48°N	245	264	275
56°N	198	214	222

Example Calculation. Middle-of-the-day winter temperatures at a certain building site are typically 50°F. The site's latitude is 40° north. The collectors are to be mounted at an angle of 50°. Compare the efficiencies of the single-glazed Grumman Sunstream 332A collector in Figure 9-4a with the double-glazed 432A in Figure 9-4b. Assume that the collector inlet temperature is 110°F.

Table 9-2 tells us that the insolation on the collector surface is 303 Btu/hr-ft^2. This leads to a fluid parameter of 0.20:

$$\frac{110 - 50}{303 \text{ Btu/hr-ft}^2} = 0.20$$

The efficiency of the single-glazed collector for this fluid parameter is 58 percent, while that of a double-glazed model is 57 percent.

If the parameter were 0.3, the efficiencies would be 49 percent and 51 percent for the single- and double-glazed panels, respectively. For a parameter of 0.35, the efficiencies would be 45 percent and 49 percent. These percentages illustrate that as the fluid parameter increases, the collector efficiency drops, because higher fluid parameters indicate greater temperature differences between collector and outside air, and heat loss is proportional to this temperature difference. The efficiencies also indicate something about the effect of multiple glazing. Notice that at the low fluid parameters indicative of mild climates, the single-glazed collector is actually more efficient than the double-glazed version. This is because the decrease in energy transmissivity due to the second glazing outweighs its added insulating value. But at higher fluid parameters, the double-glazed collector becomes slightly more efficient. The selective surface on the absorber plates of the two Grumman panels deserves some of the credit for keeping the performance of the single-glazed collector so close to that of the double-glazed one. If flat black paint rather than black chrome were used on the absorbers, they would radiate away far more energy and the extra insulating value of the second pane of glass would be somewhat more useful.

The preceding section described a method of comparing collector performances based on their efficiency graphs, which can be obtained from most collector manufacturers or distributors. It might be wondered why, since selection of the right collector is important to the design of the system, someone has not done extensive comparative testing of the panels on the market. The answer is, someone already has. That "someone" is the Solar Rating and Certification Corporation (SRCC) of Washington, DC. SRCC has examined hundreds of different collectors and published a detailed directory of the results, which is available from them at a very reasonable price (see the "Suggested Reading" at the end of this chapter for the address). For each collector in the directory, thermal performance data are listed corresponding to a variety of collector temperatures and insolation conditions. The ratings are in terms of thousands of Btu collected per day per panel.

When a collector is certified and rated by SRCC, the manufacturer is required to affix to each panel an SRCC label bearing the thermal performance test results. SRCC ratings are analogous to Btu ratings of a house's heating furnace or water heater. An increasing number of collectors bear these ratings.

If you don't have the SRCC directory, the distributors or manufacturers of the collector you are considering purchasing should be able to furnish it, assuming that the panels have been tested and certified by SRCC. Sample ratings are illustrated in Figure 9-5. The three daily insolation values on the top line (2000, 1500, and 1000 Btu/ft^2) correspond roughly to clear,

COLLECTOR A

SOLAR COLLECTOR STANDARD

100-81

Category ΔT (°F)	COLLECTOR RATING NUMBERS Thousands of BTUs per Day per Panel		
	2,000 BTU/ft²	1,500 BTU/ft²	1,000 BTU/ft²
A (-9)	34.39	25.98	17.59
B (+9)	31.97	23.55	15.17
C (+36)	27.96	19.54	11.39
D (+90)	19.89	11.81	4.63
E (+144)	-	-	-

COLLECTOR B

SOLAR COLLECTOR STANDARD

100-81

Category ΔT (°F)	COLLECTOR RATING NUMBERS Thousands of BTUs per Day per Panel		
	2,000 BTU/ft²	1,500 BTU/ft²	1,000 BTU/ft²
A (-9)	59.06	44.58	30.18
B (+9)	54.55	40.08	25.67
C (+36)	46.87	32.56	18.63
D (+90)	31.25	18.00	5.87
E (+144)	-	-	-

Figure 9-5 SRCC rating numbers. (Courtesy of Solar Rating and Certification Corporation, Washington, DC.)

mildly cloudy, and cloudy skies, respectively. The numbers in the left-hand vertical column (—9, +9, +36, +90, and +144) are Fahrenheit temperature differences between the ambient and the collector inlet. For most space heating and domestic water heating systems, the C (+36) and D (+90) lines should be the most useful.

Economic Considerations Comparing the efficiencies and Btu outputs of various collectors on the market gives us some useful information as to each one's performance but does not tell us which one is most economical to buy. One way of answering this question is to divide the Btu output of each panel by its cost per square foot, yielding an "energy collected per dollar invested" figure.

Consider the collectors in Figure 9-5. Assume that collector *A* sells for \$445 per panel, while *B* sells for \$790. Using the clear-sky insolation value (2000 Btu/ft^2) and the D (+90) rating (corresponding to a 90°F temperature difference between collector inlet and ambient), we obtain energy outputs of 19.89 thousand Btu for collector *A* and 31.25 thousand Btu for collector *B*. The energy per dollar ratios are thus as follows:

$$A: \quad \frac{19{,}890 \text{ Btu}}{\$445} = 44.70 \text{ Btu/dollar}$$

$$B: \quad \frac{31{,}250 \text{ Btu}}{\$790} = 39.56 \text{ Btu/dollar}$$

Collector *A* clearly produces more energy per dollar of cost. Remember, however, that a Btu per dollar comparison is only one factor in choosing the right solar collector for a certain application. Other considerations, such as type of glazing, absorber plate construction, type of insulation used, and so on, are also very important. The preceding sections of this chapter should be studied carefully before a decision is made.

A Low-Cost Hydronic Collector An alternative to building collectors out of materials such as copper and low-iron glass is to use much less expensive substances such as thin plastic films. As mentioned above in the glazing section, thin films often deteriorate rapidly, so in commercial collectors films are generally coated onto a stiffer FRP layer to increase their durability (and, unfortunately, their expense). It is the belief of engineers at Acme Solar Works,[10] however, that the rapid degradation of thin films has not been due mainly to intolerance of UV radiation and high collector temperatures, as it has been assumed for a number of years, but instead to the constant flexing and flapping of the plastic in the wind. This type of continuous stress "work hardens"[11] the plastic and embrittles it. What Acme has done is to develop a thin-film glazed collector in which the Tedlar outer glazing is kept unwrinkled and bowed slightly outward by a positive air pressure behind it, created by a small fan. So far, in several years of testing under wind conditions as high as 190 mph, the pressurized film glazings are holding up.

What makes the Acme collector inexpensive is that not only the glazing, but also the *absorber* and *flow passages*, are made of thin films. Their solar panel consists of four film layers: the outer Tedlar glazing, two inner films between which the water flows, and a rear outer film (see Figure 9-6).

The two inner films are made from a modification of a polyimide material called Kapton.[12] Kapton's typical use in the past has been on wiring and printed circuit boards in the electronics industry. Kapton is able to withstand high temperatures; in testing of the panels, it survived stagnation conditions of 290°F. The upper layer of Kapton is coated with a semiselective black surface, and acts as the collector's absorber, transferring heat to the water underneath. The absorber is "100 percent wetted," which means that the water comes in contact with all of its surface, and so, being only 1/1000

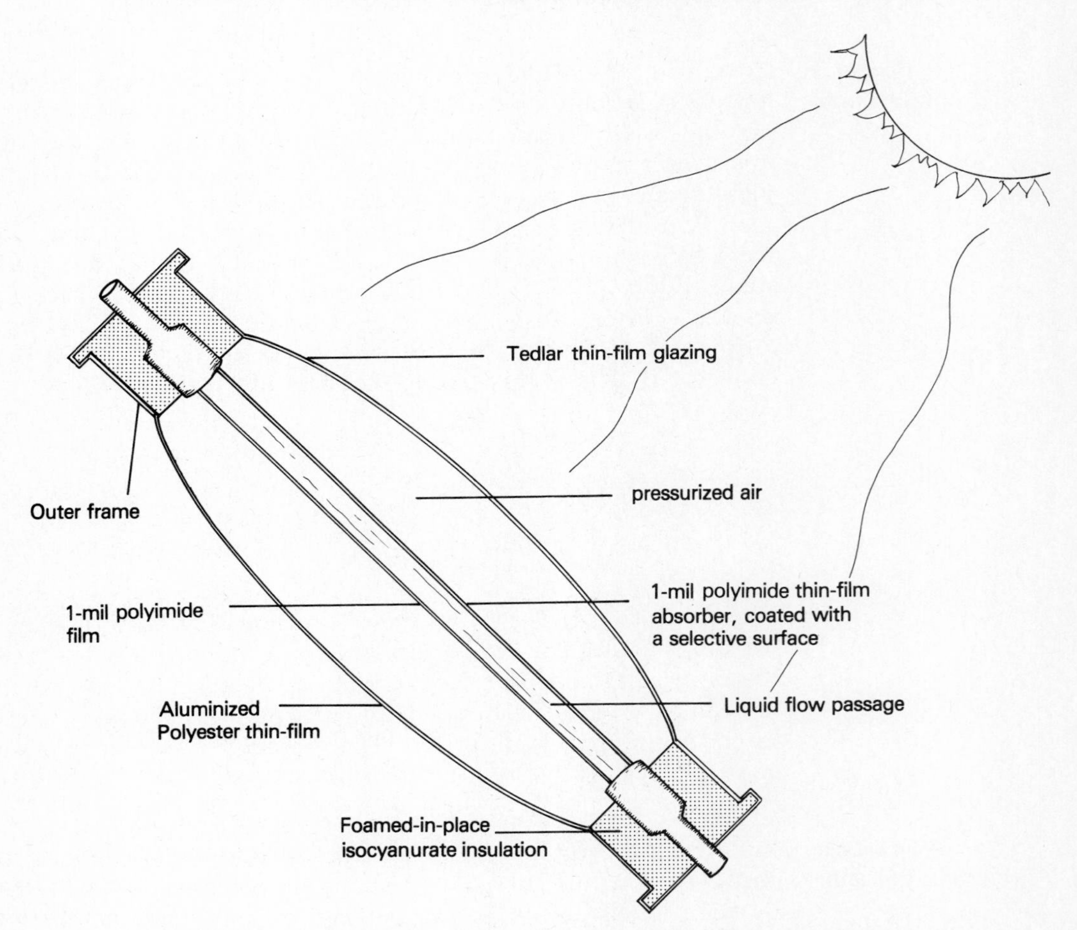

Figure 9-6 Volkspanel. Volkspanels consist of four thin film planes. Liquid flows between the inner two planes. Pressurized air keeps the two outer films taut, helping to prevent wind damage.

inch thick, absorbed solar energy is transmitted quickly and efficiently through the plastic into the water.

Because there is such little mass in the collectors, they heat up very rapidly. Acme's claim, supported by DSET test results, is that its panels are more efficient than collectors with flat black copper absorbers, but less efficient than those with black chrome selective surfaces. The 1984 cost of the panels to distributors is a very low $5 to $6.50 per square foot, depending on the quantity ordered. Because of the low price, and the belief that the panels will eventually be accepted into widespread use, Acme has named their product the Volkspanel Model T.

Volkspanels are designed for on-site replacement of the thin films should any of them degrade. The big question with this as well as most other new products is: Is it really as durable as the manufacturer claims? Although the Volkspanel has undergone DSET's 30-day stagnation test with no measurable degradation, and has received an SRCC rating, it is still new on the scene and must be looked at with a certain amount of skepticism until it has withstood the test of time. What is certain, however, is that the solar industry will benefit greatly if it can develop radically less expensive components and materials, and thus products such as the Volkspanel should be strongly encouraged.

AIR COLLECTORS

The hydronic collectors that we have been studying are used mostly for domestic hot water and swimming pool applications. Air systems, however, are often the wiser choice for space-heating purposes. Air collectors can be fabricated on the building site itself and can cost far less per square foot than most hydronic ones. Air systems are not subject to freezing problems and they continue to perform well even with minor leaks that would be unacceptable in a hydronic system. Except possibly for the circulating fans, the components of an air system should outlast those of a hydronic system.

Design considerations for the glazing, case, and insulation of an air collector are similar to those for the hydronic type. It is in absorber plate design that the differences between the two collectors are most marked.

Air collector absorber plates are simpler than those for hydronics, in that they have no need for liquid flow channels. Air flows between the absorber plate and the insulation behind it, and/or between the glazing and absorber plate (see Figure 9-7). When air flows on both sides of the absorber, both its front and back can transfer heat. This is a very efficient design for situations in which the air is no more than 30 to 40°F above ambient temperatures (those of the outside air). But in higher-temperature situations, air flowing between the absorber and a single layer of glazing would lose much of its heat to the relatively cool glass. To circumvent this problem, collectors made for high-temperature use should have the flow passage only between the absorber plate and the insulation behind it.

The absorber plate should be designed so as to induce turbulence in the air flowing across it. Turbulence is a situation in which the flow lines of a fluid are twisted, corkscrewed, and irregularly distorted (*fluid*, remember, is a term that refers to either a gas or a liquid). A turbulent fluid is constantly being mixed and tumbled. In a solar air collector, more air is brought into contact with the heated absorber plate in turbulent flow than in smooth, orderly "laminar" flow, for in laminar flow only a thin boundary layer actually touches the absorber. The rest of the air passes through the collector almost unheated.

The irregular *eddies* (whirlpool-like currents) of turbulent flow can be of all sizes. Large irregularities in the absorber plate such as fins or corruga-

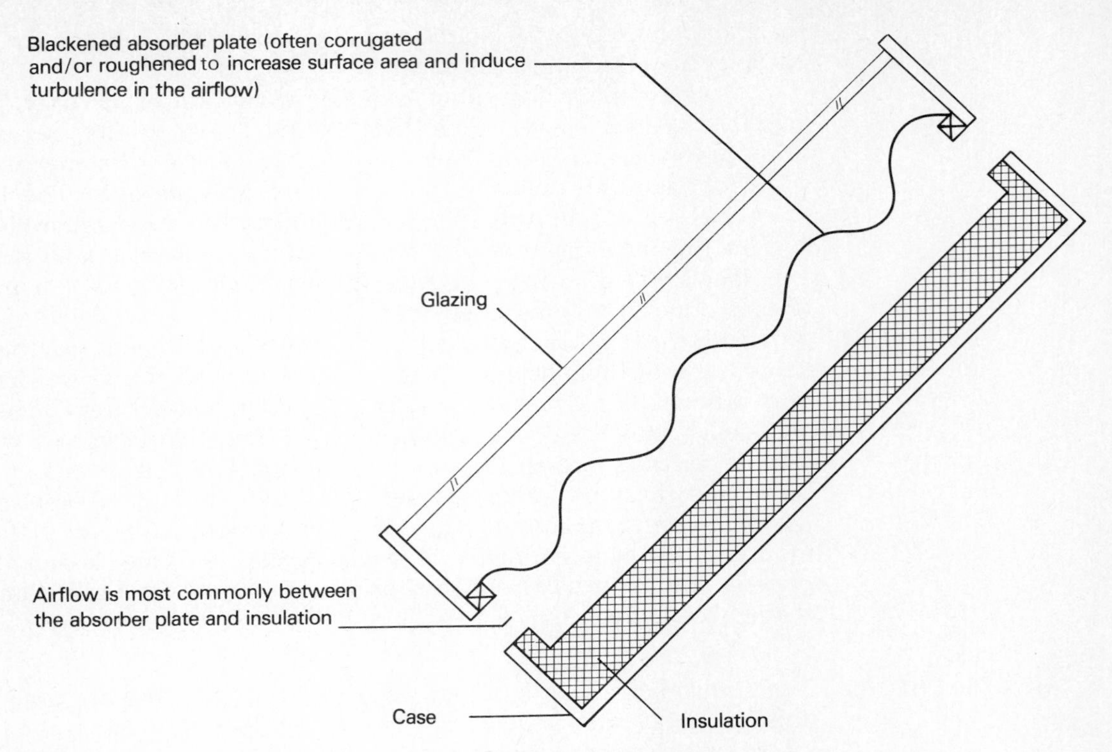

Figure 9-7 Air collector.

tions tend to induce large eddies. Smaller eddies can be set up by roughening the absorber surface. For maximum air–absorber contact, it is best if both large and small eddies are generated.

Corrugations and fins in the absorber plate have other benefits besides initiating turbulence. They also increase the total surface area of the plate, thus increasing the rate of thermal energy transfer to the air. In addition, incident sunlight that is either reflected or absorbed and reradiated by the plate will be partially reabsorbed by the fins and other protuberances (see Figure 9-8).

Absorber plates for air collectors are not always solid surfaces. Other types of materials, such as several layers of black screen, are sometimes

Figure 9-8 Reflected and reradiated incident sunlight can to some extent be reabsorbed by fins, corrugations, or other protuberances in the absorber plate.

used.[13] Screen absorbers allow some of the sunlight to penetrate to the second, third, and lower layers. These heated lower layers are somewhat insulated from thermal losses by the upper layers. Also, air flowing through screens contacts a great deal of surface area, which enhances the efficiency of heat transfer.

The most common absorber surface for a hydronic collector is metal. This is because metal's high conductivity allows heat to be quickly conducted into the flow channels and carried away to a storage tank. But it is not necessary for the absorbing surface of an air collector to have a high thermal conductivity, for the air flowing through the collector comes in contact with every square inch of the surface and thus carries away its heat no matter what the surface is made of. Concrete, glass, or even wood can be used for the absorber. Tritec Solar Industries' AirHair collector uses a mass of black painted glass fibers similar to furnace filter material for its absorbing surface. Air passes through the fibers, contacting five to seven times the amount of heated surface that it would contact with a flat absorber plate.[14]

Whereas hydronic collectors are usually self-contained units that are built in a factory, shipped to a construction site, and mounted on or near a building being erected, air collectors are frequently built *on-site* and are often made an integral part of the structure of the building itself. This is possible because efficient air collectors can be quickly fabricated from readily available construction materials using ordinary carpentry tools. When the house depicted in Figure 9-9 was being constructed, an air collector was built into the entire south-facing roof slope. The upper surface of the roof is translucent fiberglass with an air space and black painted corrugated roofing below that. Under the corrugated absorbing surface is another air space and then insulation, which both prevents heat from escaping from the house and prevents excessive heat from entering the house when the sun is shining on the collectors. Airflow in this system is both on top of and below the corrugated metal absorber.

Figure 9-9 The entire roof of this house is a solar collector. Sunlight passes through the fiberglass roofing, across an air space, and is absorbed by black-painted metal panels. Air flowing under the absorber is heated by contact with it, and then blown into a gravel thermal storage bin under the house.

NOTES

1. Available from the Chemax Corporation, 211 River Road, New Castle, DE 19720.

2. Thermatool Corporation, 280 Fairfield Avenue, Stamford, CT 06902.

3. Grumman Energy Systems, Inc., 445 Broadhollow Road, Suite 20, Melville, NY 11747.

4. American Solar King Corp., 6801 New McGregor Highway, Waco, TX 76710.

5. Filon brochure FSEB/5M—7/82. Filon Division of Vistron Corp., 12333 Van Ness Avenue, Hawthorne, CA 90250.

6. Brand names it is sometimes sold under include Thermax, Celotex, and Trymer. Sometimes other types of insulations are sold under these names, however, and the insulation should be checked to make sure that it is composed of polyisocyanurate.

7. *Solar Applications Seminar*, Pacific Sun, Inc., Palo Alto, CA, 1977, p. B-10.

8. Sunbox collectors are available from Northern Solar Power Co., 311 South Elm Street, Moorhead, MN 56560.

9. See note 3.

10. Acme Solar Works, Lodi, CA. From an article by Barbara Atkinson and Philip Caesar, "The Volkspanel Model T," *Solar Age*, Vol. 8, No. 4, pp. 32–33, as well as private communication with Acme Solar Works.

11. Ibid.

12. Ibid.

13. One company that sells such a collector is Solar Development, Inc., 11799 East 30th Avenue, Aurora, CO 80010.

14. Tritec Solar Industries, Inc., 711 Florida Road, P.O. Box 3145, Durango, CO 81301.

SUGGESTED READING

Directory of SRCC Certified Solar Collector Ratings. 1001 Connecticut Avenue NW, Washington, DC 20036; Solar Rating and Certification Corporation, 1983.

Solar Products Specifications Guide. SolarVision, Inc., Church Hill, Harrisville, NH 03450. Contains descriptions of many solar products besides collectors, including computer software, sunspaces, packaged systems, and system controls. Updated several times a year.

REVIEW QUESTIONS

1. Describe the five basic parts of a flat-plate collector and the function of each.
2. Define *hydronic* collectors.
3. Why are the absorber plates of most hydronic collectors made of either copper or aluminum?
4. Why are aluminum flow channels not used in solar collectors designed to carry water?

5. Why are header pipes in swimming pool heating collectors larger in diameter than those in collectors for heating domestic water supplies?

6. For what applications are collectors with selectively surfaced absorber plates cost effective?

7. What are the advantages and disadvantages of using multiple glazing on flat-plate collectors? Why is it that a large percentage of collector manufacturers produce single-glazed panels almost exclusively?

8. For what applications would you recommend plastic glazing for a collector array?

9. Why is insulation *outgassing* potentially detrimental to a solar collector?

10. What is the function of the collector case?

11. Define *efficiency* as applied to flat-plate collectors.

PROBLEMS

1. Calculate the January fluid parameters of collector arrays that heat the fluid they contain to a point where the inlet temperature is 110°F and that are located in the following cities:
   (a) Portland, Oregon
   (b) San Diego, California
   (c) Atlanta, Georgia
   (Ambient temperature and latitude data can be obtained from Table III-1. Collector tilt angles are equal to latitude + 10°.)

2. Use Table 9-2 to find clear-sky January insolation values for the following situations:
   (a) Site is at a latitude of 48°N, collectors are tilted at an angle of 58°.
   (b) Site is at a latitude of 32°N, collectors are tilted at an angle of 32°.

3. A family is considering the two collectors in Figure 9-5 for possible use in their solar system. Type *A* sells for $525 and type *B* for $725. Which one produces the most energy per dollar of cost? (Use the clear-sky insolation value, 2000 Btu/ft^2, and the D (+90) category, corresponding to a 90° temperature difference between collector inlet and ambient.)

4. Assume that the collectors in a domestic water heating system will be used for preheating only and will raise the temperature of incoming water to 36°F above that of the ambient temperature. Compare the energy per dollar ratios of the collectors in Figure 9-5, assuming that *A* sells for $395 and *B* for $690. (Use the clear-sky insolation value.)

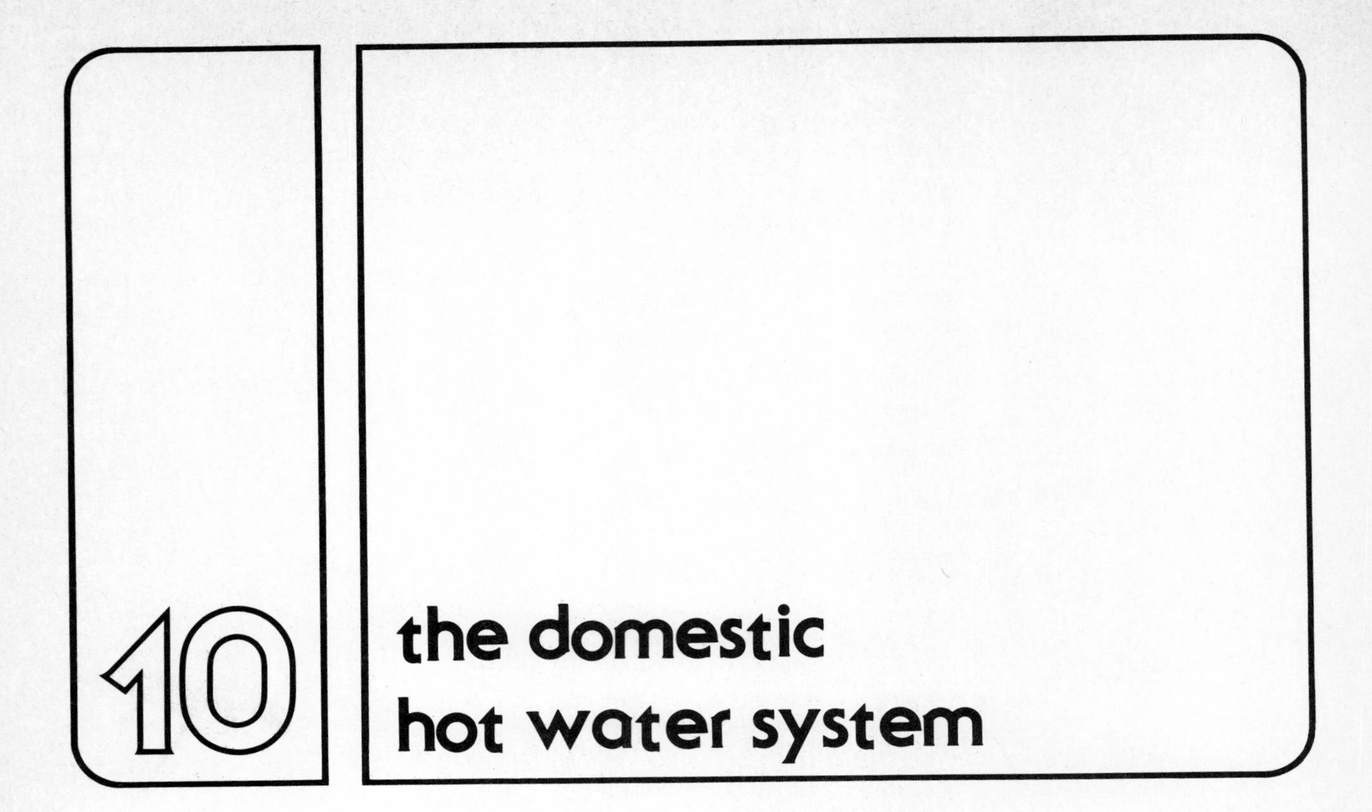

Of the many uses of solar power, domestic water heating is one of the most popular. Domestic hot water (DHW) solar systems can be used throughout the year, for hot water is needed in a house both in winter and summer. Domestic water heating differs in this respect from space heating, which is needed only during cold periods.

Domestic hot water systems are not by any means recent inventions. They were widely used in Florida and California during the early years of the century, until the advent of low-priced oil and natural gas did away with them. It has been the skyrocketing fuel costs of recent years that have led to the renewed interest in such systems.

A DHW system for a family of four typically employs 40 to 80 square feet of solar collector surface producing from 60 to 120 gallons of hot water on a sunny day. The heated water is kept in insulated storage tanks until it is needed in the house. In the United States and Canada, most systems use small electric pumps to circulate the water between tank and collector, but in other countries natural convection processes are often used instead. Many of the DHW systems sold in Japan, for instance, are of the thermosyphoning variety. Israel is another country in which thermosyphoning systems are used extensively. During much of the year solar-heated water is the only source of hot water in many Israeli dwellings. Although most of this chapter is devoted to active water heating systems that employ pumps and other externally powered devices to control the transport of thermal energy, thermosyphoning systems as well as passive systems of other types are also examined.

THE COMPONENTS OF ACTIVE DOMESTIC WATER HEATING SYSTEMS

Active water heating systems in general contain the following components:

1. An array of solar collectors
2. A storage system
3. The circulating pump
4. The automatic system controller
5. A freeze protection system
6. Vents to release trapped air
7. Safety and other valves

The Collector Array

A "ballpark" method used for rough sizing of domestic water systems is: include 14 to 20 square feet of solar collector area for each person using hot water. Use the lower figure for households without washing machines and dishwashers and the upper figure for those with them. The collectors are usually single-glazed models, although double-glazed types are sometimes employed in areas with severe winter climates.

A more accurate method than the above rule of thumb for determining solar array square footage is first to calculate the daily hot water requirements of the household, and then, with knowledge of the climate of the site, compute the square footage of the collectors needed to provide the necessary Btu.

The Solar Energy Applications Laboratory at Colorado State University has developed a method of sizing a DHW system's collector array based on the insolation received at the site. A common way to design a system is to size it to meet almost all of the domestic hot water needs during the summer months. The array will then provide typically about half of the needs during the time of the year with the worst weather (usually January). It is generally more cost-effective to do this than to purchase sufficient solar panels to provide all of the year's hot water needs. (The extra cost of such an array would be considerable, and during much of the year, the array would furnish many more gallons of hot water than could be used.) The following formula, derived from the Colorado State methodology,[1] is of help in computing the collector area needed for a domestic hot water system:

$$\text{collector area needed} = \frac{0.85 \times \text{average daily hot water load (Btu)}}{\text{mean January insolation on a horizontal surface (Btu/ft}^2\text{-day)}} \qquad (10\text{-}1)$$

Because so many variables are not considered in this method (such as the ambient temperature of the site or the type and brand of solar collectors chosen), equation (10-1) should not be expected to yield exact results. Its purpose is to provide an approximate collector sizing guide for water systems. It is assumed that well-built collectors are installed that are designed for domestic water heating applications.

To use equation (10-1), the daily hot water load as well as the mean January insolation must be known. Table 10-1 lists typical Btu of heat needed to provide hot water for various household functions, while the map in Figure 10-1 depicts average daily insolation values during the month of January for various locations throughout the country. Insolation values can also be obtained from Table III-1.

TABLE 10-1 Daily Hot Water Needs and Energy Requirements of a Typical Household

Function	*Hot Water Required (gallons/day)*	*Energy Required to Provide This Amount of Hot Water*[a] *(Btu/day)*
Baths and showers	15 per person	9,383 per person
Hand and face washing	2 per person	1,251 per person
Food preparation	3	1,877
Dishwashing		
By hand	5	3,128
Automatic	15	9,383
Washing machine	21	13,136

[a]It is assumed that the temperature at the water heater outlet is 135°F, and that the temperature of cold water from the main is 60°.

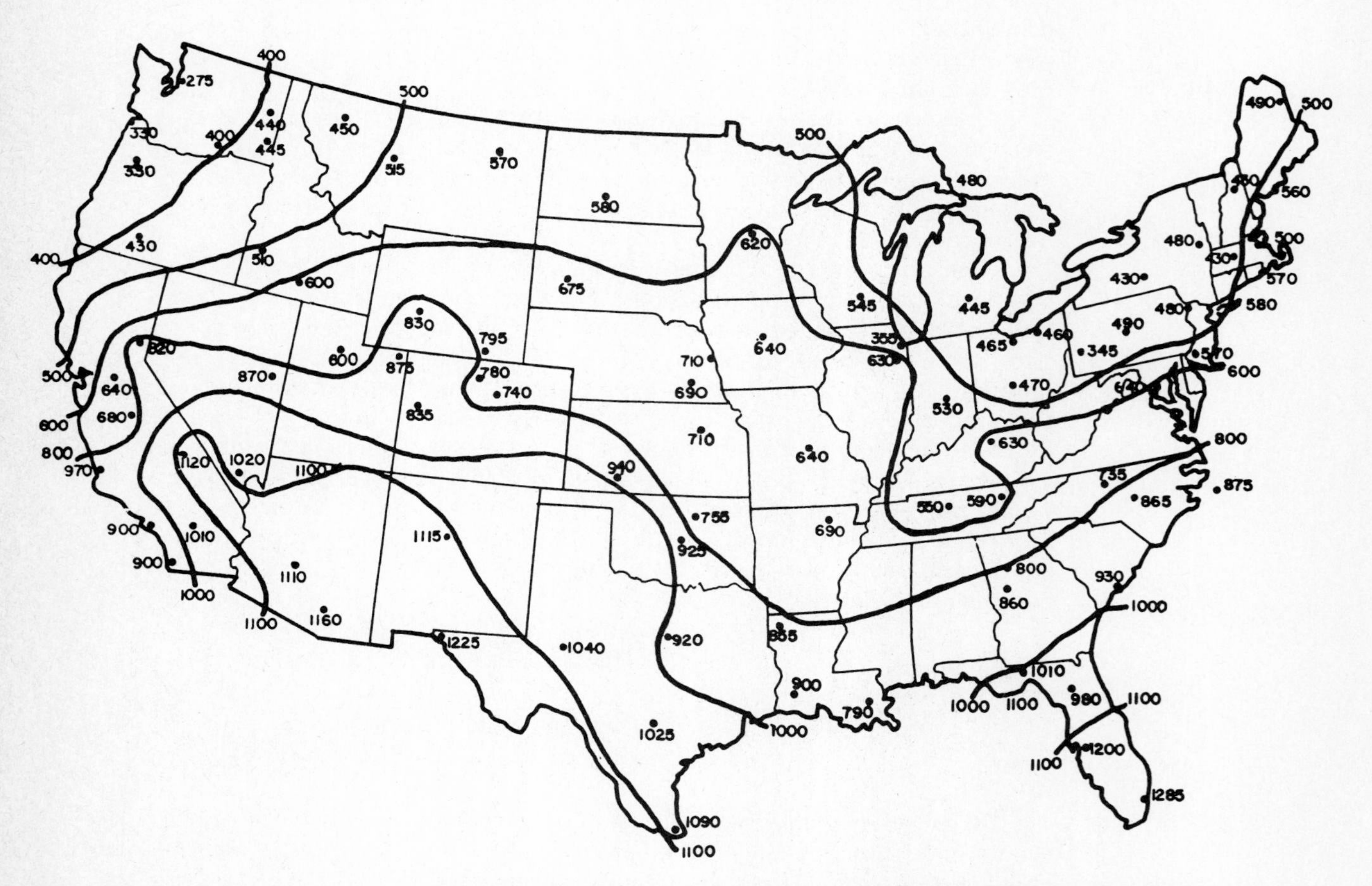

Figure 10-1 Average daily solar radiation (Btu/ft^2), month of January. (From *Solar Heating and Cooling of Residential Buildings—Design of Systems*, Solar Energy Applications Laboratory, Colorado State University, for U.S. Department of Commerce, 1980, p. 9-27.)

Example Calculation. A family of four owns a washing machine as well as a dishwasher. Their home is located in Lincoln, Nebraska. How many square feet of collectors does their solar system require?

To determine the collector area, the first step is to compute the daily hot water requirements, using Table 10-1:

Baths and showers:	9,383 Btu/day per person, × 4 people	37,532 Btu/day
Washing machine:		13,136 Btu/day
Automatic dishwasher:		9,383 Btu/day
Hand and face washing:	1,251 Btu/day per person, × 4 people	5,004 Btu/day
Food preparation:		1,877 Btu/day
		66,932 Btu/day

It is observed from Figure 10-1 that Lincoln, Nebraska, receives an average of 580 Btu per day per square foot in January. Equation (10-1) can now be solved:

$$\text{collector area needed} = \frac{0.85 \times 66{,}932 \text{ Btu/day}}{690 \text{ Btu/day-ft}^2} = 82 \text{ ft}^2$$

Collector arrays in the Northern Hemisphere should face within 30° of due south (in the Southern Hemisphere the situation is reversed). Due south is generally the best orientation, although if there is frequent early morning or late afternoon fog at the site, a 15° rotation of the collectors to the southwest or southeast can improve system efficiency.

Although flat-plate collectors are usually mounted on roofs, they are often placed on the ground and occasionally on the sides of a building. If mounted on a south-facing roof, a collector should be installed flush with the roof's pitch if its tilt angle is approximately equal to the latitude angle plus 10 to 15°. For instance, if the latitude of the site is 40°, the roof pitch should be about 50 or 55° if the collectors are to be mounted flush on top of it. Otherwise, collectors can be mounted on racks or standoffs in order to achieve the proper angle (see Figure 10-2). Racks are also used on ground-mounted collectors.

(a)

(b)

Figure 10-2 (a) Collectors can be mounted on racks if the roof is at the wrong pitch for the latitude of the site. This house is located at 39° north latitude, which means that the collectors should be tilted from horizontal at between 49 and 54°. (Photo by Christina Meltzer.) (b) In this situation, the roof is not only at the wrong pitch; it also does not face toward the south. The mounting racks compensate for both of these problems.

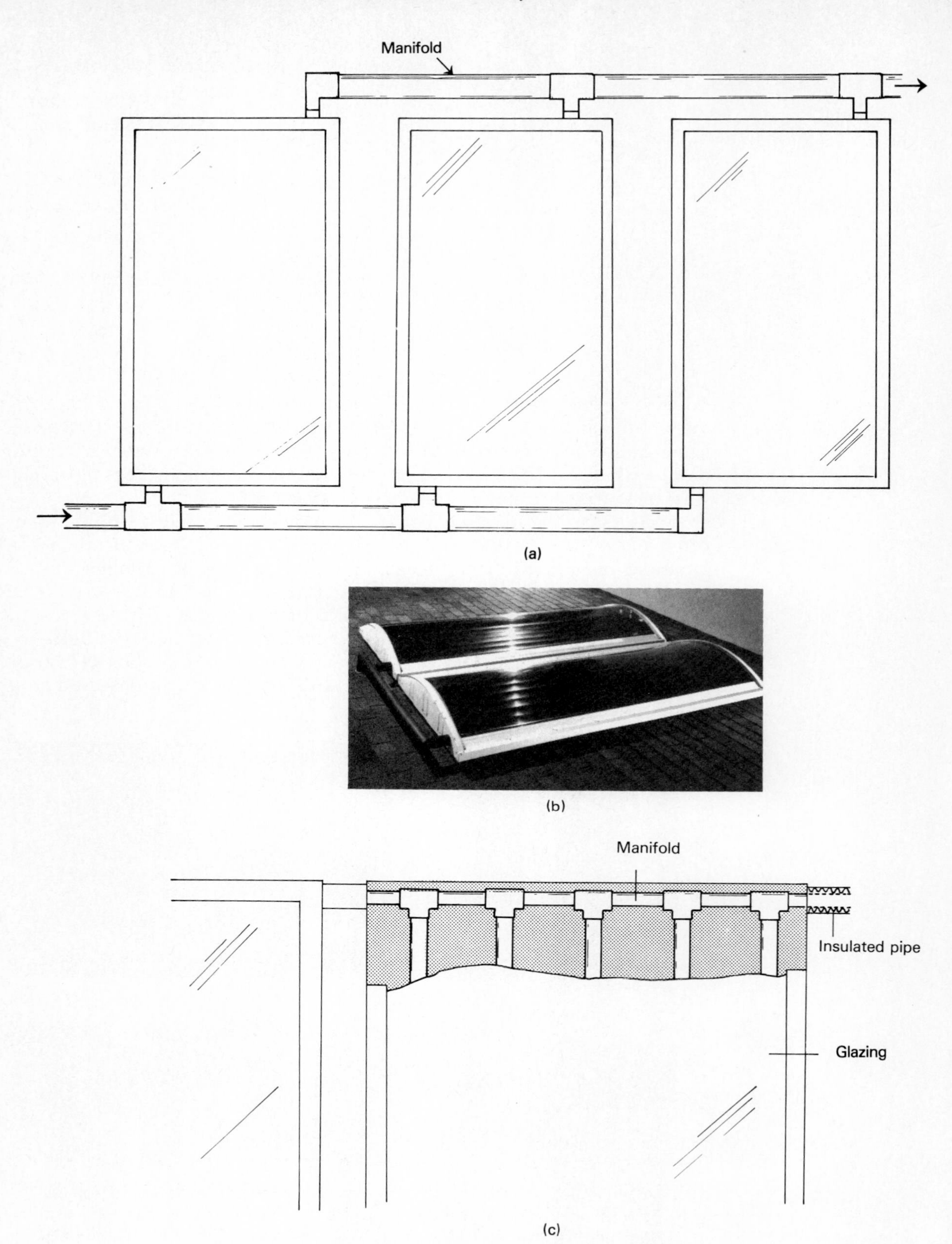

Figure 10-3 (a) Collector array with external manifolds. (b) Internal manifolds. (c) Close-up of internal manifold. [(b), Courtesy of Grumman Energy Systems Company, Melville, NY.]

The manifold pipes of a collector array—the pipes that carry liquid to and from the absorber plates—are sometimes built into the collectors themselves and at other times are installed on-site, outside the collectors (see Figure 10-3). Internal manifolds have the advantage of not needing pipe insulation. Arrays with internal manifold collectors can be put together quickly and present a neat appearance, since little or no piping is visible from the outside. Internal manifolds are of limited size, however, and this places restrictions on the way an array can be plumbed. Usually, no more than five internal manifold collectors can be connected together in series. If longer runs are attempted, the flow rates necessary to supply the panels with sufficient water will approach the erosion limit of the metal in the flow channels. External manifolds do not have this problem, for they can be made of a size to accommodate virtually any flow rate.

For maximum energy efficiency, it is necessary to keep the flow rates the same or nearly the same through each panel in an array. If the panels are plumbed together as in Figure 10-4, panel 1 will have the highest flow rate since fluid has less distance to travel and thus will encounter less friction than on paths through panels 2 and 3. The result will be different temperatures in each of the collectors, which will lower the overall efficiency of the system.

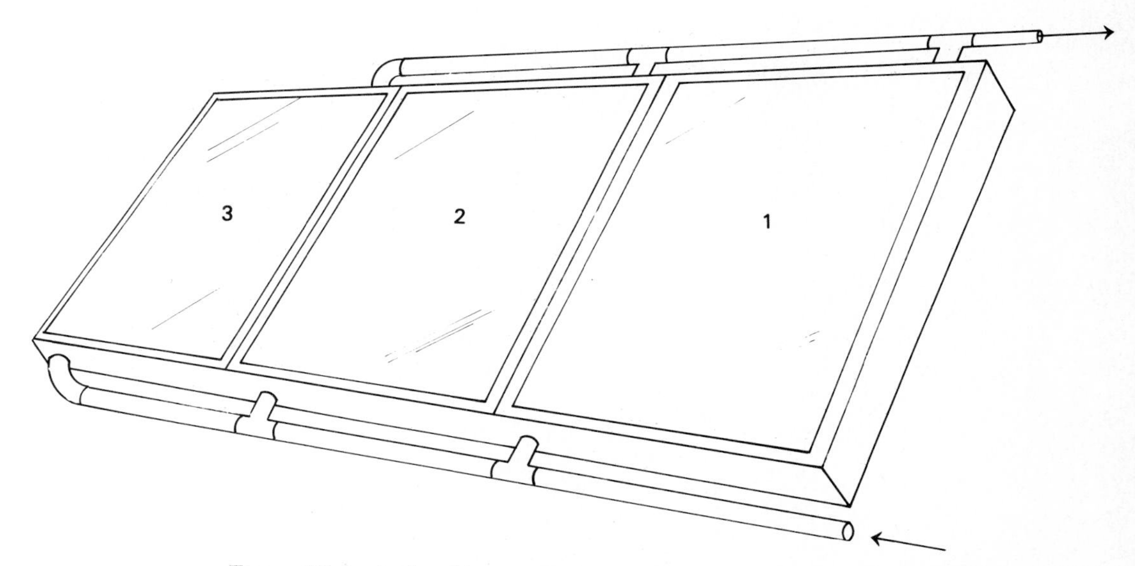

Figure 10-4 A plumbing configuration that will lead to thermal imbalance in the collector array.

A better way to plumb the system is as in Figure 10-5, for in this case the total length of travel is the same no matter which collector the fluid flows through. This leads to flow rates and temperatures that are nearly identical in each collector.

Temperature Sensors. After a collector array is installed, sensors for measuring temperature are mounted in it which signal the system controller to turn on the circulating pump or drain down the panels in the event of a freeze. Sensors whose measurements help control the circulating pump should be mounted at the top (the hottest part) of the collector array. When possible, it is best to install them inside the collector itself. If this is not practical, a sensor of the "well" or "probe" type may be mounted in the

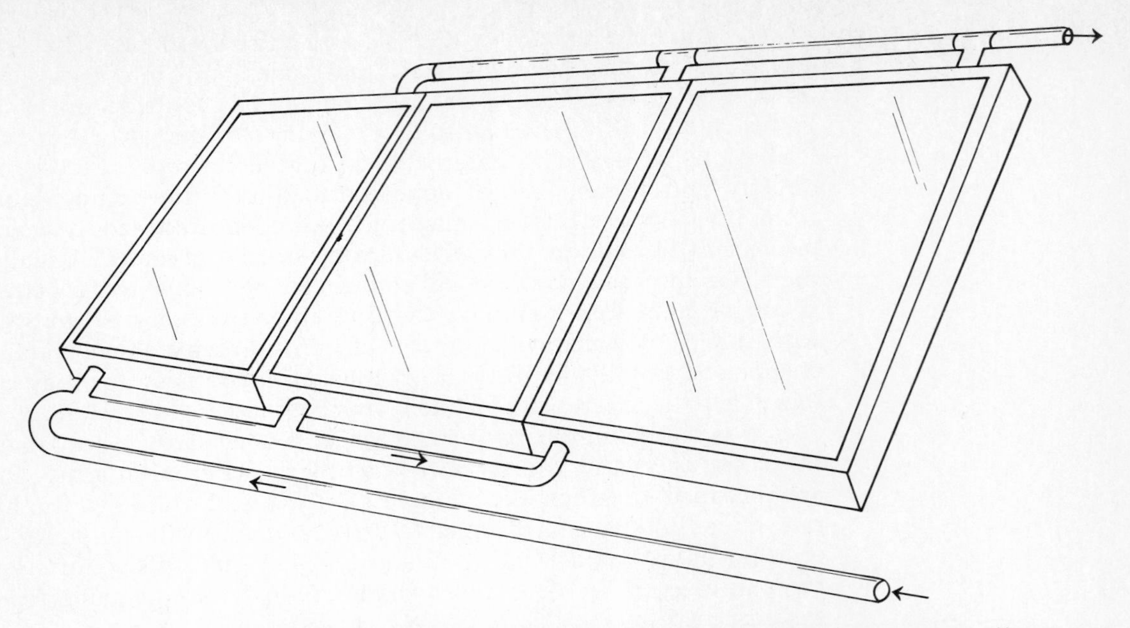

Figure 10-5 The total length of travel for the fluid is the same no matter which collector it flows through.

outlet pipe. Other types of sensors can be strapped onto the outside of this pipe (see Figure 10-6). Insulation should be put over the sensors to cut down on thermal losses and so that they read actual hot water temperatures rather than those of the outside air.

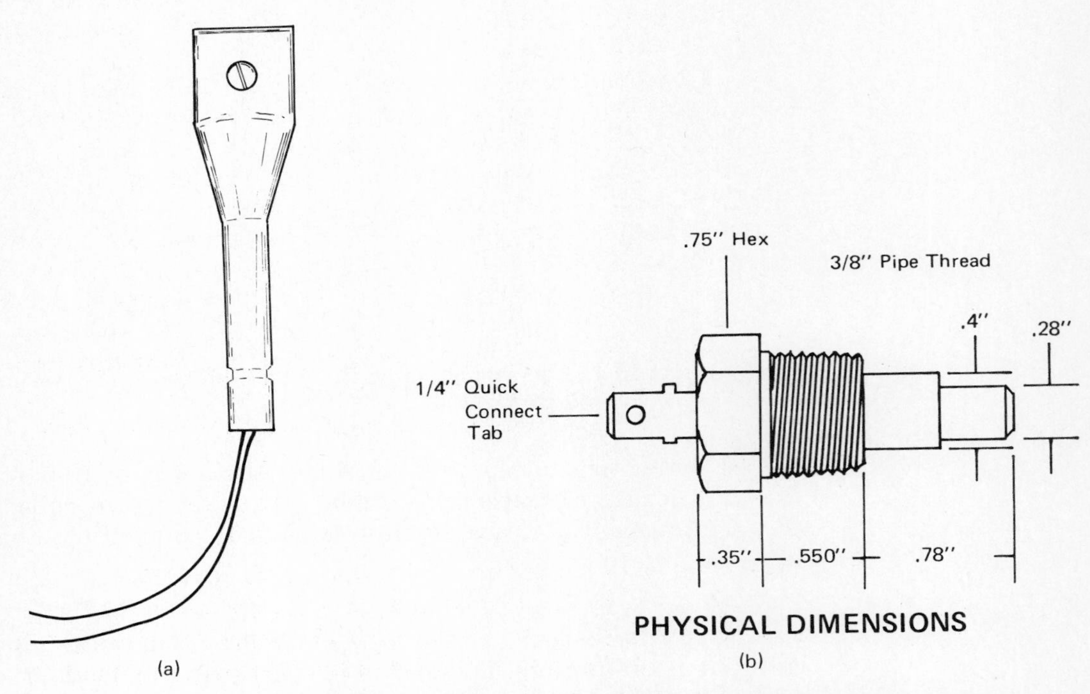

Figure 10-6 Temperature sensors. (a) This type of sensor can be screwed directly to the absorber plate. Its flattened end also allows it to be strapped tightly to a manifold pipe. (b) Made for mounting in a storage tank wall. [(b), Courtesy of Midwest Components, Inc., Muskegon, MI.]

Solar systems that need protection against freezing (those with collector flow passages filled with water rather than antifreeze or oil) have other sensors mounted in the lowest, and coldest, part of the collector array. The best location is on the bottom third of the absorber plate itself.

Freeze sensors must be mounted inside the collector, for during a clear night, absorber plate temperature can actually be over 10°F colder than that of the outside air.[2] This is due to a phenomenon called *night sky radiation*. A flat black collector absorber plate is an excellent emitter of radiative energy. On clear nights, much of this emitted energy travels through the atmosphere and into space. Outer space is at a temperature close to absolute zero and thus sends little radiant energy back toward the collector. As a result, the panel's temperature can drop below freezing when ambient air temperature is as high as 45°F. Freeze sensors must thus be located where they will measure actual absorber temperature. At least two freeze sensors should be used, mounted in two different collectors on the bottom third of each one's absorber plate.

The Storage System

A solar system should have enough thermal storage to provide at least a day's worth of hot water during periods when the sun is not shining. A good rule of thumb is to include 1½ to 2 gallons of storage for every square foot of solar collector area.

In retrofit situations for single-family dwellings, storage generally consists of an insulated glass-lined tank plus the existing water heater. In new construction, storage is provided by one large (80- to 120-gallon) water heater.

One large water heater is cheaper to buy and takes up less space than a small heater plus an additional storage tank. If an existing water heater in a retrofit situation has been in service more than seven years, it is considered by many solar contractors to be more cost-effective to replace it with one big heater at the time when a DHW system is being installed.

Figure 10-7 illustrates the methods by which the storage units and collectors of both one- and two-tank solar systems are plumbed. Note that water going to the collector array comes from the bottom and coldest point of the storage tank, while water returning from the collectors enters the tank at a depth one-third of the way down from the tank. This maintains a layer of the hottest water at the top of the tank, ready for use in the house.

Water entering the tank at a high velocity can upset the stratification in the tank, mixing cold water with hot. To prevent this from occurring, a dispersing valve should be added to the end of the inlet tube, as illustrated in Figure 10-7. This device can simply be a length of pipe with pinholes in it, which has the effect of breaking up a "slug" of water into many small streams.

In wintertime, the collectors of many systems act as *preheaters*, raising the water temperature part but not all of the way to the temperature required by the house. The water heater then raises the temperature the rest of the way. In summer, however, no auxiliary heating of the water is necessary for much or most of the time, and the water heater element can be turned off. In a two-tank system, a bypass channel should also be installed so that water can be directed from the storage tank into the house without entering the water heater (see Figure 10-9). If the water is allowed to enter the turned-off heater, it will have a chance to cool off, significantly impairing the system's performance.

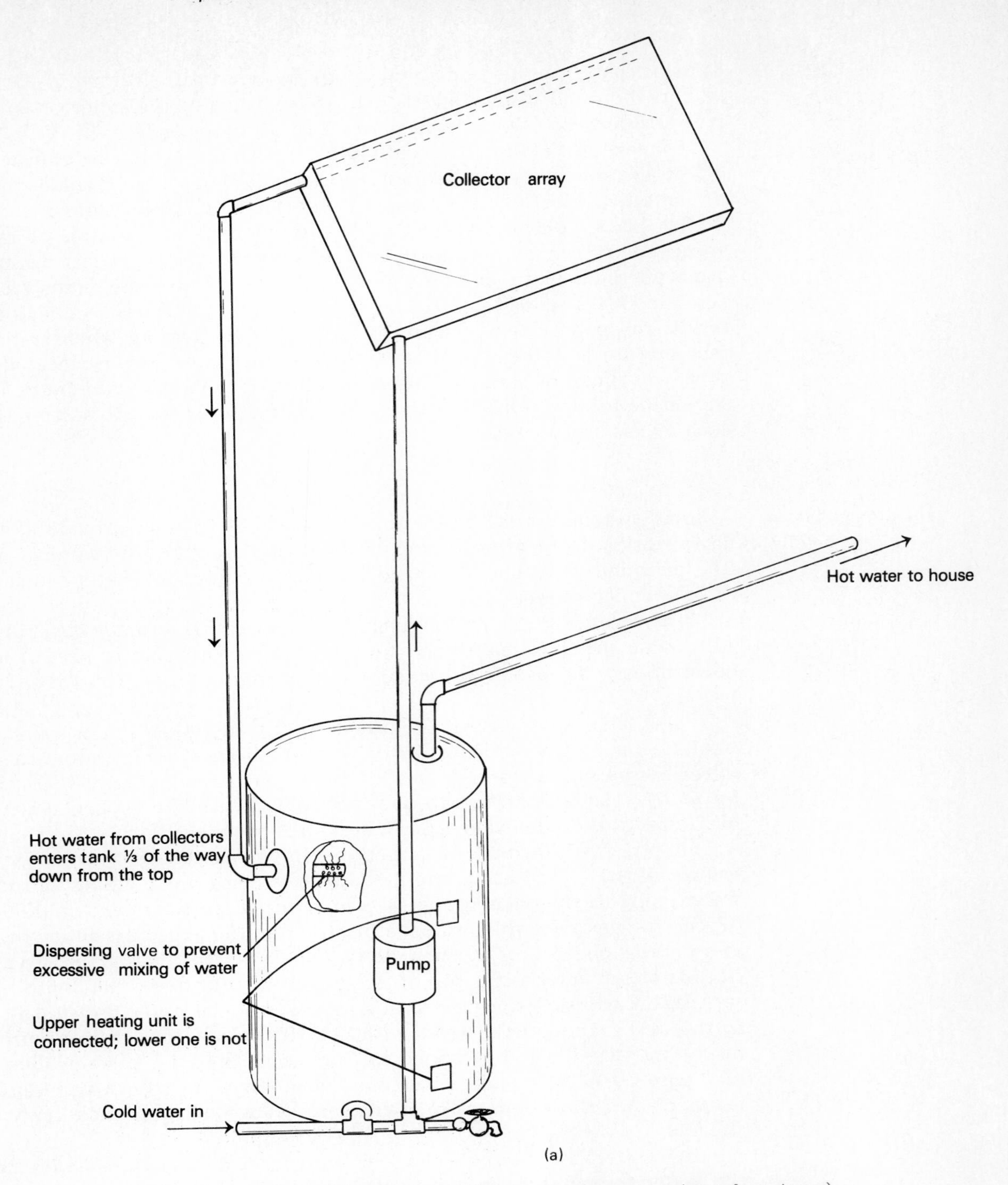

Figure 10-7 (a) One-tank solar system (open-loop type.)

All the storage vessels in a solar system should be insulated with 6 inches of fiberglass or its equivalent. Manufacturers often don't insulate their tanks to this degree. If foam insulation is to be added to the tank of a water heater, however, care must be taken to prevent a fire danger. Many foams on the market are highly flammable and release very toxic gases as they burn. They must not be used anywhere near a part of the water heater that has even a remote chance of igniting them. Fiberglass is often used for wrapping

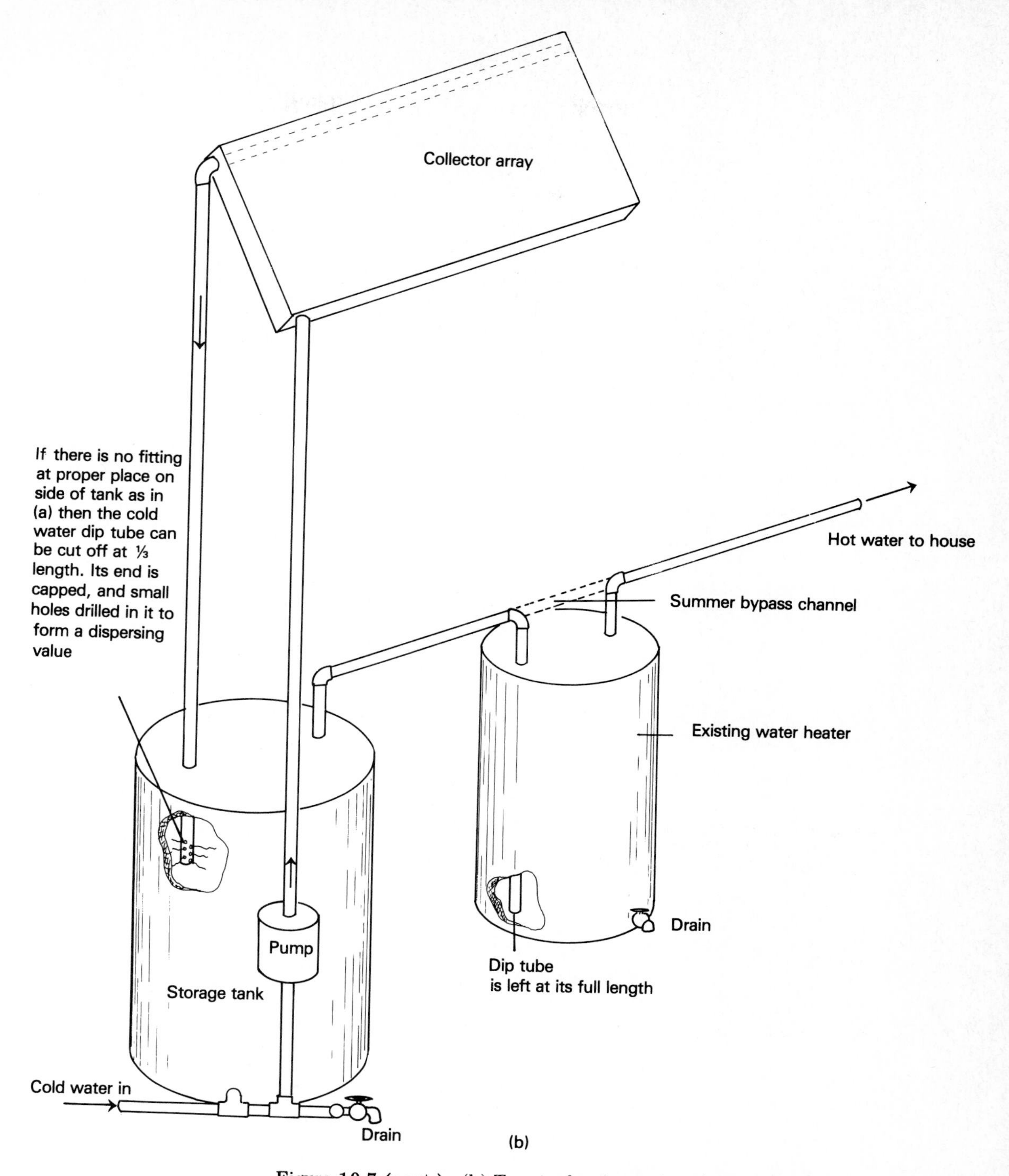
Collector array
If there is no fitting at proper place on side of tank as in (a) then the cold water dip tube can be cut off at ⅓ length. Its end is capped, and small holes drilled in it to form a dispersing value
Hot water to house
Summer bypass channel
Existing water heater
Drain
Dip tube is left at its full length
Pump
Storage tank
Cold water in
Drain
(b)

Figure 10-7 (cont.) (b) Two-tank solar system (open-loop type).

a heater because of its reduced fire danger, but what many people don't know is that the foil or paper facing attached to it *is* easily ignited. No matter what type of insulation is installed, manufacturer's specifications must first be carefully studied.

The top of an electric water heater does not have an exhaust flue as does a gas heater and so can be insulated more heavily. It is thought by some that since none of the thermal energy in an electric heater is lost through a

Figure 10-8 DHW system. DHW system collector array. (Courtesy of Grumman Energy Systems Company, Melville, NY.)

flue, it must be more energy conservative. But before jumping to this conclusion, it should be remembered that the electric power generating plant from which the energy comes is often not very efficient. At a typical plant 3200 Btu of fossil fuel is burned for every 1000 Btu of electrical energy actually delivered to the home (1000 Btu = 0.29 kilowatthour). Also, an electric water heater is considerably more expensive to operate than a gas unit of the same size.[3]

A new alternative to the choices described above is the heat pump water heater. A heat pump will deliver up to three times more thermal energy than a conventional electric resistance heater for every watt of power drawn from the wall socket. This is because a heat pump is a device that

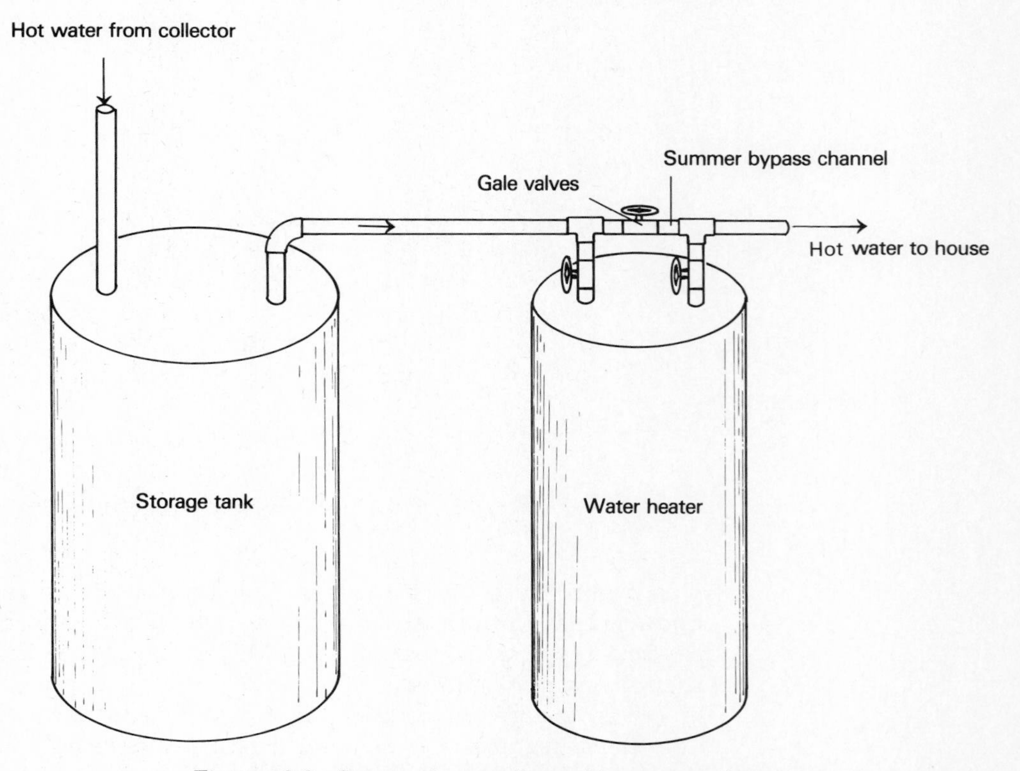

Figure 10-9 Summer bypass channel in a two-tank system.

simply moves thermal energy from one place to another (for instance, from the air around the water heater into the water itself), rather than creating the heat by forcing electric current through a resistor. Because of their greater efficiency, heat pump water heaters produce monthly power bills that are competitive with those for gas heaters.

In the diagram of a one-tank domestic hot water system (Figure 10-7a), the water tank's heating unit is located a third of the way down from the top. If the tank has an additional heating unit near its bottom, as some do, it should be disconnected. Otherwise, since convection carries heated water to the top of a tank and colder water to the bottom, the lower heating element will stay on until all the water above it is warmed up. It will thus be doing work the solar collectors are intended for and will consume much more energy than necessary. The heating element located near the top, on the other hand, heats only the water in that part of the tank, providing a ready supply of hot water but not competing with the solar panels.

Since gas water heaters have their burners located at the bottom of the tank, they are not recommended for one-tank solar systems. Heat pump units, though, are ideal for this application and should become increasingly popular as their price comes down.

A temperature sensor similar to those mounted on the collector array must also be mounted on the storage tank. The temperature it measures is compared by the differential thermostat in the automatic system controller with temperatures at the top of the collector array, and the circulating pump is turned on and off at appropriate times. Some tanks have openings designed especially for the sensors. On others, it is necessary to cut through the outer shell and insulation and fasten the sensor directly to the tank wall. It should be mounted near the bottom so that it reads the temperature of the coldest water in the tank. It should then be covered with silicone gel or some other sealant and recovered with insulation to ensure that it reads tank temperature rather than that of the outside air.

Open Loops and Closed Loops. In an open-loop solar system, the same water that passes through the collector array eventually finds its way to the sink faucets, shower heads, hose bibs, and other water outlets of the house. For this reason, open-loop systems must use potable water. Figure 10-7 depicts examples of one- and two-tank open-loop setups.

A disadvantage of an open loop is that the water in it can freeze on cold nights, rupturing pipes and the absorber plate flow passages and causing extensive repairs. To guard against this, some collector arrays are automatically drained on cold nights, whereas others are kept above freezing by circulating warm water through them or by using small electric heating devices like heat tape.

Aluminum flow passages cannot be used in open-loop systems, for water will corrode them. In some locations with high-mineral-content groundwater supplies, even copper flow passages will be ruined, especially at the temperatures produced in a solar collector.

Many systems do not use water in the collector array but instead employ a noncorrosive fluid that will also not freeze at the temperatures likely to be encountered. Systems such as these are called *closed loops* because the fluids they employ are circulated only between the collector array and a heat exchanger in which the solar-heated fluid gives up its thermal energy to the building's water supply.

Heat exchangers can either be separate units from the storage tanks or an integral part of them (see Figure 10-11). In the heat exchangers, the

Figure 10-10 DHW system collector array. (Courtesy of Grumman Energy Systems Company, Melville, NY.)

solar-heated fluid is separated from the water supply by a thin, well-conducting wall usually made of metal. The more surface area there is in the interface between the two liquids, the faster and more efficient is the energy exchange. Figure 10-11 depicts typical heat exchanger designs that achieve large interface areas. Many types of heat exchangers employ double walls of material between the two fluids, as in Figure 10-11b. Some building codes require this type of design because if one of the walls happens to leak, the potable water is still protected from contamination by the second wall.

Silicone oils and antifreeze solutions such as propylene glycol are fluids often used in closed-loop solar systems. Propylene glycol is the cheaper of the two, but it tends to break down after a year or two into corrosive acids that can eat through pipes (even "inhibited" glycol solutions can break down eventually if stagnation temperatures in the collector are high). To guard against this happening, the acidity of the fluid needs to be measured periodically and, if dangerously high, the system must be drained and the liquid replaced.

Solar contractors often prefer silicone oil over propylene glycol for use in a collector array, for although silicone is initially more expensive, it does not have to be replaced every several years and thus is more economical in the long run. It is, however, more viscous than propylene glycol or water, which means that a higher-powered pump might be necessary to circulate it. In addition, all pipe joints in a silicone system must be very tight, for the oil is extremely good at "finding leaks," especially at threaded connections.

Closed-loop systems require expansion tanks that are used to catch overflowing fluid as the collector's temperatures rise. Closed loops do not, however, require the sometimes unreliable drain-down valves or other freeze-protection devices of open loops. Although closed loops are slightly less energy efficient than open loops, the dependable protection against freezing and corrosion damage that they offer make them attractive choices for many, if not most, DHW applications. In the experience of this author, they have proved to be significantly more trouble-free than open-loop systems.

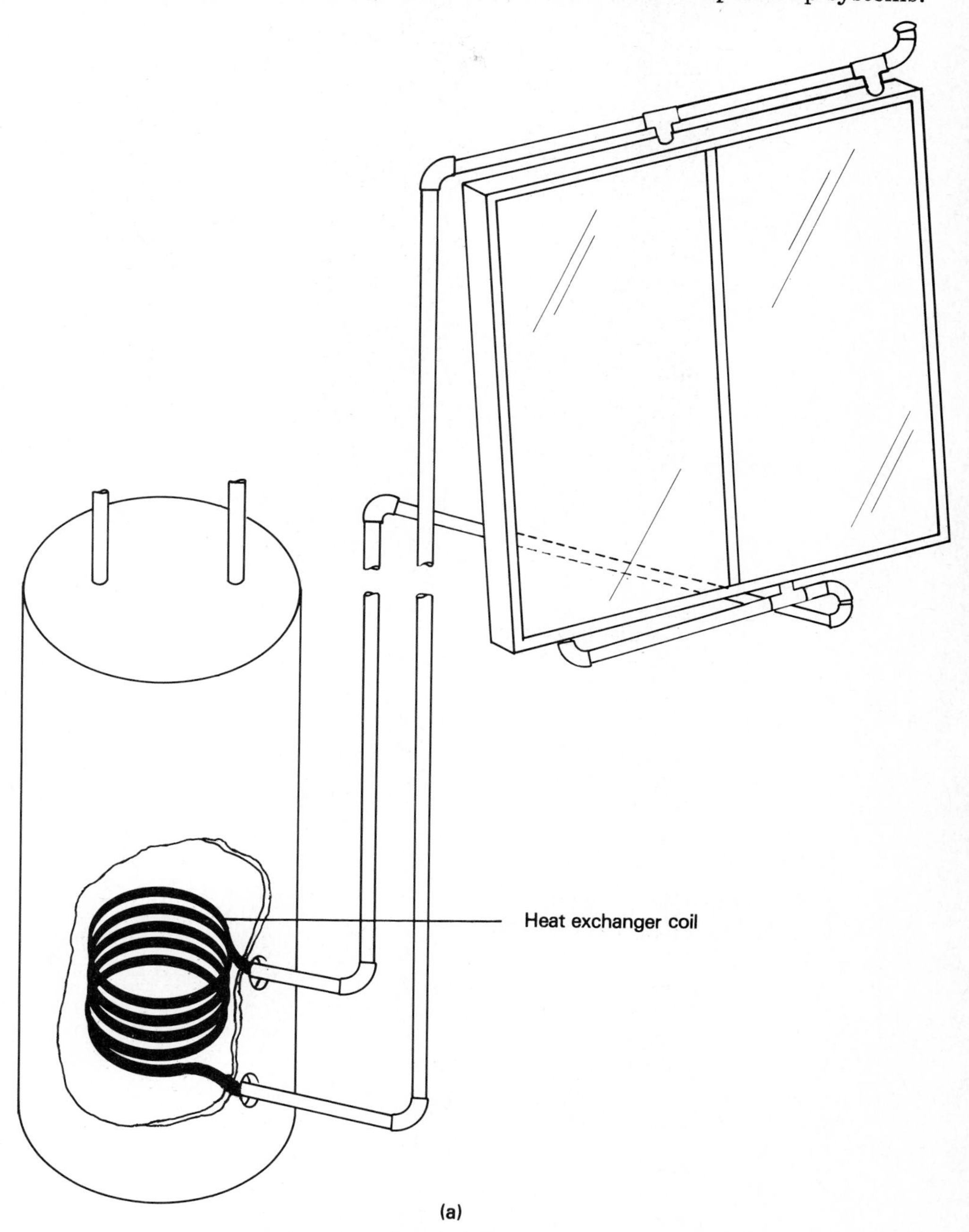

Figure 10-11 Heat exchangers. (a) The heat exchanger can be an integral part of the storage tank. Driven by convection, water in the tank flows past the solar-heated exchanger coil. The pump circulates silicon oil or antifreeze from the collector array through the heat exchanger coil and back again.

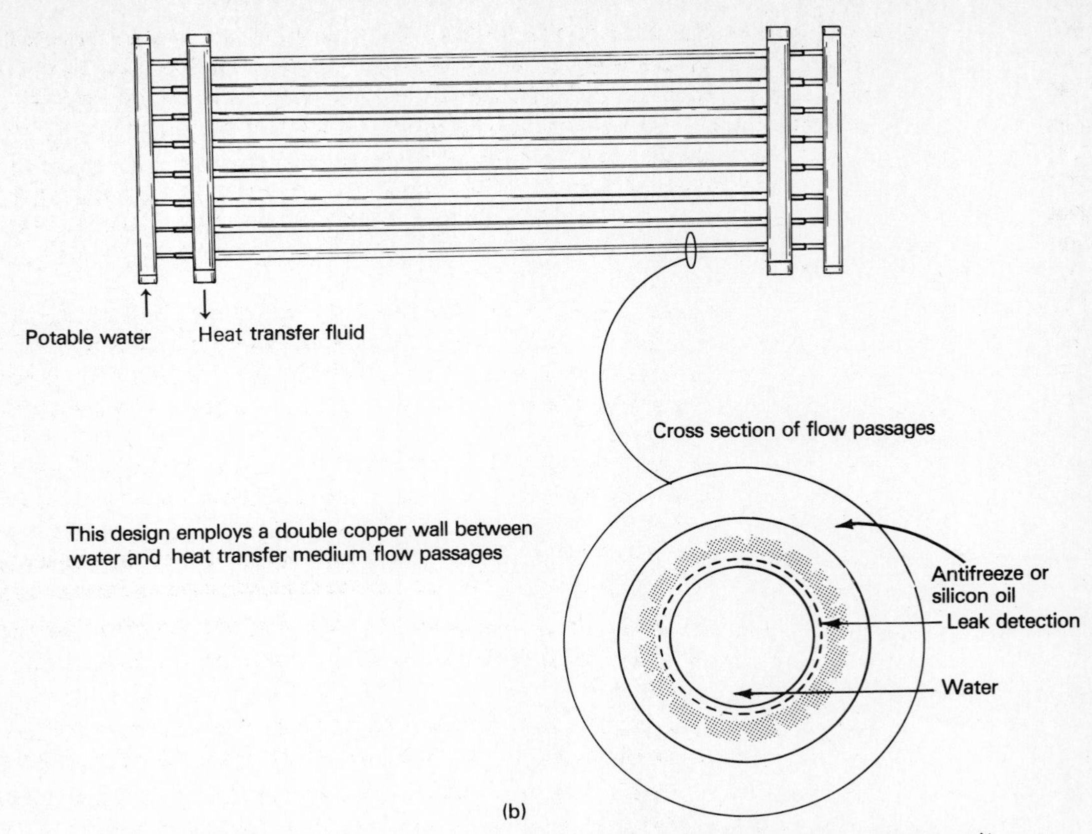

Figure 10-11 (cont.) (b) The heat exchanger can also be a separate unit from the tank.

The Circulating Pump

The electric pumps that circulate water or other liquids through an active solar system are generally of the centrifugal type. Centrifugal pumps have rotating blades in them that catch low-pressure incoming water and whirl it around until it is thrown out of the pump at a high velocity. Typical pump sizes for single-family DHW systems range from 1/12 to 1/40 horsepower.

Solar circulating pumps are rated to withstand temperatures of over 200°F. Those used in open loops should have all surfaces that come in contact with potable water made of stainless steel, bronze, or some other noncontaminating material. Those pumps used in closed loops can have their inside surfaces made of cast iron and are usually somewhat cheaper because of this. Typical electric power draws for single-family DHW pumps are roughly the same as the power draw of a 100-watt light bulb, so the pump's cost of operation is generally very low.

The Automatic System Controller

The system controller contains a differential thermostat that compares the temperatures read by sensors on the collectors and storage tank and switches the circulating pump on and off when appropriate. The pump is turned on when collector temperature is slightly higher than that of storage and off when it approaches that of storage. The exact temperatures at which the pump is turned on and off vary from system to system and from controller to controller, but the goal is always the same: to cycle the pump only when it can deliver water to storage that is warmer than the water already in the tanks.

Many system controllers also operate drain-down valves or turn the pump back on whenever sensors on the bottom of the collector array indicate that there is a freezing danger (turning the pump back on sends warm water from storage through the collectors). Closed-loop systems do not require this additional feature on their controllers, since the fluids that they use will not freeze.

Some of the more sophisticated controllers on the market provide digital readouts of temperatures at various locations in the system. These readouts can be useful and convenient in monitoring system performance.

Automatically controlled pumps sometimes turn on and off at the wrong times and/or wrong temperatures. A common cause of this malfunction is incorrect mounting of the sensors. A sensor strapped onto a water pipe or storage tank must make good thermal contact with the surface if it is to read accurately. If the surface is at all rough, it is advisable to sand it smooth before mounting the sensor. If there is a layer of paint or protective grease on the surface, that must be removed as well. It also helps to spread a thin layer of silicone heat transfer cream onto the surface before mounting the sensor.[4] After mounting the sensor, it is important that it be insulated so that it measures fluid temperatures only, rather than those of the outside air.

The Freeze Protection System

The most common type of freeze protection for open-loop systems is a valve operated by the system controller that drains down the solar panel array whenever its temperature approaches freezing (see Figures 10-12 and 10-14).

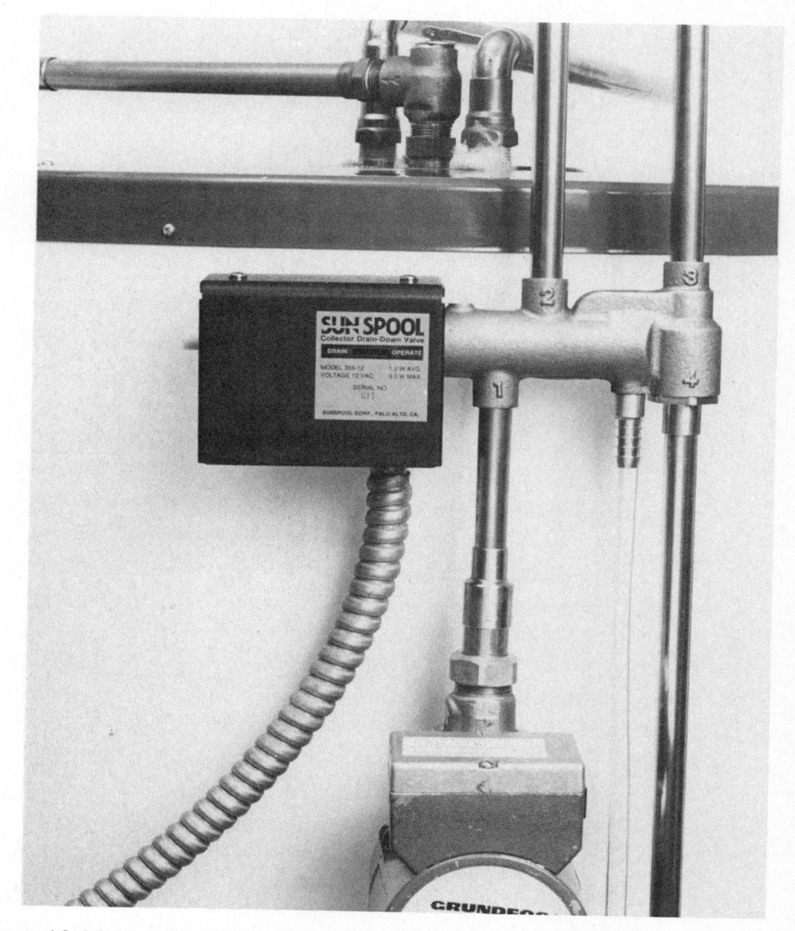

(a)

Figure 10-12 (a) Drain-down valve mounted next to the tank. Notice also the pressure/temperature valve on top of the tank, and the Grundfos pump below the drain-down valve. (Courtesy of Sunspool Corp., Menlo Park, CA.]

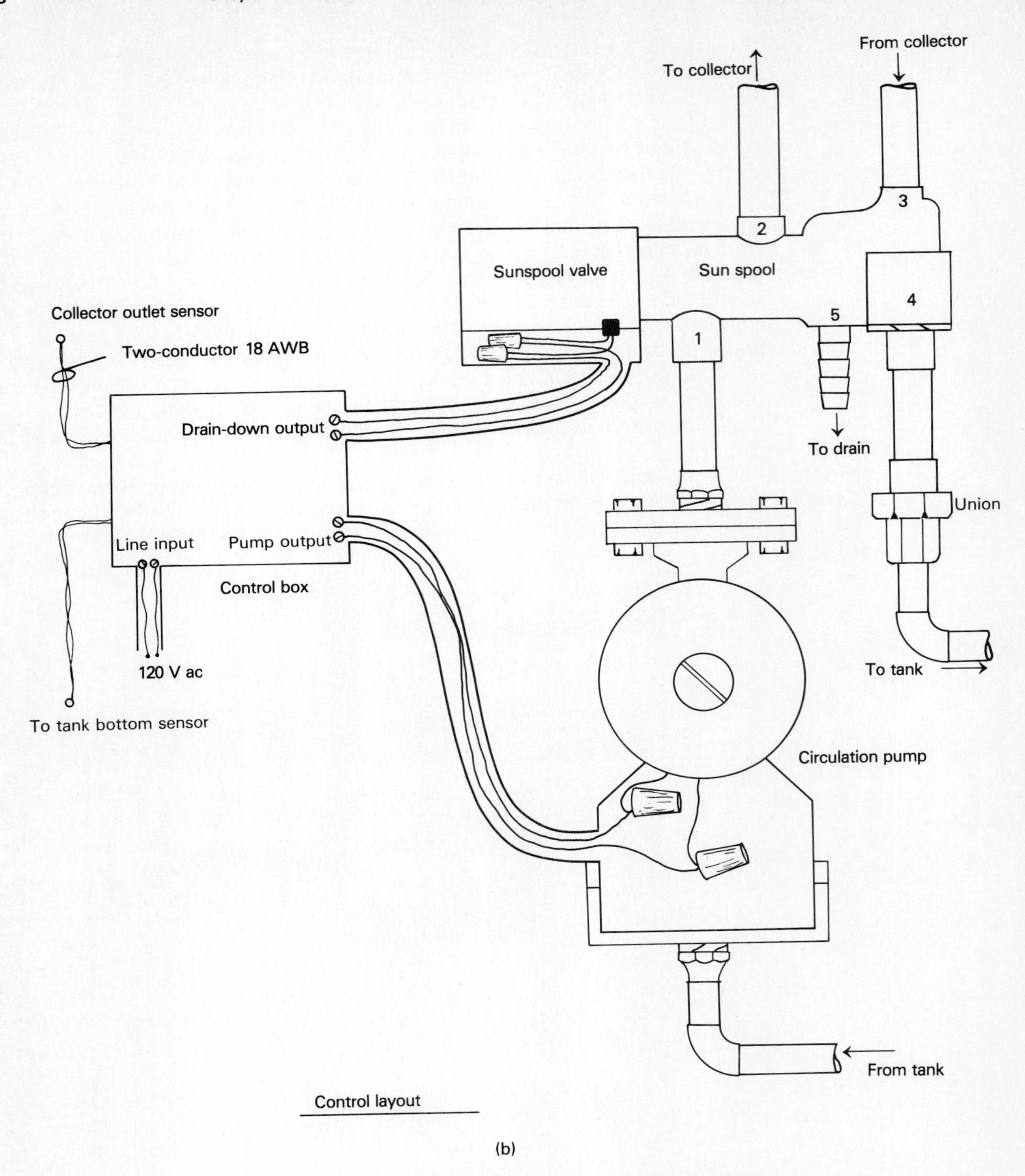

Figure 10-12 (cont.) (b) Control layout.

The disadvantage of these devices is that they sometimes get clogged with debris and fail to open. To guard against this, all the pipes in a solar system should be thoroughly flushed out with a citric acid solution immediately after the system has been plumbed. In areas with high-mineral-content water supplies, it might also be necessary to flush out the system at regular intervals to wash away scale deposits.

As mentioned above, system controllers can also be used to turn on the circulating pump whenever outside temperatures drop. Doing this, however, depletes the system's supply of hot water.

Small electric heating devices such as "heat tape" are sometimes put inside each solar panel. These devices, too, can be turned on and off with the aid of the automatic system controller. But it should be remembered that if freezing temperatures are also accompanied by electric power outages, then heat tape, circulating pumps, and system controllers all become inoperable, and this can result in thousands of dollars worth of solar panels being ruined. Drain-down valves do not suffer from this weakness, for they are made to open in the event of a power failure.

Air Vents and Vacuum Breakers

Vents that allow trapped air to escape are installed at the highest point of a solar system. Some of these vents also aid in emptying the collector array of fluid by serving as vacuum breakers whenever the system's drain-down valve is opened. Vents and vacuum breakers must be mounted in a vertical position and insulated to guard against heat loss and freeze-up. Another vacuum breaker must also be installed *inside* the building where temperatures remain above freezing. A good location for this breaker is near the storage tank.

Safety and Other Valves

A sophisticated solar water heating system employs a number of different valves in various parts of the system. Some are opened and closed automatically, whereas others are manually operated.

Pressure/temperature (P/T) relief valves are valves that open whenever temperatures or pressures in the system get dangerously high. One should be installed on the system's backup water heater and another at the top of the collector array.

The liquid in a solar collector cools down quite rapidly after the sun has set. To prevent this fluid from flowing into the storage tank and displacing the less dense, heated liquid in the tank, a one-way check valve must be included in the system. Many drain-down valves have a check valve feature built into them.

Tempering valves are located downstream of the solar heating system, between it and the water outlets of the building. The device is connected to the hot water supply line and to the cold water line (see Figure 10-13). Its function is to mix cold water with solar-heated water whenever temperatures get high enough to present a scalding danger to the user. Tempering valves are controlled by a mechanical thermostat and so do not need electric power in order to operate.

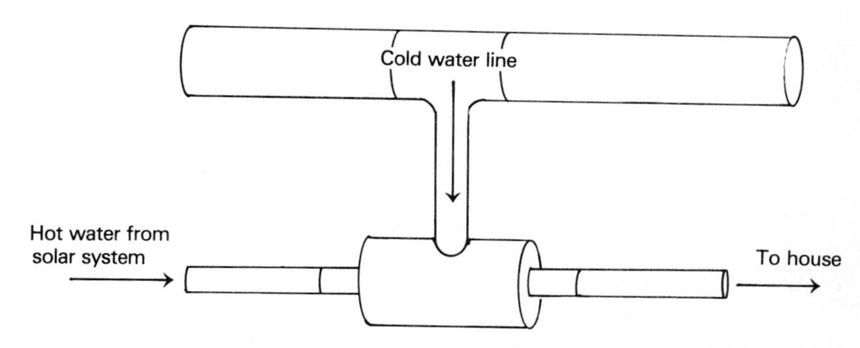

Figure 10-13 Tempering valve. Tempering valve admits cold water into the hot water line whenever there is a scalding danger.

Gate valves—valves that can be manually screwed shut to close down flow—should be installed at enough locations so that storage tank and pump can be removed for repair or replacement without having to drain the rest of the system. Some manufacturers (such as Grundfos) build gate valves into the fittings that are sold with the pumps.

See Figures 10-14 and 10-15 for illustrations of complete open- and closed-loop systems employing the whole gamut of the solar hardware discussed above.

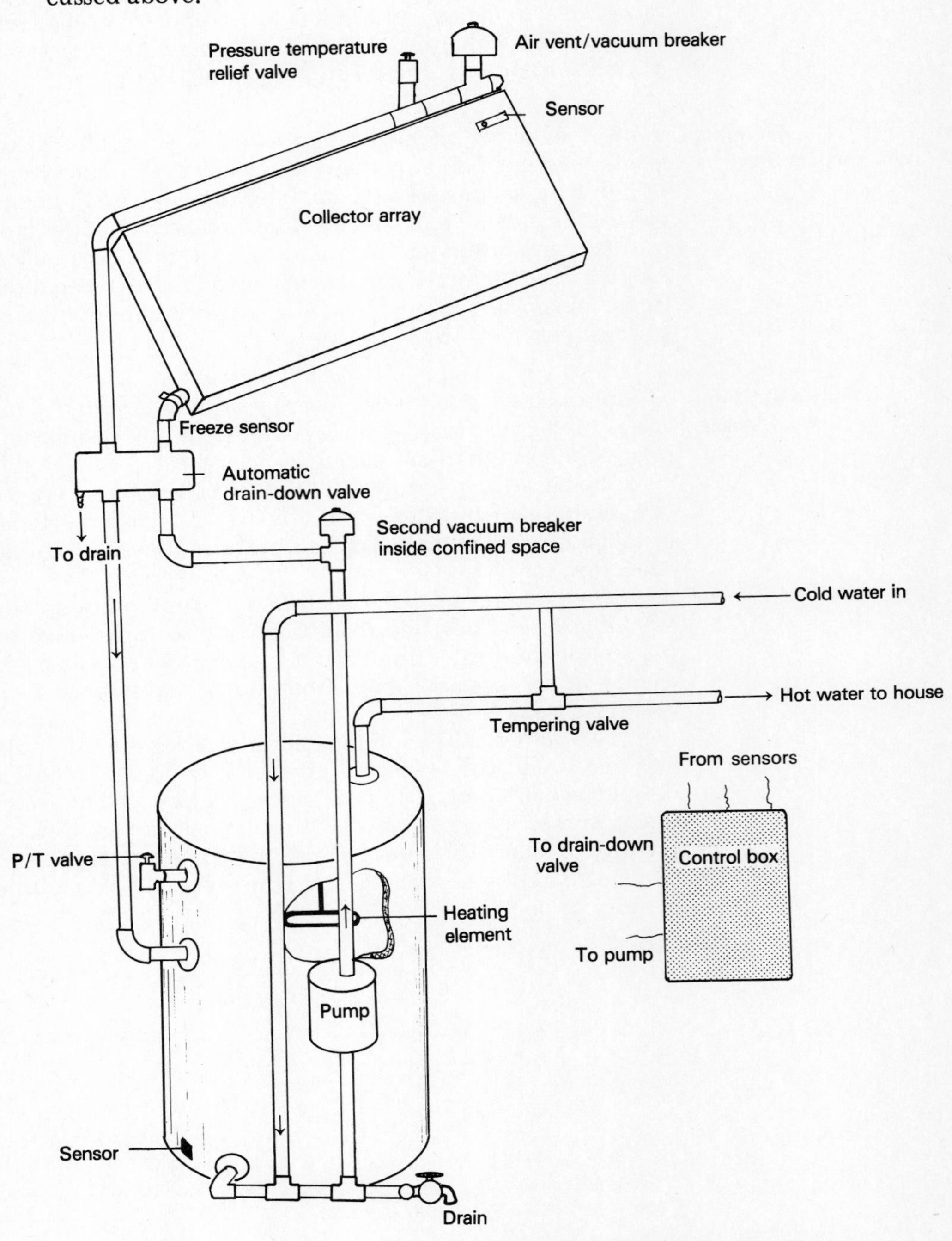

Figure 10-14 Open-loop system (one-tank configuration). In addition to the components above, gate valves are installed at enough locations so that the tank or pump can be removed without having to drain the system.

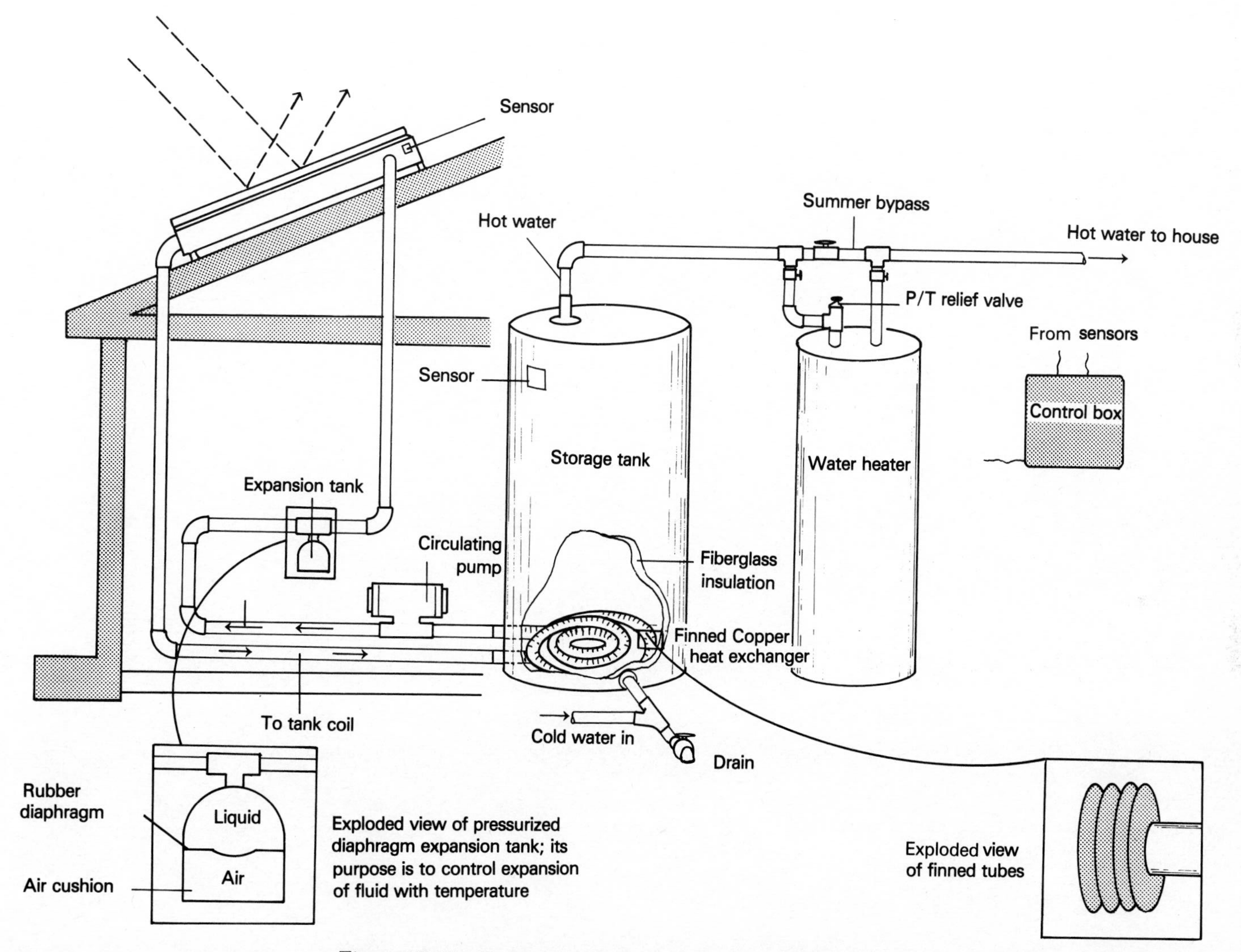

Figure 10-15 Closed-loop system (two-tank configuration).

Purchasing DHW Components

When designing a solar system, it is often necessary to contact manufacturers in order to obtain information and specifications concerning the system's components. Table 10-2 lists manufacturer's addresses and phone numbers for the components discussed above.

TABLE 10-2 Manufacturers and Distributors

Components	*Sample Manufacturers*
Temperature sensors	Midwest Components, Inc. P.O. Box 787 Muskegon, MI 49443 (616) 777-2602
	Sunspool Corp. 439 Tasso Street Palo Alto, CA 94301 (415) 324-2022
Insulated tanks and water heaters	A. O. Smith Corp. P.O. Box 28 Kankakee, IL 60901 (815) 933-8241
	Aquatherm, Inc. 541 Main Street South Weymouth, MA 02190 (617) 659-2643
	Vaughn Corp. 386 Elm Street Salisbury, MA 01950 (617) 462-6683
Closed-loop collector fluids	Sunworks Division of Sun Solector Corp. P.O. Box 3900 Somerville, NJ 08876 (201) 469-0399
	Monsanto Co. 800 North Lindbergh Boulevard St. Louis, MO 63166 (314) 694-2153
	Dow Corning Corp. Department 2314 South Saginaw Road Midland, MI 48640 (517) 496-5985
Heat exchangers that are separate from the storage tank	Aerco-Spirec Products, Inc. 159 Paris Avenue Northvale, NJ 07647 (201) 767-3333
	Alfa-Laval, Inc. 2115 Linwood Avenue Fort Lee, NJ 07024 (201) 592-7800
	Grumann Energy Systems Co.[a] 445 Broadhollow Road, Suite 20 Melville, NY 11747 (516) 454-8674

TABLE 10-2 (cont.)

Components	*Sample Manufacturers*
Expansion tanks	Taco, Inc. 1600 Cranston Street Cranston, RI 02920 (401) 942-8000
Circulating pumps	Taco, Inc. 1600 Cranston Street Cranston, RI 02920 (401) 942-8000
	Grundfos Pumps Corp. 2555 Clovis Avenue Clovis, CA 93612 (209) 299-9741
Automatic system controllers	Rho Sigma Division of Watsco, Inc. 1800 West Fourth Avenue Hialeah, FL 33010 (305) 885-1911
	U.S. Solar Corp. P.O. Drawer K Hampton, FL 32044 (904) 468-1517
	Solar Control 2524 East Vine Drive Ft. Collins, CO 80524 (303) 221-5166
Drain-down valves	Sunspool Corporation 439 Tasso Street Palo Alto, CA 94301 (415) 324-2022
Air vents and vacuum breakers	Hoffman Specialty Division of ITT 1700 West Tenth Street Indianapolis, IN 46222 (317) 632-7546
Pressure/temperature relief valves	Readily available at many hardware and building supply stores (standard equipment on water heaters)
Tempering valves	ITT Lawler Mount Vernon, NY

[a]Grumman manufactures a package that includes *all* components (except collector and tank) necessary for a DHW system.

INSULATING PIPE RUNS

The most important considerations in choosing an insulation for a pipe run are (1) that it provide a high-enough R-value for the system's needs, and (2) that it not deteriorate when exposed to the weather. The *HUD Installation Guidelines for Solar Domestic Hot Water Systems*[5] recommends that a minimum of R-4 insulation be used on all pipes in the collector loop under 1 inch in diameter and R-6 on piping 1 inch or over.

Elastomer insulation is a type commonly used, but it should be noted that it often does not provide an adequate R-value. Elastomer pipe insulation is sold in ½- and ¾-inch wall thicknesses with typical R-values of 2 and 3, respectively, requiring that it be doubled up to meet HUD guidelines.[6]

Elastomer insulation is often installed over existing piping by slitting it lengthwise, but since the insulation shrinks over time, these slits can widen and expose the pipe. Sealing the cracks with mastic or silicone at the time of installation lessens this problem.

Many types of insulation deteriorate drastically after several years of exposure to ultraviolet radiation and extremes of temperature (see Figure 10-16). Although some ultraviolet protection is provided by painting the insulation, a far better solution is to encase it in a metal or PVC jacket. Plastic pipe may be used for this purpose.

Figure 10-16 Badly weathered pipe insulation.

Rigid fiberglass pipe insulation is typically sold in 1-inch wall thicknesses and encased in weatherproof PVC jackets. One inch of rigid fiberglass provides an insulating value close to the HUD recommendations, but if water should get into the fiberglass either through a pipe leak or a gap in a seam of the jacket, its R-value will be severely reduced.

Urethane and phenolic foams are good choices for many solar applications. Both provide excellent thermal protection. One inch of urethane has an insulating value of R-7.2 and 1 inch of phenolic foam, R-5. As with other materials, though, both must be encased in weatherproof jackets if mounted on the outside of the house. Phenolic foam is less durable than urethane but will not give off noxious gases if burned.

Northeast Specialty Insulations, Inc., of Saxonville, Massachusetts, is a distributor of isocyanurate insulation, a modified foam of urethane able to withstand higher temperatures. The company sells a type designed to fit over two pipes—the incoming and the outgoing ones—at one time. It can save labor and will cost less than a separate insulation tube for each pipe. Northeast also makes an insulation with a PVC jacket thick enough to be buried directly in the ground. This feature is of great use in installations where the collectors must be mounted apart from the house, or where hot water is to be pumped to an outside pool or hot tub.

TROUBLESHOOTING YOUR DOMESTIC HOT WATER SYSTEM

The troubleshooting guide shown in Table 10-3 lists problems sometimes encountered in DHW systems. Because it is so much easier to *prevent* these problems than to rework an installed system, the guide should be studied carefully during the design and construction phase of your system, as well as being used to help correct difficulties in existing setups.

TABLE 10-3 **Troubleshooting Guide**

The Problem	*Which Components to Examine*	*What Might Be Wrong with Them*	*How to Correct the Trouble*
System does not produce enough hot water	Collector array	Are the collectors oriented properly?	Face them south and tilt them up from horizontal at an angle equal to the latitude plus 10 to 15°.
		Are they partially shaded between 9 A.M. and 3 P.M.?	Remove obstructions or move the collectors.
		Is there enough square footage of collectors?	Use the tables and method described at the beginning of this chapter to size your array.
		Is the glazing clean?	Clean if necessary. Do it early in the day before the glass gets hot.
	Temperature sensors	Are the sensors loose?	Tighten them.
		Do they make good contact with pipes, absorber plate, or storage tank?	If the surfaces of contact are rough, they must be sanded smooth. A thin layer of silicone heat transfer cream should be applied before the sensor is remounted. Also, there should be no paint between the sensor and the surface whose temperature it is measuring. If possible, mount collector array sensors directly on the absorber plate.
		Are the sensors insulated from the outside air?	Insulate them well, preferably to a value of R-11.
		Are all electric wire splices on or near the sensors protected from the weather?	You do not want water to contact these splices. Contain them in a weatherproof box.
		Are the sensors themselves at fault?	Dip each one alternately in hot and cold water. The automatic system controller that they are attached to should turn the circulating pump on and off. If this does not happen, and the controller, pump, and all electrical splices check out, replace the sensor.
	Automatic system controller	Is it wired properly?	Check instructions.
		Are its electrical connections loose?	Inspect and tighten if necessary.
		Is the controller itself working properly?	Check instruction manual on how to test for this. Do not test by jumping the common and load terminals, as this can damage the unit.
	Piping	Is the piping configuration to the collectors correct?	Improper configurations can cause hot spots and inefficiency. See Figure 10-5 for proper flow patterns.
		Are the pipes insulated sufficiently?	*HUD Installation Guidelines* specify R-4 value insulation or greater on all pipes between the collector array and storage tank, and R-6 if the piping is 1 inch in diameter or greater. If elastomer-type insulations are used, check their R-value. It is often far too low. Check also for gaps and cracks in the insulation. Seal them with silicone or replace the damaged section. Enclose all outside pipe insulation in plastic or metal jackets.
	Check valve	Does it leak, allowing reverse convection to occur at night?	Nighttime reverse convection can quickly deplete the system of its hot water. Repair or replace a faulty check valve or install a solenoid shutoff valve in its place.

TABLE 10-3 (cont.)

The Problem	*Which Components to Examine*	*What Might Be Wrong with Them*	*How to Correct the Trouble*
	Storage tank	Is the tank large enough?	You should have 1½ to 2 gallons of storage for every square foot of solar collector. If you do not, add more storage to the system.
		Is it insulated well enough?	Tanks need to be insulated with 6 inches of fiberglass or its equivalent.
		Where is the tank located?	Tanks should not be located out of doors, where they will be exposed to low temperatures. If practical, they should be placed in the heated portion of the house.
	Tempering valve	Is the valve adjusted correctly?	Set the adjustment of the temperature indicator higher, if necessary.
		Does the valve supply cold water only?	If so, it must be replaced.
Freezing damage *Open-loop systems*	Drain-down valve	Is the valve clogged?	Flush out the entire system with a citric acid solution.
		Is it operating correctly?	After flushing the system, put the freeze sensor in a bowl of ice water (leave the wires from the controller attached to it). The drain-down valve should open. If it does not, and if all wire splices are secure, then the sensor, system controller, and valve must be inspected individually following the instruction manuals.
	Collector array	Can the collectors drain totally?	Collectors must be oriented and plumbed so that they drain completely when the drain-down valve is opened, or else freezing damage is likely to occur.
	Exterior pipe runs	Can all exterior pipe runs drain completely?	Horizontal pipe runs must have a minimum slope of 2°, or about 4 inches of drop for every 10 feet of horizontal run in order to drain properly. Half-inch-diameter pipe should have an even greater slope. Sagging in the pipe runs can lead to puddles inside the pipes that are subject to freezing. Copper pipe will sometimes sag over time if not supported every few feet. Half-inch pipe is especially prone to this problem.
	Freeze sensor	Is the sensor in the right location?	Freeze sensors must be mounted inside collectors, on the bottom third of the absorber plate.
	Air vents/ vacuum breakers	Are the vacuum breakers working correctly?	If they do not open up when the drain-down valve does, they will prevent the system from draining properly. Vacuum breakers must be installed in a vertical position to work correctly. External vacuum breakers must be installed to help prevent them from freezing shut. One should also be mounted *inside* the conditioned space, near the storage tank.
Closed-loop systems	Heat transfer fluid	Is there enough antifreeze in the system?	Check manufacturer's specs to see if the freezing point of the antifreeze–water mixture in the collector loop is lower than the coldest temperatures that you are likely

TABLE 10-3 (cont.)

The Problem	*Which Components to Examine*	*What Might Be Wrong with Them*	*How to Correct the Trouble*
			to encounter in your area. If it is not, add more antifreeze. If you have been topping off your system with pure water periodically, it is very likely that you will have to add more antifreeze. A sample of the fluid in the collector loop should be analyzed by a local lab or the antifreeze manufacturer each fall to determine if the antifreeze content is high enough, as well as to test for acidity level.
Water leaks in the roof	Collector array and piping	Are the roof penetrations of the piping and the collector mounting hardware watertight?	No-caulk roof shoes are recommended for all roof penetrations of the piping. Silicone sealant around the shoes provides extra protection. Roofing cement should be applied on the roof penetrations of all collector mounting hardware.
Water leaks in the collector loop	Piping	Is there seepage from threaded or soldered fittings?	Tighten or resolder all leaking fittings.
	Pressure/ temperature valve	Is its pressure setting too low?	Adjust pressure setting or replace valve.
		Does the valve reseat properly?	Clean the seat or replace valve.
	Collectors	Are the internal flow channels leaking?	If your warranty is not yet void, call your dealer! Fixing collector leaks in the field can be tricky. If you decide to attempt it, braze rather than solder, for high collector temperatures can melt some solders.
Not enough pressure	Valves and vents	Are your P-T valves or air vents spitting liquid?	Adjust or replace them.
	Collectors and piping	Are they leaking?	Repair.
	Expansion tank	Has it lost its pressure?	Recharge it with pressure. This can often be done with a bicycle or automobile tire pump.
Noisy system	Pipes	Is air trapped in the line?	Install float air vents at all high points of the system.
		Do the pipes rattle?	Isolate them from hard surfaces such as rafters and studs against which they can bang. Mount them firmly with the proper pipe-hanging hardware. Make sure there is insulation between pipe and hanger.
	Air vent	Is its cap tight?	Tighten if necessary.
	Air purger (closed-loop systems)	Is it installed backward?	Reverse it. Some purgers have arrows that are supposed to face in the direction of flow.
	Pump	Is the pump air locked?	Many pumps have a venting screw in their bodies. Loosen it, bleed off the air, and retighten.
Antifreeze in the drinking water	Heat exchanger	Is the heat exchanger leaking?	Replace it.
No flow	Pump	Is the pump air locked?	Loosen venting screw in the pump body, bleed off air, and retighten.

TABLE 10-3 (cont.)

The Problem	*Which Components to Examine*	*What Might Be Wrong with Them*	*How to Correct the Trouble*
		Is the pump installed horizontally?	The motor on pumps that are installed horizontally must sometimes be oriented in a certain way. Check pump installation manual.
		Is the pump's impeller stuck?	Loosen impeller screw, or spin impeller with a screwdriver (not with your fingers) to get it started.
		Does the pump have more than one speed?	Turn the pump to its highest speed.
		Can the pump supply the head that the system requires?	Check the pump specifications. If more head is needed, add a second pump or replace with a more powerful pump.
	Collectors and piping	Are they air locked?	Check air vent by operating the vent plunger manually. If it seems to be functioning normally, install additional vents at all high points that don't yet have them.
		Is the piping large enough?	Replace or add a second pump.
	Gate valves	Is a closed gate valve preventing flow?	Open it.
	Flow regulator	Is it clogged?	Clean it.
		Is it installed backward?	Reverse if necessary.
Performance of the system decreases over time	Collectors	Is the glazing dirty?	Clean it.
		Are trees or new buildings shading more of the array than when it was installed?	If possible, remove obstructions or relocate the array.
		Do the collectors use plastic glazing?	The transmissivity of some plastic glazings will decrease over a period of several years (see Chapter 7). These glazings must eventually be replaced. Resurfacing can prolong their lives if performed early enough.
		Are the absorber plate surface coatings chipped and peeling?	They might have to be recoated. Check with your dealer.
	Check valve	Is the check valve getting worn and no longer stopping nighttime convection losses?	Replace, or install a solenoid shutoff valve in the line.
	Backup water heater	Is there a buildup of sediment in the bottom of the water heater?	Drain heater until the water comes out clear. Repeat every few months.
System will not turn on or turn off	Sensors	Are the sensors mounted and insulated correctly? Are their electrical connections tight and protected from the weather? Are the sensors themselves broken?	See the section in this table on what to do if the system does not produce enough hot water. Look under the subheading "Temperature sensors."
	Automatic system controller	Is the controller wired and set correctly?	Check instructions.
		Are its electrical connections loose?	Tighten them.
		Is the controller working properly?	Check instruction manual on how to test for this. Do not test by jumping the common and load terminals, as this will damage the unit.

TABLE 10-3 (cont.)

The Problem	*Which Components to Examine*	*What Might Be Wrong with Them*	*How to Correct the Trouble*
Do the collector mounting racks deteriorate or loosen up over time?	Mounting	Are they made of wood?	All wood used in the racks must be treated in some way to prevent its deterioration due to sun, rain, and extremes of temperature. Pressure-treated wood is best if wood is to be used. Metal racks, however, last longer on the average.
		If the racks are made of metal, were dissimilar metals used in their construction?	It is important that only compatible metals be in contact in the rack. Bolts and lag screws used, for instance, must be compatible with any parts of the rack that they happen to be touching. Otherwise, galvanic corrosion can occur. Your building supply store or solar system dealer can help you to choose compatible materials for your setup.
		Are the racks properly connected to the roof?	It is not enough to attach collector arrays to the roof sheathing only. They must in some way be tied into the rafters. Lag bolts can either be sunk directly into the rafters or into blocking beneath the sheathing that is firmly connected to the adjacent rafters. The bolts should be sunk at least 2 inches into the rafters or blocks. Spanners between the rafters can take the place of blocks. Mounting techniques are described in detail in the *HUD Installation Guidelines.*[a]
Water that comes out of a house tap or shower head is too hot	Tempering valve	Does the system have a tempering valve?	If not, install one. See Figure 10-13 for location.
		Is the valve adjusted correctly?	Readjust.
Snow builds up on the collectors	Collector array	Are the collectors mounted at a steep enough angle?	The tilt angle with respect to horizontal should be equal to the latitude of the site plus 10 to 15°. In areas with heavy snowfall, the collector angle should be at least 45°.

[a] *HUD Installation Guidelines for Solar Domestic Hot Water Systems*, available from the U.S. Government Printing Office, Washington, DC.

PASSIVE DOMESTIC HOT WATER SYSTEMS

A variety of passive DHW components have been introduced into the North American market in recent years. Some are new designs; others have been used in other parts of the world for a long time. The basic difference between passive and active DHW systems is that in the former electric pumps are not required, nor is much of the other hardware associated with active systems, such as automatic controllers and differential temperature sensors. Because passive systems are simpler, with fewer or no moving parts or electronic components, they can last longer and cost less money to buy, install, and maintain. As shown below, however, there are trade-offs for these advantages.

Thermosyphoning Systems

The passive DHW system most like active ones is the thermosyphoning variety, typified by the Amcor and Solahart systems depicted in Figure 10-17. In it, the storage tank is located immediately above the collector,

(a)

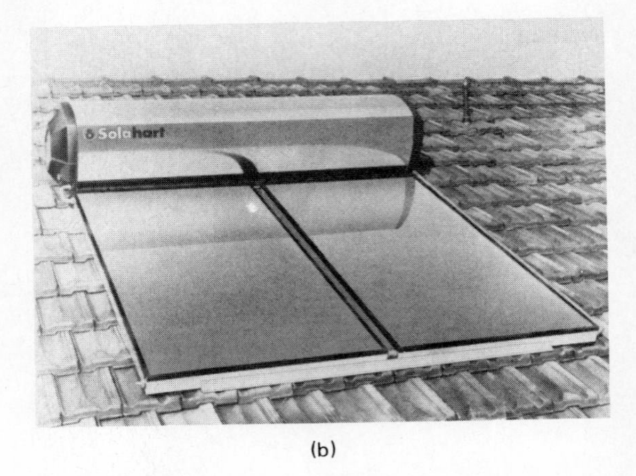

(b)

Figure 10-17 Thermosyphoning water heater systems. (a) Amcor system. (b) Solahart closed-loop system. [(a), Courtesy of Amcor, New York; (b), courtesy of Solahart USA, San Diego, CA.]

which is a flat-plate type very similar to the ones studied in Chapter 9. Water heated in the collector rises into the tank through the process of convection.

The storage tank is mounted horizontally to distribute its weight across several of the roof rafters. Storage tanks full of water are heavy; a full 40-gallon tank weighs about 350 pounds. If mounted on the roof it is a good idea to locate the tank over a load-bearing wall if possible. If must also be ascertained before installation whether or not the roof framing is sufficient to support the weight.

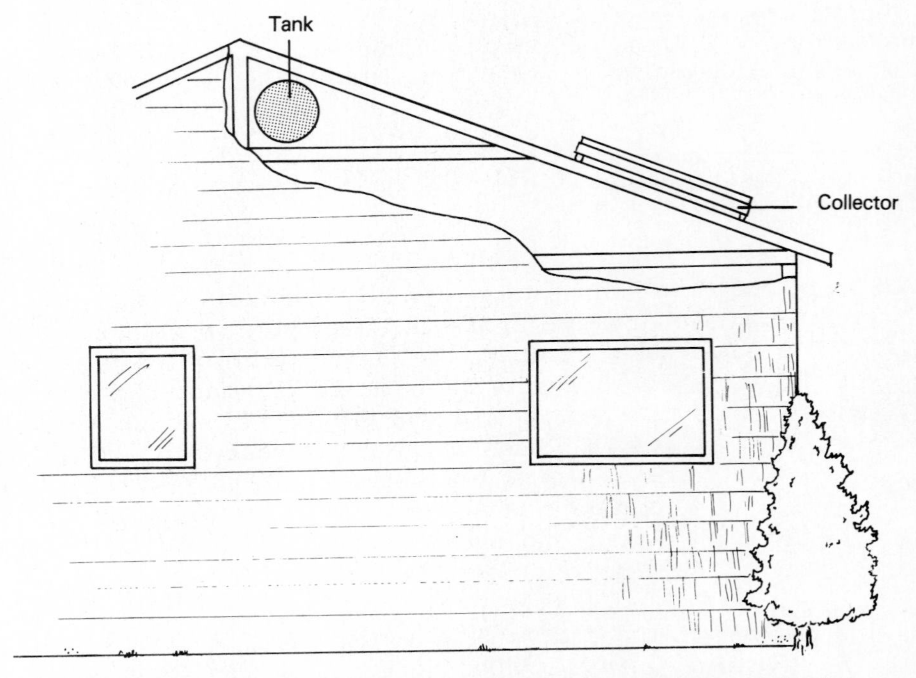

Figure 10-18 Attic installation of storage tank in a thermosyphoning system.

In the Amcor system, as well as in others, there is the option of locating tanks in the attic, but above the tops of the collectors (see Figure 10-18). Electric backup heating elements may also be added to the tanks.

A problem with thermosyphon systems is how to provide reliable freeze protection for them while maintaining the simplicity and lack of moving parts that makes them so appealing. Amcor employs a "dripper valve" in their systems that releases water at an increasing rate as temperatures drop below 38°F. The company is confident enough of its product to offer a five-year warranty against freezing damage. E/One and Solarhart are two other companies that use temperature-activated-water releasing valves in their thermosyphon systems.[7] Solarhart also manufactures a closed-loop system in which nonfreezing propylene glycol fluid in the collector thermosyphons between the outer and inner jackets of the storage tank, transferring its heat to the water within the tank. A cutaway view of this configuration is depicted in Figure 10-19.

Another method of freeze protection is to locate the panels inside a tempered area such as a sunspace. Still another choice is to employ one of the closed-loop thermosyphon systems available that use Freon as the collector fluid. In these systems, the Freon undergoes a phase change from liquid to gas as the sun heats the collector (the boiling point of Freon at atmospheric pressure is 39°F). The Freon vapor rises into a heat exchanger

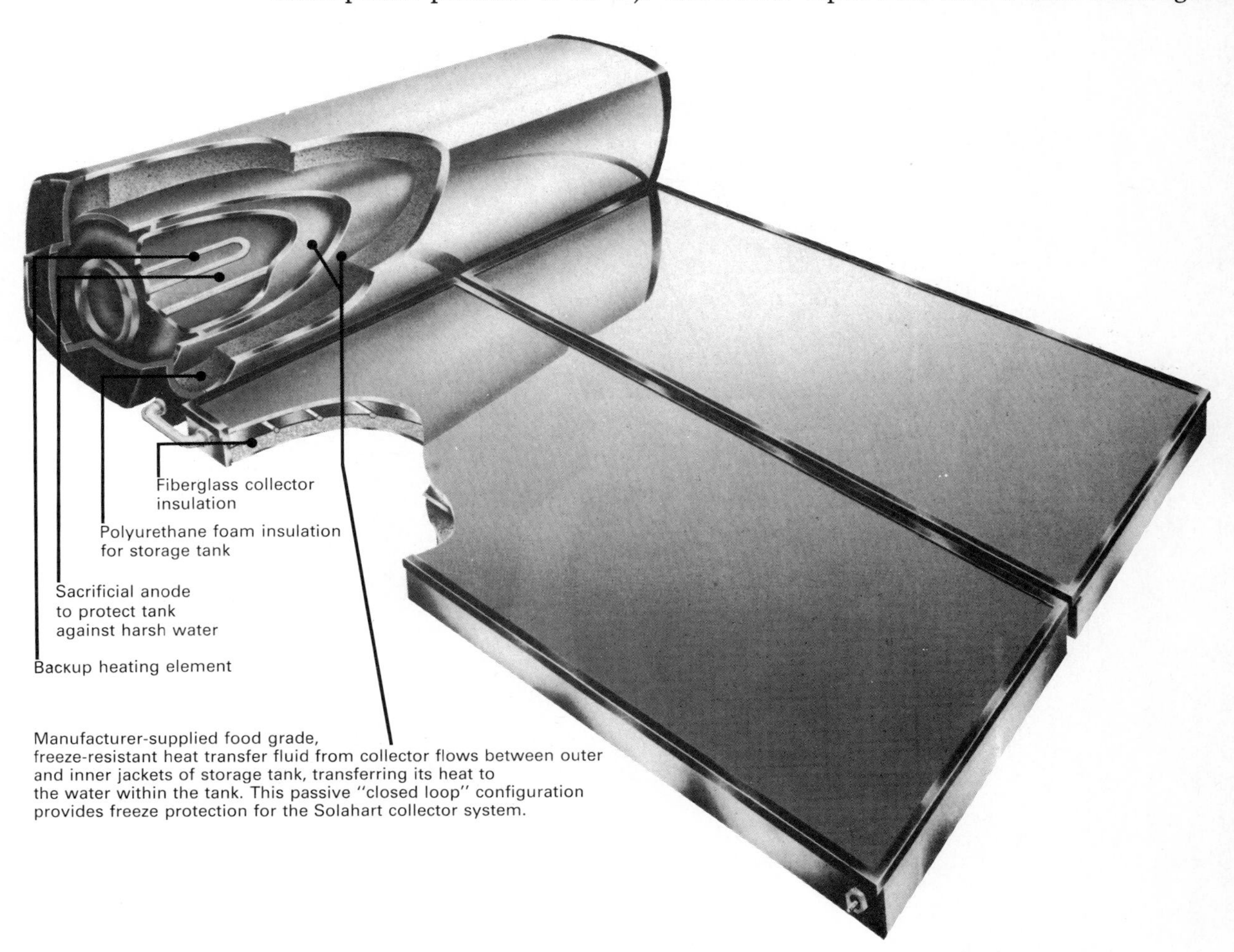
Fiberglass collector insulation
Polyurethane foam insulation for storage tank
Sacrificial anode to protect tank against harsh water
Backup heating element
Manufacturer-supplied food grade, freeze-resistant heat transfer fluid from collector flows between outer and inner jackets of storage tank, transferring its heat to the water within the tank. This passive "closed loop" configuration provides freeze protection for the Solahart collector system.

Figure 10-19 Solahart closed-loop thermosyphon system. (Courtesy of Solahart USA, San Diego, CA.)

located in the water tank. The vapor gives up latent heat at this point to the water in the tank, in the process condensing back into a liquid and flowing with the aid of gravity back to the collectors. Systems of this type are generally referred to as either phase-change or refrigerant-charged designs. Two companies that manufacture such systems are Suntime, Inc., in Bridgton, Maine, and Ying Manufacturing in Gardena, California. Refrigerant-charged water heaters have the advantage of offering extremely reliable freeze protection while maintaining mechanical simplicity.

Integrated Systems

Integrated "batch" collectors are so named because the vessels that they contain are both solar collecting surfaces and thermal storage tanks. As seen from the several designs in Figure 10-20, integral systems are basically water tanks contained in glazed boxes that are usually insulated on all their opaque walls.

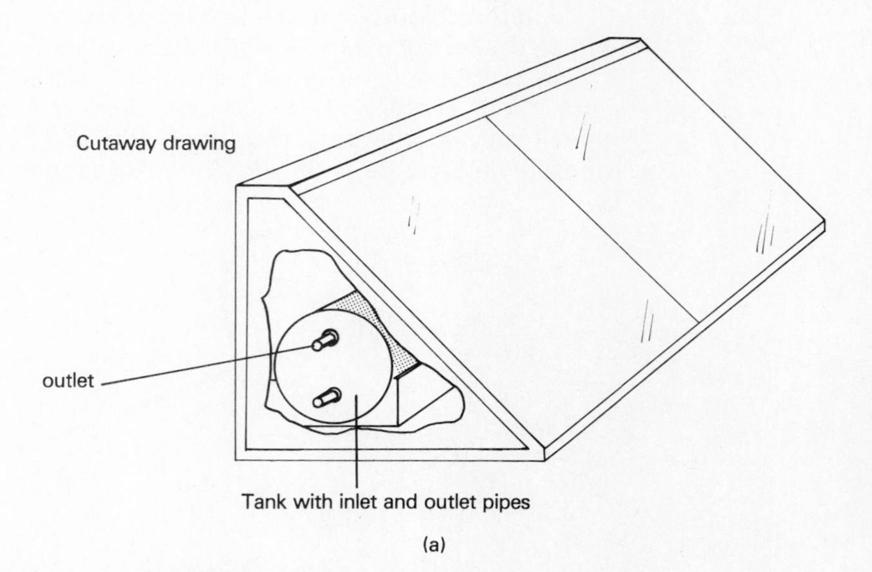

(a)

(b)

Figure 10-20 Integrated systems. (a) Owner-built integral system being installed near author's home. Water pipes are insulated and then enclosed in waterproof plastic drain pipe before being buried in the ground. A 40-gallon tank is enclosed within the insulated box. (b) Owner-built system retrofitted into existing house.

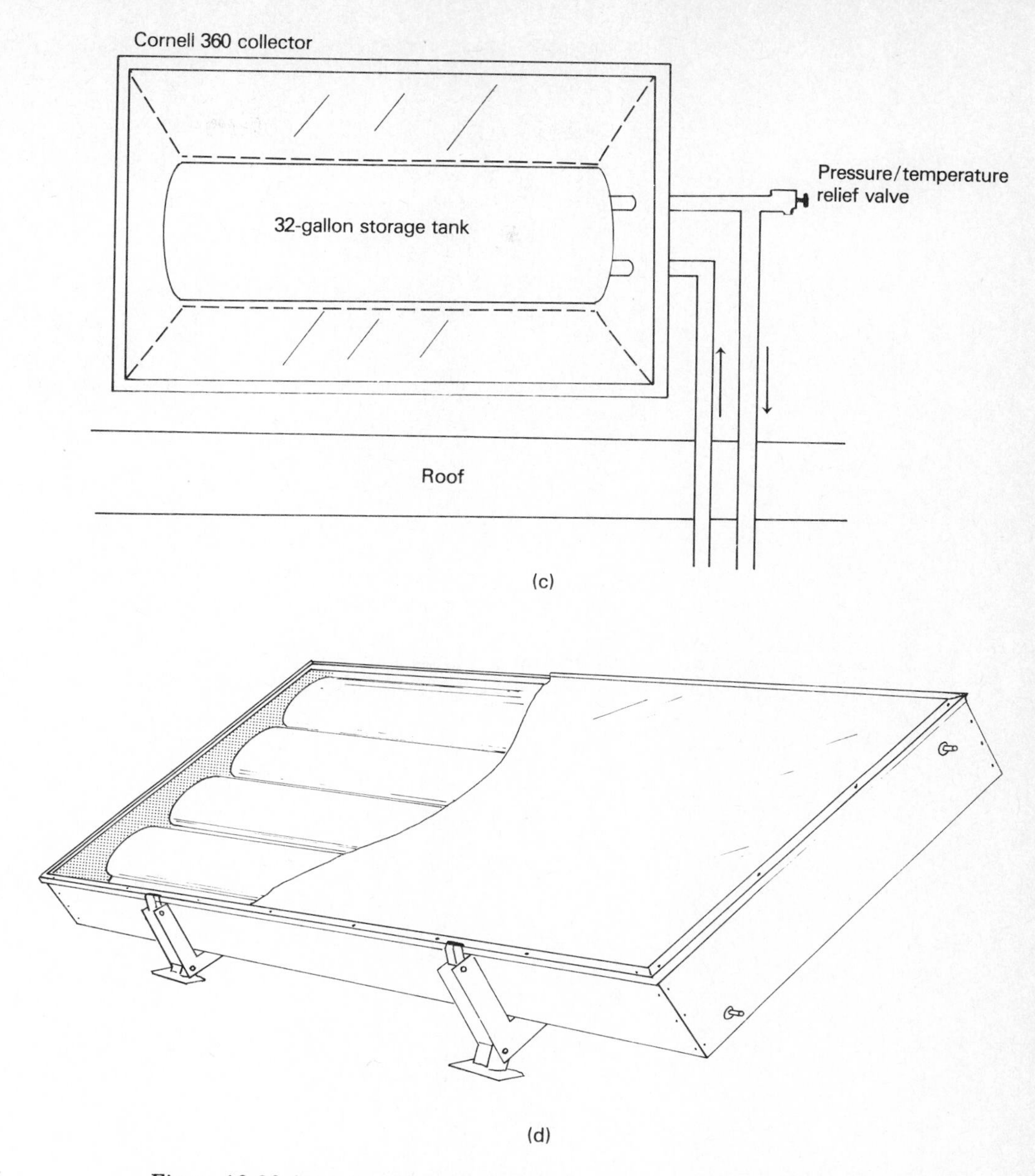

Figure 10-20 (cont.) (c) System manufactured by Cornell Energy, Inc. (d) This collector, made by Gulf Thermal Corp., contains four 6-inch-diameter stainless steel tanks. The hottest water rises to the upper tanks, where it is kept from mixing with cooler water in the lower tanks. [(c), Courtesy of Cornell Energy, Inc., Tucson, AZ; (d) courtesy of Gulf Thermal Corp., Sarasota, FL.]

Integral systems are frequently constructed on-site. By being built into the structure of the house, as depicted in Figure 10-21, they present a less obtrusive profile than if they simply sit on the roof or in the yard.

The popular site-built design illustrated in Figure 10-22 employs two water tanks, one mounted above the other. Cold water enters the lower tank first and after being heated, rises by convection to the upper one, where it

Figure 10-21 The roof gable of this home contains a built-in integral solar water heating system. This house also features a hybrid space heating system made up of solar air collectors on the roof as well as direct-gain apertures in the roof and south wall. Thermal mass in the living room and in a gravel bin under the house store the captured energy.

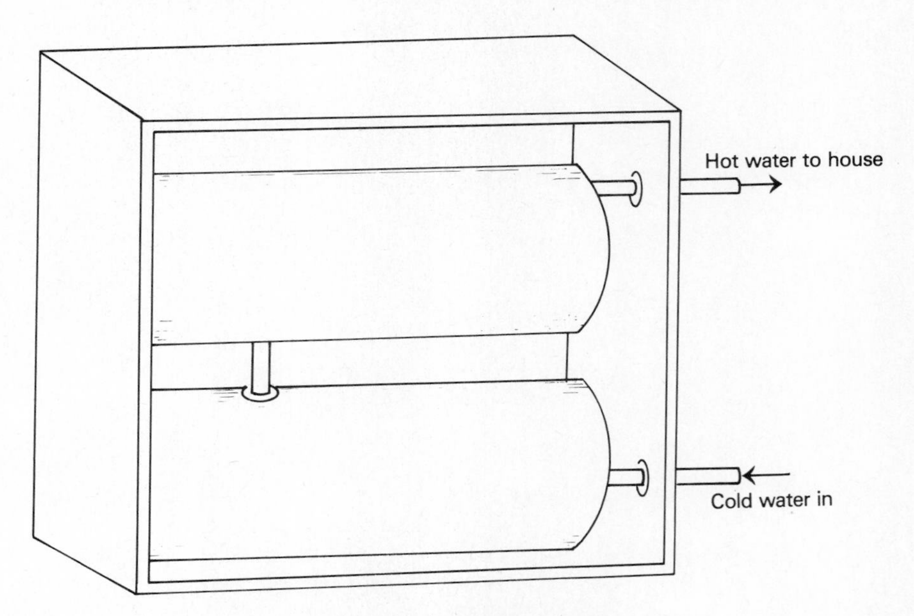

Figure 10-22 Double-tank design.

absorbs still more heat. Water for the building is drawn off the top of the upper tank, where the highest temperatures are.[8] The series arrangement of the tanks has the advantage of preventing cold incoming water from mixing with the system's hottest water.

The water tanks in integral systems are usually painted flat black for solar collection purposes. An alternative is to apply to them selective surface tape such as that sold by Berry Solar Products.[9] This helps the tanks retain more of their heat (see Chapter 6 for a discussion of the properties of the selective surfaces).

The inside walls of the boxes enclosing the tanks are frequently made to be highly reflective and are oriented at such angles as to direct incoming light toward the tanks. Tests by J. Burton, however, indicate that performances of batch collectors with black interior walls are quite close to those with reflective interiors.[10]

The biggest problems encountered by integral systems are severe nighttime heat losses. The boxes enclosing the tanks sometimes have insulating lids on them that can be closed at night, conserving warmth. A surprisingly large percentage of system owners, however, fail to do this regularly, so it is perhaps better to design systems that do not need daily attention. Double or triple glazing helps in this regard, as does a selective surface on the tanks. Another heat-conserving method currently being experimented with is to install translucent insulation (such as fiberglass furnace filter material) around the tanks. This material transmits most of the incident sunlight but retards heat loss in the same way that fiberglass wall insulation does. Although tests at the Farallones Institute[11] have indicated that collectors with selective surface-covered tanks perform somewhat better than those with transmissive insulation, the insulation has the advantage of costing much less per square foot.

In moderate climates, the mass of water contained in a batch collector is sufficient to prevent freezing. In severe climates the system is sometimes put in a tempered space such as an attic or sunspace. Dripper valves and other temperature-activated valves may also be used.

In addition to the site-built integrated systems, some manufactured units are now being marketed. One of these, the TEF SunRunner,[12] features an automatically closing internal shutter to conserve tank heat after sunset (see Figure 10-23).

The SunWizard is an integral system with a unique and rather aesthetically pleasing design. Illustrated in Figure 10-24, it is glazed on all sides and meant to be placed on light-colored south-facing surfaces that will reflect

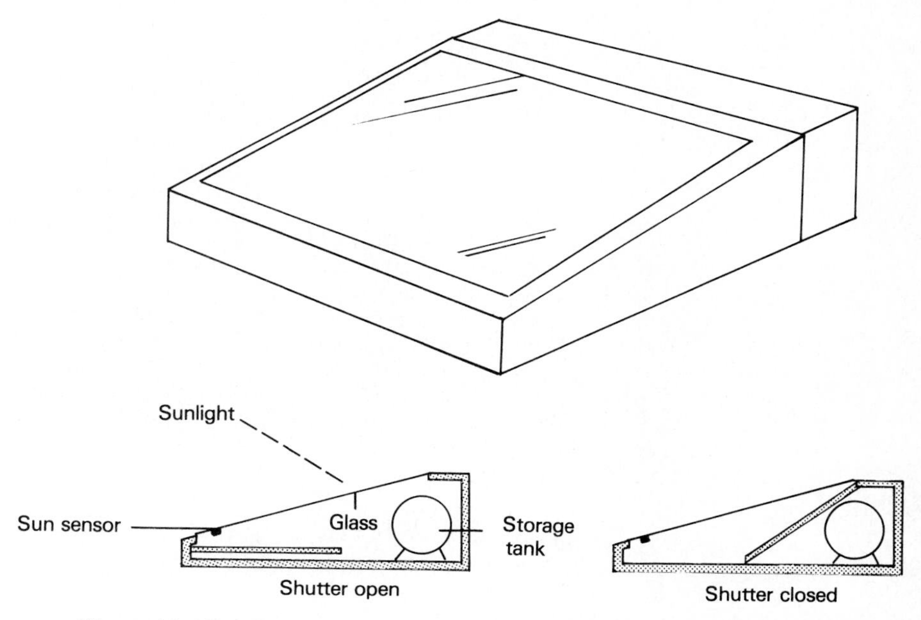

Figure 10-23 The Sun Runner integrated system, made by TEF Manufacturing, contains a shutter that automatically closes when the sun is not shining, insulating the water storage tank. The shutter is electrically operated. (Courtesy of TEF Manufacturing, Fresno, CA.)

incoming energy into the collector, thus enhancing energy gain. The SunWizard system is comparatively low in price and features an excellent "Btu collected per dollar invested" record in tests run on it.[13]

The Glazing/Tank Ratio

A critical factor for all types of passive DHW systems is the glazing area/tank volume ratio. One to 1½ square feet of glazing is recommended for every 2 gallons of water.

In summary, passive DHW systems are simpler in design than active systems and generally easier and quicker to install. There is less that can go wrong with them. But they are heavy, and this can cause problems in mounting them on roofs. Since they depend on natural convection to circulate the water, the tanks in thermosyphon systems must be located nearby and above the collector. Active system tanks, on the other hand, may be located wherever convenient.

Integral passive systems suffer from heat losses through the glazing at night and thus have lower thermal performance statistics than those of well-constructed active systems. This problem can be minimized with insulated covers, selective surface coatings for the tanks, and translucent insulation installed between tank and glazing.

(a)

Figure 10-24 SunWizard integral system. (a) Tank is glazed on all sides and meant to be placed on a light-colored surface that will reflect incoming solar energy into the collector. The SunWizard features an excellent "Btu collected per dollar invested" record in independent tests run on it.

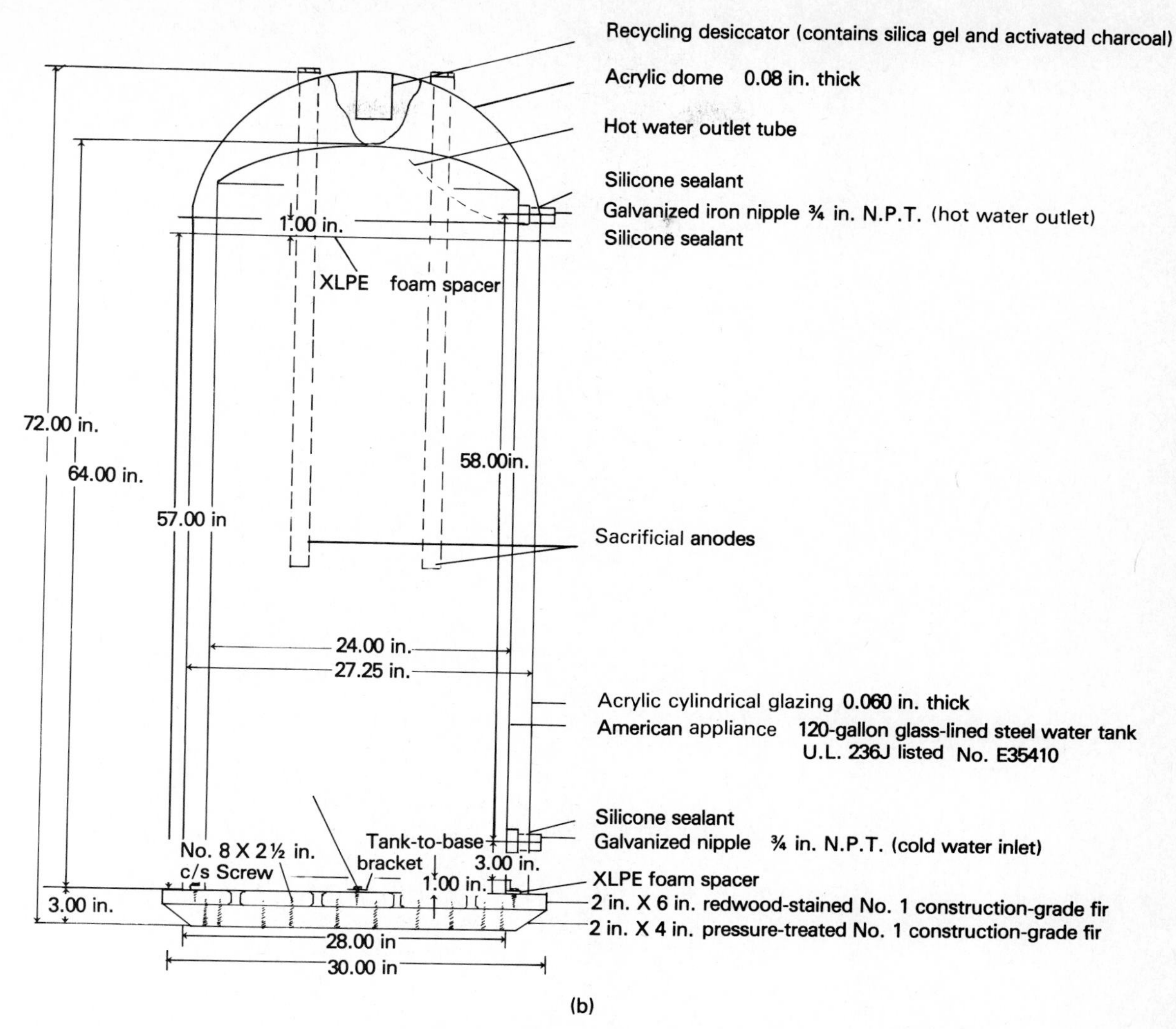

Figure 10-24 (cont.) (b) Cutaway view of the system. (Courtesy of Sun-Wizard, Inc., Harbor City, CA.)

Although there are some disadvantages to passive DHW systems, their advantages suggest that their use in North America will increase markedly in the years to come.

NOTES

1. Solar Energy Applications Laboratory, Colorado State University, *Solar Heating and Cooling of Residential Buildings: Design of Systems*, U.S. Department of Energy, 1980, pp. 9-21 to 9-29.

2. Douglas Beaman and John Rigter, "Freeze Protection: What You Should Know," *Northern California Sun*, Vol. 8, No. 6, p. 10.

3. Bruce Anderson, *The Solar Home Book*, Brick House Publishing Company, Andover, MA, 1976, pp. 196-197.

4. Many hardware or building supply stores carry silicone heat transfer cream. Radio and TV repair stores also generally have, or can get it. It is also called "heat sink cream," as it is used in mounting power transistors onto heat sinks.

5. *HUD Installation Guidelines for Solar Domestic Hot Water Systems.* Available from the U.S. Government Printing Office, Washington, DC.

6. Useful statistics and calculations concerning pipe insulation are contained in the following article: Lew Boyd and John Pesce, "The Costs and Benefits of Solar Pipe Insulation," *Solar Age*, Vol. 6, No. 11, p. 57.

7. John Burton, "Passive Solar Water Heating," *Northern California Sun*, Vol. 9, No. 1, pp. 4–7. Amcor Solar Energy Division, The Amcor Group Ltd., 350 Fifth Avenue, Suite 1907, New York, NY 10118; (212) 736-7711. Environmental/One Corp. (E/One), 2773 Balltown Road, Schenectady, NY 12309; (518) 346-6161. Solahart USA, San Diego, CA.

8. See Appendix I, Lab Project B, for instructions on building an integral batch collector.

9. Berry Solar Products, 2850 Woodbridge Avenue, Edison, NJ 08337. (201) 549-0700.

10. John W. Burton and Peter R. Zweig, "Side by Side Comparison Study of Integral Passive Solar Water Heaters," *Proceedings of the 6th National Passive Solar Conference, American Section of the International Solar Energy Society*, 1981, pp. 136–140.

11. Ibid.

12. TEF Manufacturing, 1550 North Clark, Fresno, CA 93703; (209) 441-1833.

13. SunWizard, Inc., 1424 West 259th Street, Harbor City, CA 90710; (213) 539-8590.

SUGGESTED READING

Directory of SRCC Certified Solar Water Heating System Ratings. Washington, DC: Solar Rating and Certification Corporation, 1983.

REVIEW QUESTIONS

1. Describe what is meant by *internal* and *external* manifolds in relation to a collector array.
2. Explain why the plumbing arrangement in Figure 10-4 will lead to uneven flow rates through the collectors.
3. Do closed-loop systems generally need freeze sensors mounted on the collectors? Do open-loop systems? Explain your answers.
4. What is the function of a two-tank system's summer bypass valve? How does it add to the efficiency of the system?
5. What features should a pump have that will transport solar-heated potable water?
6. What are the functions of a system's air vent and vacuum breaker?
7. Compare the characteristics of various types of pipe insulations.
8. One of the common causes of insufficient hot water is trouble with the temperature sensors. What are the types of things that can go wrong with them?
9. What problems can cause freezing damage in an open-loop system?
10. Compare the characteristics of thermosyphoning systems with those of integrated systems.

PROBLEMS

1. A family of five living in Boise, Idaho, has a home with a washing machine but no automatic dishwasher. How many square feet of collectors does their DHW system require?

2. A three-person family that lives in Houston, Texas, owns a home with a washing machine and an automatic dishwasher. How many square feet of collectors does their DHW system require?

3. At what angle with respect to horizontal should a DHW collector array located in New York City be mounted?

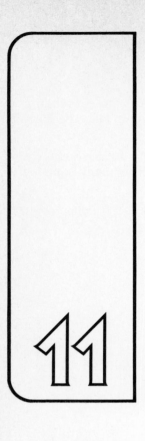

space-heating systems

AN INTRODUCTION TO SPACE HEATING

Earlier in the book, passive methods of heating and cooling buildings were studied. Although very useful, there are times when those methods are not applicable—for instance, in many retrofit situations. The components of active solar systems are more applicable to installation on existing structures. They can also be used on some buildings that are too shaded for a passive system to be practical. This is because an active system's collectors do not have to be placed on the building they will heat. They can be installed in a nearby yard or on a structure such as a garage or barn.

Active systems are either of the air or the hydronic type. *Air systems* use collectors such as those described in Chapter 9, thermal storage bins filled with gravel, and ductwork to transport the heated air. The dampers and blowers of the system's air handler direct the air to the right place at the right time and are operated by an automatic control box and sensors.

Hydronic systems are composed of hydronic collectors (see Chapter 9), piping, one or more pumps to circulate the water or other liquid through the system, a thermal storage tank, and a control box and sensors.

Air and Hydronic Systems Compared

Both air and hydronic systems have been used with good results in every climatic zone of the United States, but each type has its own special advantages. An air system's main advantage is its durability. There is little that can go wrong with its collector array, ductwork, or rock storage bin. The system will not freeze and can still function even with minor leaks. Air collectors begin producing useful heat earlier in the day than those of the hydronic type and can thus minimize the use of conventional backup heaters on cold but clear mornings. Air systems also continue to produce heat later in the afternoon than hydronic systems.[1] Data from a Colorado State University study indicates that because of the longer daily operating time, air systems produce more usable energy over an entire heating season than do similarly sized hydronic systems.[2]

The collectors and storage bins of an air system can be fabricated on the building site itself, which is generally much less expensive than buying manufactured components. The equivalent components of a hydronic system are considerably more time consuming to construct on site, and it is rarely cost-effective to do so. Air systems are easily adapted for summer cooling of the living space; hydronic systems are not.

A possible drawback of an air system is the cost of operating its blowers. This cost is greatly affected by the storage bin, ductwork design, and type of air handler chosen. A well-designed system for a 1500-square-foot house might use 100 kilowatthours of energy per month for its operation during the heating season. A not so carefully designed system can use two or three times this amount.[3] Blower energy requirements are related to pressure drops across the system's components, which can be minimized by the sizing techniques summarized later in the chapter.

A hydronic system's pumps cost far less than air blowers to operate. Pumps, piping, and storage also take up much less space than air handlers, ductwork, and rock bins. And it is easier to retrofit a compact hydronic system than an air system into an existing structure.

There are a number of heat distribution options with a hydronic system. Baseboard radiators may be used, or a radiant slab or a forced-air system. An air-based solar heating system, however, is usually limited to a forced-air distribution system.

AIR SYSTEMS

The Modes of Operation

Air systems operate in several different modes. The heat from their collector arrays can be blown directly into the building's living space or it can be directed into the rock storage bin, raising its temperature. During cloudy periods or in the evening, the storage bin's thermal energy may be used to heat the living area. When there is insufficient thermal energy in both the collectors and the rock storage, conventional heaters are turned on.

The system can also preheat the building's water supply, and in summer it can function as a cooler, blowing night air through the rock bin and then passing living space air through the cold rocks the next day.

The system's modes are governed by an automatic control box acting on temperature information received from sensors on the collectors, in the storage bin and living space, and if the system performs DHW and cooling functions, on the water tank and outside the house as well.

Mode 1: Space Heating Directly from the Collectors. In this mode, the air handler's blowers transport heated air from the solar panels directly to the living space (see Figure 11-1). The automatic controller initiates this mode when the solar collector temperature rises above a predetermined point and the living space thermostat indicates that heat is needed. The mode persists until the living space thermostat signals the controller that the room temperature has risen to a comfortable level. If the solar panels are still collecting heat at this time, the controller shifts the system into mode 2.

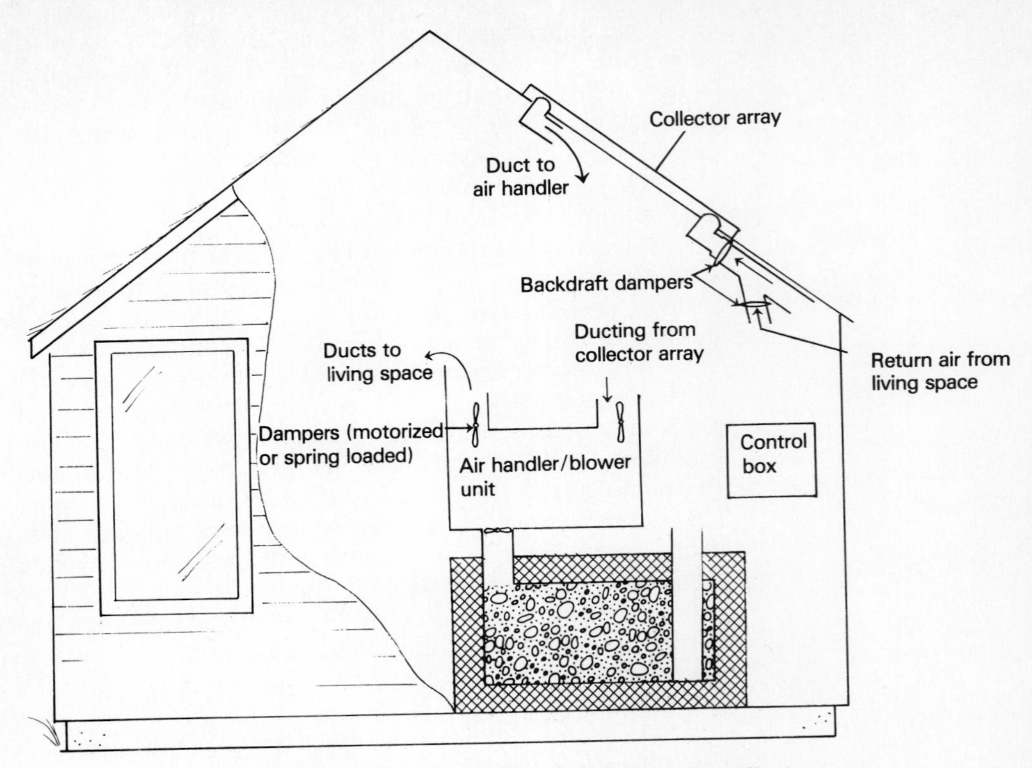

Figure 11-1 Space heating directly from the collectors.

Mode 2: Storing Heat. In this mode, the air handler directs heated air from the panels to the storage bin. As the air enters the bin its temperature can be 140°F or even higher. As it travels through the bin the air transfers thermal energy to the rocks and should return to the collectors at about 70°F (see Figure 11-2). The cool return air allows the panels to operate at lower temperatures. This is advantageous, because lower operating temperatures increase the collector's efficiency.

Mode 3: Heating the Living Space from Energy Stored in the Rock Bin. When the sun has set or is behind clouds, the collector temperature drops until it is too cold to provide heat. If the storage bin is at a high enough temperature and the house still needs heat, the air handler circulates living space air through the bin. Notice from Figure 11-3 that air is blown through the rock in the opposite direction as in mode 2. Air now enters the *cool* end of the bin and emerges at the hot end at a temperature almost equal to that of the hottest rocks.

Mode 4: Backup Heating. Conventional heating equipment is used to warm the house during times of insufficient sunlight. Any type of heater can be employed as the backup, although forced-air furnaces fit conveniently into the solar system's ductwork. They should be installed so that return air from the living space first passes through the rock storage, picking up any thermal energy remaining there, and then passes through the forced-air furnace (see Figure 11-4).

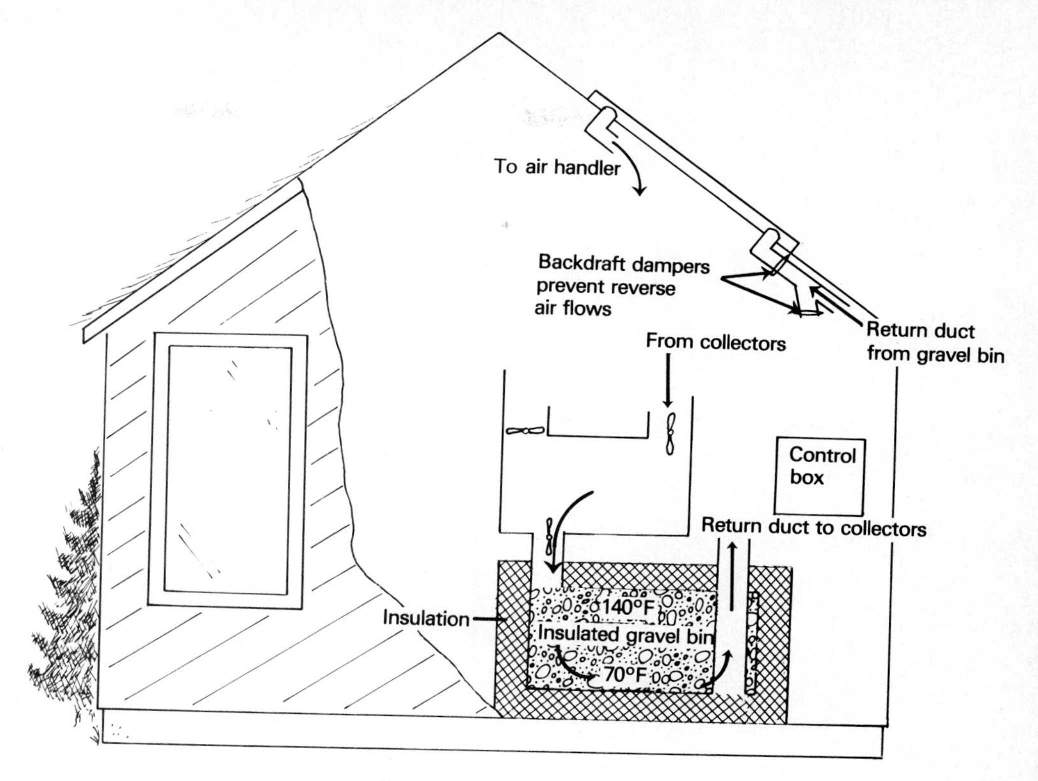

Figure 11-2 Storing heat.

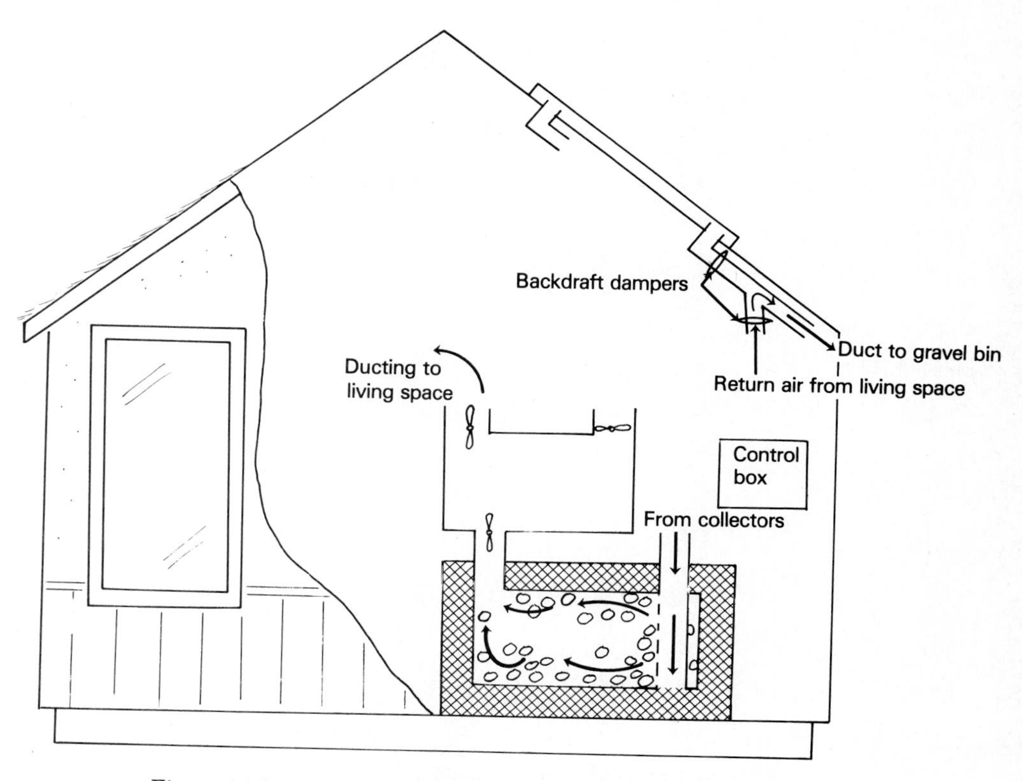

Figure 11-3 Living space heated by energy stored in the gravel bin.

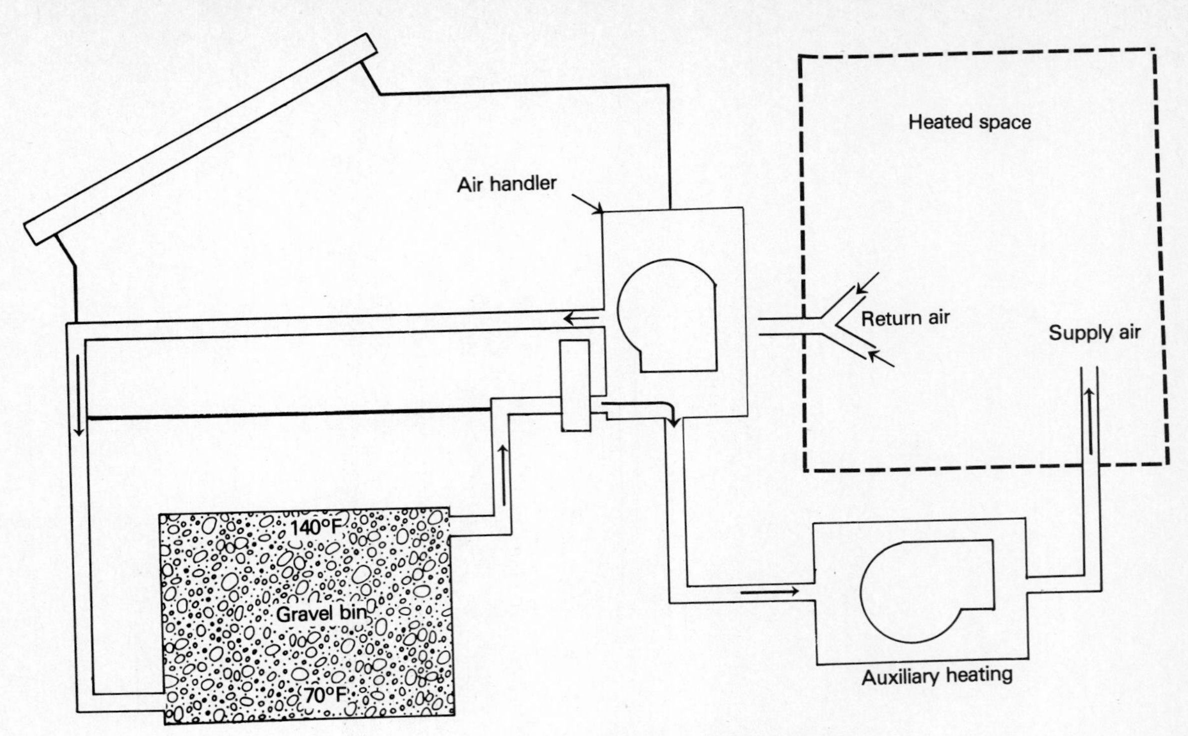

Figure 11-4 Auxiliary heating.

Mode 5: Domestic Water Heating. A solar air system can preheat the building's water supply if an air-to-water heat exchanger is installed inside the ductwork. A common location for it is on the vent of the air handler that leads to the rock bin (see Figure 11-5). Heated air will not pass through an exchanger in this location during operational mode 1 (space heating directly

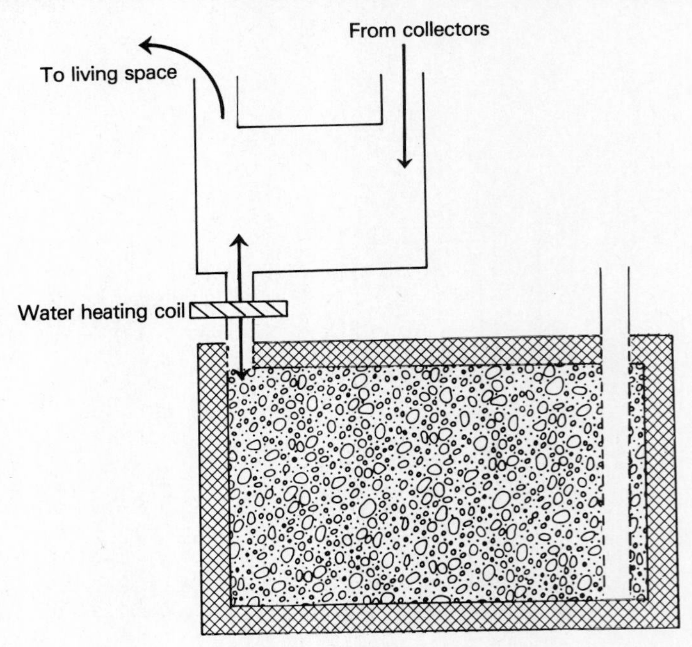

Figure 11-5 Domestic water heating. Hot air can pass through water heating coil in both mode 2 (storing heat) and mode 3 (heating living space from rock storage).

from the collector array). This is fitting, because mode 1's top priority is to bring the living-space temperature rapidly up to a predetermined level, and draining heat away for DHW purposes could slow down this process. Once the desired room temperature has been reached, the system switches to mode 2 (storing heat) and domestic water heating can begin. Warm air also passes through the air-to-water heat exchanger during mode 3 (heating living space from storage).

A disadvantage of using air systems to preheat water is that the cost of operating the space-heating system's main blowers during summer months when room heating is not needed can be somewhat high compared to the economic return. TEF Manufacturing of Fresno, California, is one company that circumvents this problem by offering an additional low-powered blower specifically designed for DHW applications as an optional feature of its system.

Mode 6: Space Cooling. Installing a "cooling tee" on the discharge side of the collector, as illustrated in Figure 11-6, will allow the air handler to draw outside air into the system. A cooling tee shuts off flow from the collector while it opens the system to the outside. Its dampers are motor driven and governed by the system's control box. The air intake for the tee should be located on the north side of the building and in the shade, if possible. The cooling mode is put into effect whenever interior temperatures are uncomfortably high and outside air temperature is sufficiently low.

The outside air that flows through the cooling tee is either directed into the living area or through the rock bin, storing "coolth" for future use. As cool air is blown into the living space from outside, it pushes exhaust air through the return ducts to the collectors, where the air escapes through vents that are opened in summer.

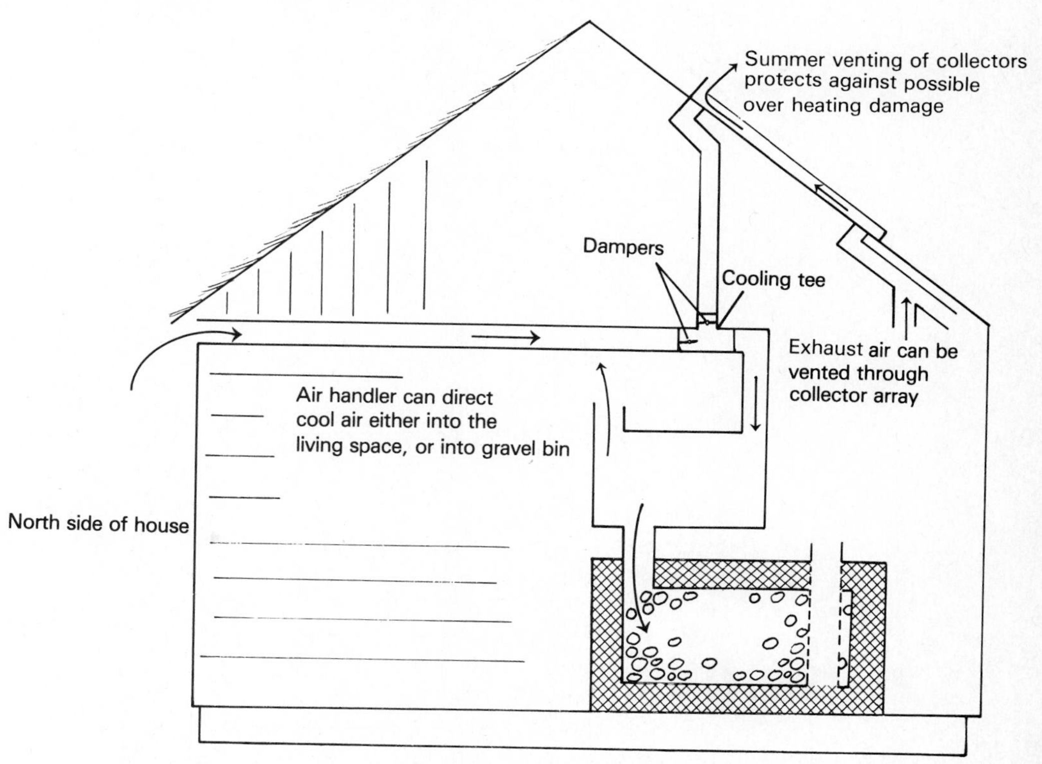

Figure 11-6 Cooling tee.

System Components

The Rock Storage Bin. The rock storage bin of an air system is constructed of either reinforced concrete or wood and is insulated on all sides, including the bottom, to a value of R-11 if in a heated space, or to R-30 if in an unheated location. Concrete bins should be insulated on the inside with rigid fiberglass.[4] Wood bins must be lined on the inside with a noncombustible material able to withstand 200°F temperatures, such as gypsum board. Rock bin construction details are outlined in Figure 11-7.

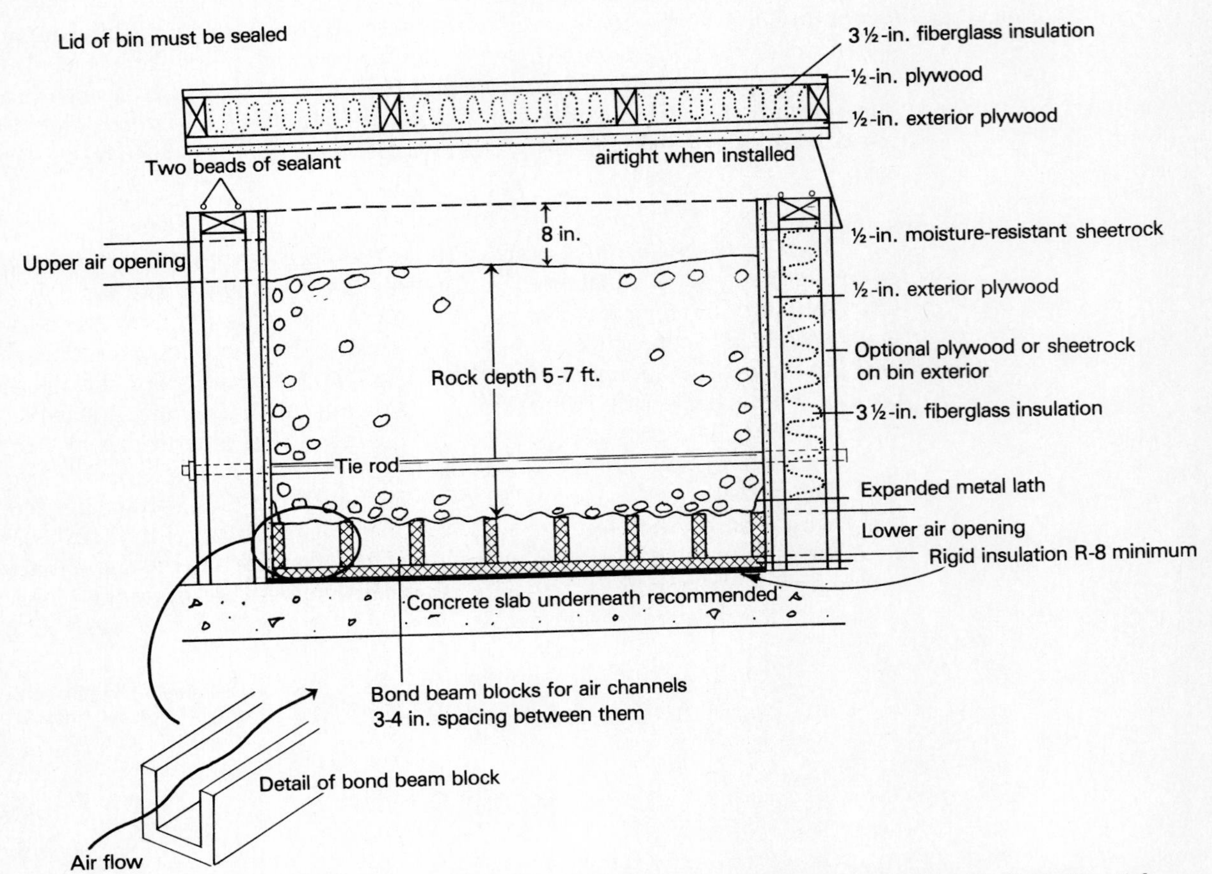

Figure 11-7 Rock bin details. Diagram constructed with the aid of *Residential Heat Storage Unit Installation Manual*, Solaron Corporation, Englewood, CO.

An 8-inch airflow space is left above the rock and an 8-inch-high opening is put in the bin wall to service it. Bond beams below the rock form air channels to the bin's lower opening, which must be twice as wide as the upper one due to air restrictions caused by the beams.

Crawl spaces and basements are popular locations for storage bins. Warmth escaping from them will rise into the living space, helping to heat it. If, instead, the bin is buried in the earth, all of its surfaces that are in contact with the ground must be thoroughly waterproofed.

The bin should be filled with clean, dry river gravel or hard crushed rock, ¾ to 1½ inches in diameter. Very fine rock such as pea gravel should be avoided, as it will fill in the air spaces and restrict the flow. Rock depths are typically 5 to 7 feet.

Bins should be designed so that airflow through the rocks is *vertical.* In horizontal bins there is the possibility of "channeling"—in other words, the heated air flowing only through the top layers of rock (see Figure 11-8). If a horizontal rock bin must be used due to space limitations, vertical heat separation problems can be reduced by making the bin no more than 32 inches deep. It should be no less than 24 inches deep or the system's blower will have to work excessively hard to force air through, leading to a heavy power draw.[5]

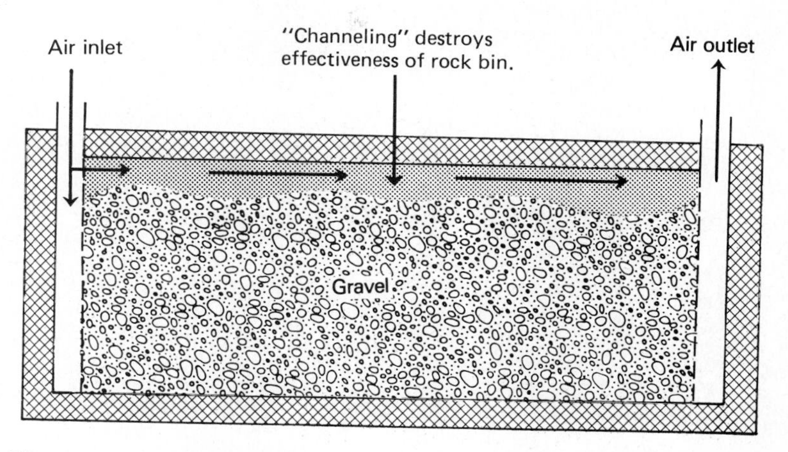

Figure 11-8 If a horizontal flow rock bin is used, there must be no space on top of the gravel that can "channel" air from inlet to outlet aperture without ever passing through the rock.

Temperature Stratification in Rock Storage Bins. Figure 11-9 depicts a typical temperature profile of a vertical storage bin at various times during a clear winter day. The system's air handler directs solar-heated air downward through the rock bin throughout the day. What is interesting is that even though the upper levels of rock get hotter and hotter until they reach temperatures of up to 140°F, the bottom levels remain at room temperature.

The advantage of sharply stratified storage bins is that they allow air used for space heating (mode 3) to be warmed to almost the temperature of the bin's hottest rocks, while air taken from the bottom of the bin during the heat storage mode (mode 2) can be returned to the collectors at a far lower temperature, enabling them to operate at high efficiency.

The reason that incoming air temperature affects a collector's efficiency is because efficiency is heavily dependent on the rate of heat loss from the collector. Heat loss, in turn, is proportional to the temperature difference between the inside and outside of the solar panel. The lower the panel's operating temperature, the less the temperature difference. The more efficient a collector array is, the more usable energy it produces to heat the living space or recharge the rock bins.

After the sun has set, stored heat in the rock bin is used to keep the house at a comfortable temperature. Air blown upward through the rock picks up energy from the top layers and carries it to the living space, reducing the rock's temperature. Bins are generally designed so that most of their energy is used up by the next morning.

Rock bin temperature stratifications are much more stable than those in a solar hydronic system's water storage tanks. Inlet water that enters a hydronic storage tank at too high a velocity, for instance, can cause mixing of hot and cold layers. This problem cannot occur, though, in the solid storage medium of a rock bin.

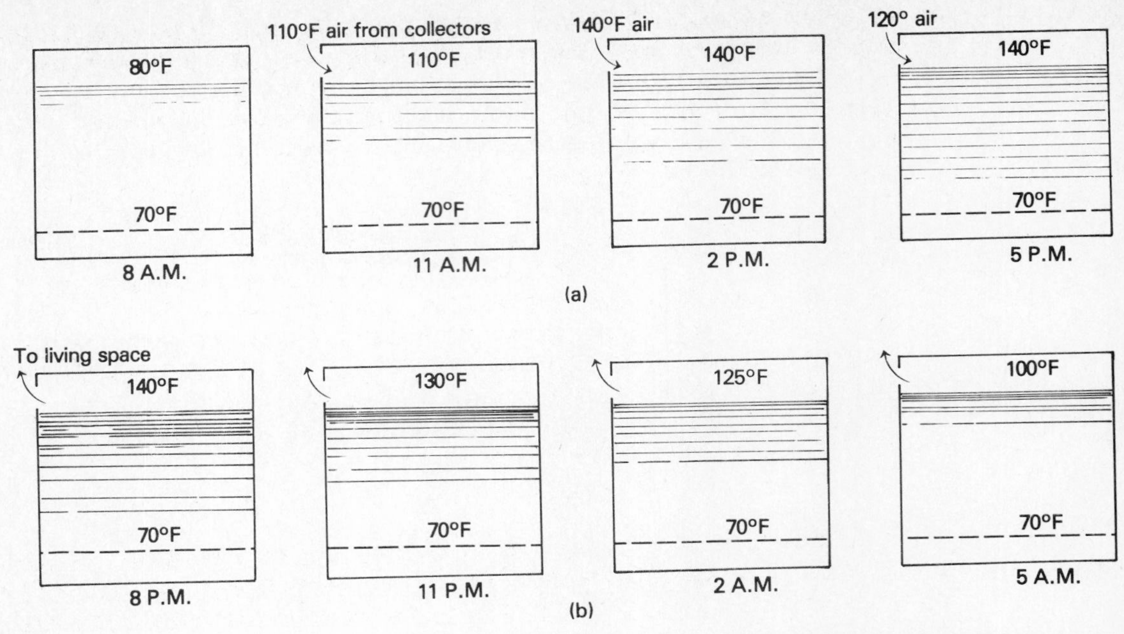

Figure 11-9 Temperature stratification in rock storage bins. (a) Charge cycle. A typical solar system during the winter starts in the morning after having used most of the stored solar energy for the nighttime heating requirement. The collector outlet temperatures (and temperatures into storage) typically peak during the middle of the day and are lower in the early morning and late afternoon. This results in stratified layers of heat such as 120 to 140°F at the top, approximately 140°F near the middle, 100 to 120°F near the bottom, and 70°F at the bottom. (b) Discharge cycle. Heat is extracted from the storage unit in reverse of the charge cycle. The energy stored in the late afternoon is used first. Energy stored at the highest temperatures during the middle of the day is used at night during the peak heating hours. Energy stored from the early morning hours is used toward the end of the peak nighttime heating hours. Propagation of the temperature profile through the pebble bed during a typical charge-discharge cycle is as measured in actual system operation at Colorado State University. (Courtesy of Solaron Corporation, Englewood, CO.)

The Air Handler and Its Dampers. The air handler is a sheet metal box containing one or more fans and several motor-driven or spring-loaded dampers, whose purpose is to direct air through the various ducts, collectors, and storage bins of the solar system.

The differential operational modes of the system can be put into effect by opening different dampers on the air handler. Figure 11-1, for instance, illustrates the damper combination for mode 1 (space heating directly from collectors). Figures 11-2 and 11-3 illustrate damper combinations for modes 2 and 3, respectively.

The fans in an air handler's blowers are either directly connected to the shaft of an electric motor or driven by belts. Direct-drive units are more energy efficient, whereas belt drives have the advantage of being quieter. If the motors are in the airstream, they need to be built to withstand system temperatures above 140°F.

It is important that an air handler's dampers be fitted with rubber seals and close tightly in order to prevent air leakages. Back draft dampers are installed in the system to permit airflow in one direction only (see Figure 11-10). They must be mounted in straight sections of ductwork away from

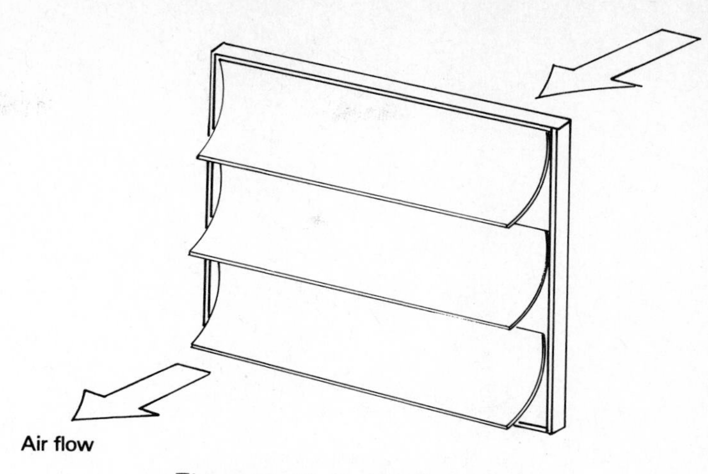

Figure 11-10 Backdraft damper.

elbows and tees to prevent leakage due to uneven airflow and velocities higher than they were designed for.[6]

Ductwork. Either metal or fiberglass ductwork is acceptable for solar systems, but if the latter is used it should be protected by plasterboard or paneling, as it is rather brittle. Ducts should be insulated to a value of at least R-4 and must be taped well at all joints to prevent leakage. Flexible connections between ducting and the air handlers will greatly reduce the noise level. All bends in the ductwork should be fitted with turning vanes to minimize pressure losses.[7]

The Control Box and Its Sensors. The control box is the "brain" of an active solar system. Based on information received from sensors in various locations, the control box automatically designates the appropriate operational mode. It signals the air handler to open a damper or turn on a blower. It is designed to optimize the performance of the collector array and storage unit as well as to operate the auxiliary heaters when necessary.

A typical controller receives information from five temperature sensors (see Figure 11-11). One of these is mounted in or near the collector outlet manifold and measures the highest air temperatures in the array. Another is in the duct running from storage bin to collector inlet. Two other sensors are mounted near the top and bottom of the storage bin (in a horizontal bin, they would be located on the hot and cold ends). The final sensor is mounted outside the building within a foot of the fresh-air intake and is used in cooling modes. It should not be mounted under the building eaves where trapped hot air can cause fallacious readings. All sensors must be accessible for repair or replacement.

The controller also receives information from an adjustable thermostat, usually of the two-stage variety, located in the living space. When room temperature falls to a certain preset level, the thermostat signals the controller and heat is supplied from collectors or storage. If the heat supplied is not sufficient, the temperature continues to drop, and at a point a few degrees below the first level the thermostat sends another signal to the controller and the backup heating unit is turned on.

To understand the control box functions, it is instructive to study the sequence of operations performed by Solaron Corporation's system controllers.[8]

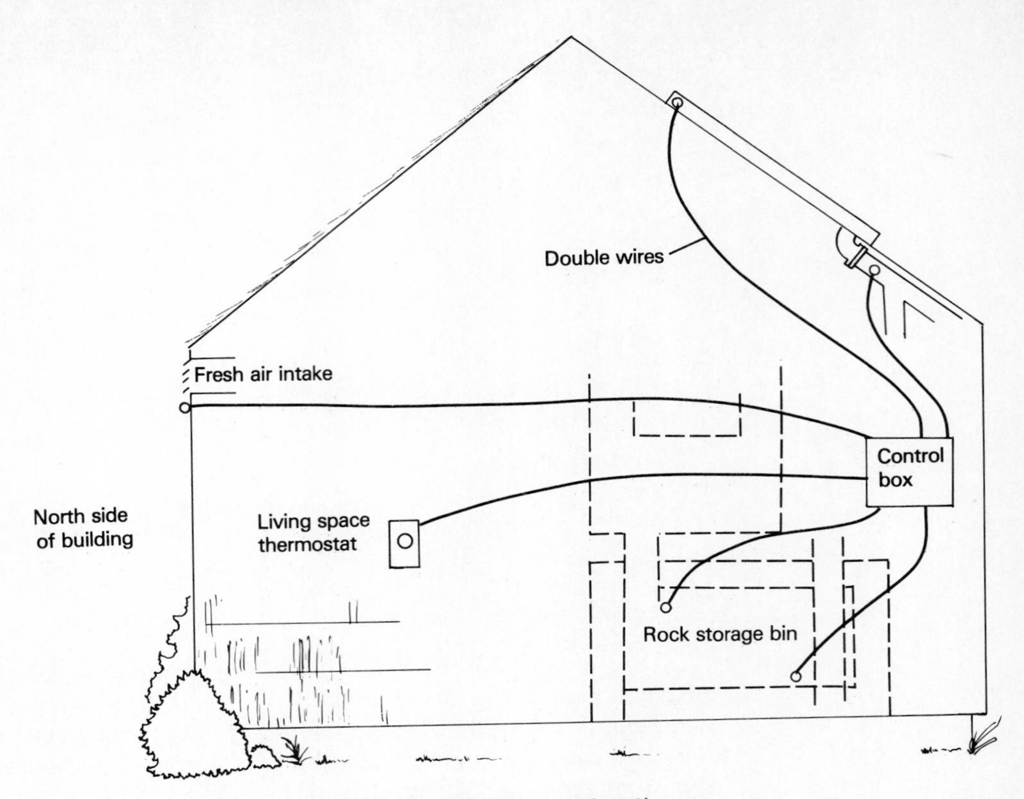

Figure 11-11 Sensor locations.

First Situation: Solar Energy Is Available. When the sensors indicate that the collector outlet temperature is 40°F above the collector inlet temperature and the room thermostat is calling for heat, the controller turns on the air handler blowers and opens the appropriate damper to transport heated air from the collectors directly to the house (see Figure 11-1). If the living-space temperature continues to fall, the controller turns on the auxiliary heating units as well. These may be gas, electric, oil, or heat pump types. If the room thermostat does not indicate that living space heating is necessary, the dampers are adjusted to direct heated air from the collector to the storage bin (this is mode 2). Auxiliary heaters are turned off.

Second Situation: Solar Energy Is Not Available. When the collector outlet temperature drops to 25°F above the collector inlet temperature and the room thermostat indicates that heat is needed in the living space, one of two situations occurs. If the temperature on the hot end of the rock storage bin is at least 90°F, air is blown through the bin and then into the living space (see Figure 11-3). If the temperature at the hot end of the bin is below 90°F, air is still transported through the storage so it can pick up the thermal energy that remains there, but it then passes through the turned-on auxiliary heating unit before it reaches the living space. If the controller were not to turn on the auxiliary unit, and air only slightly higher than room temperature were to be blown into the living space, there would be the uncomfortable sensation of drafts.[9]

The system controller can also be employed to govern the summer cooling and domestic water heating modes.

Sizing Calculations for Air Systems This section provides quantitative tools for sizing collector arrays, ductwork, and rock storage bins.

Sizing a Collector Array. The first step in sizing a collector array is to obtain a "ballpark" estimate of the required square footage from which to begin the calculation. One method of doing this is as follows:[10]

$$\text{approximate collector area} = \frac{\text{square footage of the heated living-space floor area}}{4}$$

For example, a house with a heated living-space floor area of 1600 square feet requires roughly 400 square feet of collector.

Once the initial estimate is obtained, the more accurate and very popular *f-chart method* of sizing a collector array may be employed.[11] This method was developed using computer simulations of many different situations. It involves, first, calculating two terms, called X and Y, that depend on quantities such as average monthly insolation at the site being examined, the heating load of the house, the average outside temperature, and the characteristics of the solar panels used. A trial collector square footage is also included in the X and Y terms. X may be thought of as the collector array's heat loss divided by the building's monthly heat load, while Y is the collector array's solar energy gain divided by the building's monthly heating load. After X and Y are computed, the *solar heating fraction* is obtained from the *f*-chart graph (see Figure 11-12). The solar heating fraction is the part of the building's heat load that will be supplied by the trial collector square footage for the time of year being considered. If this is not the solar heating fraction desired, the collector square footage is adjusted until a satisfactory heating fraction is achieved.

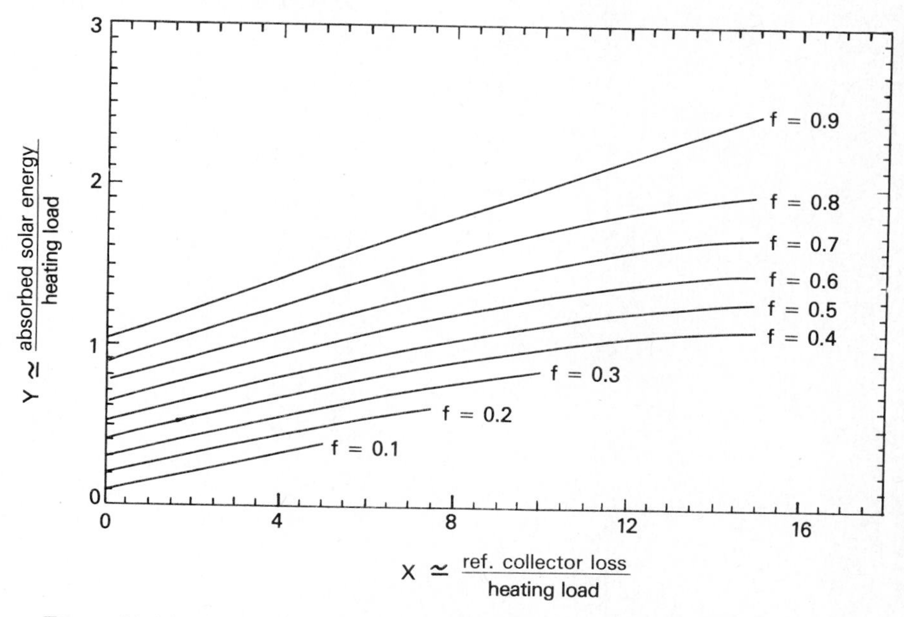

Figure 11-12 *f*-Chart for air systems. (From William A. Beckman, Sanford A. Klein, and John A. Duffie, *Solar Heating Design by the f-Chart Method*, John Wiley & Sons, Inc., New York, 1977. Reprinted by permission of the publisher.)

The terms X and Y are computed as follows:

$$X = \frac{SAh(212 - T)}{L}$$

$$Y = \frac{(0.93)VAI}{L}$$

where V = vertical-axis intercept of the ASHRAE efficiency graph of the collectors considered

A = gross collector area

h = number of hours in the month being considered

T = average temperature for the month

L = space-heating load for the month (if the system is to heat domestic water, the combined space and water heating load should be used; Chapter 2 dealt in detail with methods of calculating space-heating loads, and Chapter 10 gave a method of estimating domestic water heating loads)

S = slope of the ASHRAE efficiency graph for the collector being considered

I = average insolation on the collector surface (Btu/ft^2-month) for the month in question

The multiplier 0.93 in the Y term corrects for the fact that not all of the sunlight hits the collector surface at a perpendicular angle. The 212 in the X term is a reference temperature.

Figure 11-13 is an efficiency graph for a solar air collector manufactured by Rom-Aire. The vertical-axis intercept, V, is the point at which the

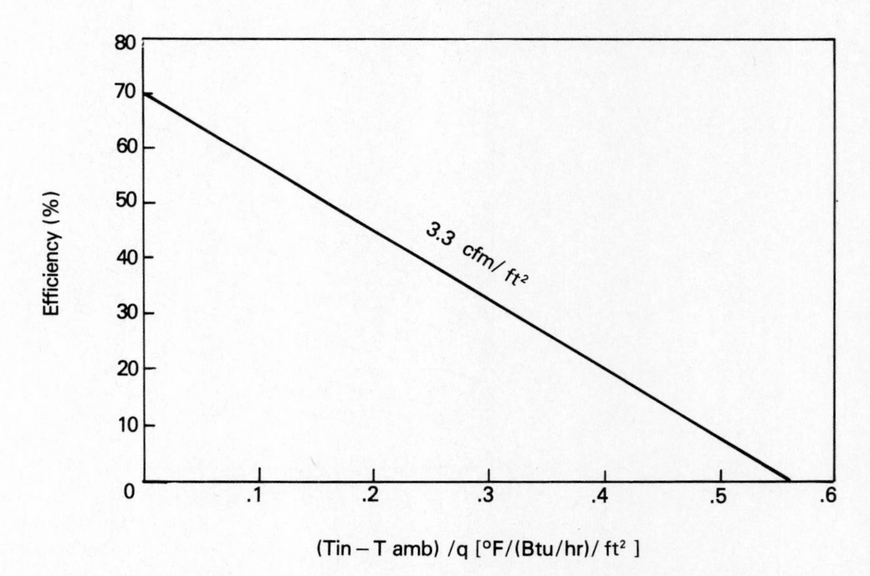

Figure 11-13 Solar air collector efficiency graph. Test data for Rom-Aire solar collectors. Tested at independent test laboratories for NASA, DOE, and HUD according to ASHRAE 93-77 and NBSIR 78-1035A. (Courtesy of Rom-Aire Solar Corporation, Lorain, OH.)

curve crosses the vertical axis and is equal to 0.70. The slope, S, of the graph can be obtained by dividing its efficiency drop by the range of fluid parameters it covers. For instance, the graph in Figure 11-13 drops from 0.70 at a fluid parameter of zero to 0.20 at a fluid parameter of 0.40. In other words, its efficiency drops 0.50. Its slope is thus calculated as follows:

$$S = \frac{0.50}{0.40} = 1.25$$

For more information on efficiency graphs, refer to Chapter 9.

Manufacturers' literature for many solar panels lists the vertical-axis intercept and slope of the graph and thus saves the designer the trouble of obtaining them from the efficiency graph. The *Directory of SRCC Certified Solar Collector Ratings*[12] and the *Solar Products Specifications Guide*[13] are two very helpful listings that include vertical intercept (also called Y intercept) and slope information for many different collectors on the market. The abbreviation for the vertical intercept, V, listed in the *Guide*, as well as in many other sources, is $F_R(\tau\alpha)$. The abbreviation for the slope is $F_R(U_L)$.

The average monthly insolation on the collector surface, I, can be obtained from Table 11-1. To use this table it is necessary to know the tilt angle of the collector and the site's latitude and K_T value, where K_T is a "cloudiness index." K_T values for various locations are listed in Table III-1.

To better illustrate how to use the *f*-chart method, let us work out an example. We wish to add enough collector area to a certain house to provide a 50 percent solar heating fraction during the month of January, which is usually the worst month of the year. The house is in Atlanta, Georgia, has a January heat load (L) of 12×10^6 Btu and a floor area of 2000 square feet. Rom-Aire collectors are used, with efficiencies described by the graph in Figure 11-13, and are at a correct tilt of latitude + 15°.

To calculate our first trial collector area, we employ the ballpark method described at the beginning of this section:

$$\text{approximate collector area } (A) = \frac{2000 \text{ ft}^2 \text{ of floor space}}{4} = 500 \text{ ft}^2$$

The other factors of the X and Y terms are as follows:

$$V = 0.70$$

$$h = 31 \text{ days in January} \times 24 \text{ hours/day} = 744 \text{ hours}$$

$$S = 1.25$$

From Table III-1, it is seen that Atlanta, Georgia, has an average January temperature of 42°F, a cloudiness index, K_T, of 0.44 in that month, and is at a latitude of 33.6. Thus the collector tilt is about 49°. These numbers are used in Table 11-1 to obtain average daily insolation on the collector surface itself. The closest corresponding values in the table are: latitude, 35°; collector tilt, 50°; K_T = 0.4 or 0.5. These values yield an insolation of 948 Btu/ft^2-day for K_T = 0.4 and 1266 for K_T = 0.5. Interpolating between the two values leads to the following calculations:

$$\text{insolation} = 948 + 0.4\,(\text{difference between two insolation values})$$
$$= 948 + 0.4(1266 - 948)$$
$$= 948 + 0.4(318)$$
$$= 1075 \text{ Btu/ft}^2\text{-day}$$

$$\text{January insolation} = 1075 \times 31 \text{ days} = 33{,}300 \text{ Btu/ft}^2\text{-month}$$

TABLE 11-1 Insolation on a Collector Surface

Tilt Angle	K_t	Jan.	Feb.	Mar.	Apr.	May	June	July	Aug.	Sept.	Oct.	Nov.	Dec.
30° North Latitude													
15°	0.3	633	767	880	991	1045	1054	1048	1011	913	798	656	604
	0.4	874	1049	1185	1322	1394	1405	1397	1349	1229	1083	905	833
	0.5	1120	1334	1508	1668	1725	1738	1729	1686	1551	1377	1161	1076
	0.6	1379	1630	1828	2004	2072	2088	2077	2025	1880	1684	1418	1324
30°	0.3	683	800	888	962	982	989	984	962	905	819	702	656
	0.4	963	1113	1215	1283	1312	1305	1298	1383	1229	1208	984	930
	0.5	1259	1447	1549	1620	1620	1595	1605	1602	1565	1462	1278	1215
	0.6	1568	1793	1894	1946	1925	1894	1907	1926	1898	1814	1606	1532
45°	0.3	700	800	856	895	888	882	879	883	861	812	714	682
	0.4	1008	1130	1173	1193	1171	1147	1157	1177	1171	1141	1015	979
	0.5	1326	1482	1509	1493	1448	1399	1413	1472	1494	1475	1338	1303
	0.6	1680	1847	1861	1811	1716	1658	1674	1747	1828	1842	1688	1647
90°	0.3	561	597	559	510	460	441	445	485	530	575	556	557
	0.4	822	850	768	642	557	516	536	608	707	815	812	819
	0.5	1103	1132	975	787	645	574	600	711	899	1068	1084	1112
	0.6	1413	1426	1202	906	690	624	636	814	1079	1338	1383	1439
35° North Latitude													
20°	0.3	587	725	848	975	1034	1054	1043	996	886	654	611	555
	0.4	821	999	1152	1313	1378	1405	1391	1328	1203	1040	723	782
	0.5	1066	1290	1466	1641	1723	1756	1739	1679	1518	1333	1194	1014
	0.6	1107	1598	1807	1990	2070	2088	2088	2014	1857	1641	1378	1271
35°	0.3	635	762	848	938	982	978	979	948	878	780	658	608
	0.4	910	1073	1172	1263	1294	1289	1291	1276	1203	1093	931	876
	0.5	1201	1402	1517	1594	1619	1612	1597	1595	1532	1421	1232	1161
	0.6	1520	1757	1869	1934	1923	1914	1918	1916	1874	1772	1551	1474
50°	0.3	654	762	818	872	877	869	873	870	828	774	668	630
	0.4	948	1089	1142	1175	1156	1144	1149	1159	1137	1093	965	923
	0.5	1266	1444	1480	1486	1429	1396	1403	1450	1465	1433	1284	1243
	0.6	1625	1818	1822	1784	1714	1653	1662	1740	1791	1798	1633	1589
90°	0.3	544	597	571	534	491	478	479	518	547	384	545	538
	0.4	814	869	792	700	613	579	582	651	751	830	809	811
	0.5	1107	1169	1030	860	714	652	675	782	953	1105	1089	1110
	0.6	1434		1266	1014	794	718	746	919	1160	1404	1408	1447
40° North Latitude													
25°	0.3	536	687	814	951	1026	1058	1043	974	851	717	566	504
	0.4	763	966	1114	1281	1369	1412	1388	1312	1165	995	800	725
	0.5	1007	1252	1429	1631	1728	1765	1738	1656	1483	1293	1051	960
	0.6	1273	1569	1759	1977	2076	2098	2087	2008	1828	1613	1323	1211
40°	0.3	584	730	814	915	964	982	968	927	843	741	609	554
	0.4	853	1038	1134	1245	1285	1296	1291	1248	1165	1050	881	816
	0.5	1139	1369	1476	1571	1606	1601	1596	1576	1496	1382	1173	1098
	0.6	1463	1730	1831	1923	1930	1923	1917	1913	1843	1730	1495	1413
55°	0.3	600	735	779	844	860	873	862	850	796	735	617	580
	0.4	890	1059	1095	1148	1147	1135	1135	1134	1103	1050	909	863
	0.5	1203	1415	1430	1451	1418	1402	1403	1437	1432	1393	1232	1177
	0.6	1560	1805	1787	1778	1702	1661	1662	1722	1766	1765	1582	1528
90°	0.3	524	606	578	562	578	513	521	535	562	582	530	518
	0.4	800	894	819	737	664	640	653	701	780	847	800	792
	0.5	1100	1209	1072	922	796	747	763	861	1003	1138	1088	1099
	0.6	1440	1558	1344	1107	913	830	852	1013	1234	1448	1417	1441

TABLE 11-1 (cont.)

45° North Latitude

Tilt Angle	K_t	Jan.	Feb.	Mar.	Apr.	May	June	July	Aug.	Sept.	Oct.	Nov.	Dec.
30°	0.3	485	604	769	929	1023	1058	1037	956	821	676	515	470
	0.4	710	860	1078	1262	1364	1412	1383	1299	1139	957	743	684
	0.5	951	1129	1391	1606	1722	1764	1728	1639	1454	1247	993	927
	0.6	1218	1421	1723	1947	2069	2120	2097	1987	1790	1571	1263	1192
45°	0.3	532	636	776	886	952	971	963	909	807	702	557	525
	0.4	794	915	1095	1215	1269	1295	1255	1224	1124	1018	822	790
	0.5	1083	1221	1434	1548	1586	1601	1587	1462	1454	1342	1110	1076
	0.6	1402	1560	1788	1895	1926	1923	1907	1895	1819	1705	1424	1399
60°	0.3	551	631	743	816	849	851	825	826	763	697	571	556
	0.4	837	928	1060	1111	1132	1121	1129	1113	1066	1012	855	846
	0.5	1152	1246	1392	1420	1417	1384	1395	1409	1395	1352	1170	1166
	0.6	1504	1597	1749	1756	1701	1639	1652	1709	1732	1736	1524	1527
90°	0.3	504	539	581	573	563	546	550	557	574	583	511	523
	0.4	778	799	834	776	723	699	705	742	804	860	780	809
	0.5	1084	1092	1099	985	888	819	847	929	1043	1163	1070	1128
	0.6	1434	1412	1396	1199	1024	939	974	1114	1310	1499	1419	1488

From *More Other Homes and Garbage*, by Jim Leckie, Gil Masters, Harry Whitehouse, and Lily Young. Copyright © 1981, Sierra Club Books. Reprinted by permission.

Notice that we did not interpolate between tilt angles or latitude. This is because insolation is much less sensitive to these variables than it is to changes in K_T.

The calculations above yield values for X and Y of

$$X = \frac{(1.25)(500)(744)(212 - 42)}{12 \times 10^6} = 6.6$$

$$Y = \frac{(0.93)(0.70)(500)(3.33 \times 10^4)}{12 \times 10^6} = 0.90$$

Referring to the f-chart graph for air systems (Figure 11-12), it is seen that the foregoing values for X and Y yield a solar heating fraction of 0.46. In other words, our trial collector area of 500 square feet will supply 46 percent of the home's heating needs in January. To supply 50 percent of the heat as we originally desired, we will have to augment the collector area somewhat. A 600 square foot array will yield X and Y values as follows:

$$X = 7.9$$

$$Y = 1.08$$

According to the f-chart graph, our new trial square footage will supply 53 percent of the building's January need. Interpolating between the two values, it is found that 557 square feet of collectors yields a January solar heating fraction of 50 percent.

A collector array that supplies 50 percent of the heating load in January, usually the coldest and worst weather month of the year, generally will supply 70 to 80 percent of the building's *annual* heating needs, which is the percentage that active space-heating systems are typically designed to provide.

It is a much trickier problem to size collector systems that are fabricated on site, since their efficiencies depend heavily on the materials used and how well they are constructed. For tightly built single-glazed panels with the airflow behind the absorber plate, however, vertical intercepts of $V = 0.53$ and slopes of $S = 1.1$ are typical,[14] and can be used as approximate design guidelines.

Duct Sizing. Several factors must be considered when sizing ductwork for a solar air system. The ducts must fit the space available, they must be designed to minimize air noise and also to minimize the power requirements of the system's blowers. The ducting must also be large enough to allow air to move fast enough through the collectors to maximize their efficiency (remember that the greater the amount of air that moves through the collectors per second, the cooler will be their operating temperature and the higher their efficiency).

The SMACNA *Fundamentals of Solar Heating* manual[15] notes that the system is generally designed to allow an air flow rate 2 to 3 cubic foot per minute (abbreviated cfm) for every square foot of solar collector used. In other words, a 350-square-foot array would require a 700- to 1050-cfm air flow rate. The SMACNA manual also recommends that the *air speed* in the ducts be 700 to 1000 linear feet per minute (abbreviated fpm).[16]

Through extensive testing, the American Society of Heating, Refrigerating and Air-Conditioning Engineers (ASHRAE) has developed a "standard friction chart" as a means of sizing ductwork. The chart originally applied to round, smooth, galvanized ducting carrying air heated to a temperature of 50 to 90°F, but it can also be applied with good results to fiberglass ducting and to the 120 to 140°F temperatures found in solar systems.

The standard friction chart illustrated in Figure 11-14 relates four system variables: cfm (on the vertical axis), friction loss per hundred feet of duct (on the horizontal axis), fpm, and duct diameter in inches (on the diagonal lines). If any two of these parameters are known, the remaining two can be read from the charts. For example, if we wish to size ductwork for a system with a 450-square-foot array, our rule of thumb of 2 to 3 cfm per square foot of array indicates that 900 to 1350 cfm is needed. We also know from above that an air velocity of 700 to 1000 fpm through the ducts is recommended. Taking minimum values of 900 cfm and 700 fpm, we obtain a required duct diameter of 15.4 inches.

Round ducts are often not the most convenient to use, since they take up more space than do rectangular ones of the same capacity. Table 11-2 provides a way of converting from round ducting to equivalent sizes of rectangular ducts. We see that a 15.3-inch round duct is equivalent to a 14-inch by 14-inch, a 15-inch by 13-inch, or an 18-inch by 11-inch rectangular duct. A 15.6-inch round duct is equivalent to a 17-inch by 12-inch, a 24-inch by 9-inch, a 28-inch by 8-inch, or a 42-inch by 6-inch rectangular size.

Rock Storage Sizing. The storage bin should contain three-fourths of a cubic foot of rock for every square foot of solar collector area. Clean river gravel or hard crushed rock, ¾ to 1½ inches in diameter, should be used as the storage medium.[17] The recommended rock depth for a single family residence is 4½ to 5½ feet,[18] and for multifamily or commercial applications it is 5 to 7 feet. The minimum dimension for the length or width of the bin is 3 feet.

The cross-sectional area of the bin (its length times its width) determines to a great extent the drop in air pressure across the rock. If this drop is too small, there can be uneven movement of air through the bin, leading to hotspots that will detract from the system's performance. The pressure drop is also dependent on air velocity. Values of air velocity between 20 and 30 fpm generally produce pressure profiles in the acceptable range.[19] The actual air velocity in the bin can be determined from the following formula:

$$\text{air velocity (fpm)} = \frac{\text{cfm}}{\text{cross-sectional area}}$$

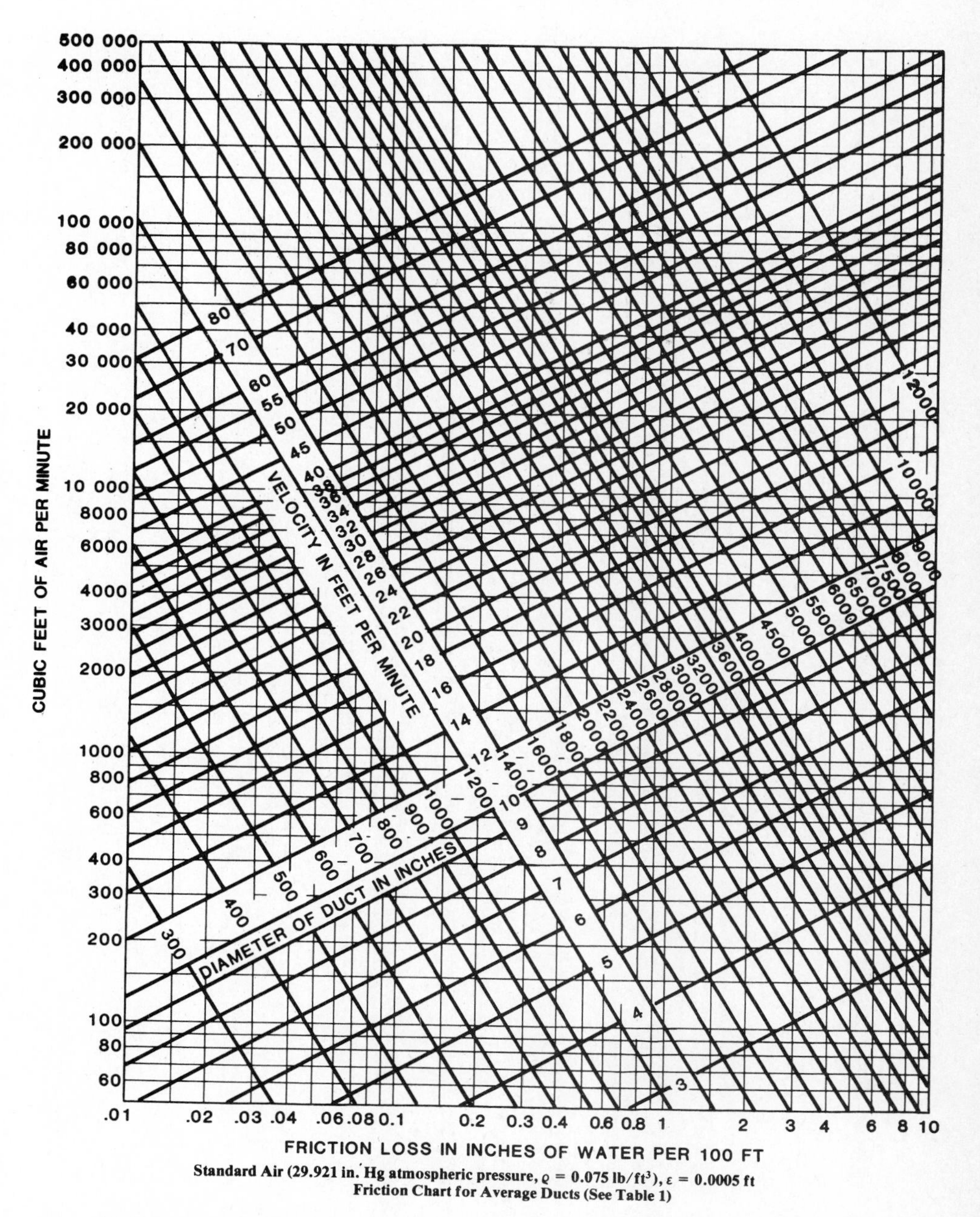

Figure 11-14 Friction chart for average ducts. (Reprinted with permission from *ASHRAE Handbook—1981 Fundamentals*, American Society of Heating, Refrigerating and Air-Conditioning Engineers, Inc., Atlanta, GA.)

TABLE 11-2 Converting Round Duct to Regular Duct Sizes (Dimensions in Inches)

Side Rectangular Duct	4.0	4.5	5.0	5.5	6.0	6.5	7.0	7.5	8.0	9.0	10.0	11.0	12.0	13.0	14.0	15.0	16.0
3.0	3.8	4.0	4.2	4.4	4.6	4.7	4.9	5.1	5.2	5.5	5.7	6.0	6.2	6.4	6.6	6.8	7.0
3.5	4.1	4.3	4.6	4.8	5.0	5.2	5.3	5.5	5.7	6.0	6.3	6.5	6.8	7.0	7.2	7.5	7.7
4.0	4.4	4.6	4.9	5.1	5.3	5.5	5.7	5.9	6.1	6.4	6.7	7.0	7.3	7.6	7.8	8.0	8.3
4.5	4.6	4.9	5.2	5.4	5.7	5.9	6.1	6.3	6.5	6.9	7.2	7.5	7.8	8.1	8.4	8.6	8.8
5.0	4.9	5.2	5.5	5.7	6.0	6.2	6.4	6.7	6.9	7.3	7.6	8.0	8.3	8.6	8.9	9.1	9.4
5.5	5.1	5.4	5.7	6.0	6.3	6.5	6.8	7.0	7.2	7.6	8.0	8.4	8.7	9.0	9.3	9.6	9.9

Side Rectangular Duct	6	7	8	9	10	11	12	13	14	15	16	17	18	19	20	22	24	26	28	30	Side Rectangular Duct
6	6.6																				6
7	7.1	7.7																			7
8	7.6	8.2	8.7																		8
9	8.0	8.7	9.3	9.8																	9
10	8.4	9.1	9.8	10.4	10.9																10
11	8.8	9.5	10.2	10.9	11.5	12.0															11
12	9.1	9.9	10.7	11.3	12.0	12.6	13.1														12
13	9.5	10.3	11.1	11.8	12.4	13.1	13.7	14.2													13
14	9.8	10.7	11.5	12.2	12.9	13.5	14.2	14.7	15.3												14
15	10.1	11.0	11.8	12.6	13.3	14.0	14.6	15.3	15.8	16.4											15
16	10.4	11.3	12.2	13.0	13.7	14.4	15.1	15.7	16.4	16.9	17.5										16
17	10.7	11.6	12.5	13.4	14.1	14.9	15.6	16.2	16.8	17.4	18.0	18.6									17
18	11.0	11.9	12.9	13.7	14.5	15.3	16.0	16.7	17.3	17.9	18.5	19.1	19.7								18
19	11.2	12.2	13.2	14.1	14.9	15.7	16.4	17.1	17.8	18.4	19.0	19.6	20.2	20.8							19
20	11.5	12.5	13.5	14.4	15.2	16.0	16.8	17.5	18.2	18.9	19.5	20.1	20.7	21.3	21.9						20
22	12.0	13.0	14.1	15.0	15.9	16.8	17.6	18.3	19.1	19.8	20.4	21.1	21.7	22.3	22.9	24.0					22
24	12.4	13.5	14.6	15.6	16.5	17.4	18.3	19.1	19.9	20.6	21.3	22.0	22.7	23.3	23.9	25.1	26.2				24
26	12.8	14.0	15.1	16.2	17.1	18.1	19.0	19.8	20.6	21.4	22.1	22.9	23.5	24.2	24.9	26.1	27.3	28.4			26
28	13.2	14.5	15.6	16.7	17.7	18.7	19.6	20.5	21.3	22.1	22.9	23.7	24.4	25.1	25.8	27.1	28.3	29.5	30.6		28
30	13.6	14.9	16.1	17.2	18.3	19.3	20.2	21.1	22.0	22.9	23.7	24.4	25.2	25.9	26.6	28.0	29.3	30.5	31.7	32.8	30
32	14.0	15.3	16.5	17.7	18.8	19.8	20.8	21.8	22.7	23.5	24.4	25.2	26.0	26.7	27.5	28.9	30.2	31.5	32 .7	33.9	32
34	14.4	15.7	17.0	18.2	19.3	20.4	21.4	22.4	23.3	24.2	25.1	25.9	26.7	27.5	28.3	29.7	31.0	32.4	33.7	34.9	34
36	14.7	16.1	17.4	18.6	19.8	20.9	21.9	22.9	23.9	24.8	25.7	26.6	27.4	28.2	29.0	30.5	32.0	33.3	34.6	35.9	36
38	15.0	16.5	17.8	19.0	20.2	21.4	22.4	23.5	24.5	25.4	26.4	27.2	28.1	28.9	29.8	31.3	32.8	34.2	35.6	36.8	38
40	15.3	16.8	18.2	19.5	20.7	21.8	22.9	24.0	25.0	26.0	27.0	27.9	28.8	29.6	30.5	32.1	33.6	35.1	36.4	37.8	40
42	15.6	17.1	18.5	19.9	21:1	22.3	23.4	24.5	25.6	26.6	27.6	28.5	29.4	30.3	31.2	32.8	34.4	35.9	37.3	38.7	42
44	15.9	17.5	18.9	20.3	21.5	22.7	23.9	25.0	26.1	27.1	28.1	29.1	30.0	30.9	31.8	33.5	35.1	36.7	38.1	39.5	44
46	16.2	17.8	19.3	20.6	21.9	23.2	24.4	25.5	26.6	27.7	28.7	29.7	30.6	31.6	32.5	34.2	35.9	37.4	38.9	40.4	46
48	16.5	18.1	19.6	21.0	22.3	23.6	24.8	26.0	27.1	28.2	29.2	30.2	31.2	32.2	33.1	34.9	36.6	38.2	39.7	41.2	48
50	16.8	18.4	19.9	21.4	22.7	24.0	25.2	26.4	27.6	28.7	29.8	30.8	31.8	32.8	33.7	35.5	37.2	38.9	40.5	42.0	50
52	17.1	18.7	20.2	21.7	23.1	24.4	25.7	26.9	28.0	29.2	30.3	31.3	32.3	33.3	34.3	36.2	37.9	39.6	41.2	42.8	52
54	17.3	19.0	20.6	22.0	23.5	24.8	26.1	27.3	28.5	29.7	30.8	31.8	32.9	33.9	34.9	36.8	38.6	40.3	41.9	43.5	54
56	17.6	19.3	20.9	22.4	23.8	25.2	26.5	27.7	28.9	30.1	31.2	32.3	33.4	34.4	35.4	37.4	39.2	41.0	42.7	44.3	56
58	17.8	19.5	21.2	22.7	24.2	25.5	26.9	28.2	29.4	30.6	31.7	32.8	33.9	35.0	36.0	38.0	39.8	41.6	43.3	45.0	58
60	18.1	19.8	21.5	23.0	24.5	25.9	27.3	28.6	29.8	31.0	32.2	33.3	34.4	35.5	36.5	38.5	40.4	42.3	44.0	45.7	60
62		20.1	21.7	23.3	24.8	26.3	27.6	28.9	30.2	31.5	32.6	33.8	34.9	36.0	37.1	39.1	41.0	42.9	44.7	46.4	62
64		20.3	22.0	23.6	25.1	26.6	28.0	29.3	30.6	31.9	33.1	34.3	35.4	36.5	37.6	39.6	41.6	43.5	45.3	47.1	64
66		20.6	22.3	23.9	25.5	26.9	28.4	29.7	31.0	32.3	33.5	34.7	35.9	37.0	38.1	40.2	42.2	44.1	46.0	47.7	66
68		20.8	22.6	24.2	25.8	27.3	28.7	30.1	31.4	32.7	33.9	35.2	36.3	37.5	38.6	40.7	42.8	44.7	46.6	48.4	68
70		21.1	22.8	24.5	26.1	27.6	29.1	30.4	31.8	33.1	34.4	35.6	36.8	37.9	39.1	41.2	43.3	45.3	47.2	49.0	70
72			23.1	24.8	26.4	27.9	29.4	30.8	32.2	33.5	34.8	36.0	37.2	38.4	39.5	41.7	43.8	45.8	47.8	49.6	72
74			23.3	25.1	26.7	28.2	29.7	31.2	32.5	33.9	35.2	36.4	37.7	38.8	40.0	42.2	44.4	46.4	48.4	50.3	74
76			23.6	25.3	27.0	28.5	30.0	31.5	32.9	34.3	35.6	36.8	38.1	39.3	40.5	42.7	44.9	47.0	48.9	50.9	76
78			23.8	25.6	27.3	28.8	30.4	31.8	33.3	34.6	36.0	37.2	38.5	39.7	40.9	43.2	45.4	47.5	49.5	51.4	78
80			24.1	25.8	27.5	29.1	30.7	32.2	33.6	35.0	36.3	37.6	38.9	40.2	41.4	43.7	45.9	48.0	50.1	52.0	80
82				26.1	27.8	29.4	31.0	32.5	34.0	35.4	36.7	38.0	39.3	40.6	41.8	44.1	46.4	48.5	50.6	52.6	82
84				26.4	28.1	29.7	31.3	32.8	34.3	35.7	37.1	38.4	39.7	41.0	42.2	44.6	46.9	49.0	51.1	53.2	84
86				26.6	28.3	30.0	31.6	33.1	34.6	36.1	37.4	38.8	40.1	41.4	42.6	45.0	47.3	49.6	51.7	53.7	86
88				26.9	28.6	30.3	31.9	33.4	34.9	36.4	37.8	39.2	40.5	41.8	43.1	45.5	47.8	50.0	52.2	54.3	88
90				27.1	28.9	30.6	32.2	33.8	35.3	36.7	38.2	39.5	40.9	42.2	43.5	45.9	48.3	50.5	52.7	54.8	90
92					29.1	30.8	32.5	34.1	35.6	37.1	38.5	39.9	41.3	42.6	43.9	46.4	48.7	51.0	53.2	55.3	92
96					29.6	31.4	33.0	34.7	36.2	37.7	39.2	40.6	42.0	43.3	44.7	47.2	49.6	52.0	54.2	56.4	96

Reprinted with permission from *ASHRAE Handbook—1981 Fundamentals*, American Society of Heating, Refrigerating and Air Conditioning Engineers, Atlanta, Georgia, p. 33.46.

Once the collector array has been sized and a tentative cfm flow rate for the ducting has been arrived at, the rock bin may be sized. For example, consider a 250-square-foot collector array connected to ducting designed for a flow of 600 cfm. The system requires 250 $ft^2 \times 3/4 = 187.5$ ft^3 of rock. Assume that the rock depth in the storage bin is 5 feet (midway between the recommended extremes of 4½ and 5½ feet). The cross-sectional area of the bin can be found as follows:

$$\text{length} \times \text{width} \times \text{depth} = \text{volume} = 187.5 \text{ ft}^3$$

$$\text{cross-sectional area} = \frac{\text{length} \times \text{width}}{\text{depth}} = \frac{187.5 \text{ ft}^3}{5 \text{ ft}}$$

$$= 37.5 \text{ ft}^2$$

To determine whether this cross section is acceptable, calculate the air velocity through the bin:

$$\text{air velocity} = \frac{600 \text{ cfm}}{37.5 \text{ ft}^2} = 16 \text{ fpm}$$

This is too low a velocity, but it can be increased by designing the ducting to accommodate a higher flow rate. Since flow rates of 2 to 3 cfm per square foot of collector are recommended, our 250-square-foot array can have flow rates of 500 to 750 cfm. If we design the ducts to handle 750 cfm, the air velocity through the rock bin becomes

$$\frac{750 \text{ cfm}}{37.5} = 20 \text{ fpm}$$

It is thought by some designers that oversizing a bin (increasing its volume without changing the collector square footage) improves the system's performance by providing several days' worth of storage. Studies performed by Solaron Corporation of Colorado, however, indicate the impracticality of doing this. The cost of the system will be significantly raised, while the total energy stored over a heating season will usually be only slightly increased. This is because if the rock bin size is increased without also adding to the solar collector square footage, it will take a long series of sunny, warm days to charge up the storage bin. To use all the stored heat, it will then take a number of cloudy, cold days. These ideal conditions do not occur that often in winter. The result is that the extra storage volume is rarely utilized.[20]

HYDRONIC SYSTEMS

Hydronic space-heating systems in some ways resemble the domestic hot water system of Chapter 10. Both use hydronic collectors, storage tanks, and pumps governed by automatic control boxes that receive signals from sensors mounted in various parts of the system. But there are important differences between the two types of systems, the most obvious being the larger sizes of the space-heating components. Another difference is that hydronic space-heating systems are very often of a *drain-back* type. In this type of system the tank is unpressurized, which allows water in the collector array to empty back into it *whenever the pump is shut off* (see Figure 11-15). This provides a simple and effective freeze-protection mechanism for the

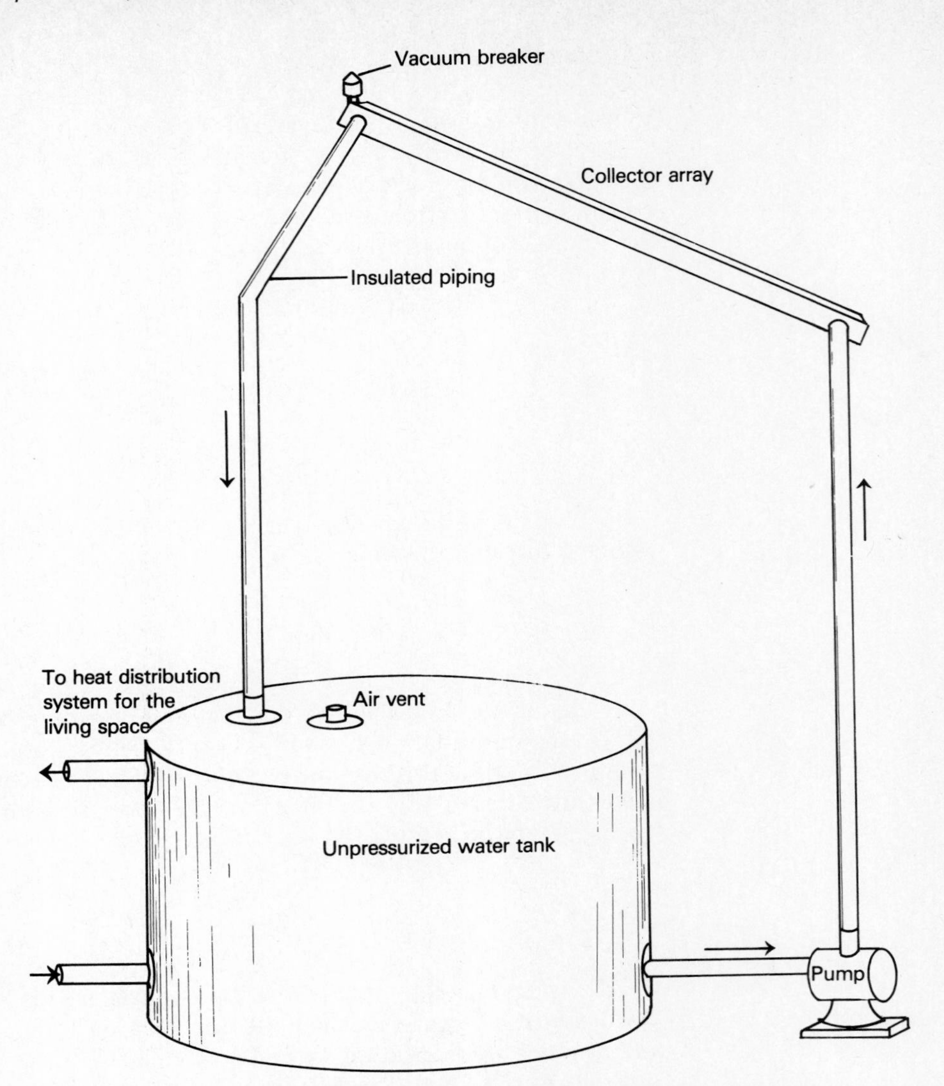

Figure 11-15 Hydronic drain-back system. Because the system is not air-tight, water in the collector array will drain back into the tank whenever the pump is shut off.

collector array. Either water is being circulated through the collectors—in which case they will not freeze—or the array is totally empty of water.

A space-heating system's storage tanks are unpressurized not only by design but by necessity. It is simply too expensive for most applications to buy a 500- to 1000-gallon pressurized tank. One disadvantage of unpressurized systems, however, is that the pump must lift water without the aid of the syphoning effect that occurs in pressurized systems that are totally filled with liquid. In those systems, liquid descending from the collector array helps to "pull" the ascending liquid up (see Figure 11-16). In unpressurized systems, air in the piping will prevent this from occurring, necessitating larger pumps and higher electrical power needs.

Unlike air systems, hydronic setups are not usually able to pump thermal energy directly from the collector array to the living area. The heated water must first go through the storage tank. This is done because to keep a hydronic collector's operating temperature as low as possible and thus maximize its performance, water must be pumped through it fast enough

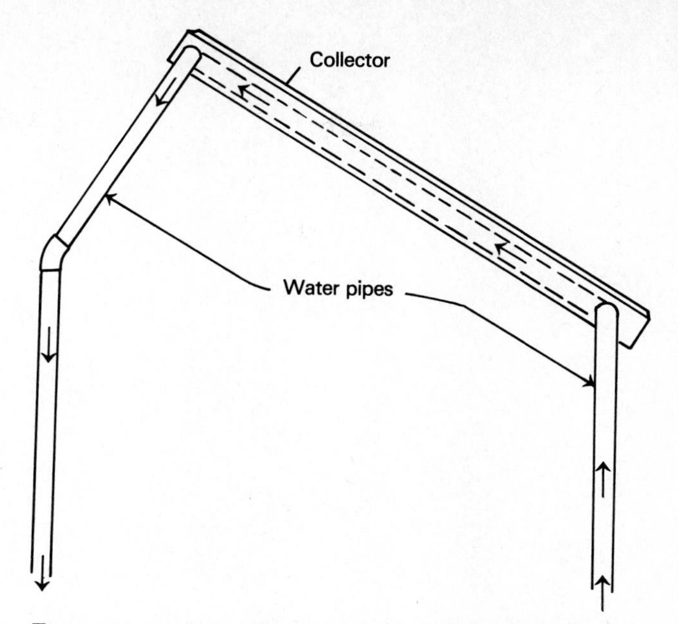

Figure 11-16 Siphoning effect. In an airtight hydronic system completely filled with liquid, fluid descending from the collector array helps "pull" ascending fluid up to the array. In unpressurized drain-back systems, this syphoning effect does not occur, and thus the pump must be somewhat powerful.

so that the water only receives a temperature increase of 10 to 20°F on each pass. The water in the collectors is at best only slightly hotter than the water in the storage tank, so a collector-to-living-space circulation loop offers little advantage.

Thermal Energy Distribution

Thermal energy can be transferred from the system's storage tank to the living area of the house in several ways. Hot water may be pumped through pipes in a slab floor, warming the slab and eventually the space above it, or the water may be circulated through a system of baseboard radiators. It can also be pumped through a coil in the ductwork or air handler of a forced-air system, transferring its energy to the air passing through the coil (see Figure 11-18).

Radiant Slabs. A properly designed radiant slab system provides even, comfortable heat throughout the house. It is perhaps the most effective system for heating a room right down to floor level, but this type of system limits the amount of floor area that can be covered with rugs, which greatly slow down the rate of heat transfer. Also, there is a significant waiting period from the time hot water is pumped through the slab to the time it begins giving off heat to the house. This can be a problem on cold mornings when heat is needed in a hurry, and might lead to large backup heater operating costs.

Baseboard Radiators. Baseboard radiators do not have the time-lag problems of pipes in a slab, and they have the advantage of being retrofittable into an existing structure. Solarized baseboard radiator systems resemble

Figure 11-17 Solar hydronic space heating in a subdivision in Fort Bragg, California. Question: Can you tell what time of day it is from the shadows in this picture? (Photo by Ed Sander.)

those used with conventional heating equipment, but the number of linear feet of radiator required in solarized systems is considerably greater. This is because typical solar-heated water temperatures are 100 to 120°F, whereas for conventional heaters they are 180°F. In cold climates the cost of providing sufficient solar-heated baseboard radiator footage might be prohibitively high, and thus in those situations (as well as in more moderate temperature zones) forced-air systems offer an attractive alternative.

Forced-Air Systems. Figure 11-18 diagrams a solarized forced-air system. In it, air returning from the living space is preheated as it goes through a solar hot water coil. The air then passes through the furnace and, if necessary, receives more heat. The heating coil must be large enough so that it transfers sufficient thermal energy to the air even when using water at tem-

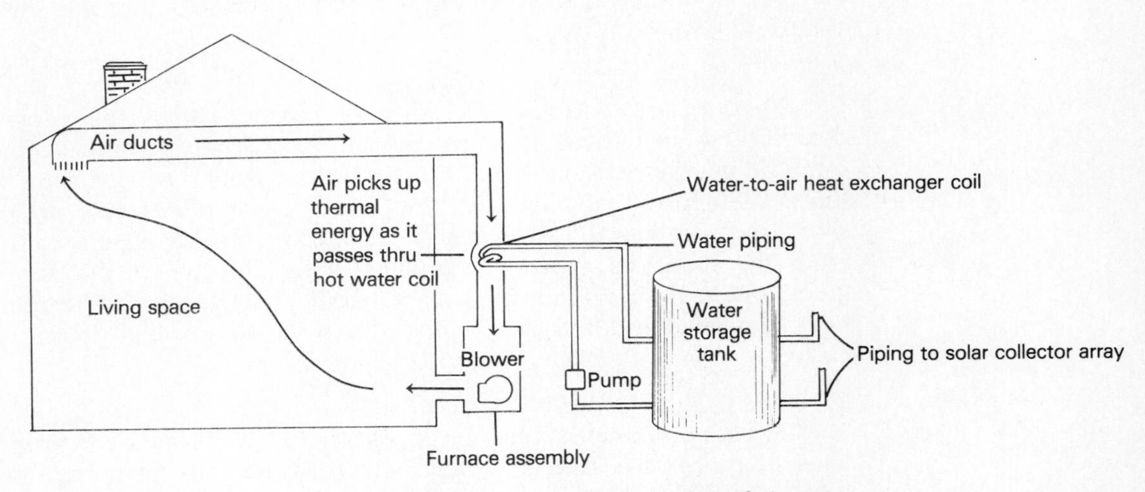

Figure 11-18 Hydronic forced-air system.

peratures as low as 90°F. This allows the collector array and storage tank temperatures also to be low, which improves the system's efficiency. The Magic Aire Division of United Electric Company in Wichita Falls, Texas, is a company that manufactures coils designed specifically for the solar industry. They are made with two-, three-, or four-coil rows and come in a variety of sizes for different applications.

A disadvantage of forced-air hydronic systems is that the ducting they require is far more expensive and takes up more space than the water pipes used in other hydronic systems. Also, forced-air blowers make more noise than water pumps and cost considerably more to operate.

The Chill Chaser System. A product that combines the advantages of baseboard radiators and forced-air systems is the Chill Chaser, manufactured by Turbonics, Inc.[21] The unit is 19 inches by 21 inches by 4 inches deep and contains its own water pump and fan. The pump circulates hot water from any source—a solar storage tank, a boiler, or even a hot water heater—through the Chill Chaser's 21-pass copper coil. The fan blows air over the heated coil and into the room. Two gallons per minute of 120°F water pumped through the unit delivers 6000 Btu per hour of heat—enough for one to two rooms. For convenience of operation, the Chill Chaser can be controlled by a wall-mounted room thermostat or by the solar system's differential thermostat.

Low-Cost Domestic Water Heating. Figure 11-19 depicts a simple way of preheating a domestic water supply from a hydronic space heating system's storage tank. Cold water is passed through a heat exchanger coil in the storage tank before it enters the water heater. The beauty of this arrangement is that pump, sensors, and control box are not required. Notice, however, that water only flows through the heat exchanger coil during the times when hot water is also being used in the house.

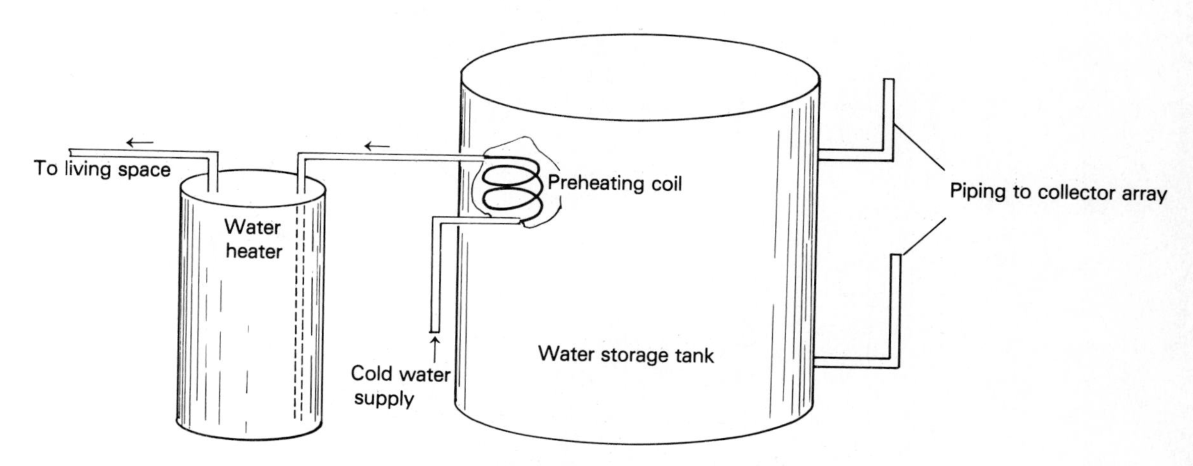

Figure 11-19 Domestic water heating.

Sizing Hydronic Systems A hydronic system's collector array is sized using the same method as for air systems. The *f*-chart corresponding to hydronic solar systems is depicted in Figure 11-20. It is recommended that the thermal storage tank be designed to hold 2 gallons of water for every square foot of solar collector.

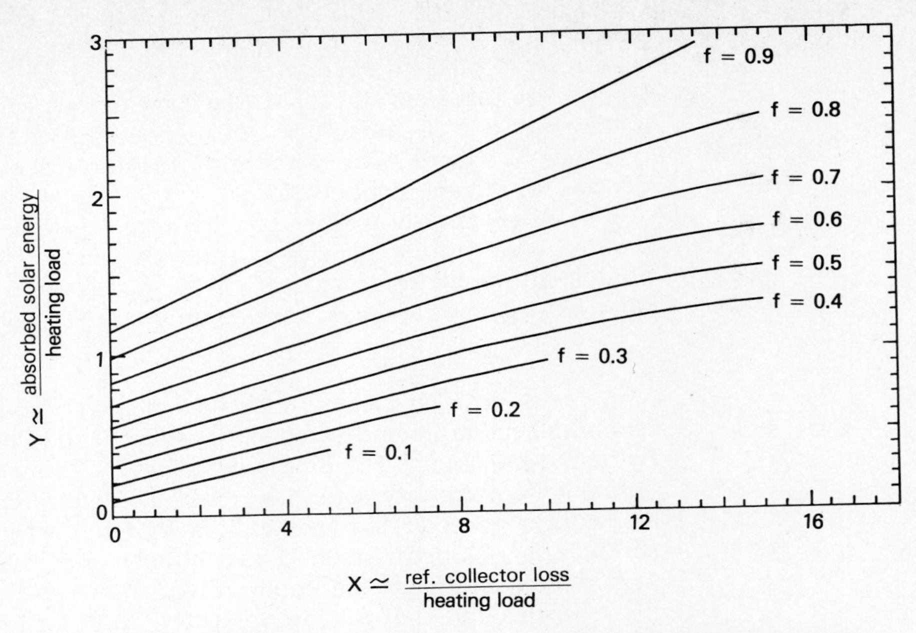

Figure 11-20 *f*-Chart for liquid systems. (From William A. Beckman, Sanford A. Klein, and John A. Duffie, *Solar Heating Design by the f-Chart Method*, John Wiley & Sons, Inc., New York, 1977. Reprinted by permission of the publisher.)

HYBRID SYSTEMS

Many space-heating systems in use today are not purely active or passive in design, but are hybrids of both. Of the large number of ways the two types of systems can be combined, some employ passive methods of thermal energy collection and storage, but an active method of energy delivery to the living space. Other designs do just the reverse. Some are almost totally of one type or another, such as a fan-assisted solar wall or sunspace system that is in all other respects completely passive (fans are sometimes employed in these designs to improve the distribution of thermal energy throughout the building). Described below are several hybrid designs and the reasons why they were engineered as they were.

The Spath House: A Sunspace with Active Energy Delivery

The Spath house, illustrated in Figure 7-4, employs a sunspace containing water-filled oil drums to fulfill collection and thermal storage functions, but circulates the heated sunspace air throughout the house by means of ductwork and a blower. It was felt by the designer, Ed Sander, that more even temperatures throughout the house could be achieved by use of the forced-air system.

The blower transports heated sunspace air to the rest of the house through a duct under the ground floor. Air is returned to the sunspace for reheating by a duct whose opening is located at the top of the stairwell, in the highest part of the house.

Active Collection with Passive Delivery

In one type of active collection/passive delivery design, heated water or other fluid from an array of solar panels is pumped through pipes in a thermally massive floor slab. The slab stores heat and releases it gradually to the living space in the same way that it would in a passive direct-gain system—through radiation and natural convection.

Because solar-heated fluid is circulated through the slab using pumps, the amount of heat gained by the slab can be carefully controlled. Summer overheating, for instance, is easily prevented.

Independent Systems

Also common in hybrid buildings are independent designs, in which two or more types of systems operate separately of each other. Figure 10-21 depicts a house whose living space is heated both by passive direct gain and by an active forced-air system.

The skylights and south windows allow direct solar radiation to enter the house. Part of its energy is stored in the tile floor and the 3 inches of concrete below it.

Heated air from the roof's flat-plate solar panels is blown through ducting into a rock bin under the house. Conduction, natural convection, and radiation then distribute its thermal energy throughout the house. In the center of the roof is a built-in integral passive water heater system for the house's domestic needs (see Chapter 10 for a discussion of these systems). Insulated sliding shutters are just visible at the tops of the skylights. Notice also the louvered overhang for summertime shading.

Low-Mass Sunspaces with Separate Thermal Storage

Movable insulation is generally advisable on sunspace glazing to reduce nighttime heat losses. If, however, the sunspace is not needed daily as a living area, and if its thermal storage mass is moved to another location, it is not as necessary to include movable insulation in its design. What is needed instead is a means of efficiently transporting energy collected by the sunspace to the storage mass. Natural convection usually does not provide a sufficiently rapid energy transfer, so forced convection is employed. Figure 11-21 depicts such a setup, in which the thermal storage mass is a rockbed underneath the house.

Figure 11-21 Low-mass second-story sunspace. (Photo by Ed Sander.)

NOTES

1. *Application Engineering Manual*, Solaron Corporation, Englewood, CO, 1980, p. 100.5.

2. Ibid.

3. Private communication with TEF Manufacturing Corporation, Fresno, CA.

4. If concrete bins are insulated on the outside, the concrete can act as a heat sink, preventing rapid charging and discharging of the bin. Fiberglass should be used on the inside, for the high temperatures might damage urethane or Styrofoam.

5. *Engineering and Applications Manual*, TEF Manufacturing Corporation, Fresno, CA, 1981, p. 27.

6. *Application Engineering Manual*, p. 300.4.

7. *Fundamentals of Solar Heating*, Sheet Metal and Air Conditioning Contractors National Association, Vienna, VA, for the U.S. Department of Energy, 1978, p. 7-6.

8. *Application Engineering Manual*, p. 500.4.

9. Ibid.

10. Ibid., p. 800.1.

11. William A. Beckman, Sanford A. Klein, and John A. Duffie, *Solar Heating Design by the f-Chart Method*, John Wiley & Sons, Inc., New York, 1977, p. 75.

12. *Directory of SRCC Certified Solar Collector Ratings*, Solar Rating and Certification Corporation, Washington, DC, 1983.

13. *Solar Products Specifications Guide*, SolarVision, Inc., Harrisville, NH, 1983.

14. Based on a survey of the characteristics of manufactured panels.

15. *Fundamentals of Solar Heating*, p. B-1.

16. Ibid., p. 7-6. Feet per minute (fpm) expresses the *speed* of air in the ducts. Cubic feet per minute (cfm) expresses the *volume* of air flowing through the ducts each minute.

17. *Application Engineering Manual*, p. 400.6.

18. Calculated from information in the *Residential Heat Storage Unit Installation Manual*, Solaron Corporation, Englewood, CO, 1978, pp. 1–15.

19. Ibid.

20. *Application Engineering Manual*, p. 400.5.

21. Turbonics, Inc., 11200 Madison Avenue, Cleveland, OH 44102.

SUGGESTED READING

The following two manuals contain practical, "hands-on" information, and are available by writing or calling the companies.

Application Engineering Manual. Solaron Corp., 1885 West Dartmouth Avenue, Englewood, CO 80110. (303) 762-1500.

Engineering and Applications Manual. TEF Manufacturing Corp., 1550 North Clark, Fresno, CA. (209) 441-1833.

REVIEW QUESTIONS

1. Compare the characteristics of air and hydronic space heating systems.
2. Describe the *storing heat* mode of an air system. Under what circumstances is this mode put into operation?
3. Why is pea-sized gravel not recommended for thermal storage bins?
4. What is the function of the air handler?
5. What do cfm and fpm stand for?
6. Contrast drain-back with drain-down systems.
7. Define a *hybrid system.*

PROBLEMS

For Problems 1 through 3, calculate the required collector area to provide 50 percent of the solar heating fraction in January.

1. The house is located in Milwaukee, Wisconsin, has a January heat load of 14×10^6 Btu and a floor area of 1900 square feet, and uses *air* collectors with efficiency characteristics described by Figure 11-13 and a tilt equal to the latitude + 10°.
2. The home is in Spokane, Washington, has a January heat load of 8×10^6 Btu and a floor area of 1400 square feet, and uses *air* collectors with vertical-axis intercept $V = 0.70$ and slope $S = 0.91$. The collectors are tilted at an angle equal to the latitude + 15°.
3. The home is in Amarillo, Texas, has a January heat load of 1×10^7 Btu and a floor area of 1600 square feet, and employs *hydronic* collectors with tilt equal to the latitude + 12° and with $V = 0.73$ and $S = 0.83$. (Remember to use the *f*-chart for *hydronic* collectors.)
4. What range of cfm is recommended for an air system with a 500-square-foot array? What range of fpm is recommended?
5. What is the required round duct diameter for a system with 1100 cfm and 900 fpm? For a system with 950 cfm and 700 fpm?
6. What sizes of rectangular ducts are equivalent to circular ducts with a
   (a) 13-inch diameter?
   (b) 7.5-inch diameter?
7. What volume of rock storage bin should be used with a 350-square-foot air collector array?
8. A 300-square-foot air collector array is connected to ducting designed for a flow of 750 cfm. The rock bin is 5 feet deep. What is the air velocity through it? Is it in the recommended range?
9. A 425-square-foot air collector array is connected to ducting designed for 850 cfm. The rock bin is 5.5 feet deep. What is the air velocity through it? Is it in the acceptable range?

swimming pools, hot tubs, and spas

SWIMMING POOLS

Swimming pools, both of the residential and the community variety, are one of North America's most popular recreational installations. Because of the volume of water they contain and because the preferred temperature for them is usually 80°F or higher, heating them requires considerable amounts of energy, especially in the nonsummer months. As fuel prices continue to rise, heating costs are becoming prohibitively high for the owners of many home pools, as well as for community and school pools. Solar heating offers an attractive alternative to fuel oil, gas, or electric heaters. Solar pool systems are relatively inexpensive, and payback periods in some locations can be as short as two to three years.[1]

Residential solar pool systems typically require 200 to 700 square feet of open, sunny area close to the pool for mounting of the collectors (commercial installations need up to 3000 square feet).[2] Roofs of buildings are usually used for this purpose, but other open locations, such as yards, are also employed. Sometimes backup heaters are included in the system, while at other times homeowners prefer to rely totally on solar. During the pool season it generally takes seven to ten days to solar-heat the water up to its desired temperature after a severe cold spell has occurred.[3]

How Pools Lose Heat

Swimming pools lose heat through the three basic methods of thermal energy transport discussed in Chapter 1—conduction, convection, and radiation—as well as through a fourth method—evaporation.

Conduction, or heat transport by contact between molecules, occurs through the walls of the pool into the surrounding ground. Conductive heat loss rates are usually low compared to convective, radiative, and evaporative losses, for it is generally a slower method of thermal energy transport.

Natural convection, or energy transfer by thermally generated currents in a fluid, carries the warmest water in the pool to the top, where its heat is carried off by the air. The more breezy the day, the faster heat is lost. Forced convection due to the pool pump mixes the warm and cool water layers, reducing surface temperatures and, as a result, surface heat losses as well.

Radiative losses, those due to emission of infrared energy from the warm water, are balanced to some extent during a sunny day by absorption of solar energy but can be quite large during nighttime hours, especially if the sky is clear (see Figure 12-1 for an explanation of this phenomenon, called "night sky radiation").

A gallon of water that evaporates from a heated pool takes with it thermal energy equivalent to 2½ kilowatthours of electricity.[4] Evaporation can account for up to 75 percent of a pool's thermal energy losses. Pools not only lose heat through evaporation, they lose water and expensive chemicals as well. Evaporation is much more severe in dry climates such as the American Southwest than in humid areas. Evaporative losses are also much greater in windy locales.

Both convective and evaporative losses can be reduced by providing windbreaks around the pool in the form of solid fences, dense shrubs, and trees or the house itself. Windbreaks should be designed so that the pool receives as much direct sunlight as possible between 8 A.M. and 4 P.M.

Pool covers are another effective way of reducing heat losses. They also can convert the pool into an efficient collector of solar radiation.

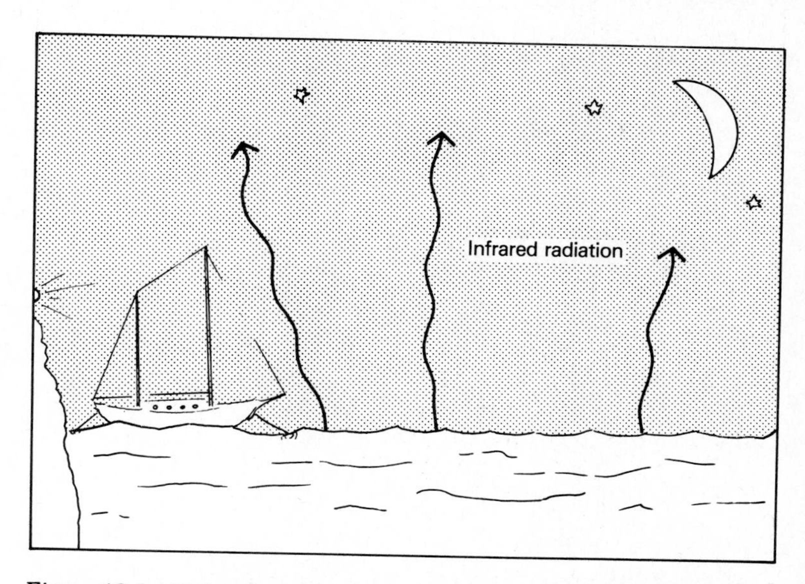

Figure 12-1 Night sky radiation. A pool is continuously radiating away thermal energy in the infrared range. During clear nights, much of this energy travels through the atmosphere and into space. Outer space is at a temperature close to absolute zero and thus sends little radiant energy back to the pool. As a result, the pool can lose considerably more heat than on an overcast night, even if air temperatures are the same. This is because a cloud layer is considerably warmer than outer space, and thus radiates more infrared energy back toward the pool.

Pool Covers: Conservation Plus Passive Collection

A clear or translucent swimming pool cover dramatically reduces evaporative, convective, and radiative losses but allows sunlight through that can help warm the pool. Experiments indicate that a clear cover raises the average temperature of a pool in a sunny location about 10°F.[5] It is often possible to maintain the temperature of a covered, unshaded pool at over 80°F during the summer months without heat from any other sources.

The simplest pool cover is a single sheet of plastic. Other types of covers are constructed of multiple layers of plastic or some sort of foam material that contains insulating air pockets. Figure 12-2 depicts several pool blanket designs. Some of the manufacturers that produce pool covers are:

Foam blankets: MacBall Industries, Inc., 1820 Embarcadero Street, Oakland, CA 94606. (415) 534-0274.

Covers containing air bubbles: Sealed Air Corporation, Park 80 Plaza East, Saddle Brook, NJ 07662. (201) 791-7600.

Covers with perimeter held in tracks: Cover Pools, Inc., 117 West Fireclay Avenue, Murray, VT 84107. (801) 262-2724.

For long life, covers should be made of ultraviolet-resistant material.

Besides helping to control energy losses, pool covers also reduce evaporative chemical losses and cut down on pool cleaning time and expense by helping to keep leaves and dirt out of the water.

Figure 12-2 Pool covers. (a) Thin films of plastic. (b) Layers of foam.

(c)

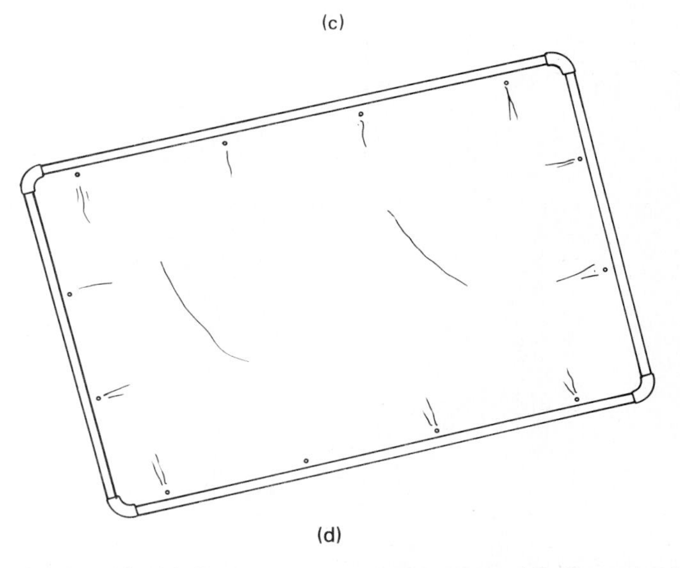

(d)

Figure 12-2 (cont.) (c) Covers containing air bubbles. (d) Covers held taut with frames or in tracks along the pool perimeter.

Dark-colored covers are inferior to transparent ones for both collecting and retaining heat, for although they get quite warm when in the sun, the heat penetrates only the upper few inches of water. Transparent covers allow the sunlight to penetrate deep into the pool. In addition, since dark colors are excellent emitters as well as absorbers of energy, they radiate away slightly more thermal energy at night than do either clear or lighter-colored covers.[6]

Painting a Pool. Painting the walls and bottom of a pool a dark color to improve its solar collection potential has been tried on numerous occasions without much success. A pool's water absorbs 75 to 85 percent of the sunlight passing through it,[7] leaving little energy that even reaches the walls. In one California study,[8] a black pool bottom reduced the annual heating requirement by an average of only 10 percent.

Another problem with painting the pool to increase solar absorption is that it is difficult to find chlorine-resistant dark paints. Most suitable coatings for pools are available in light colors only.

Active Solar Collection Systems for Pools

The amount of solar energy collected by a pool's surface is frequently insufficient to keep the pool at a comfortable temperature. In these situations solar collector arrays can provide the additional heat. To avoid excessively large panel areas, however, it is important that solar collectors be used *in addition to*, rather than instead of, a pool cover.

Covering the pool with a transparent blanket is a method of passively heating the pool's water as well as conserving its thermal energy. A flat-plate collector array constitutes an active system, since it uses a pump for water circulation. Like DHW and space heating systems, active pool systems also employ an automatic controller, temperature sensors, and pressure relief and check valves. A thermal storage tank, though, is not needed because heat can be stored in the pool itself. Figure 12-3 depicts a typical active solar pool heating system.

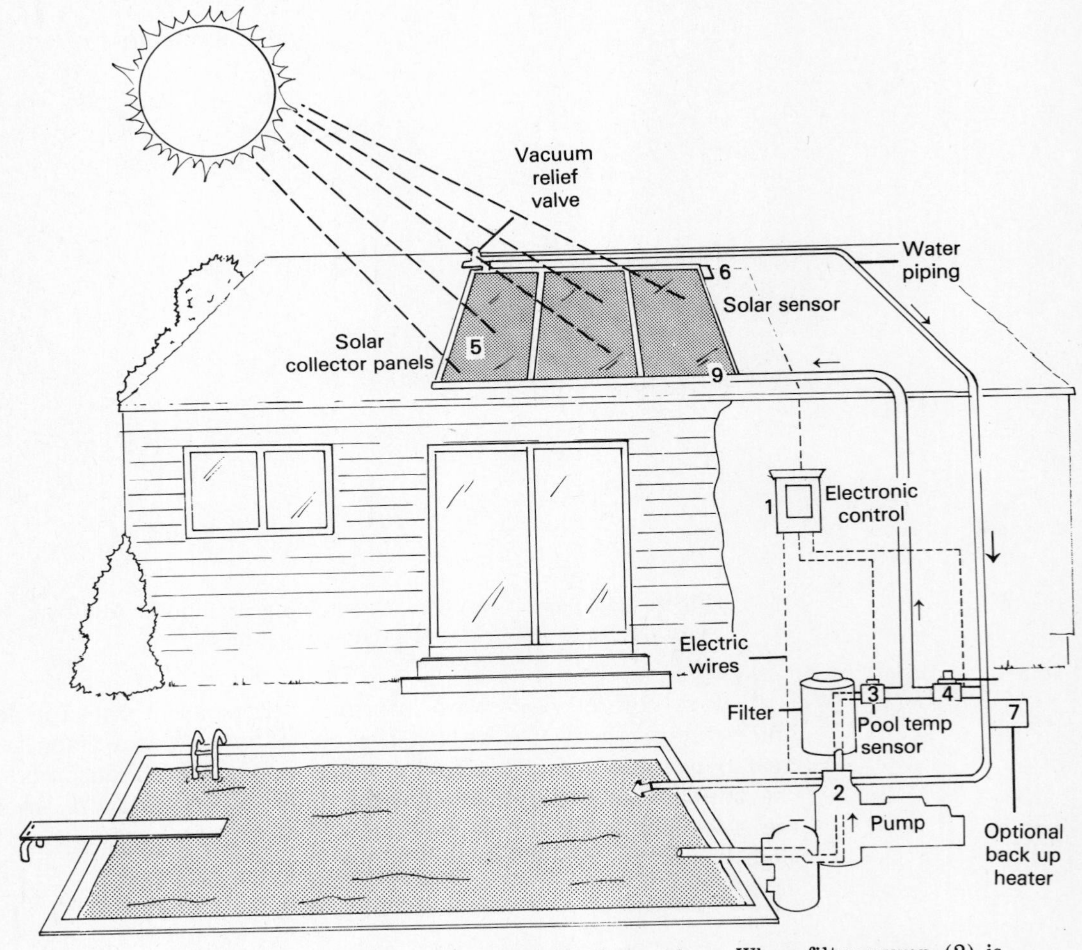

Figure 12-3 Active solar pool heating system. When filter pump (2) is running and pool water sensor (3) signals that temperature is below selected setting on the electronic control (1), the system will automatically activate valve (4) and direct the water flow through solar collector panels (5)—if the solar sensor (6) finds that there is sufficient solar energy available to warm water. A conventional pool heater (7) can be used to supplement the solar heating system. (Courtesy of Solar Industries, Inc., Manasquan, NJ.)

The domestic water system that was studied in Chapter 10 was designed to heat a small volume of water (approximately 100 gallons) to a temperature of 130°F or higher. The purpose of a swimming pool system, on the other hand, is to heat a large volume of water (15,000 gallons) to a temperature of 80°F. While temperatures in domestic water heating collectors frequently reach 160°F and higher, 90°F is a typical pool collector outlet temperature.

Because of the lower temperatures involved, pool collectors can be constructed of plastic or synthetic rubber instead of the more expensive metals used in DHW and space-heating panels. Pool collectors are usually uninsulated and unglazed, which greatly reduces their cost. Because the thermal conductivities of most plastics are as much as 1000 times lower than those of metals such as copper, pool collector flow channel design differs markedly from those of other types of collectors.

The flow channels in panels with metal absorber plates are spaced 5 to 8 inches apart (see Figure 9-1). Because of the high thermal conductivity of the metal between the flow channels, absorbed heat is quickly conducted through the metal plate into the liquid in the channel. If this same design were used for plastic collectors, however, thermal energy would take excessively long to flow through the poorly conducting plastic and much of the energy would be radiated away before it ever reached the flow channel. For this reason, plastic and rubber collectors are designed with little or no space between the flow channels. In addition, the flow channel walls are made as thin as possible to lower the transmission time of thermal energy through them. Figure 12-4 depicts common designs for swimming pool collectors.

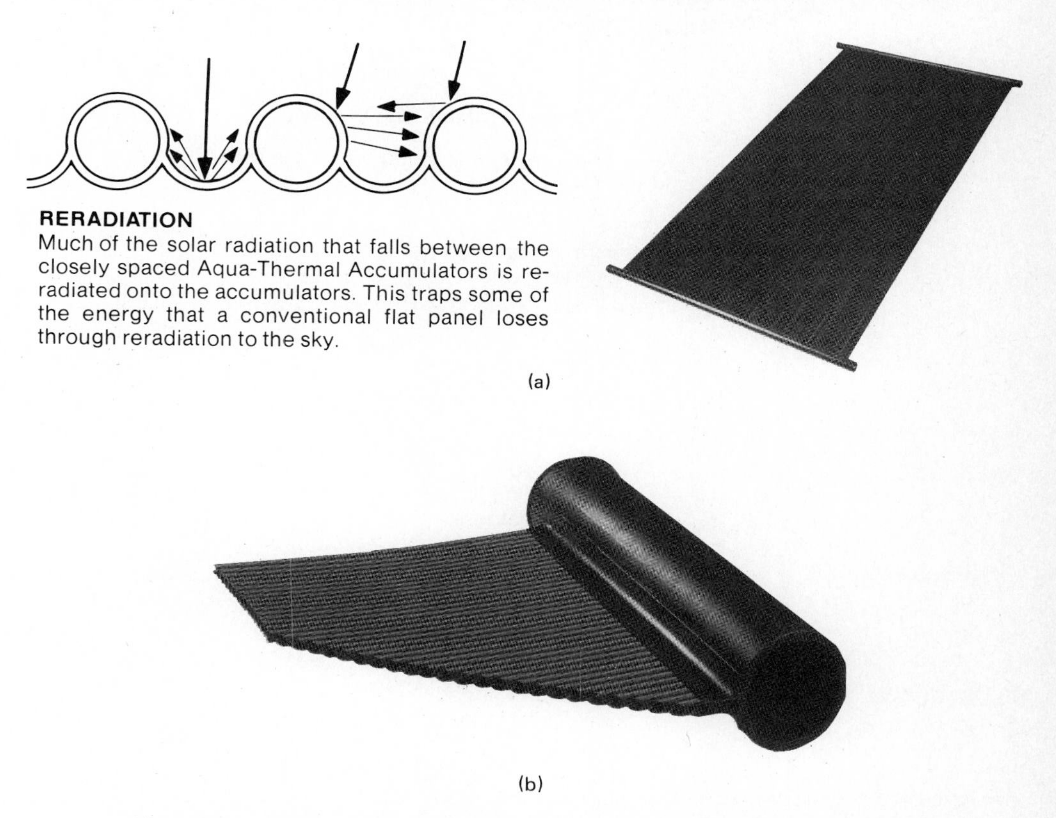

RERADIATION
Much of the solar radiation that falls between the closely spaced Aqua-Thermal Accumulators is reradiated onto the accumulators. This traps some of the energy that a conventional flat panel loses through reradiation to the sky.
(a)
(b)

Figure 12-4 Swimming pool collector designs. (a) Cross section of Solar Industries' Aqua-Thermal Accumulator. (b) Fafco's design. [(a), Courtesy of Solar Industries, Inc., Manasquan, NJ; (b) courtesy of Fafco, Inc., Menlo Park, CA.]

Carbon black is often added to the plastic of a pool collector during its molding process. This chemical will make the plastic more resistant to ultraviolet degradation, as well as improving its thermal conductivity to some extent. Unlike aluminum, galvanized steel, and even copper to some extent, plastics are not sensitive to corrosion from pool chemicals such as chlorine.

Since the absorbing surface of an unglazed collector is in direct contact with the atmosphere, wind currents tend quickly to carry away its thermal energy. The lower the air temperature and/or the higher the wind speed, the faster the rate of heat loss. In windy areas it is advisable to install windbreaks around the collector arrays (making sure that they do not shade the panels) or use glazed collectors.

Most plastic collectors cannot be glazed, for the stagnation temperatures that would then occur during warm, sunny days when no water was circulating through them would rapidly deteriorate the material. This is why metal absorbing plates are used in most glazed collectors. Two of the plastic

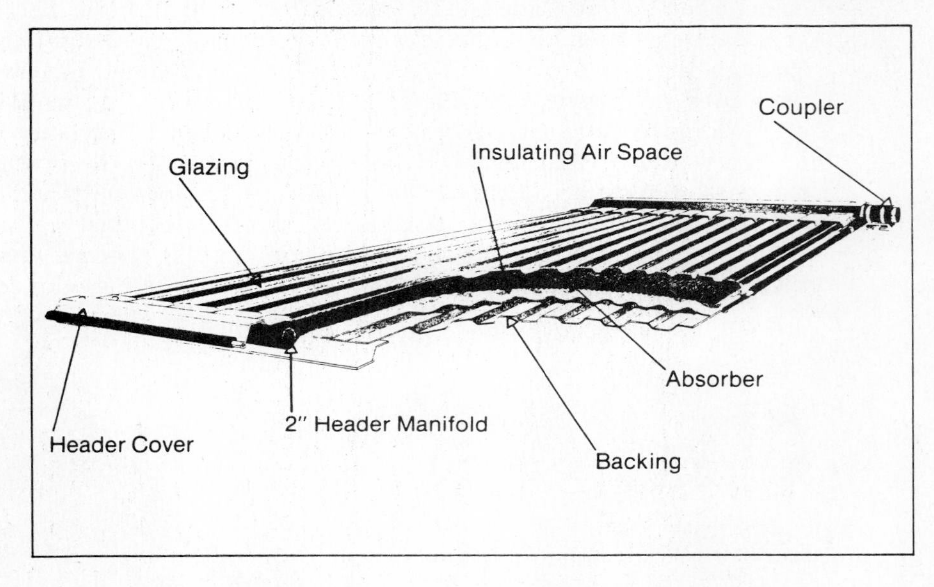

(a)

MATERIALS:

Absorber Plate—FAFCO ultrastablized black polypropylene copolymer.

Glazing—Special profile fiberglass reinforced polyester resin with ultraviolet and weather resistant surface.

Back Plates—Fiberglass reinforced polyester resin.

Header Covers—High temperature ABS with ultraviolet and weather resistant acrylic surface.

PRESSURE:

Collector hydrostatically pressure tested @ 30 psi.
Operating pressure of 10 psi at outlet coupler is recommended maximum.
Vented atmospheric draindown systems, or closed systems with adequate expansion tank volume and pressure relief are recommended.

TEMPERATURE:

Recommended Fluid Operating Temperature = 60° F to 130° F (16° C to 55° C).
Maximum Intermittent Fluid Operating Temperature = 212° F (100° C).

Absorber—Melt Temperature 338° F (170° C)
Stagnation—Detail Recommendations, see Fafco IV Product Description Report.

FLUID FLOW:

Recommended: 1.25 gpm/panel
Design Range: 1-gpm per panel (minimum)
6-gpm per panel (maximum)

CERTIFICATIONS: CALIFORNIA TIPSE, FLORIDA SOLAR ENERGY CENTER, IAPMO

FAFCO reserves the right to change specifications. * Pat. Pending

Figure 12-5 (a) Fafco IV specifications.

(b)

Figure 12-5 (cont.) (b) Fafco IV panels heating the Steinhart Aquarium in San Francisco's Golden Gate Park. (Courtesy of Fafco, Inc., Menlo Park, CA.)

collectors that are glazed are the Fafco IV and the Solar Industries Aqua-Therm GL models. Their absorbers are made of high-temperature polymers enclosed in a glazed case. Unlike the collectors studied in Chapter 9, no foam or fiberglass insulation is added. Some insulating value is provided by the air spaces in the case (see Figure 12-5).

Sizing a Swimming Pool Solar Panel Array

Solar Industries of Manasquan, New Jersey,[9] has developed a simple method of sizing pool collector arrays, using empirical data from their 800-square-foot test pool and later verifying these data with extensive computer modeling techniques. The tables they have developed were designed to be used with the Solar Industries' unglazed, uninsulated polypropylene panels but should also be applicable to the similarly constructed panels sold by other manufacturers, such as Fafco.

It is assumed when using the above-mentioned tables that the collector array is facing south and oriented at the proper tilt angle for the site's latitude. If the pool is intended mainly for summertime use, the array should be tilted at an angle with respect to the horizontal equal to the latitude of the site minus 10 to 15°. If the pool season includes the spring and/or fall months, the tilt angle should equal the latitude angle, and if the pool is intended for year-round use, the tilt should equal the latitude angle plus 10 to 15° (see Figure 12-6).

The first step in the Solar Industries sizing method is to calculate a *basic panel area* using the amount of insolation that the site receives. If the average daily insolation on a horizontal surface is not known for the site, refer to insolation Table III-1, and if the site location is not included in this

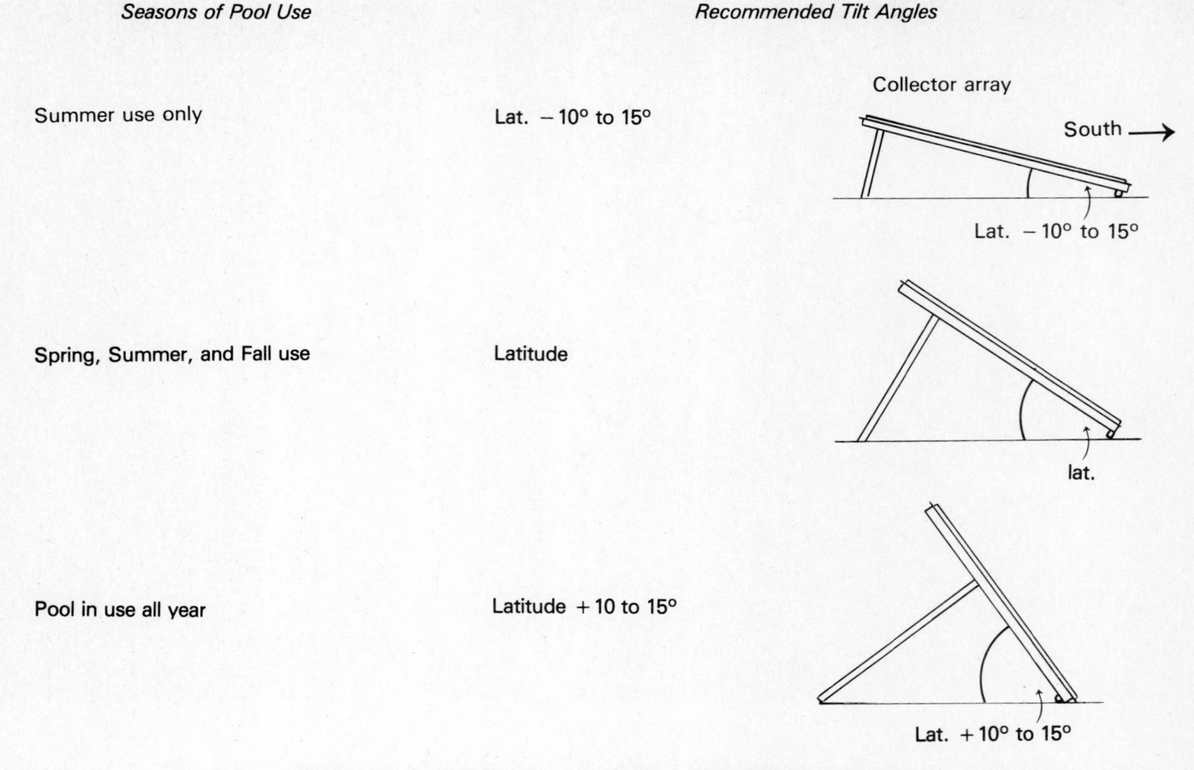

Figure 12-6 Tilt angles of south-facing collector arrays. For good drainage, the minimum tilt of any collector should be 15°.

table, to the final section in Chapter 4, which explains how to select a city with a similar climate. Once insolation data have been obtained, select the lowest figure for the months that the pool is in operation, and use Table 12-1 to find the *basic area factor.* This factor is the ratio of solar panel area to swimming pool surface area and must be multiplied by the pool's square footage to obtain the basic panel area.

As an example, a 750-square-foot pool located in Denver, Colorado, with a season that runs from April through September, receives less daily

TABLE 12-1 Basic Area Factors

Average Daily Insolation (Btu/ft^2-day) on a Horizontal Surface	*Basic Area Factor*
1100	0.85
1200	0.80
1300	0.75
1400	0.70
1500	0.66
1600	0.62
1700	0.59
1800	0.56

Source: This table was computed, with permission, from information in *Application Manual Number 1,* Solar Industries, Inc., Manasquan, NJ, p. 2.

insolation in September (1727 Btu/ft^2) than in any other month that it is in operation. From Table 12-1, a basic area factor of 0.59 is obtained. The basic panel area is then computed as follows:

$$\text{basic panel area} = 0.59 \times 750\ ft^2 = 442\ ft^2$$

The basic panel area must be modified by several factors in order to arrive at a final solar panel area. The first factor is dependent on the site's average wind velocity and is listed in Table 12-2. If the wind speed is not known and instruments for measuring it are not available, the qualitative descriptions in Table 12-2 of the effects of different winds will help the designer to select the proper factor.

TABLE 12-2 Wind Velocity Factors

Average Wind Velocity at Pool (mph)	*Definition*	*Factor without Pool Blanket*	*Factor with Transparent Pool Blanket*
0–10	Calm to light	1	0.85
10–15	Moderate	1.33	1.11
15–20	High	1.64	1.35
Wind Velocity (mph)	*Apparent Effect of Wind*		
8–10	Leaves and twigs in motion, wind starts to extend light flag		
10–15	Raises dust and loose paper; small branches are moved		
15–20	Small trees in leaf begin to sway		

Source: Reprinted with permission from *Application Manual Number 1,* Solar Industries, Inc., Manasquan, NJ, p. 13.

A pool blanket significantly decreases heat losses due to wind. Table 12-2 lists wind factors for both covered and uncovered pools. These factors are to be multiplied by the basic pool area. For example, an uncovered pool with a 450-square-foot basic panel area and located on a site with 10- to 15-mph average winds has a wind factor of 1.33. Multiplying this factor by the basic panel area yields a modified area of 598.5 square feet. If the pool described above were covered by a transparent blanket, the correction factor would be 1.11, leading to a 499.5-square-foot panel area, almost 100 square feet less than for the uncovered pool.

Knowledge of the atmospheric humidity is also important for sizing a solar system. The lower the humidity, the higher the pool's evaporation rate. Since each gallon of water that evaporates takes with it 2½ kilowatthours' worth of energy, the atmospheric humidity can greatly affect a pool's thermal energy losses as well as the solar panel square footage it requires. The correction factors for atmospheric humidity are listed in Table 12-3.

Since a large fraction of a swimming pool's heating requirement can be met by sunlight impinging directly on the pool's surface, the amount of shading the pool receives has a dramatic effect on the solar collector area needed to supply the additional heat. Table 12-4 lists correction factors for various shading situations. The solar panel area required for a pool that is enclosed in an insect screen is the same as if the pool were 25 percent shaded all day.

TABLE 12-3 Atmospheric Humidity Corrections

The amount of moisture in the air (relative humidity) has an effect on the evaporation rate from the pool. The higher the relative humidity, the lower the evaporation rate. Since this means that the heat loss from the pool will be reduced, less solar panel area is required. In unusually dry regions additional solar panel area must be provided to compensate for the added heat losses.

Average Relative Humidity (%)	*Climate*	*Factor*
60+	Moist	0.75
30–60	Moderate	1.0
0–30	Dry	1.25

Source: Reprinted with permission from *Application Manual Number 1,* Solar Industries, Inc., Manasquan, NJ, p. 13.

TABLE 12-4 Shading Corrections

If all or part of the pool is shaded during the daylight hours, the solar panel area must be increased to compensate for the heat the pool does not receive from direct solar radiation.

Amount of Pool Surface in Shade	*Shade All Day*	*Shade 75% of Day*	*Shade 50% of Day*	*Shade 25% of Day*
100	2	1.75	1.50	1.25
75	1.75	1.56	1.38	1.19
50	1.50	1.38	1.25	1.13
25	1.25	1.19	1.13	1.06
0	1	1	1	1

Source: Reprinted with permission from *Application Manual Number 1,* Solar Industries, Inc., Manasquan, NJ, p. 13.

Example Calculation. Calculate the collector area needed for a pool covered with a transparent blanket and located on a site with an average wind velocity of 8 miles per hour, a humidity of 50 percent, and subject to 25 percent shading for 50 percent of each day. The basic solar panel area for the pool is 525 square feet.

From Tables 12-2, 12-3, and 12-4 it is found that the wind velocity factor is 0.85, the humidity level factor is 1.0, and the shading factor is 1.13. Multiplying these factors by the basic area yields a final solar panel area:

$$\text{final solar panel area} = 525 \text{ ft}^2 \times 0.85 \times 1.0 \times 1.13 = 504 \text{ ft}^2$$

HOT TUBS AND SPAS

Hot tubs and spas are becoming very popular residential installations. Their numbers may one day rival those of home swimming pools. Using traditional energy sources to heat tubs and spas is expensive and wasteful. Solar energy offers a less expensive and very promising alternative.

Although hot tubs and spas have similar energy requirements, their construction is somewhat different. Hot tubs are self-supporting structures that are made out of wood such as oak, redwood, cedar, teak, or mahogany, or of other materials, such as concrete or metal. Spas are usually made of fiberglass, polyethylene, or gunnite and are not self-supporting, requiring a surrounding structure such as a bed of sand to prevent them from breaking under the weight of the water.

It is thought by some that use of hot tubs began in the Orient, where communal bathing is a centuries' old tradition. Spas, on the other hand, receive their name from the town of Spa in Belgium. The mineral springs there have been visited by bathers since the early thirteenth century for the purposes of healing and relaxation.

Hot tubs and spas are sold with circulation systems consisting of a pump, heater, filter, and often an air blower. The blower powers air jets in the walls and/or floor that create a relaxing massage effect and an enjoyable sense of buoyancy. Tubs and spas lose heat up to four times as fast with the blower turned on, however, as they do when it is off. This is because the air blown through the water carries thermal energy out of the spa at a very rapid rate.

The heating unit, which is usually powered by gas or electricity, is designed to raise the water temperature to between 95 and 106°F,[10] 15 to 26°F warmer than a typical swimming pool temperature. Solar heating systems can readily supply some or all of the energy necessary for maintaining the water at the proper temperature. The energy characteristics of tubs and spas differ somewhat from pools, however, and need to be studied before solar heating systems designed for them are examined.

A Comparison of the Energy Characteristics of Hot Tubs, Spas, and Swimming Pools

Hot tubs and spas designed for home use have surface areas that range from 13 to 14 square feet to 75 square feet, whereas pool areas are 300 to 400 square feet or higher. The small surface area of tubs and spas allows quick installation and removal of an insulated cover. Covers for pools are more time consuming to install and remove and, as a result, many home owners neglect to use them every day, leading to unnecessary energy losses.

Swimming pools are usually used during daylight hours so that the pool and the collector arrays receive solar energy during the times when the pool is uncovered and losing heat at its maximum rate. Spas and tubs, on the other hand, are normally used sometime between 4:00 and 11 P.M.,[11] which means that they receive little or no solar energy replenishment during their period of maximum heat loss. This coupled with the greater temperatures that they must be heated to implies that solar energy is generally not able to provide as high a percentage of a spa or hot tub's total heating needs as it can for a swimming pool system. If the tub or spa system includes thermal storage tanks, however, that can hold energy captured during the day for use in the evening, the solar heating fraction increases considerably.

If an air blower is not installed in a spa that is heated by an array of solar panels, it is likely that during sunny periods of the year the backup heater will not be necessary. The temperature of a covered 400-gallon spa drops only a fraction of a degree per hour when surrounded by air at an ambient temperature of 60°F or higher. When uncovered and in use but with the air jets turned off, the spa loses only 2 to 3°F per hour.[12]

The second type of solar system for tubs and spas—the one that employs not only a collector array but a storage tank as well—employs a bank of solar collectors to charge both the tub and tank with thermal energy. As the tub cools down during use, an automatic system controller connected to sensors in various parts of the system turns on the pump and delivers hot water from storage.

Figure 12-7 depicts systems of both types discussed above. Note that the same storage tank that is used for a DHW or space heating system can also be used to provide thermal energy for a spa or hot tub.

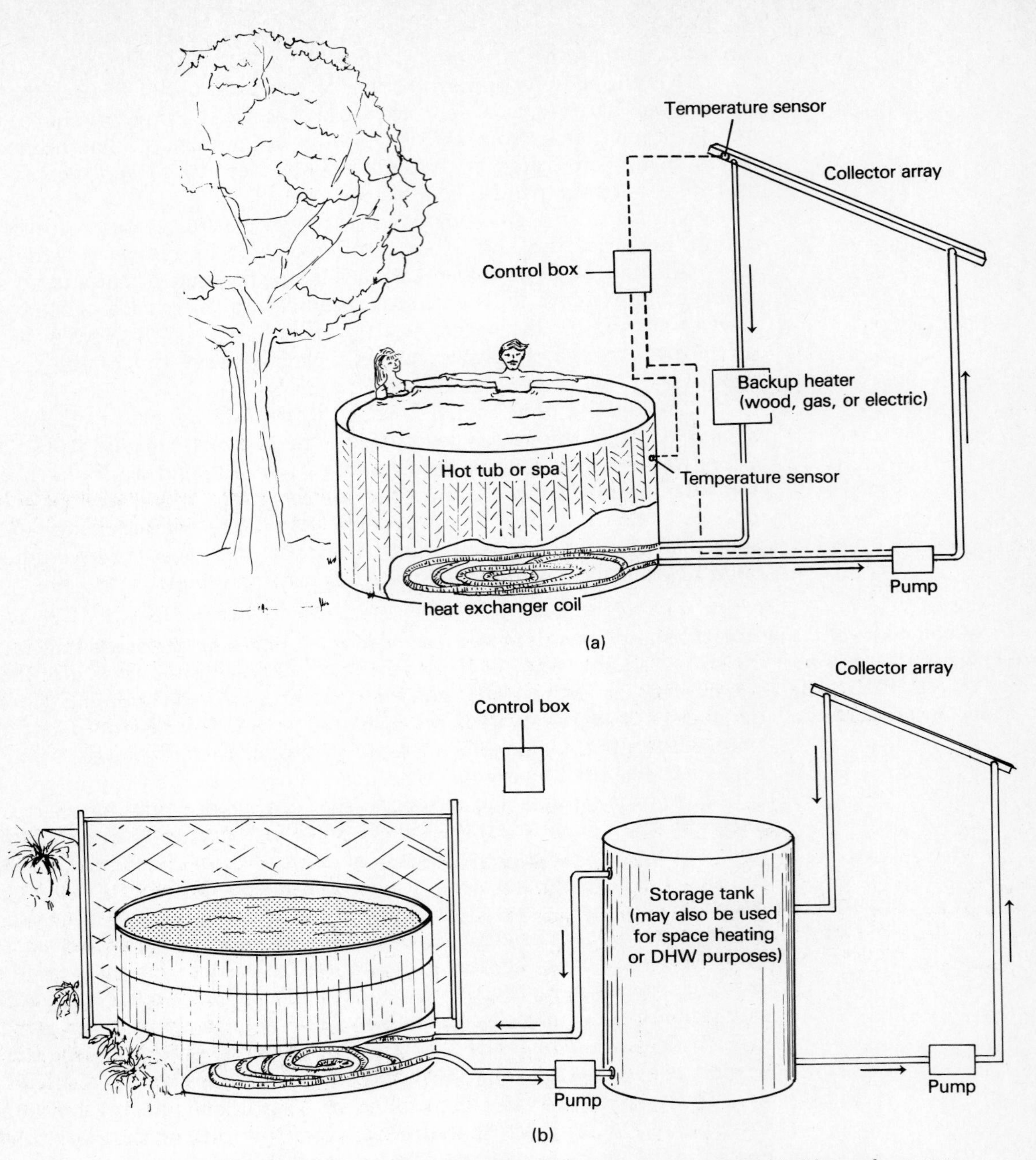

Figure 12-7 Solar-heated hot tub and spa systems. (a) Thermal energy stored in hot tub or spa only. (b) Thermal energy stored in hot tub/spa, as well as in additional storage tank.

To Glaze or Not to Glaze. Both glazed and less expensive unglazed solar panels are used for spa and hot tub heating. Unglazed panels are appropriate when the weather is warm but become extremely inefficient as ambient air temperatures drop to 40°F or more below the collector operating temperature (which should be at least 100°F and preferably higher). This disqualifies unglazed panels for use in winter in all but the mildest climates, such as are found in Florida or southern California.

To determine whether to purchase glazed or unglazed panels for a particular installation, the designer must first decide which months of the year

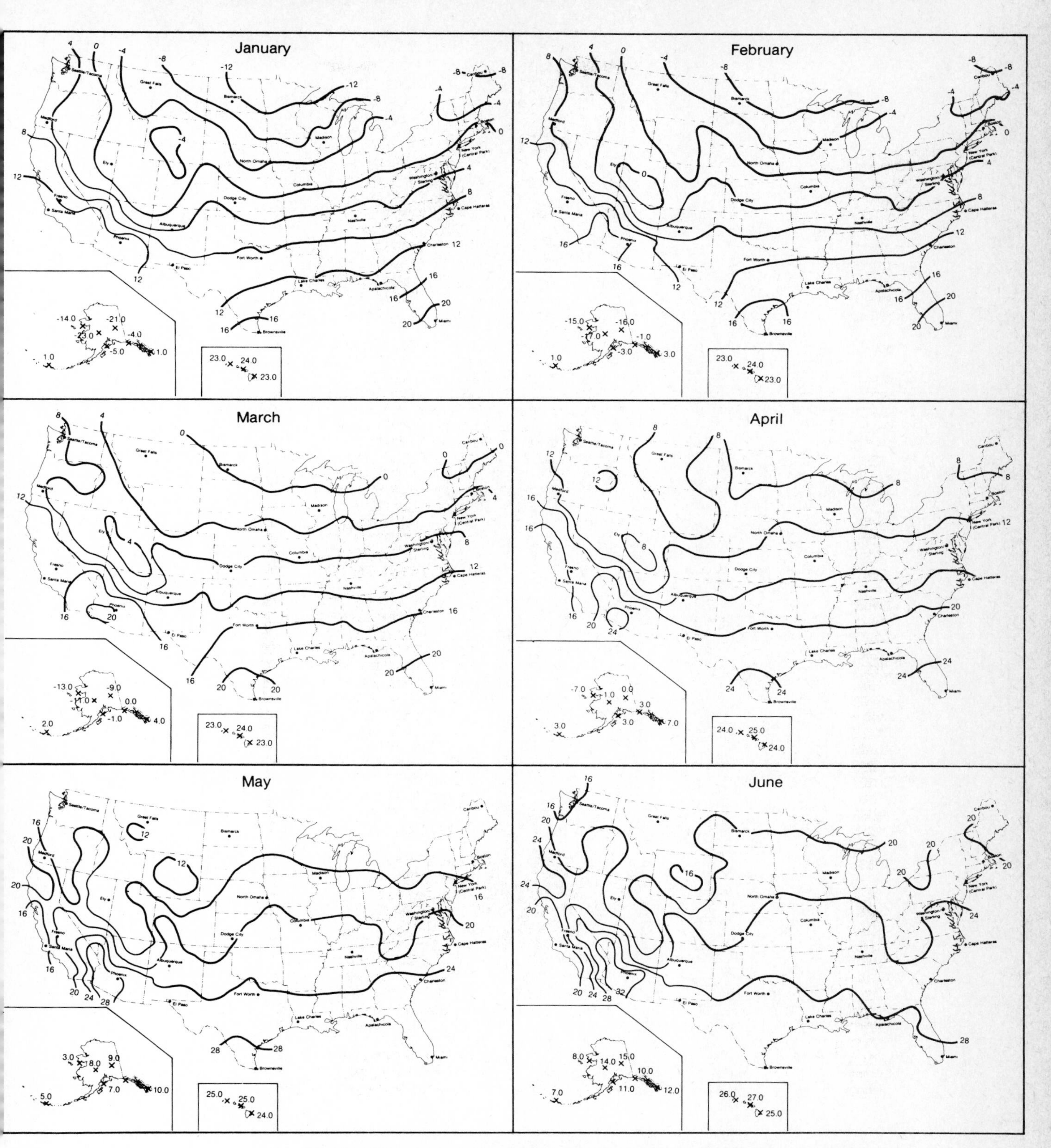

Figure 12-8 Average daytime dry bulb temperatures. To convert temperatures to Fahrenheit, multiply by 1.8 and add 32. For instance, $4^{\circ}C \times 1.8 + 32 = 39.2^{\circ}$ Fahrenheit. (From *Solar Radiation Energy Resource Atlas of the United States*, SERI/SP-642-1037, Solar Energy Research Institute, Golden, CO, 1981.)

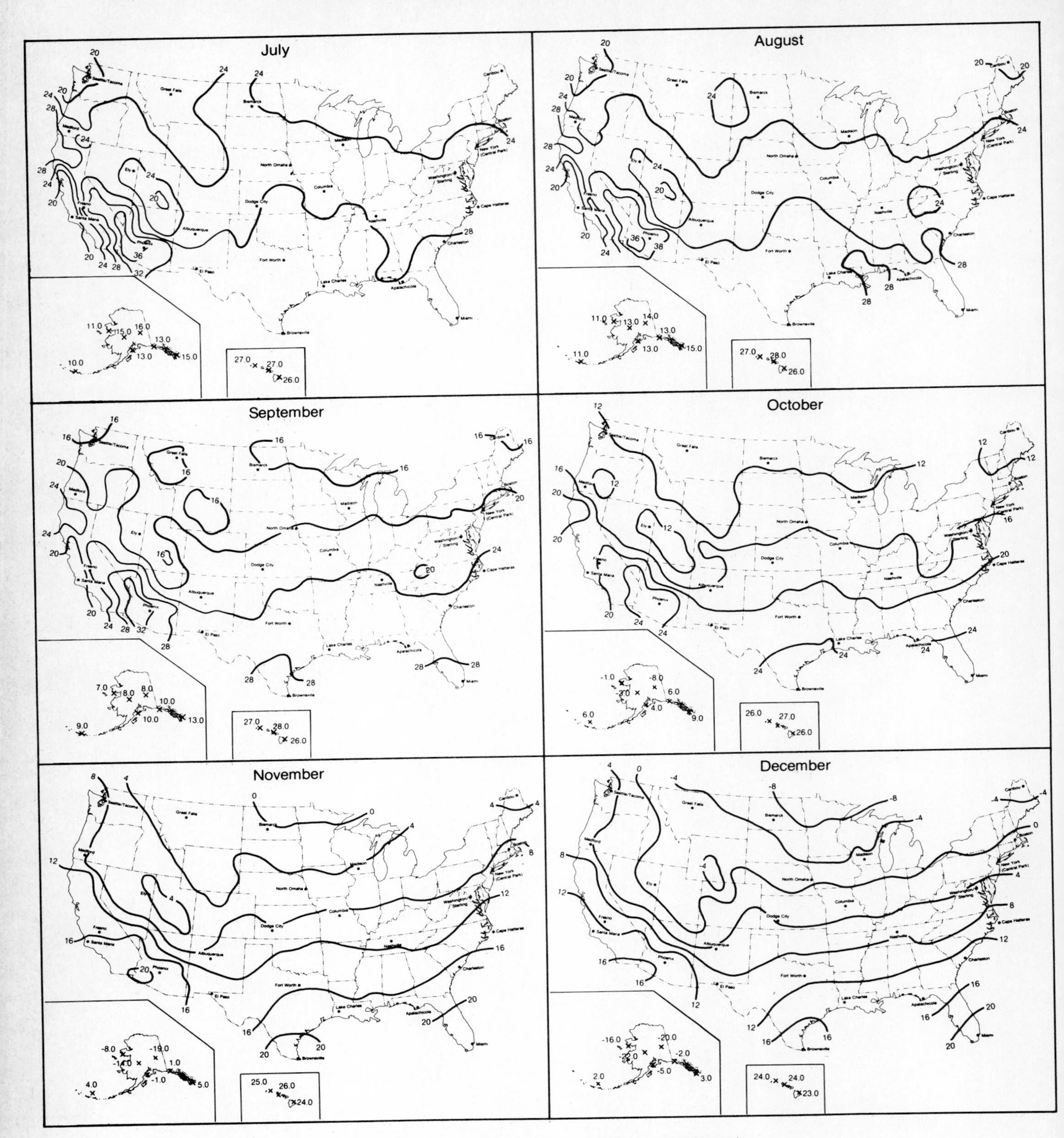

Figure 12-8 (cont.)

the spa will be used and then determine the average daytime temperatures for those months. If the average daytime temperature of the coldest month of the spa season is less than 60°F (16°C) then glazed panels should be used. Figure 12-8 will be a help in estimating daytime temperatures of various locations. It is the daytime rather than 24-hour average temperature that is significant, for it is only during daylight hours that the collector array functions.

Glazed collectors are advisable if storage tanks that are also part of a DHW or space heating system are used, for the desired storage temperature for those systems is 130°F or above, and that is beyond the efficient range of unglazed panels.

The Hot Tub or Spa Cover

As with swimming pools, a snugly fitting cover on a spa or hot tub dramatically lowers thermal energy losses and operating expenses. For example, a 50-square-foot spa located in an area with an ambient temperature of 60°F and a humidity of 70 percent loses only 2200 Btu per hour when covered by an 8-inch-thick layer of foam with an insulating value of R-1 but loses from 7400 to 12,100 Btu per hour if uncovered.[13]

Sizing a Hot Tub or Spa System

Solar Industries has developed an empirical formula for sizing solar panels for a tub or spa. Their method is for *glazed* solar panels and is based on the volume of water rather than surface area as was done for swimming pools. The first step in the sizing procedure is to determine the site's average insolation for the month of the spa season with the least amount of solar radiation (this is done in the same way as for swimming pools). If the spa is used all year long, January will usually be the month chosen. After finding an insolation value, refer to Table 12-1 for the basic area factor.

Determine the volume of the spa in gallons. This information can often be obtained from the distributor or manufacturer of the spa. Or alternatively, the following method may be used to find volume:

For a rectangular spa (Figure 12-9a):

1. Multiply the length times the width times the depth, in feet, which gives you the volume in cubic feet.
2. Multiply by 7.48 to obtain the volume in gallons.

For a cylindrical-shaped spa or tub:

1. Divide diameter by 2 to obtain radius.
2. Square the radius.
3. Multiply by $\pi \times$ height (π is about 3.14) to obtain volume in cubic feet.
4. Multiply by 7.48 to obtain volume in gallons.

For example, if height is 4 feet and diameter is 6 feet,

(1) 6 ft/2 = 3-ft radius

(2) $(3\text{ ft})^2 = 9$

(3) $\pi \times 4\text{ ft} \times 9 = 113\text{ ft}^2$

(4) $7.48 \times 113 = 845$ gallons

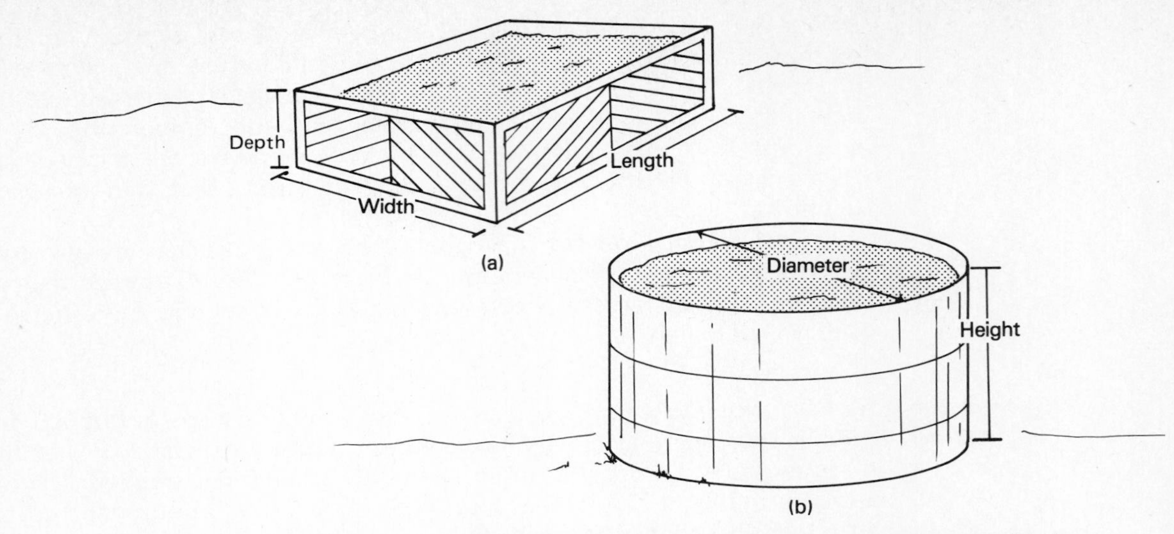

Figure 12-9 Determining volume in gallons. (a) Rectangular spa. (b) Cylindrical spa or tub.

After determining the spa's volume, divide it by 4 and multiply by the basic area factor found in Table 12-1.

Example Calculation. A 500-gallon spa located in Houston, Texas, is to be used from March through October. Calculate the solar panel area it requires.

It is seen from insolation Table III-1 that for Houston, October is the month of the spa season with the least amount of solar radiation (1276 Btu per day). Using this value in Table 12-1, a basic area factor of 0.75 is obtained. The basic panel area is then determined as follows:

$$\text{basic panel area} = \text{basic area factor} \times \frac{\text{spa volume (gallons)}}{4}$$

$$= 0.75 \times 125$$

$$= 94 \text{ ft}^2$$

Figure 12-10 Author (left) and friends engaged in a typical California recreational activity. What's not so typical about this hot tub is that it's solar heated, using glazed flat-plate collectors. A small wood furnace provides backup heating.

After obtaining the basic panel area, utilize Tables 12-2, 12-3, and 12-4 in the same way as was done for swimming pools to calculate a final panel area that takes into account wind velocity, humidity, shading, and whether or not the spa is covered during standby periods (see Figure 12-10).

NOTES

1. Douglas E. Root, Jr., William M. Partington, Jr., and Lowell Lotspeich, *Solar Heating for Swimming Pools*, Florida Conservation Foundation, Inc., Winter Park, FL, 1982, p. 5.
2. Ibid., p. 19.
3. Ibid.
4. Ibid., p. 7.
5. Ibid., p. 8.
6. Ibid.
7. Ibid., p. 9.
8. *Solar Applications Seminar*, Pacific Sun, Inc., Palo Alto, CA, 1977, p. D-31.
9. *Application Manual Number One*, Solar Industries, Inc., Manasquan, NJ, 1979, pp. 2–13.
10. Douglas E. Root, *Heating Alternatives for Spas and Hot Tubs*, Florida Solar Energy Center, Installation Note 9, Publication FSEC-IN-9-81, 1981, pp. 1, 8.
11. Ibid., p. 3.
12. Douglas E. Root, "Heating Options for Spas and Hot Tubs," *Solar Age*, Vol. 7, No. 4, p. 48.
13. Calculated from information in *Heating Alternatives for Spas and Hot Tubs*, Appendix I.

SUGGESTED READING

ROOT, DOUGLASS E., PARTINGTON, WILLIAM M., AND LOTSPEICH, LOWELL, *Solar Heating for Swimming Pools*. Winter Park, FL: Florida Conservation Foundation, 1982. Available by writing or calling the Environmental Information Center of the Florida Conservation Foundation, Inc., 935 Orange Avenue, Winter Park, FL 32789. (305) 644-5377.

REVIEW QUESTIONS

1. Describe the four ways that pools can lose heat.
2. Describe how light-transmissive pool covers transform the pool into a passive collector.
3. Contrast flat-plate collectors for pools with those for DHW systems.
4. How do hot tubs differ from spas?
5. What effect do air jets have on a spa's rate of heat loss?
6. What is a typical temperature of a swimming pool? Of a spa or hot tub? During what times of day are these installations typically used?
7. When should glazed panels be used for spa or hot tub heating?

PROBLEMS

1. A 925-square-foot swimming pool located in San Diego, California, is used from March through November. Calculate the basic solar panel area for the pool.
2. Calculate the final panel area for the system, assuming that the pool is covered when not in use. The average wind speed is 7 mph, humidity is less than 30 percent, and the pool is totally shaded from 1:30 P.M. on (in other words, for 25 percent of the prime daylight hours of 9 A.M. to 3 P.M.).
3. A 675-square-foot pool in Las Vegas, Nevada, is used from April through October. No pool cover is used, average wind speed is 13 mph, humidity is less than 30 percent, and the pool averages 50 percent shade for half of each day and total sunshine the rest of the time. Calculate the final panel area.
4. A 1000-square-foot pool in Columbus, Ohio, is used from May through September. A plastic cover is installed when no one is swimming, the average wind speed is 12 mph, humidity averages 66 percent, an insect screen is installed around the pool, and in addition, 25 percent of the pool is shaded throughout the day. What is the final panel area?
5. A 375-gallon spa located in Raleigh, North Carolina, is used from April through October. Calculate the collector area it requires. Wind velocity is 9 mph, the spa is unshaded all day, and covered during standby periods and the mean humidity at 7 P.M., during a popular time of use, is 68 percent.

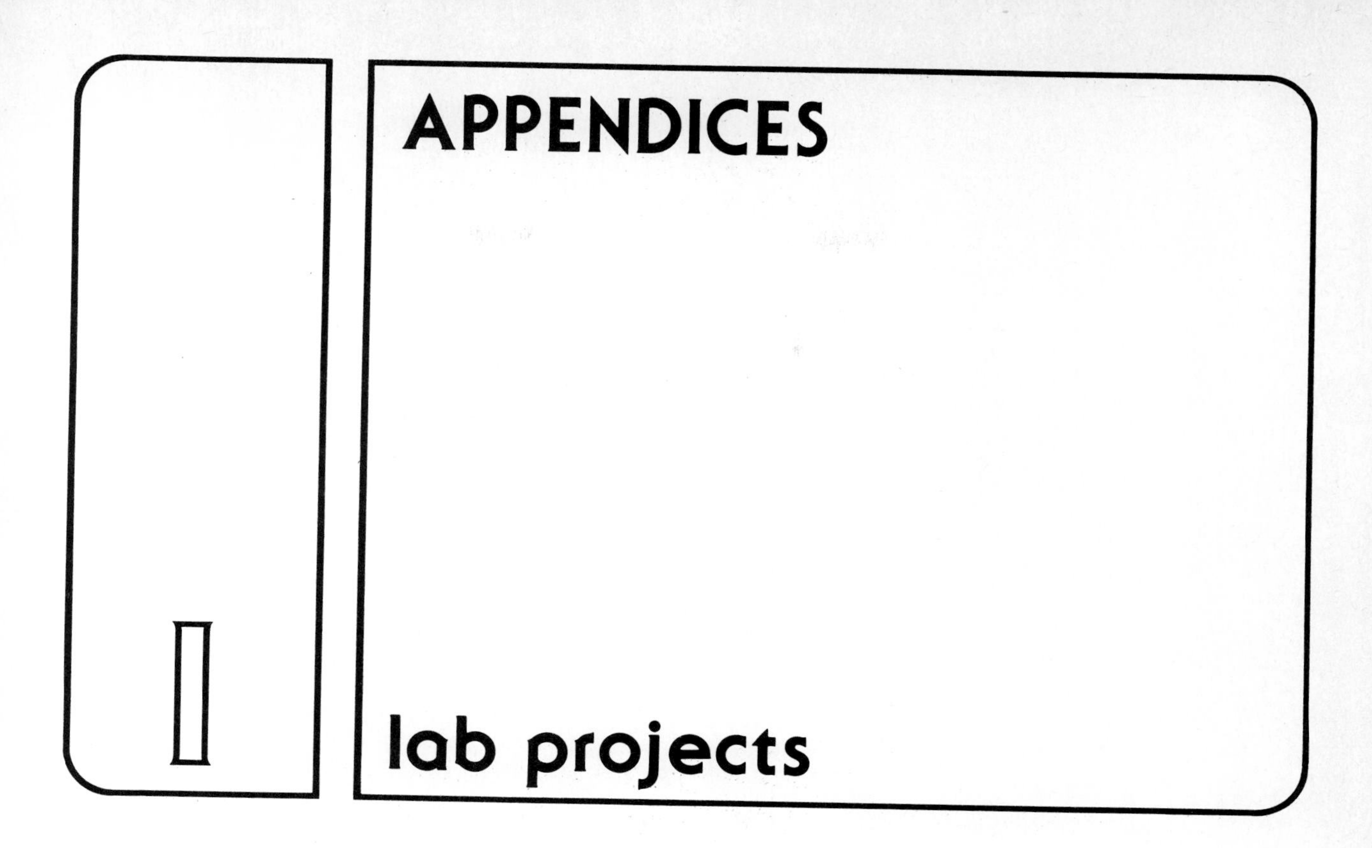

There are two objectives to the lab projects described in this appendix. The first is to provide students with "hands-on" experiences in solar energy. The second is to offer an inexpensive way of obtaining workable solar energy components. The collector whose construction is described in Lab Project A, for instance, can be built for approximately 25 percent of the cost of a "store-bought" model of the same size.

LAB PROJECT A: THE OWNER-BUILT FLAT-PLATE COLLECTOR

The do-it-yourself "recipe" for flat-plate collector construction presented below was originally inspired by a design of Roy Irving,[1] a former instructor at Sonoma State University in California. But in the five years that the author of this textbook has been conducting collector construction classes, Irving's design has been modified and remodified, with the aim of improving efficiency and making construction less difficult. If carefully constructed, the collector should provide many years of useful and satisfying service.

Table I-A-1 lists the tools and materials needed for an 18-square-foot collector. Notice that only common hand tools are required, and that all materials, with the exception of the thermal cement, are readily available from building supply stores.

The first step in collector construction is to put together the pipe grid. The length and position of each piece of pipe is depicted in Figure I-A-1. Pipes can be cut with a hacksaw, although a pipecutter makes a neater slice and is safer to use. The ends of each cut section should be reamed out (pipecutters often have a tool on them to do this) and the last 1 inch on each end of the pipe sanded until shiny.

TABLE I-A-1

(a) Materials List

Quantity	*Description*
40 ft	Type M ½-inch rigid copper pipe
2	½-inch copper 90° elbows
12	½-inch copper T fittings (tees)
2	½-inch copper threaded fittings
18 ft²	3½-inch fiberglass insulation with a foil face or 1½-inch polyisocyanurate rigid foam insulation
18 ft	1 × 6 pine or fir (1 × 4 wood if foam insulation is used)
18 ft	1 × 4 pine or fir (1 × 2 if foam insulation is used)
18 ft²	⅜-inch plywood (more than one piece may be used if joints are well sealed, although one piece is recommended)
18 ft	1-inch-wide wood lath
0.2 lb	50-50 or 60-40 plumber's solder
	Flux
	Waterproof glue
	Silicone sealer
30 ft	16- to 18-gauge galvanized wire
1 pt	Aluminum primer
1 pt	Flat black paint able to withstand 250°F
½ lb	Number 4 galvanized nails
42 ft	8-inch-wide 16-gauge aluminum flashing
18 ft²	Glazing (see the text for suggestions)
1 qt	Chemax Tracit 1100 nonhardening thermal cement available from Chemax Corp., 211 River Road, New Castle, DE 19720, or equivalent
110	Pop rivets or sheet metal screws

(b) Tool List

Hammer
Hand or power saw
Drill
Pipecutter or hacksaw
Fine sandpaper
Propane torch
Sheet metal shears
Sawhorses
Pliers
Paintbrush
Pop-rivet gun or screwdriver
Carpenter's square
5-foot piece of ⅝-inch-outer-diameter steel rod, galvanized pipe, or electrical conduit
Staple gun
Work gloves

Notice from Figure I-A-1 that four of the pipe sections need to be "cut to fit." Their lengths should be adjusted so that they are just long enough to bottom out in the fittings they go into.

Soldering of the pipes is best done on sawhorses over a nonflammable surface like a slab floor or the ground. Notice that there are two flames issuing from the nozzle of the torch: a bright inner one and a dimmer outer one. Figure I-A-2 illustrates where the hottest point of the flames is—just in front of the tip of the inner flame. It is this spot that should touch the copper piping. Be careful not to get burned! The outer flame, although hard to see, is hot to a distance of 1½ feet or more from the nozzle.

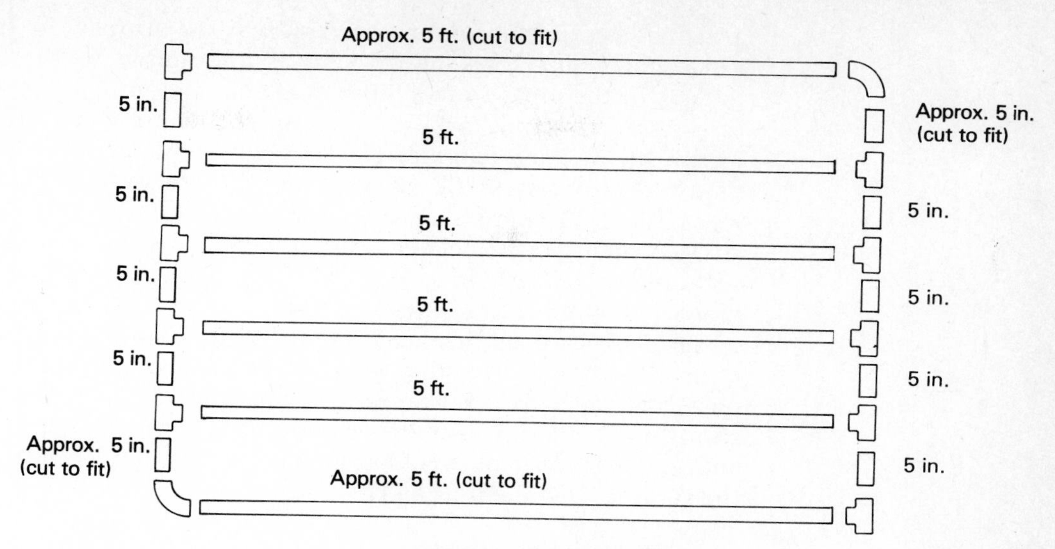

Figure I-A-1 Pipe grid.

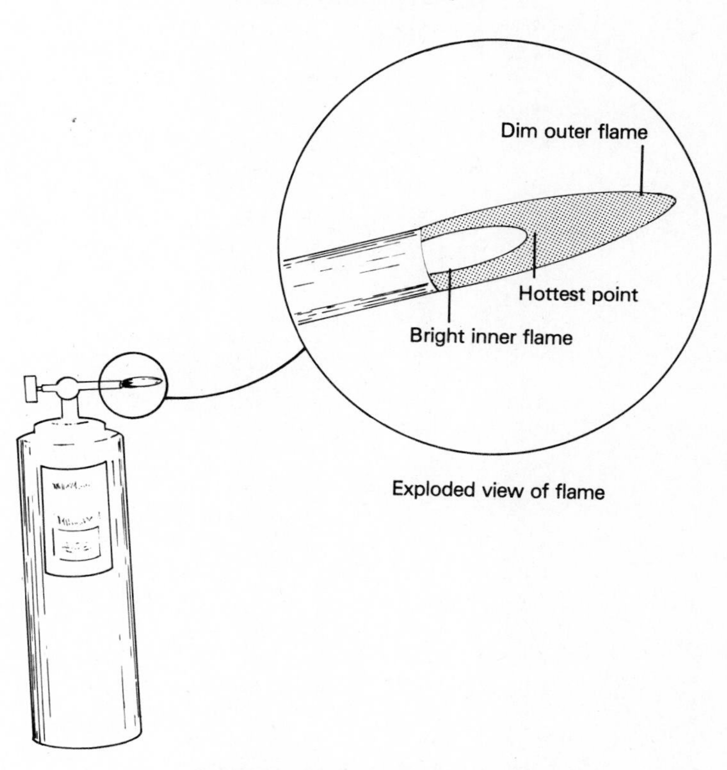

Figure I-A-2 Use of the torch.

Apply an ample amount of flux to the sanded pipe ends as well as to the insides of the tees and elbows. Heat each joint separately with the torch. After 15 to 20 seconds, touch the solder to the copper, and see if it will melt. When it does, it will be drawn into the fitting and around the pipe. Heat for just a few seconds more and then go on to another pipe. Prolonged heating can ruin the joint! And don't use too much solder. If you can see globs hanging from a cooled joint, you have used too much.

After you have soldered the pipe grid and let it cool, pressure test it. The easiest way is to plug up one end with a cap or your finger and attach a water hose to the other end. You will see dripping or spurting from any leaks. Make sure that the dripping is not simply water running down the outside of the pipes from a loose hose connection.

After repairing leaks, cut pieces of flashing to fit the pipe grid, as per Figure I-A-3. Then hammer a groove down the center of each strip, for the riser pipes to rest in. To do this, lay two pieces of wood (you can use the wood that will later be made into the collector case) next to each other with an 11/16-inch gap in between, as in Figure I-A-4. Tack the wood together with pieces of lath so that it will not spread apart when hammering the grooves. Lay your long pieces of flashing down on this jig with the centerline of the flashing over the gap between the pieces of wood, and lay a 5/8-inch-outer-diameter rod or pipe on the flashing. Copper is too soft for this purpose, but steel rod, galvanized pipe, or metal electrical conduit are tough enough. Place a 2 × 4 piece of wood lengthwise on top of the pipe or rod, and hammer on it until a U-shaped groove about 3/8 inch deep is formed in the flashing (see Figure I-A-4).

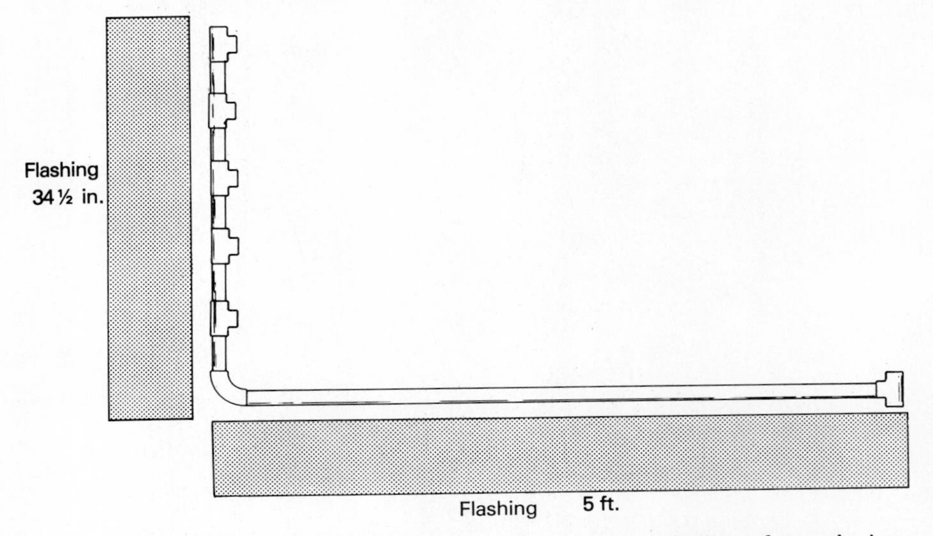

Figure I-A-3 Flashing dimensions. Cut a piece of flashing for each riser pipe and for each header pipe.

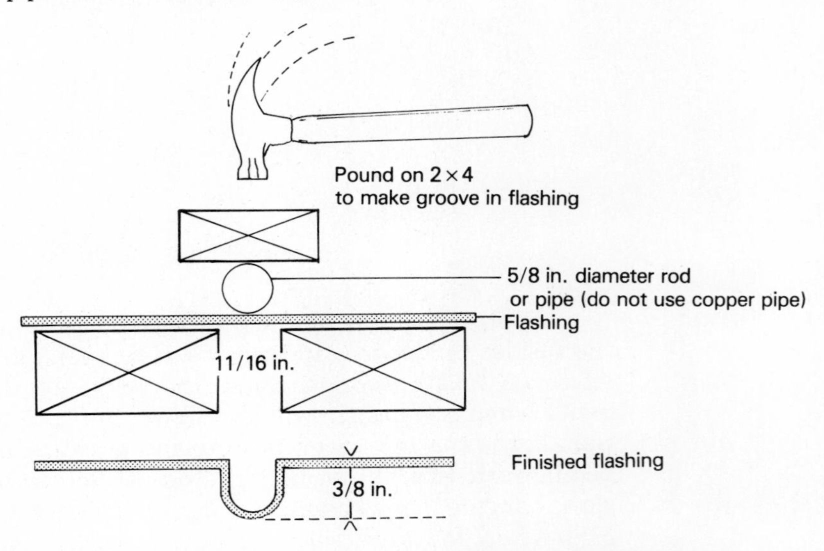

Figure I-A-4 Grooving the flashing for the riser pipes.

Next, prepare both of the short pieces of flashing. These will fit on the header pipes at each end on the grid. Cut slots in the flashing pieces to accommodate the elbows and tees, as shown in Figure I-A-5. Pound a groove along the edge of the flashing piece for the header to nest in.

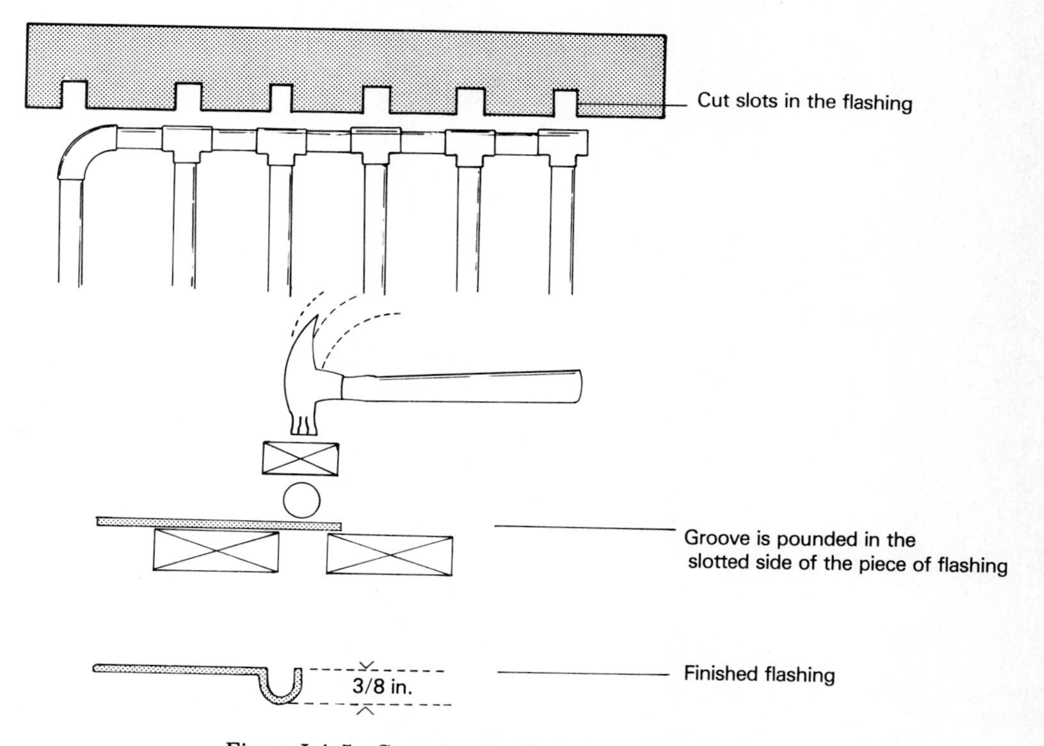

Figure I-A-5 Grooving the flashing for the header pipes.

You are now ready to drill holes in the flashing for the wires that will hold the pipe grid in place. Make the holes just big enough for the galvanized wire that you have bought. Holes should be spaced every 6 inches on either side of the groove, as in Figure I-A-6.

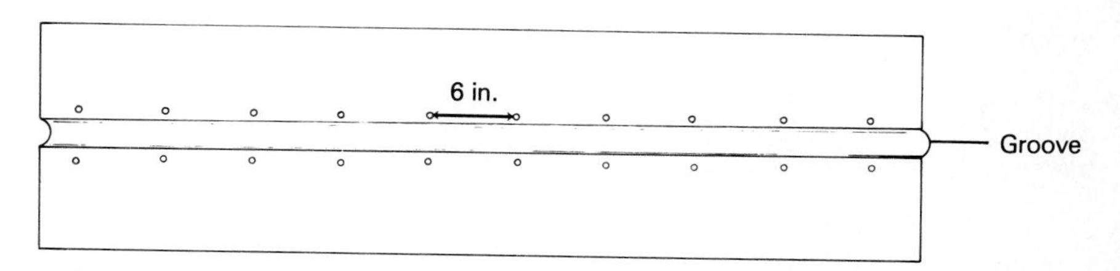

Figure I-A-6 Locating wire holes in the flashing. Space holes every 6 inches on each side of groove.

Flashing is sold with a layer of protective grease on it. Clean this off using a solvent (apple cider vinegar works quite well). Then lay a strip of thermal cement down the center of each groove. Besides making a good thermal bond, the cement will provide a dielectric between the copper pipe and aluminum flashing that will prevent electrolysis. Use the wood jig that you built to press the pipe grid into place. Do not hammer on the 2 × 4, however; just walk on it and it will press the pipe grid tightly into the grooves. Wipe off the excess thermal cement that has squeezed out of the sides of the grooves and *save it* for reuse; it's expensive. After each section of pipe has

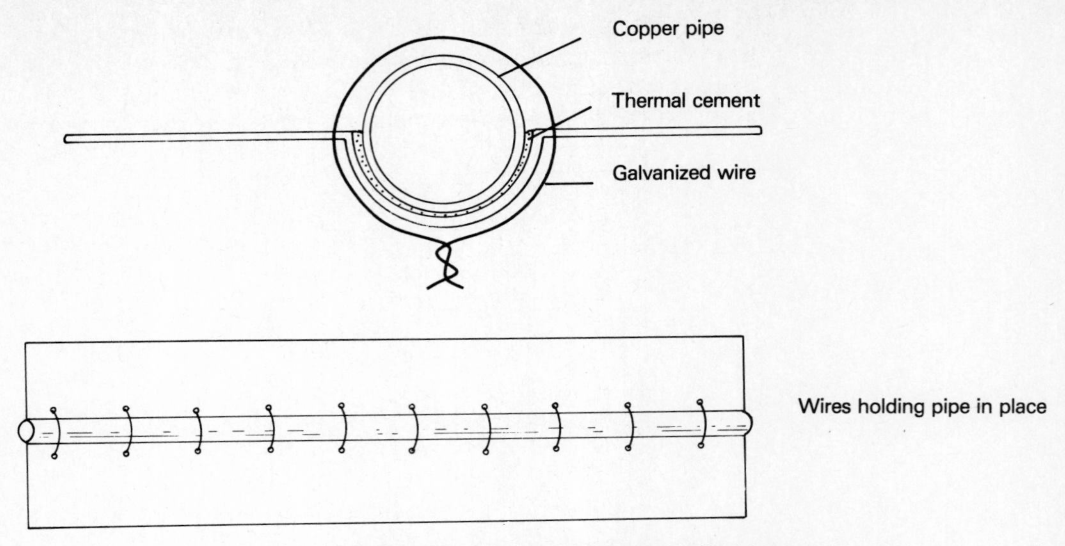

Figure I-A-7 Attaching pipes to flashing.

been pressed into the flashing, wire it in place before going on to another section (see Figure I-A-7).

Notice that pieces of flashing on adjacent riser pipes overlap each other. Fasten the pieces together with double rows of sheet metal screws or pop rivets. Increasing the thickness of the absorber plate in this way will speed up thermal energy transfer into the flow channels (see Figure I-A-8).

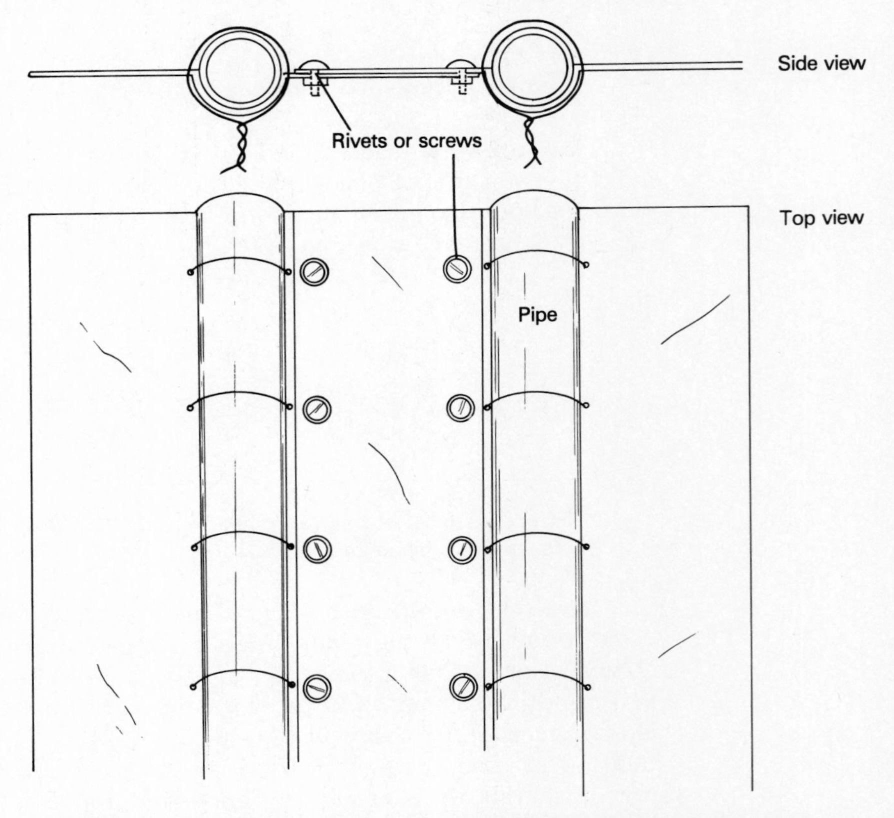

Figure I-A-8 Fastening pieces of flashing together. Fasten adjacent flashing pieces together with pop rivets or screws.

Now it is time to paint the absorber plate. Prepare the surface with a good primer and then paint it with a flat black paint that will withstand 250°F temperatures. Rustoleum paint for wood stoves works quite well. Paint only the side facing the sun, and cover everything.

The wooden case should be built to just barely enclose the absorber plate (see Figure I-A-9). Use waterproof glue as well as nails on all joints. If a high-temperature foam insulation such as polyisocyanurate is used instead of fiberglass, the case can be made 2 inches thinner.

The collector must be seasoned before the permanent glazing is installed. Staple an inexpensive layer of clear plastic such as 4-mil polyethylene over the collector and leave it in the sun for a week. Resins and gases will bake out of the paint, glue, thermal cement, and insulation, and be deposited on the clear plastic.

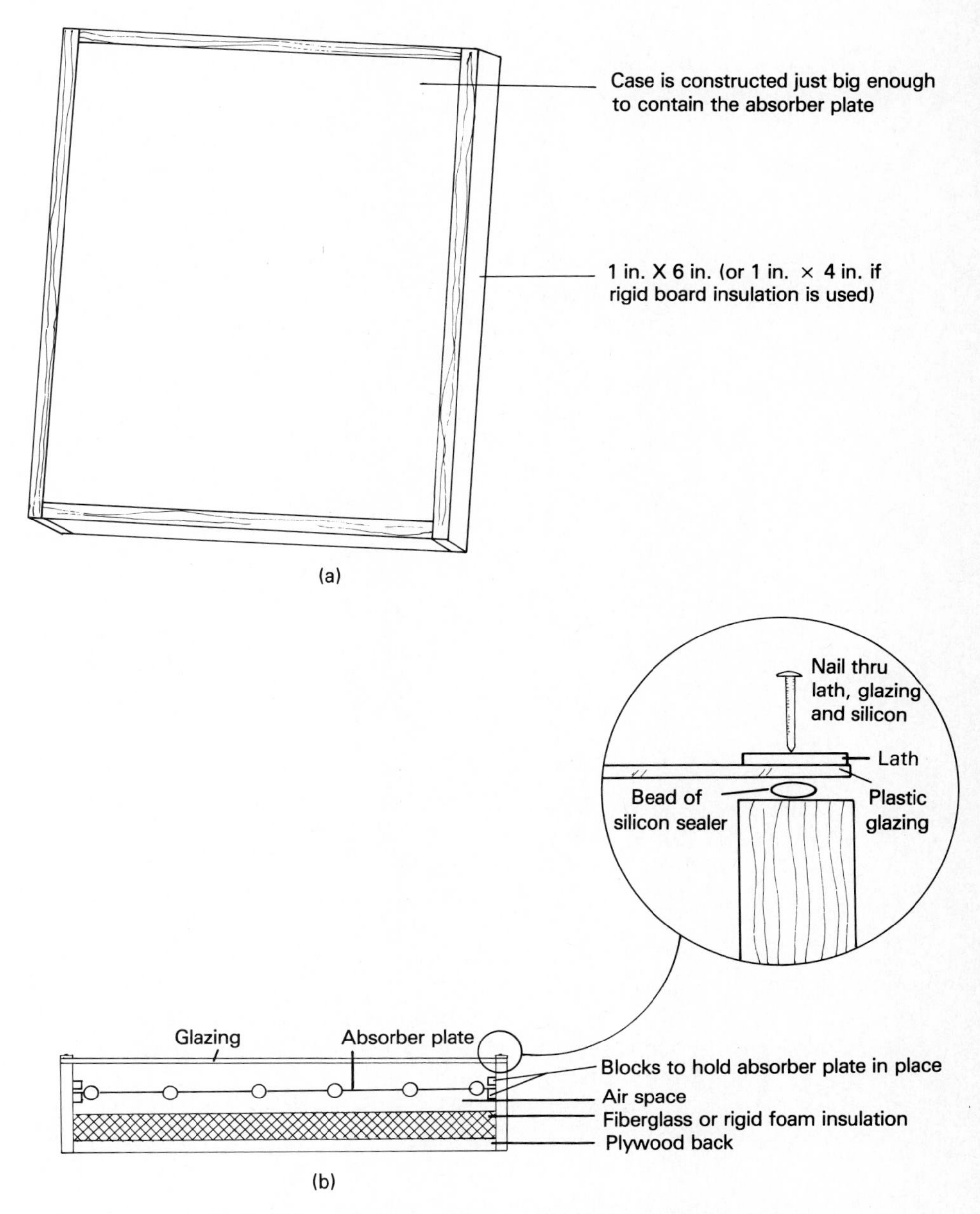

Figure I-A-9 (a) Case. Part (b) shows the cross section of (a).

Tedlar and other ultraviolet-resistant thin plastic films are inexpensive choices for the permanent glazing, but sagging and wind damage can be a problem with them. A more durable alternative that is still quite a bit cheaper than tempered glass is a fiberglass-reinforced plastic such as Filon. To install either of the glazings described above, lay a bead of silicon sealer around the edge of the collector, and place the glazing on top of it. If Tedlar is used, it should be stretched over the collector on a warm day or in a heated room (to reduce sagging), and held in place temporarily with tape. A support wire strung from end to end down the center of the collector will also minimize sagging.

After the glazing is in place, lay strips of lath along the edges of the collector frame and nail them down through the plastic glazine and silicone. The lath will exert an even pressure on the glazing, making a tight seal as well as reducing the chance of rips.

Coat the case with a couple of layers of water seal, and your collector is ready for use. The construction process should have taken you about 20 hours.

LAB PROJECT B: THE OWNER-BUILT INTEGRAL COLLECTOR

More carpentry and less plumbing is involved in building an integral collector as compared to a flat-plate collector. As discussed in Chapter 10, integral solar collectors have several advantages over flat plates, such as simplicity and an inherent resistance to freezing. The trade-off for this is a lower level of performance, due largely to nighttime heat losses from the storage tank.

The integral collector described below consists basically of a box glazed on one side to receive sunlight and insulated on all other sides, and enclosing a water storage tank (see Figure 10-20). The tank is black in color for collection purposes while the interior walls are highly reflective.

Several steps are involved in construction of the collector:

1. Construction of the case
2. Insulation installation
3. Mounting the tank
4. Glazing installation

Table I-B-1 lists the necessary tools and materials for this project.

Construction of the Case

The collector case is framed with 2 × 4's as detailed in Figure I-B-1. Two of the 6-foot pieces must be ripped (cut lengthwise) at a 45° angle to accommodate mounting of the glazing. Sheath the case with 3/8-inch plywood on all sides except the sloped 45° face, as in Figure I-B-2.

Insulation Installation

Cut insulation to fit *snugly* into the spaces between studs. All sides of the box with plywood on them should be insulated. Any gaps between studs and insulation should be filled. Scraps of fiberglass roll insulation make good filler material.

Rigid insulation can be held in place with small blocks of wood nailed to the studs, as in Figure I-B-3. Insulation with at least one side faced with reflective foil should be used, and that side should face the interior of the collector.

TABLE I-B-1

(a) Materials List

Amount	*Description*
60 ft	Economy grade 2 × 4's
Three 4-ft × 8-ft sheets	3/8-inch plywood CDX
25	2-inch wood screws (of a size to fit shelf brackets below)
35	2½-inch wood screws (for glazing frame)
Two 4-ft × 8-ft sheets	Foil-faced isocyanurate rigid insulation 1½ inches thick (or other high-temperature rigid insulation, with thermal resistance of R-10 or higher)
½ lb	No. 16 box nails
½ lb	No. 7 box nails
2 ft	2-inch × 12-inch fir or pine
4	Right-angle shelf brackets
1	40- to 50-gallon glass- or stone-lined water heater core
20 ft	1-inch × 1-inch fir or pine
20 ft	1-inch lath
1 tube	Silicone caulk
½ pint	High-temperature flat black paint
8-16 ft^2	Selective surface tape (optional). Ordering information can be obtained from Berry Solar Products, 2850 Woodbridge Avenue, Edison, NJ 08337. (201) 549-0700.
2	¾-inch galvanized nipples; length varies depending on size of tank

(b) Tool List

Circular saw (hand or table model)
Saber saw
Drill
Screwdriver
Hammer
Pipe wrench
Paintbrush
Caulk gun

Mounting the Tank

Cut the 2-foot piece of 2 × 12 wood into two 1-foot squares. With the saber saw, cut an arc out of each piece for the tank to nest in. The size of arc will differ depending on the tank diameter. When mounted, the center of the tank should be 1 foot above the bottom of the piece of wood (see Figure I-B-4). Position the stand as in Figures I-B-4 and I-B-5, and bracket it to the base of the case.

Glass or stone-lined water heater cores make fine tanks for integral collectors. Before installing, make sure that the lining is not cracked, or rusting will result. Insert a flashlight or torch in one of the tank's holes and inspect the interior carefully by looking through the other holes.

If a tank of a different size than 40 to 50 gallons is used, the case size must be adjusted so that there is at least 1 square foot of aperture for every 2 gallons of water. Tanks should have dip tubes in them that will transport incoming water to the far end of the tank. Without these tubes, incoming water is likely to find a short path to the outflow hole, returning to the house with only minimal heating.

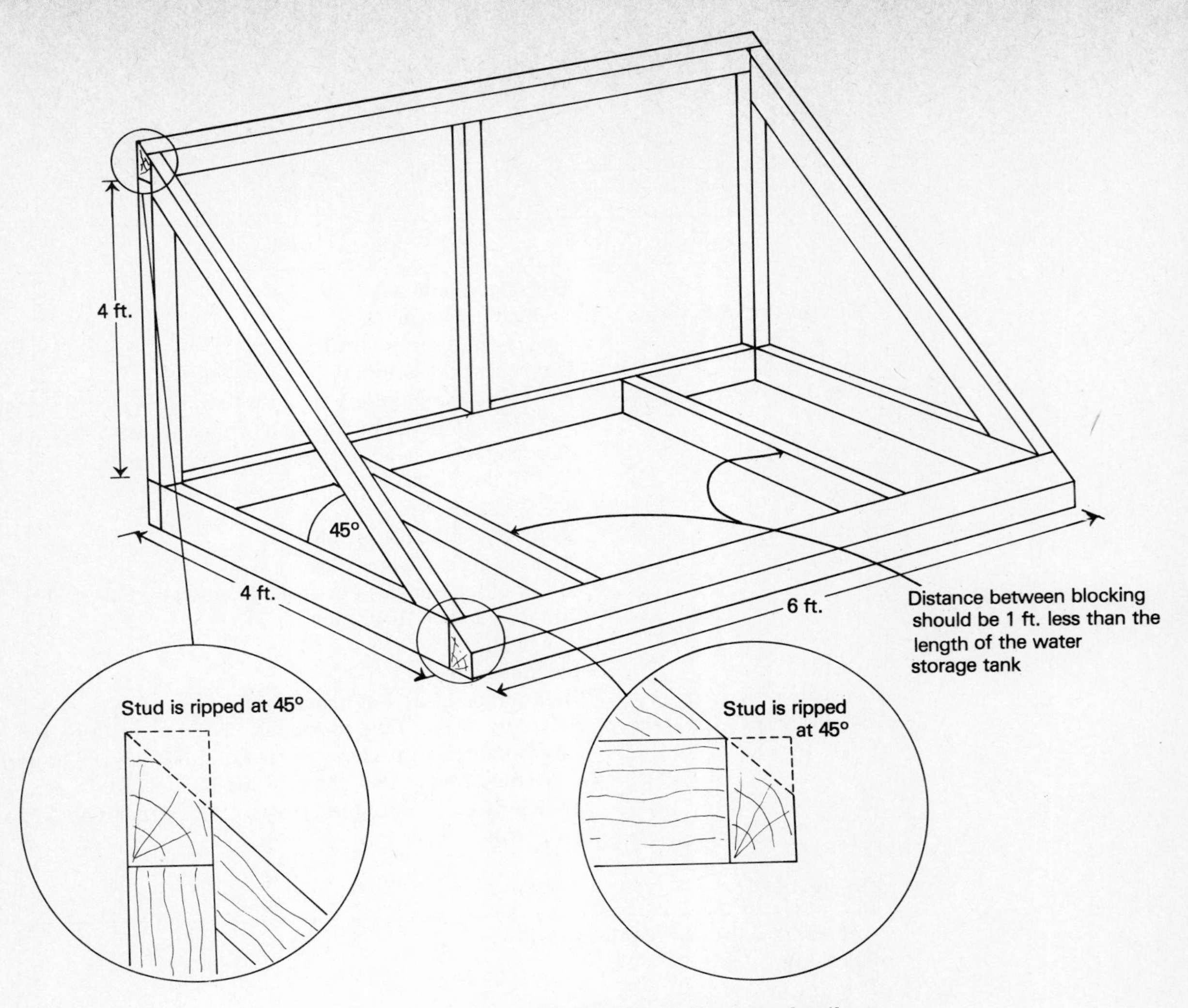

Figure I-B-1 Framing details.

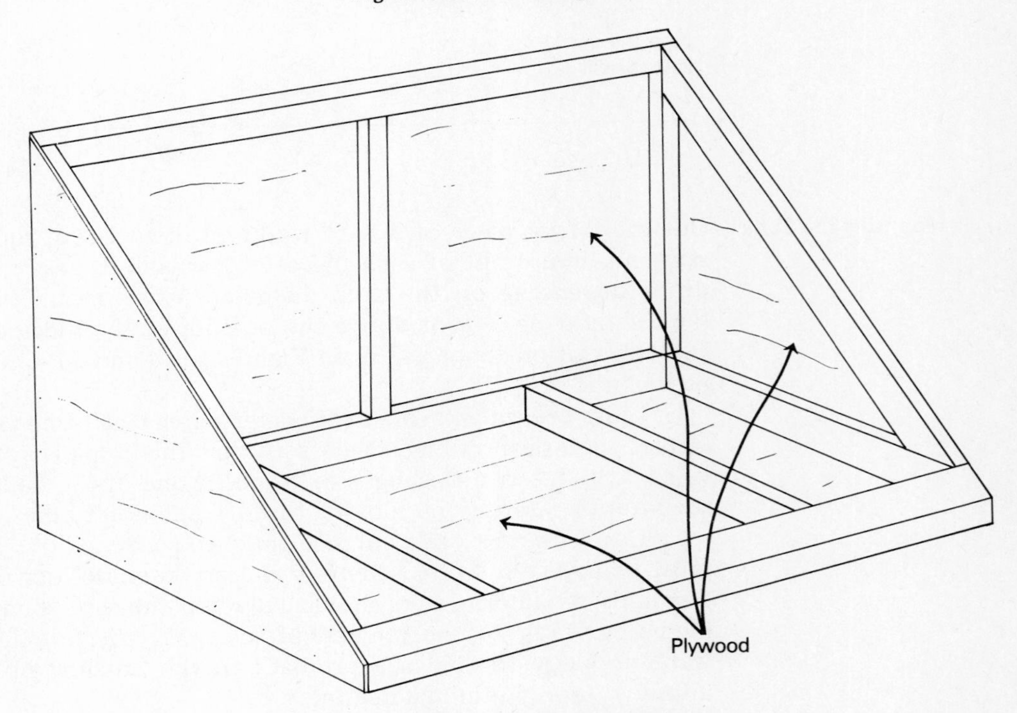

Figure I-B-2 Plywood sheathing.

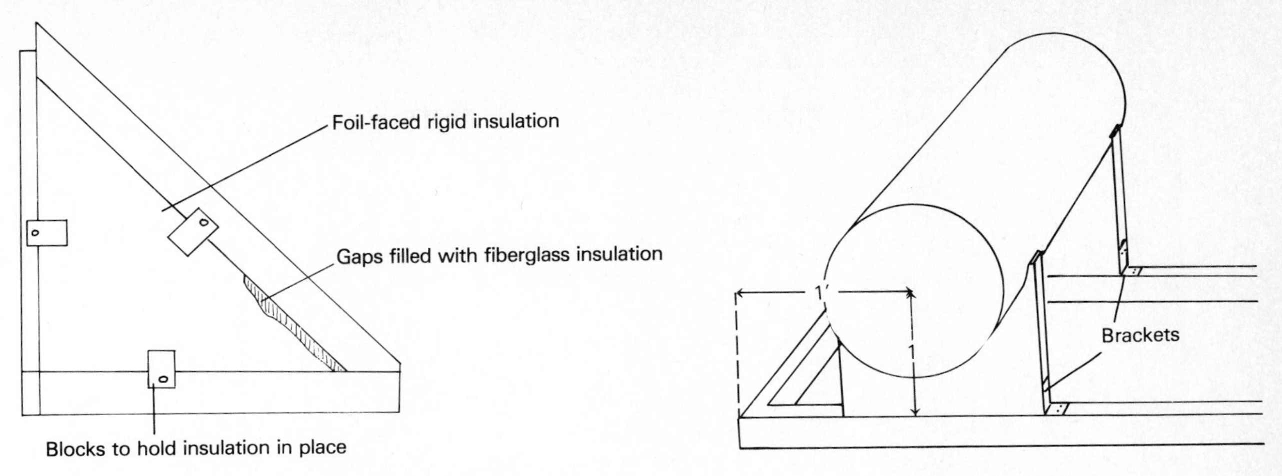

Figure I-B-3 Blocking and insulation details.

Figure I-B-4 Tank stand.

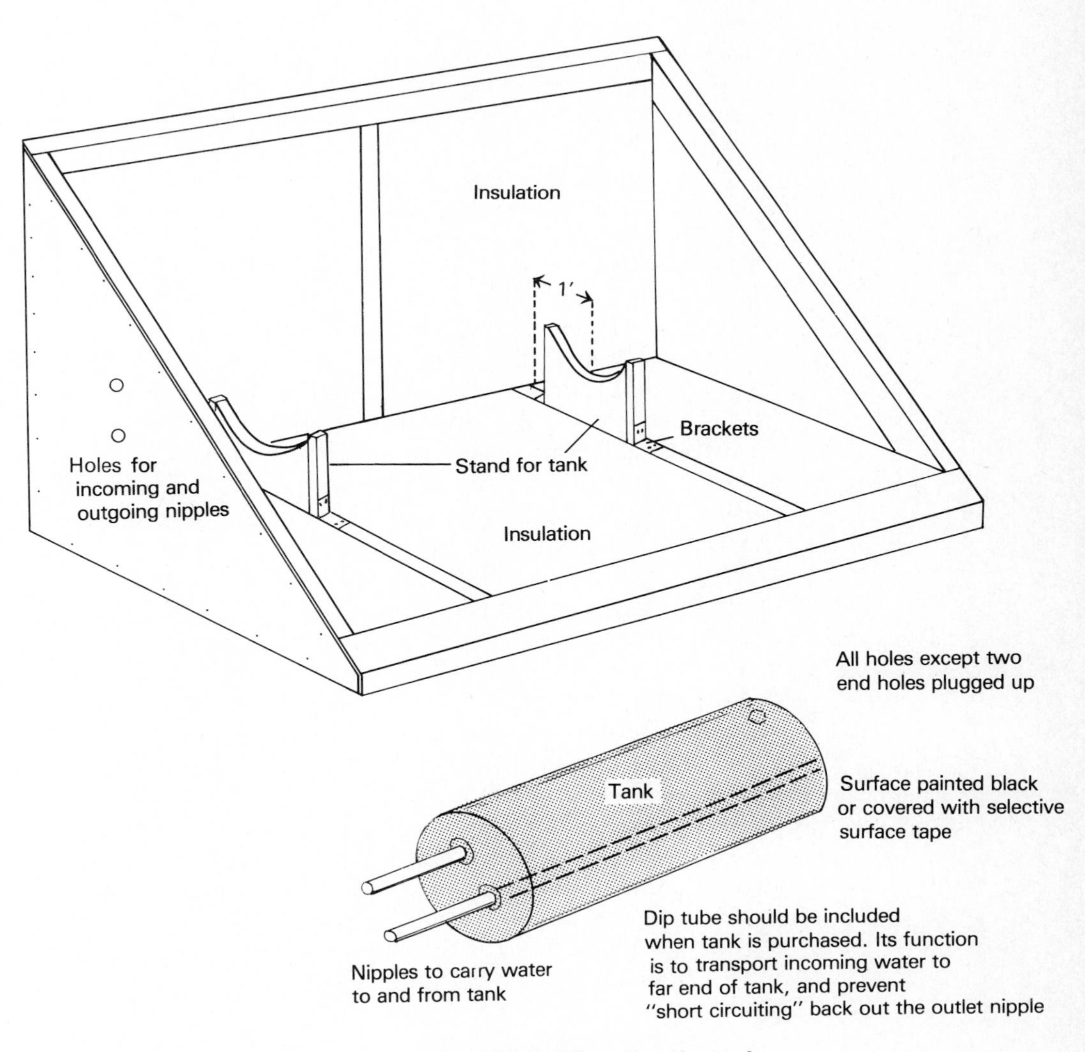

Figure I-B-5 Mounting the tank.

The tank should also contain a "sacrificial anode" to inhibit corrosion of the tank walls. New tanks as well as many used ones are generally sold with dip tube and anodes already installed.

Install galvanized nipples on the "cold water in" and "hot water out" holes on the end of the tank. All other holes must be plugged. The nipples must be long enough to go through the side of the case. Nipple length differs depending on tank size. After the nipples are installed, sit the tank on the tank stand (which has already been mounted in the case) and position the nipples so that one is directly above the other. See where the nipples hit the side of the case, and drill holes for them.

Paint tank and nipples with two coats of high-temperature flat black. Alternatively, apply selective surface tape to the side of the tank receiving direct sunlight, or even better (but slightly more expensive) to the whole tank (see Chapter 6 for a discussion of selective surfaces). Any part of the tank or nipples not covered by selective surface should be painted flat black.

Glazing Installation

Unless the builder has experience with glass it is recommended that plastic glazing be used, as it is reasonably unbreakable and easier to work with (see Chapter 7 for a discussion of the various options). Except in areas with mild climates and warm nights, at least two layers of glazing should be used. In severe climates, a third layer is recommended. The outer layer should be durable and ultraviolet resistant. The higher-quality Filons and other FRP materials are popular choices. The inner layer(s), which can be a thin film, must be able to withstand high temperatures.

Figure I-B-6 depicts detailing for a double layer of plastic glazing. If a third glazing layer is employed, two spacer frames are used.

The spacer frame should be made of high-quality wood that will stand up to the weather. A bead of silicone caulk should be applied to the edge of the aperture, and then the glazings and spacer frame taped in place until the lath exterior frame can be screwed down.

If thin-film glazings are used, sagging tends to be a problem. The glazings can be supported by wires to reduce their span.

Water vapor condensation and resultant fogging of the glazing is often a problem in the air space between multiple glazing layers. Even if the spacer frame is carefully sealed, condensation still frequently occurs. Besides impairing the glazing's transmissivity, condensation can lead to wood rot in the frame. To relieve the problem, two or three 1/8-inch holes should be drilled in the top of the frame and the same number in the bottom, allowing between-glazings moisture to evaporate. Notice in Figure I-B-6 that the upper vent holes are angled so that a minimum of rain enters them. Fine mesh screen on the outside of the vents also cuts down on rain leaks as well as keeps bugs out. If between-glazings condensation still does not readily evaporate, more vent holes might be necessary.

Seal all cracks and around protruding pipes with silicone caulk, protect the plywood with paint or some other coating, and your integral collector is ready for use.

Some integral collectors have insulated covers on them that can be closed at night to retain heat. But it has been found that most collector owners do not usually take the trouble to close the covers each night at sunset and open them early the next morning. This is why the discussion above described a collector that operates well either with or without night insulation.

As with other collectors, integral units are mounted on roofs of buildings as well as on the ground nearby. Because they are fairly heavy, it is often easiest to build integral collectors *in the location* where they will stay. Or they can be transported in sections to their permanent location. If the collector must be moved in one piece to the top of a building, it is recommended that a hand or motor-driven winch be used to pull it up skids to the roof (see Figure I-B-7).[1]

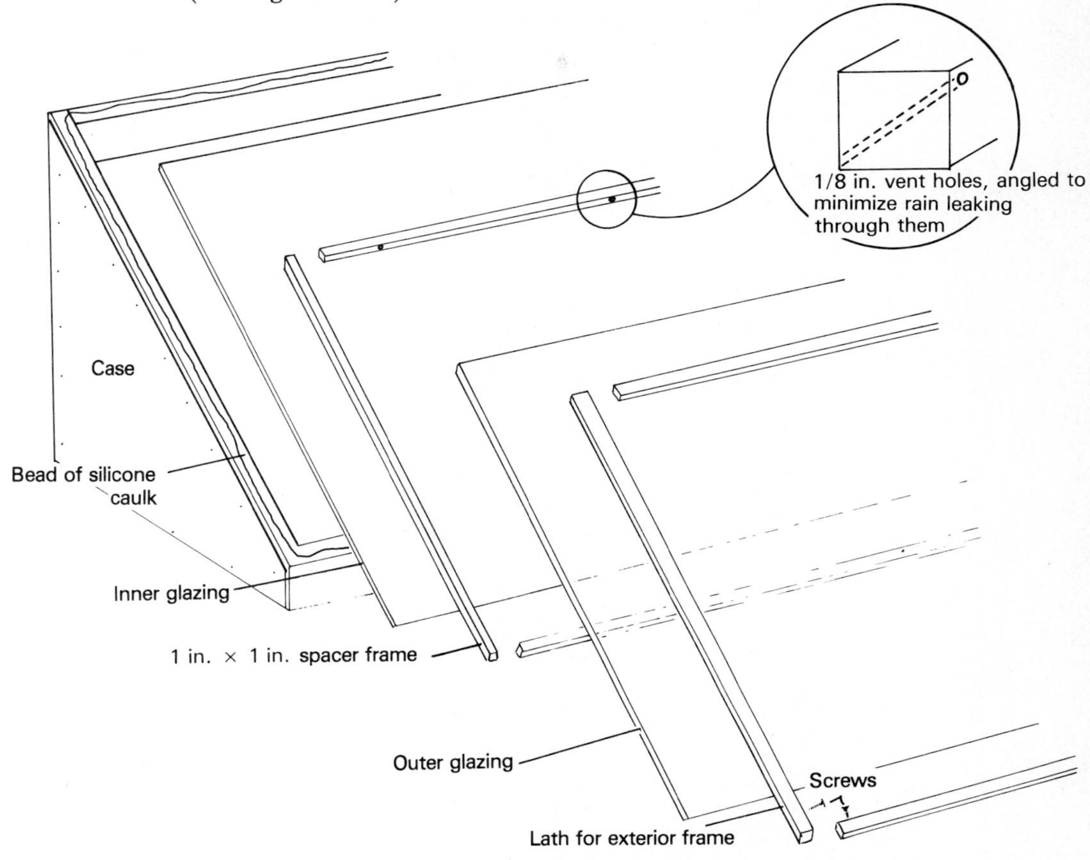

Figure I-B-6 Glazing detailing.

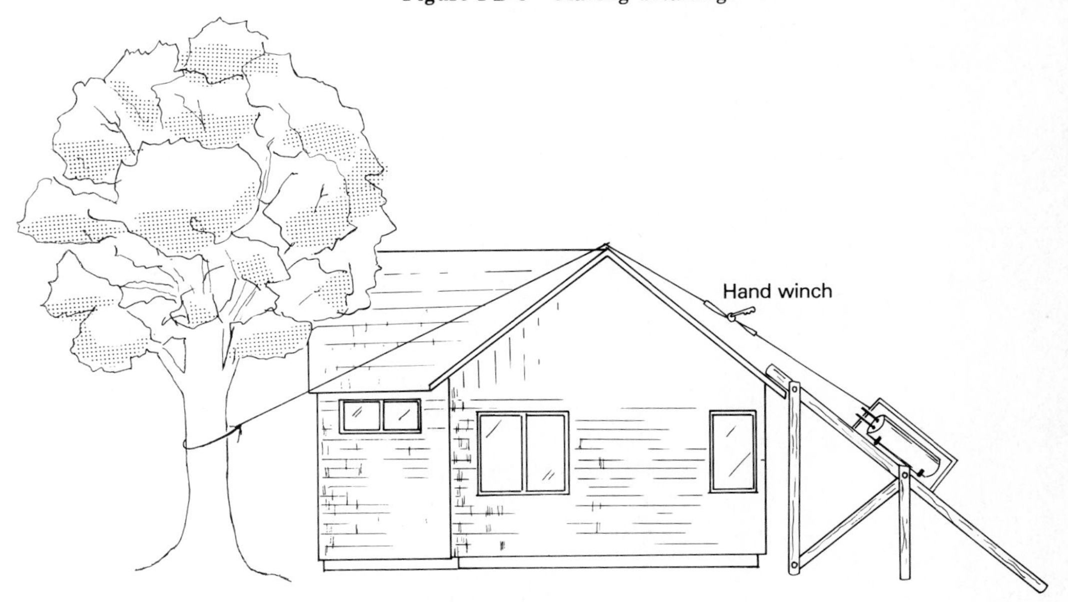

Figure I-B-7 Raising the collector to the roof.

Note: The most common complaint concerning integral collector systems encountered by the author has been that in winter, the water simply does not get hot enough. The problem often turns out to be that there is not enough glazing per gallons of water to be heated. One to 1½ square feet of glazing should be included for each 2 gallons of water.

NOTES

Lab Project A

1. Roy Irving's design is described in the booklet *Alternative Energy Fair Program and Consumer Guide*, Alternative Energy Association of Sonoma State University, Rohnert Park, CA, 1979, pp. 13-18.

Lab Project B

1. An excellent reference on both owner-built and commercial integral collectors is David A. Bainbridge's *The Integral Passive Solar Water Heater Book*, 1981. Available from the Passive Solar Institute, P.O. Box 722, Davis, CA 95617. Chapter 8, "Building Your Own," is of special interest to the owner-builder.

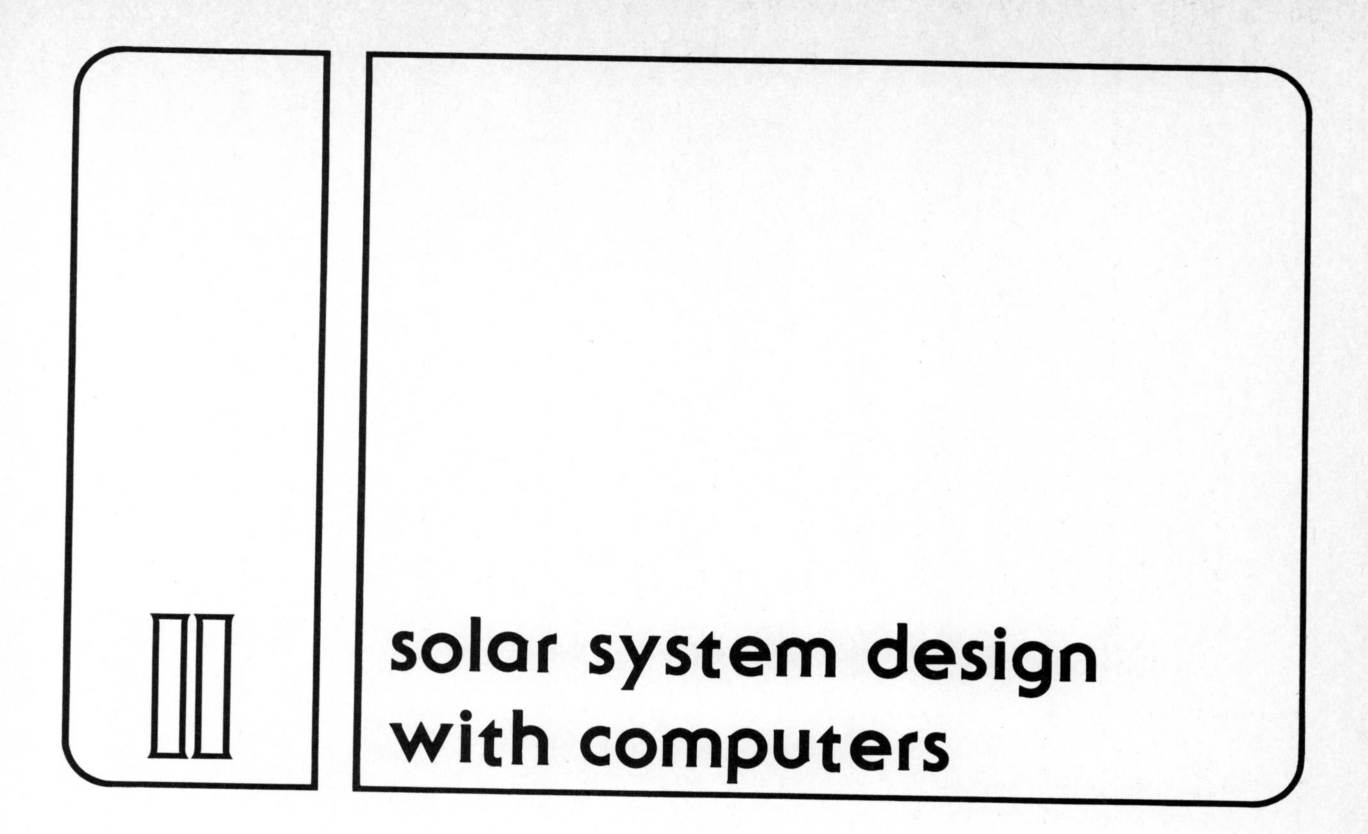

II solar system design with computers

Personal computers and programmable calculators are becoming increasingly popular and reasonably priced. Among their many other benefits, they can be of great help in designing solar systems of all types. Some areas in which they are particularly useful are as follows:

1. Calculation of building heat loss
2. Calculation of solar collection areas for passive as well as active space heating, domestic water heating, and swimming pool heating systems
3. Heat storage system sizing
4. Economic analysis of solar systems

There are a variety of software packages on the market designed to handle the above-mentioned areas of calculation, varying in price from several dollars to several hundred dollars. These programs are fast becoming the tools of the trade for many professional designers, so it is important to make them accessible to others who want to use them. Table II-1 lists and describes some of the systems available.

Milburn-Sparn is a Colorado-based architectural firm that employs a number of computers and software packages in their design work. They use an Apple II Plus with 48K storage and two disks, a Vector Graphic System, and the services of Timeshare, a San Jose, California, company whose services are accessed via Milburn-Sparn's Apple computer.

TABLE II-1 Computer and Calculator Programs for Solar System Design

Name of Program	*Available from:*	*For Use with:*	*Functions*	*Average Data Preparation Time*	*Cost*
F-CHART	Solarsoft, Inc. Division of Kinetic Software, Inc. Box 124 Snowmass, CO 81654	Apple II Plus 48K storage, two disk drives	For active liquid or air-based systems. Besides monthly and yearly solar heating fractions and heat losses, output data includes annual cash flow analysis. Output data are presented graphically.	1–5 min	$400
IMPSLR	Princeton Energy Group 575 Ewing Street Princeton, NJ 08540	Apple II (32K), Commodore PET, TRS-80 with disk	Predicts annual auxiliary energy needed by buildings using Trombe wall, water wall, or direct-gain passive systems.	10 min to 8 hr, depending on detail	$250
MICRO FIX	Princeton Energy Group 575 Ewing Street Princeton, NJ 08540	Apple II, PET, TRS-80, HP-85, 16K storage	Hourly performance of direct-gain or *attached sunspace* systems. Outputs include auxiliary heat demand and hourly temperature of room air and storage mass.	10 min	$150
DHW-2 SOLCOST	Solar Energy Corp. P.O. Box 3065 Princeton, NJ 08540		For domestic and small commercial solar water heating systems. Typical outputs include collector and storage system sizing, tilt angle, percentage of energy requirements met by solar, and a complete economic analysis. The program is run on Solar Energy Corp.'s computer.		$25 for a small computer run
SEEC VI Passive Design	Solar Environment Eng. Co. 2524 East Vine Drive Fort Collins, CO 80524	TI-59, HP-67, HP-97, HP-41C calculators; TRS-80 I, II, or III computer	This program uses Los Alamos's design methods to compute solar savings fractions and auxiliary heating requirements for direct-gain, water wall, and Trombe wall passive solar systems. It will also perform a life-cycle cost analysis of the system.	10 min	$125 for calculator, $200 for micro-computer
SEEC VII Shading Computation	Solar Environment Eng. Co. 2524 East Vine Drive Fort Collins, CO 80524	TI-59, HP-67, HP-97, HP-41C calculators	A shading analysis. The program computes the effect of different window overhangs and side fins on the percentage of the aperture that receives shade.	3 min	$75

SEEC VIII for swimming pool systems	Solar Environment Eng. Co. 2524 East Vine Drive Fort Collins, CO 80524	TI-59	A design program for active solar swimming pool systems and for pool blankets. Outputs include monthly average pool temperature and solar heating fraction.	5 min	$125
SUNPAS	Solarsoft, Inc. Box 124 Snowmass, CO 81654	Apple II Plus 48K storage, two disk drives	Performance analysis of Trombe and water walls, direct-gain and sunspace systems, based on the design techniques in Los Alamos's Passive Solar Design Handbook. Features graphic output of data. Example: Bar graph of the solar savings fraction versus aperture area.	10 min to 2 hr	$250
SUNOP	Solarsoft, Inc. Box 124 Snowmass, CO 81654	Apple II Plus 48K storage, two disk drives	Performs an economic analysis of passive solar designs.	1 min	$250
T-SWING	Solarsoft, Inc. Box 124 Snowmass, CO 81654	Apple II Plus 48K storage, two disk drives	Calculates the temperature swings in a building. It is especially effective for sizing and locating thermal storage mass in a passively heated building in such a way as to keep the temperature as stable as possible. Program is sold with SOLGAIN, a program for calculating the solar energy transmitted through glazing oriented at different angles and partially shaded.	10 min to 2 hr	$700 with SOLGAIN
F-CHART PLUS	F-Chart Software 4406 Fox Bluff Road Middleton, WI 53562	Apple II, TRS-80 Model III disk drive, 48K storage	Analyzes active liquid or air-based systems, as well as passive direct-gain and solar wall systems. Outputs include fraction of energy needs met by the solar system and life-cycle-cost and analysis.	1 hr	$400
13 Solar Energy Programs	F-Chart Software 4406 Fox Bluff Road Middleton, WI 53562	TI-59, HP-41C	A book of 13 programs for hand-held calculators that includes an *f*-chart program, one for direct-gain systems, and another for life-cycle-cost estimates.	1 hr	$50
Solar Energy Calculations	Solarcon Dept. SA-69 607 Church Ann Arbor, MI 48104	TI-59 and other programmable calculators	This 200-page book lists over 20 solar energy energy calculations, and includes 10 program descriptions, although actual program steps are not included. (They can be ordered separately.)		$28

TABLE II-1 Computer and Calculator Programs for Solar System Design (cont.)

Name of Program	*Available from:*	*For Use with:*	*Functions*	*Average Data Preparation Time*	*Cost*
		Systems and Services That Are Time-Share-Accessed[a]			
Boeckh Cost Estimating System	Boeckh Systems 615 East Michigan Street Milwaukee, WI 53202	Most microcomputers	A job cost estimating system originally designed for insurance appraisers that is useful in providing quick, inexpensive construction cost approximations.		
Orr Cost Estimating System	Cost Systems Engineers 131 East Exchange Suite 222 Ft. Worth, TX 76106		A considerably more detailed construction cost data base.		
Quickee	Energy Systems Engineers Boeing Computer Services 8000 East Girard Avenue Suite 508 Denver, CO 80231		An energy analysis program geared for commercial rather than residential building projects.		
CALPAS 3	Berkeley Solar Group 3140 Grove Street Berkeley, CA 94073		For use in modeling the performance of passive solar homes. Does hourly calculations for an entire year. Program can be purchased as well as time-share accessed.		

[a]These systems are not purchased, but accessed via computer terminals.

The architects do initial job cost estimates using Boeckh Systems's services, and more complete economic analyses with the Orr system. The Boeckh System is especially helpful for closing deals, in that it provides a quick idea of the project expense.

Milburn-Sparn uses Solarsoft's programs (such as SUNPAS, SUNOP, and T-SWING) for their residential passive design projects, and Quickee for commercial applications. They report that although the capital investment for their new systems was considerable, there is no doubt that it was worthwhile. Actual building costs are typically within 1 or 2 percent of the computer estimates, and the thermal analysis programs provide immediate feedback on energy-saving ideas, rather than waiting until the building is completed and occupied.

INSOLATION AND WEATHER DATA

The following parameters are listed in Table III-1 for 219 American and Canadian locations:

Parameter	*Description*
Elev	Elevation (feet).
Lat	Latitude.
HS	Average daily hemispheric radiation on a horizontal surface (Btu/ft^2-day).
VS	Average daily solar radiation on a vertical, south-facing surface (Btu/ft^2-day).
TA	(T min + T max)/2, where T min and T max are monthly or annual averages of daily minimum and maximum ambient temperatures, in degrees Fahrenheit.
DXX	Monthly or annual degree-days. D65 corresponds to degree-days calculated with a base temperature of 65°F. The other DXX's correspond to other base temperatures. D60, for instance, corresponds to a base temperature of 60°F, which is useful for energy calculations in buildings with lowered thermostat settings (see Chapter 2 for a discussion of degree days and base temperatures).
KT	Clearness ratio, also called "cloudiness index."
LD	Stands for "Latitude − Declination" and is calculated by subtracting the midmonth solar declination from the latitude of the site.

TABLE III-1 Insolation and Weather Data

BIRMINGHAM, ALABAMA ELEV 630 LAT 33.6

	HS	VS	TA	D50	D55	D60	D65	D70	KT	LD
JAN	707	864	44	216	346	493	654	800	.43	55
FEB	967	995	47	144	246	372	517	647	.47	47
MAR	1296	986	53	61	135	237	389	520	.49	36
APR	1673	921	63	6	20	60	116	232	.53	24
MAY	1857	852	71	1	3	12	20	100	.53	15
JUN	1918	843	77	0	0	0	0	24	.53	10
JUL	1810	821	80	0	0	0	0	14	.51	12
AUG	1724	874	79	0	0	0	0	16	.53	20
SEP	1455	990	74	0	1	5	6	52	.52	32
OCT	1211	1179	63	6	20	61	137	237	.54	44
NOV	858	1056	52	75	151	258	391	539	.49	53
DEC	661	857	45	193	318	463	614	769	.43	57
YR	1346	936	62	702	1240	1961	2844	3950	.51	

MOBILE, ALABAMA ELEV 220 LAT 30.7

	HS	VS	TA	D50	D55	D60	D65	D70	KT	LD
JAN	828	964	51	97	180	294	451	585	.46	52
FEB	1100	1066	54	54	119	204	337	452	.50	44
MAR	1407	993	59	19	55	125	221	343	.52	33
APR	1722	884	68	3	8	26	40	145	.54	21
MAY	1872	828	75	0	0	0	0	51	.54	12
JUN	1868	819	80	0	0	0	0	15	.52	7
JUL	1715	778	82	0	0	0	0	11	.49	9
AUG	1641	799	82	0	0	0	0	12	.50	17
SEP	1449	909	78	0	0	0	0	28	.50	29
OCT	1299	1164	69	2	7	21	39	133	.55	41
NOV	955	1096	59	22	63	135	211	356	.51	50
DEC	759	928	53	73	149	252	385	533	.45	54
YR	1386	935	67	270	581	1056	1684	2664	.51	

MONTGOMERY, ALABAMA ELEV 203 LAT 32.3

	HS	VS	TA	D50	D55	D60	D65	D70	KT	LD
JAN	752	896	48	148	256	394	556	698	.44	54
FEB	1013	1011	51	88	166	277	419	544	.48	46
MAR	1341	987	57	30	85	166	299	424	.50	35
APR	1729	920	65	3	12	39	76	186	.54	23
MAY	1897	849	72	1	2	7	8	72	.54	13
JUN	1972	848	79	0	0	0	0	16	.55	9
JUL	1841	820	81	0	0	0	0	10	.52	11
AUG	1746	861	81	0	0	0	0	11	.53	19
SEP	1468	963	76	0	0	0	0	32	.52	30
OCT	1262	1188	66	3	11	35	93	180	.55	42
NOV	915	1099	55	41	103	191	306	454	.51	52
DEC	719	916	49	131	231	365	512	667	.45	56
YR	1390	946	65	445	866	1474	2269	3295	.52	

PHOENIX, ARIZONA ELEV 1112 LAT 33.4

	HS	VS	TA	D50	D55	D60	D65	D70	KT	LD
JAN	1021	1462	51	78	162	285	428	584	.62	55
FEB	1374	1608	55	29	85	168	292	419	.66	47
MAR	1814	1471	60	9	36	100	185	327	.69	36
APR	2355	1236	68	1	4	16	60	130	.74	24
MAY	2676	1034	76	0	0	0	0	24	.77	14
JUN	2739	973	85	0	0	0	0	2	.76	10
JUL	2486	964	91	0	0	0	0	0	.70	12
AUG	2293	1080	89	0	0	0	0	1	.70	20
SEP	2015	1415	84	0	0	0	0	3	.72	32
OCT	1576	1674	72	0	1	5	17	64	.70	43
NOV	1150	1606	60	9	34	96	182	314	.66	53
DEC	932	1407	53	60	137	250	388	544	.61	57
YR	1371	1326	70	187	459	919	1552	2411	.71	

PRESCOTT, ARIZONA ELEV 5023 LAT 34.6

	HS	VS	TA	D50	D55	D60	D65	D70	KT	LD
JAN	1016	1533	37	401	555	710	865	1020	.64	56
FEB	1335	1623	41	270	407	546	686	826	.66	48
MAR	1777	1502	44	193	335	487	642	797	.69	37
APR	2275	1251	52	49	127	248	394	540	.72	25
MAY	2629	1060	60	3	18	73	165	302	.75	16
JUN	2762	998	69	0	1	5	33	91	.76	11
JUL	2309	953	76	0	0	0	0	18	.65	13
AUG	2092	1043	73	0	0	0	0	38	.64	21
SEP	1954	1430	68	0	1	7	23	107	.71	33
OCT	1543	1709	57	10	48	127	254	398	.71	45
NOV	1140	1674	46	152	281	427	576	726	.68	54
DEC	927	1477	39	356	509	663	818	973	.63	58
YR	1815	1352	55	1436	2282	3294	4456	5837	.69	

TUCSON, ARIZONA ELEV 2556 LAT 32.1

	HS	VS	TA	D50	D55	D60	D65	D70	KT	LD
JAN	1099	1539	51	80	166	292	442	593	.64	53
FEB	1432	1615	54	41	106	201	333	463	.67	46
MAR	1864	1447	58	15	56	133	243	388	.70	35
APR	2363	1186	66	2	7	26	81	169	.74	22
MAY	2671	1001	74	0	0	0	0	44	.77	13
JUN	2730	953	82	0	0	0	0	4	.76	9
JUL	2341	922	86	0	0	0	0	1	.66	11
AUG	2183	1007	84	0	0	0	0	3	.67	19
SEP	1979	1322	80	0	0	0	0	7	.70	30
OCT	1602	1623	70	0	2	8	29	91	.70	42
NOV	1208	1631	59	12	44	114	221	350	.67	51
DEC	996	1464	52	64	144	262	403	559	.62	55
YR	1874	1307	68	214	525	1036	1752	2673	.70	

WINSLOW, ARIZONA ELEV 4882 LAT 35.0

	HS	VS	TA	D50	D55	D60	D65	D70	KT	LD
JAN	985	1490	33	540	694	849	1004	1159	.63	56
FEB	1327	1637	39	309	446	585	725	865	.67	49
MAR	1780	1528	45	184	321	472	626	781	.69	37
APR	2283	1272	54	33	101	205	348	490	.73	25
MAY	2595	1064	63	2	10	45	124	238	.74	16
JUN	2712	1001	72	0	1	3	14	55	.75	12
JUL	2347	968	78	0	0	0	0	9	.66	14
AUG	2141	1074	76	0	0	0	0	17	.66	22
SEP	1928	1429	70	0	1	6	19	88	.70	33
OCT	1513	1691	57	12	50	129	252	396	.70	45
NOV	1119	1658	43	217	357	505	654	804	.67	54
DEC	894	1424	34	503	657	812	967	1122	.62	58
YR	1804	1351	55	1800	2639	3611	4733	6025	.69	

YUMA, ARIZONA ELEV 207 LAT 32.7

	HS	VS	TA	D50	D55	D60	D65	D70	KT	LD
JAN	1096	1575	55	28	87	177	308	455	.65	54
FEB	1443	1675	59	8	33	93	192	303	.69	46
MAR	1919	1534	64	3	11	41	97	212	.72	35
APR	2413	1231	71	0	1	5	24	73	.76	23
MAY	2728	1023	79	0	0	0	0	11	.78	14
JUN	2814	966	86	0	0	0	0	1	.78	9
JUL	2453	948	94	0	0	0	0	0	.69	11
AUG	2329	1070	93	0	0	0	0	0	.71	19
SEP	2051	1406	87	0	0	0	0	1	.73	31
OCT	1623	1694	76	0	0	1	5	25	.72	43
NOV	1215	1690	64	3	12	44	108	215	.68	52
DEC	1000	1515	56	22	74	158	276	427	.64	56
YR	1925	1358	74	64	218	520	1010	1724	.72	

Source: J. Douglas Balcomb et al., *Passive Solar Design Handbook*, Vol. 3, Los Alamos National Laboratory, Los Alamos, NM, for U.S. Department of Energy, DOE/C5-0127/3, 1982, pp. 264–292.

TABLE III-1 (cont.)

FORT SMITH, ARKANSAS ELEV 463 LAT 35.3

	HS	VS	TA	D50	D55	D60	D65	D70	KT	LD
JAN	744	996	39	346	497	651	806	961	.48	57
FEB	999	1108	43	204	331	468	608	748	.50	49
MAR	1312	1056	50	85	177	308	471	611	.51	38
APR	1616	929	62	3	15	56	132	246	.52	26
MAY	1912	892	70	0	1	7	17	88	.55	16
JUN	2089	898	78	0	0	0	0	12	.58	12
JUL	2065	908	82	0	0	0	0	3	.58	14
AUG	1877	973	81	0	0	0	0	4	.58	22
SEP	1501	1080	74	0	0	0	0	36	.55	33
OCT	1201	1242	63	3	11	46	135	228	.56	45
NOV	851	1119	50	81	169	295	438	588	.52	55
DEC	682	963	42	274	421	574	729	884	.48	59
YR	1406	1013	61	996	1622	2405	3336	4409	.54	

LITTLE ROCK, ARKANSAS ELEV 266 LAT 34.7

	HS	VS	TA	D50	D55	D60	D65	D70	KT	LD
JAN	731	947	40	331	482	636	791	946	.46	56
FEB	1003	1089	43	212	342	479	619	759	.50	48
MAR	1313	1037	50	83	175	308	470	611	.51	37
APR	1611	913	62	4	16	61	139	259	.51	25
MAY	1929	888	70	0	1	7	21	91	.55	16
JUN	2106	896	78	0	0	0	0	11	.58	11
JUL	2032	891	81	0	0	0	0	4	.57	13
AUG	1860	952	81	0	0	0	0	5	.57	21
SEP	1518	1074	73	0	0	2	5	41	.55	33
OCT	1228	1251	62	3	14	54	143	248	.56	45
NOV	847	1084	50	81	169	298	441	591	.50	54
DEC	674	922	42	270	418	571	725	880	.46	58
YR	1406	995	61	984	1617	2414	3354	4447	.54	

BAKERSFIELD, CALIFORNIA ELEV 492 LAT 35.4

	HS	VS	TA	D50	D55	D60	D65	D70	KT	LD
JAN	766	1044	48	131	246	391	543	698	.49	57
FEB	1102	1276	52	50	120	225	353	493	.56	49
MAR	1595	1350	57	18	66	149	266	418	.62	38
APR	2095	1191	63	3	13	50	140	233	.67	26
MAY	2509	1057	70	0	2	7	22	92	.72	16
JUN	2749	1013	77	0	0	0	0	16	.76	12
JUL	2683	1037	84	0	0	0	0	2	.76	14
AUG	2421	1198	82	0	0	0	0	4	.75	22
SEP	1992	1508	77	0	0	0	0	18	.73	34
OCT	1458	1632	67	1	4	17	55	143	.68	45
NOV	942	1305	56	21	72	156	276	422	.57	55
DEC	677	957	48	124	236	379	530	685	.47	59
YR	1752	1213	65	348	759	1372	2185	3224	.68	

DAGGETT, CALIFORNIA ELEV 1929 LAT 34.9

	HS	VS	TA	D50	D55	D60	D65	D70	KT	LD
JAN	958	1422	47	144	257	398	549	704	.61	56
FEB	1281	1549	52	63	135	239	371	505	.64	49
MAR	1772	1514	57	23	74	155	271	416	.69	37
APR	2274	1263	64	3	11	41	118	200	.73	25
MAY	2591	1061	72	0	1	5	14	65	.74	16
JUN	2766	1004	80	0	0	0	0	9	.76	12
JUL	2603	1012	87	0	0	0	0	1	.73	14
AUG	2383	1164	86	0	0	0	0	2	.73	21
SEP	2008	1493	79	0	0	0	0	12	.73	33
OCT	1516	1688	68	1	4	16	57	130	.70	45
NOV	1085	1575	56	30	88	174	296	438	.65	54
DEC	876	1374	48	132	238	378	527	682	.60	58
YR	1845	1342	66	396	809	1406	2203	3165	.71	

FRESNO, CALIFORNIA ELEV 328 LAT 36.8

	HS	VS	TA	D50	D55	D60	D65	D70	KT	LD
JAN	657	886	45	176	308	457	611	766	.45	58
FEB	1012	1195	50	81	167	288	423	563	.53	50
MAR	1566	1389	54	37	107	209	344	500	.62	39
APR	2093	1246	60	6	25	81	182	298	.68	27
MAY	2484	1092	67	1	3	14	51	132	.71	18
JUN	2733	1045	74	0	0	2	9	36	.75	13
JUL	2685	1076	81	0	0	0	0	5	.76	15
AUG	2423	1258	78	0	0	0	0	11	.75	23
SEP	1985	1587	74	0	0	0	0	37	.74	35
OCT	1429	1681	64	2	8	34	90	202	.68	47
NOV	888	1271	54	40	110	213	345	496	.57	56
DEC	574	800	46	165	293	442	595	750	.42	60
YR	1714	1210	62	507	1021	1741	2650	3797	.67	

LONG BEACH, CALIFORNIA ELEV 56 LAT 33.8

	HS	VS	TA	D50	D55	D60	D65	D70	KT	LD
JAN	928	1291	54	17	79	189	339	490	.57	55
FEB	1215	1373	56	9	53	140	273	406	.59	47
MAR	1610	1289	57	5	35	116	247	397	.61	36
APR	1938	1056	61	1	10	55	148	284	.61	24
MAY	2064	913	64	0	2	18	71	192	.59	15
JUN	2140	894	67	0	1	5	23	110	.59	10
JUL	2300	939	72	0	0	0	0	35	.65	12
AUG	2100	1024	73	0	0	0	0	24	.64	20
SEP	1701	1186	72	0	0	1	7	39	.61	32
OCT	1326	1342	67	0	1	6	48	122	.60	44
NOV	1003	1335	61	1	10	55	155	284	.58	53
DEC	847	1241	56	10	59	155	295	450	.56	57
YR	1600	1156	63	46	250	740	1606	2834	.61	

LOS ANGELES, CALIFORNIA ELEV 105 LAT 33.9

	HS	VS	TA	D50	D55	D60	D65	D70	KT	LD
JAN	926	1293	55	21	83	186	331	481	.57	55
FEB	1214	1377	56	13	59	143	270	404	.59	48
MAR	1619	1303	57	11	53	138	267	419	.62	36
APR	1951	1066	59	5	26	91	195	338	.62	24
MAY	2060	914	62	2	9	47	114	257	.59	15
JUN	2119	891	65	1	4	20	71	180	.59	11
JUL	2307	942	69	0	1	5	19	99	.65	13
AUG	2079	1019	70	0	1	4	15	83	.64	20
SEP	1681	1174	69	0	1	5	23	93	.60	32
OCT	1317	1335	65	0	3	17	77	168	.59	44
NOV	1004	1343	61	2	15	65	158	289	.58	53
DEC	848	1249	57	9	47	129	279	407	.56	57
YR	1596	1157	62	64	299	849	1819	3216	.61	

MOUNT SHASTA, CALIFORNIA ELEV 3586 LAT 41.3

	HS	VS	TA	D50	D55	D60	D65	D70	KT	LD
JAN	561	857	34	509	663	818	973	1128	.46	63
FEB	857	1128	38	344	482	622	762	902	.51	55
MAR	1250	1217	40	303	454	608	763	918	.54	44
APR	1756	1200	46	143	268	412	561	711	.59	32
MAY	2186	1129	53	37	109	221	371	518	.63	22
JUN	2436	1102	60	4	22	78	178	304	.67	18
JUL	2577	1201	68	0	2	9	37	118	.73	20
AUG	2213	1356	66	1	3	17	64	154	.70	28
SEP	1735	1595	61	3	15	62	145	271	.68	39
OCT	1155	1492	51	61	146	274	422	577	.62	51
NOV	659	998	42	257	400	549	699	849	.49	61
DEC	505	814	36	451	605	760	915	1070	.46	65
YR	1494	1174	50	2112	3169	4430	5890	7520	.62	

TABLE III-1 (cont.)

NEEDLES, CALIFORNIA — ELEV 886 LAT 34.8

	HS	VS	TA	D50	D55	D60	D65	D70	KT	LD
JAN	985	1477	52	80	161	278	421	572	.62	56
FEB	1353	1669	57	25	74	147	261	382	.68	48
MAR	1825	1565	62	8	27	80	150	278	.71	37
APR	2317	1280	70	1	3	10	42	94	.74	25
MAY	2652	1070	80	0	0	0	0	13	.76	16
JUN	2791	1005	88	0	0	0	0	1	.77	11
JUL	2541	1000	95	0	0	0	0	0	.72	13
AUG	2278	1121	93	0	0	0	0	0	.70	21
SEP	2015	1494	87	0	0	0	0	2	.73	33
OCT	1537	1714	74	0	1	4	10	46	.71	45
NOV	1124	1654	61	9	32	90	163	292	.67	54
DEC	913	1457	53	65	141	249	381	538	.63	58
YR	1863	1373	73	188	439	858	1428	2219	.71	

OAKLAND, CALIFORNIA — ELEV 7 LAT 37.7

	HS	VS	TA	D50	D55	D60	D65	D70	KT	LD
JAN	708	1028	49	77	202	354	508	663	.50	59
FEB	1017	1247	52	25	101	228	367	507	.55	51
MAR	1456	1307	54	12	75	199	350	505	.59	40
APR	1922	1178	56	4	36	128	270	417	.62	28
MAY	2211	1035	59	1	11	72	193	344	.64	19
JUN	2350	994	62	0	2	26	114	244	.65	14
JUL	2322	1018	63	0	1	16	80	216	.65	16
AUG	2053	1122	64	0	1	13	74	205	.64	24
SEP	1701	1359	65	0	1	8	59	170	.64	36
OCT	1212	1380	61	0	4	37	135	277	.59	48
NOV	822	1180	55	5	47	148	291	441	.54	57
DEC	647	993	50	55	164	314	468	623	.50	61
YR	1538	1152	57	179	646	1543	2909	4613	.61	

RED BLUFF, CALIFORNIA — ELEV 354 LAT 40.1

	HS	VS	TA	D50	D55	D60	D65	D70	KT	LD
JAN	570	831	45	189	316	462	614	769	.44	61
FEB	892	1138	50	92	175	290	420	561	.51	54
MAR	1354	1296	53	58	132	236	366	523	.57	43
APR	1910	1266	60	12	42	107	218	325	.63	30
MAY	2375	1164	67	2	6	22	64	146	.69	21
JUN	2600	1111	76	0	1	3	8	33	.71	17
JUL	2672	1184	82	0	0	0	0	6	.75	19
AUG	2311	1356	80	0	0	0	0	12	.73	27
SEP	1845	1649	75	0	0	0	0	35	.71	38
OCT	1227	1549	65	3	11	39	82	194	.63	50
NOV	706	1047	54	50	120	217	339	491	.50	59
DEC	511	782	46	165	283	426	577	732	.44	63
YR	1585	1198	63	572	1086	1802	2688	3827	.65	

SACRAMENTO, CALIFORNIA — ELEV 26 LAT 38.5

	HS	VS	TA	D50	D55	D60	D65	D70	KT	LD
JAN	597	829	45	183	315	464	617	772	.43	60
FEB	939	1149	50	86	172	292	426	566	.52	52
MAR	1458	1348	53	50	125	235	372	528	.60	41
APR	2004	1263	58	12	45	116	227	355	.65	29
MAY	2435	1131	64	2	9	37	120	202	.70	19
JUN	2684	1082	71	0	1	6	20	82	.74	15
JUL	2688	1131	75	0	0	0	0	29	.76	17
AUG	2368	1310	74	0	0	0	0	38	.74	25
SEP	1907	1615	72	0	1	5	5	68	.72	37
OCT	1315	1602	63	3	12	47	101	227	.65	48
NOV	782	1135	53	49	121	227	360	511	.53	58
DEC	538	784	46	168	295	442	595	750	.43	62
YR	1646	1198	60	554	1097	1871	2843	4129	.66	

SAN DIEGO, CALIFORNIA — ELEV 30 LAT 32.7

	HS	VS	TA	D50	D55	D60	D65	D70	KT	LD
JAN	976	1325	55	9	58	160	314	459	.58	54
FEB	1266	1392	57	4	33	111	237	373	.60	46
MAR	1632	1261	58	3	23	95	219	372	.61	35
APR	1937	1024	61	1	7	49	144	281	.61	23
MAY	2003	882	63	0	2	20	79	213	.57	14
JUN	2062	870	66	0	1	8	52	147	.57	9
JUL	2186	902	70	0	0	1	6	68	.62	11
AUG	2057	981	71	0	0	0	0	41	.63	19
SEP	1717	1155	70	0	0	1	16	61	.61	31
OCT	1373	1349	66	0	1	6	43	137	.61	43
NOV	1063	1387	61	1	7	48	140	278	.60	52
DEC	904	1302	57	5	37	123	257	413	.58	56
YR	1600	1151	63	23	170	623	1507	2841	.60	

SAN FRANCISCO, CALIFORNIA — ELEV 16 LAT 37.6

	HS	VS	TA	D50	D55	D60	D65	D70	KT	LD
JAN	708	1023	48	82	210	363	518	673	.49	59
FEB	1009	1229	51	32	117	247	386	526	.54	51
MAR	1455	1301	53	17	88	219	372	527	.59	40
APR	1920	1173	55	5	47	148	291	441	.62	28
MAY	2226	1038	58	1	15	82	210	363	.64	19
JUN	2377	998	62	0	3	30	120	253	.65	14
JUL	2392	1034	63	0	2	21	93	234	.67	16
AUG	2116	1149	63	0	1	17	84	219	.66	24
SEP	1742	1394	64	0	1	10	66	181	.65	36
OCT	1226	1397	61	0	4	39	137	280	.60	48
NOV	821	1172	55	5	47	148	291	441	.54	57
DEC	642	977	50	58	170	320	474	629	.49	61
YR	1556	1156	57	202	705	1643	3042	4768	.62	

SANTA MARIA, CALIFORNIA — ELEV 236 LAT 34.9

	HS	VS	TA	D50	D55	D60	D65	D70	KT	LD
JAN	854	1198	51	51	150	296	450	605	.54	56
FEB	1141	1313	52	28	103	226	364	504	.57	49
MAR	1582	1312	53	22	97	226	378	533	.61	37
APR	1921	1082	55	9	58	161	303	453	.61	25
MAY	2141	953	57	3	30	112	245	400	.61	16
JUN	2349	947	60	1	10	63	167	313	.65	12
JUL	2341	965	62	0	3	30	112	247	.66	14
AUG	2106	1057	62	0	3	27	102	241	.65	21
SEP	1730	1255	63	0	3	24	94	225	.63	33
OCT	1353	1440	60	1	7	53	159	299	.62	45
NOV	974	1341	56	5	41	131	270	417	.58	54
DEC	804	1207	52	33	118	256	409	564	.55	58
YR	1610	1172	57	155	624	1604	3053	4801	.62	

COLORADO SPRINGS, COLORADO — ELEV 6171 LAT 38.8

	HS	VS	TA	D50	D55	D60	D65	D70	KT	LD
JAN	891	1534	29	663	818	973	1128	1283	.65	60
FEB	1178	1619	31	524	664	804	944	1084	.65	52
MAR	1550	1477	35	457	611	766	921	1076	.64	41
APR	1931	1227	46	145	271	415	564	714	.63	29
MAY	2129	1035	56	19	74	166	301	451	.61	20
JUN	2369	1021	65	1	5	25	103	181	.65	15
JUL	2212	1012	71	0	1	4	9	71	.62	17
AUG	2025	1145	69	0	1	6	13	96	.63	25
SEP	1759	1475	61	3	17	66	155	279	.67	37
OCT	1359	1702	51	74	166	300	456	605	.68	49
NOV	944	1529	38	377	525	675	825	975	.64	58
DEC	782	1399	31	589	744	899	1054	1209	.63	62
YR	1596	1346	48	2853	3896	5098	6473	8023	.64	

TABLE III-1 (cont.)

DENVER, COLORADO — ELEV 5331 LAT 39.7

	HS	VS	TA	D50	D55	D60	D65	D70	KT	LD
JAN	840	1465	30	623	778	933	1088	1243	.64	61
FEB	1127	1577	33	482	622	762	902	1042	.64	53
MAR	1530	1503	37	406	559	713	868	1023	.64	42
APR	1879	1227	48	130	240	379	525	675	.62	30
MAY	2135	1061	57	18	63	143	253	406	.62	21
JUN	2351	1037	66	1	5	23	80	158	.65	16
JUL	2273	1053	73	0	0	0	0	50	.64	18
AUG	2044	1188	72	0	0	0	0	69	.64	26
SEP	1727	1491	63	3	14	51	120	232	.66	38
OCT	1300	1657	52	63	143	261	408	559	.67	50
NOV	883	1441	39	324	469	618	768	918	.62	59
DEC	732	1323	33	540	695	849	1004	1159	.61	63
YR	1570	1334	50	2592	3588	4733	6016	7535	.64	

EAGLE, COLORADO — ELEV 6512 LAT 39.6

	HS	VS	TA	D50	D55	D60	D65	D70	KT	LD
JAN	754	1234	18	992	1147	1302	1457	1612	.57	61
FEB	1078	1470	23	748	888	1028	1168	1308	.61	53
MAR	1502	1461	31	586	741	896	1051	1206	.63	42
APR	1933	1261	42	245	393	543	693	843	.64	30
MAY	2255	1103	51	38	129	271	425	580	.65	21
JUN	2509	1075	59	1	13	72	190	333	.69	16
JUL	2384	1085	66	0	0	5	43	139	.67	18
AUG	2084	1207	64	0	1	14	79	200	.65	26
SEP	1767	1530	56	6	46	142	285	432	.68	38
OCT	1307	1663	45	168	317	471	626	781	.67	50
NOV	869	1400	31	573	723	873	1023	1173	.61	59
DEC	691	1203	20	921	1076	1231	1386	1541	.58	63
YR	1597	1306	42	4278	5474	6849	8426	10147	.65	

GRAND JUNCTION, COLORADO — ELEV 4839 LAT 39.1

	HS	VS	TA	D50	D55	D60	D65	D70	KT	LD
JAN	791	1296	27	726	880	1035	1190	1345	.59	60
FEB	1119	1520	34	460	599	739	879	1019	.62	53
MAR	1553	1498	41	283	430	583	738	893	.64	42
APR	1986	1276	52	63	142	260	404	550	.65	29
MAY	2380	1132	62	4	16	58	133	255	.69	20
JUN	2598	1081	71	0	1	5	20	69	.71	16
JUL	2465	1094	79	0	0	0	0	10	.70	18
AUG	2182	1240	75	0	0	0	0	26	.68	26
SEP	1834	1573	67	1	3	15	60	133	.70	37
OCT	1345	1698	55	30	91	186	324	470	.68	49
NOV	918	1486	40	312	457	606	756	906	.63	58
DEC	731	1280	30	636	791	946	1101	1256	.60	62
YR	1661	1346	53	2514	3412	4434	5605	6931	.67	

PUEBLO, COLORADO — ELEV 4721 LAT 38.3

	HS	VS	TA	D50	D55	D60	D65	D70	KT	LD
JAN	894	1504	30	617	772	927	1082	1237	.64	60
FEB	1172	1572	35	429	569	708	848	988	.64	52
MAR	1564	1467	40	315	466	620	775	930	.64	41
APR	1956	1223	52	56	136	257	405	549	.64	29
MAY	2162	1034	61	3	17	67	148	283	.62	19
JUN	2434	1025	71	0	1	4	28	70	.67	15
JUL	2312	1030	76	0	0	0	0	15	.65	17
AUG	2102	1167	75	0	0	0	0	27	.66	25
SEP	1779	1468	66	1	3	16	55	146	.67	36
OCT	1361	1669	55	27	91	191	335	481	.67	48
NOV	954	1518	41	282	427	576	726	876	.64	58
DEC	782	1364	33	527	682	837	992	1147	.62	62
YR	1625	1335	53	2258	3163	4203	5394	6751	.65	

HARTFORD, CONNECTICUT — ELEV 180 LAT 41.9

	HS	VS	TA	D50	D55	D60	D65	D70	KT	LD
JAN	477	694	25	781	936	1091	1246	1401	.40	63
FEB	715	891	27	650	790	930	1070	1210	.43	56
MAR	978	900	36	447	601	756	911	1066	.42	44
APR	1315	888	48	109	226	370	519	669	.44	32
MAY	1568	859	58	5	30	101	226	364	.45	23
JUN	1686	853	68	0	1	6	24	107	.46	19
JUL	1649	861	73	0	0	0	0	36	.47	21
AUG	1422	880	70	0	0	2	12	67	.45	28
SEP	1154	967	63	1	6	34	106	224	.46	40
OCT	853	991	53	37	114	237	384	540	.46	52
NOV	497	679	41	265	412	561	711	861	.38	61
DEC	385	562	28	676	831	986	1141	1296	.36	65
YR	1060	835	49	2971	3948	5075	6350	7841	.44	

WILMINGTON, DELAWARE — ELEV 79 LAT 39.7

	HS	VS	TA	D50	D55	D60	D65	D70	KT	LD
JAN	571	819	32	558	713	868	1023	1178	.43	61
FEB	827	1006	34	460	599	739	879	1019	.47	53
MAR	1149	1032	42	268	417	571	725	880	.48	42
APR	1480	952	52	47	123	240	381	531	.49	30
MAY	1710	887	62	2	11	47	128	246	.49	21
JUN	1883	895	71	0	0	0	0	59	.52	16
JUL	1823	898	76	0	0	0	0	18	.51	18
AUG	1615	947	74	0	0	0	0	29	.51	26
SEP	1318	1064	68	0	2	9	32	113	.51	38
OCT	984	1113	57	11	50	129	254	399	.50	50
NOV	645	901	46	156	285	430	579	729	.45	59
DEC	489	720	35	475	629	784	939	1094	.41	63
YR	1210	935	54	1978	2829	3818	4940	6295	.49	

WASHINGTON, DC — ELEV 289 LAT 38.9

	HS	VS	TA	D50	D55	D60	D65	D70	KT	LD
JAN	572	793	32	555	710	865	1020	1175	.42	60
FEB	815	956	34	454	594	734	874	1014	.45	53
MAR	1125	979	42	262	411	564	719	874	.46	41
APR	1459	919	53	38	109	219	357	507	.48	29
MAY	1718	877	63	2	10	45	131	240	.50	20
JUN	1901	890	71	0	1	3	5	63	.52	16
JUL	1817	883	75	0	0	0	0	20	.51	18
AUG	1617	929	74	0	0	0	0	34	.51	25
SEP	1340	1058	67	0	2	12	43	131	.51	37
OCT	1004	1111	56	17	68	156	291	438	.50	49
NOV	651	883	45	179	313	460	609	759	.45	58
DEC	481	678	34	497	651	806	961	1116	.39	62
YR	1210	912	54	2004	2869	3864	5010	6372	.49	

APALACHICOLA, FLORIDA — ELEV 20 LAT 29.7

	HS	VS	TA	D50	D55	D60	D65	D70	KT	LD
JAN	853	967	54	53	125	225	368	508	.46	51
FEB	1126	1061	56	30	84	161	290	401	.50	43
MAR	1474	1016	61	10	34	94	175	302	.53	32
APR	1879	928	68	1	5	18	30	128	.58	20
MAY	2091	870	75	0	0	0	0	41	.60	11
JUN	1998	844	80	0	0	0	0	12	.56	6
JUL	1814	801	81	0	0	0	0	8	.52	8
AUG	1688	804	82	0	0	0	0	8	.51	16
SEP	1535	938	79	0	0	0	0	16	.53	28
OCT	1371	1204	71	1	3	10	22	92	.57	40
NOV	1040	1188	61	9	30	85	158	282	.54	49
DEC	818	991	55	38	101	191	318	462	.47	53
YR	1475	967	69	142	382	783	1361	2261	.54	

TABLE III-1 (cont.)

DAYTONA BEACH, FLORIDA — ELEV 39 LAT 29.2

	HS	VS	TA	D50	D55	D60	D65	D70	KT	LD
JAN	958	1116	58	19	60	133	241	368	.51	50
FEB	1213	1150	60	13	42	103	210	302	.54	43
MAR	1548	1056	64	5	17	54	120	222	.56	32
APR	1884	919	70	1	4	14	17	108	.58	19
MAY	1968	840	75	0	0	0	0	42	.56	10
JUN	1826	807	79	0	0	0	0	15	.51	6
JUL	1784	792	81	0	0	0	0	10	.51	8
AUG	1682	796	81	0	0	0	0	10	.51	16
SEP	1478	892	80	0	0	0	0	14	.51	27
OCT	1251	1054	73	0	2	6	5	61	.52	39
NOV	1035	1157	65	4	12	40	97	189	.53	49
DEC	870	1060	60	14	46	114	212	334	.50	53
YR	1459	969	71	57	183	463	902	1677	.53	

JACKSONVILLE, FLORIDA — ELEV 30 LAT 30.5

	HS	VS	TA	D50	D55	D60	D65	D70	KT	LD
JAN	900	1076	55	47	114	207	348	481	.50	52
FEB	1164	1141	56	29	80	155	282	389	.53	44
MAR	1522	1079	61	10	33	91	176	291	.56	33
APR	1856	936	68	2	6	21	24	135	.58	21
MAY	1956	847	74	0	0	0	0	50	.56	11
JUN	1885	822	79	0	0	0	0	16	.53	7
JUL	1802	800	81	0	0	0	0	11	.51	9
AUG	1694	816	81	0	0	0	0	11	.51	17
SEP	1442	900	78	0	0	0	0	20	.50	29
OCT	1223	1070	71	1	3	12	19	101	.52	40
NOV	996	1153	61	10	32	88	161	281	.53	50
DEC	818	1024	55	40	102	190	317	457	.48	54
YR	1439	971	68	138	372	762	1327	2241	.53	

MIAMI, FLORIDA — ELEV 7 LAT 25.8

	HS	VS	TA	D50	D55	D60	D65	D70	KT	LD
JAN	1057	1121	67	1	4	18	53	142	.52	47
FEB	1314	1130	68	1	3	14	67	118	.55	39
MAR	1603	990	71	0	1	6	17	76	.56	28
APR	1859	853	75	0	0	0	0	31	.57	16
MAY	1844	800	78	0	0	0	0	15	.53	7
JUN	1708	787	81	0	0	0	0	6	.48	2
JUL	1763	787	82	0	0	0	0	4	.51	4
AUG	1630	751	83	0	0	0	0	4	.49	12
SEP	1456	811	82	0	0	0	0	5	.49	24
OCT	1303	991	78	0	0	0	0	15	.51	36
NOV	1119	1131	72	0	1	4	13	61	.53	45
DEC	1019	1157	68	1	3	13	56	122	.53	49
YR	1474	941	76	3	14	55	206	599	.52	

ORLANDO, FLORIDA — ELEV 118 LAT 28.5

	HS	VS	TA	D50	D55	D60	D65	D70	KT	LD
JAN	999	1151	60	13	42	105	197	316	.52	50
FEB	1243	1158	62	9	29	79	184	256	.54	42
MAR	1582	1058	66	3	11	36	94	181	.57	31
APR	1898	910	71	1	3	10	13	88	.59	19
MAY	1989	840	76	0	0	0	0	33	.57	9
JUN	1831	809	80	0	0	0	0	13	.51	5
JUL	1801	795	81	0	0	0	0	10	.51	7
AUG	1673	786	82	0	0	0	0	9	.51	15
SEP	1497	887	80	0	0	0	0	13	.51	27
OCT	1304	1084	74	0	0	0	0	51	.53	38
NOV	1096	1219	67	3	9	30	75	162	.55	48
DEC	926	1126	62	10	32	87	170	284	.52	52
YR	1488	984	72	39	126	348	733	1415	.54	

TALLAHASSEE, FLORIDA — ELEV 69 LAT 30.4

	HS	VS	TA	D50	D55	D60	D65	D70	KT	LD
JAN	877	1033	53	73	150	256	408	542	.48	52
FEB	1138	1104	55	42	102	185	323	429	.52	44
MAR	1479	1042	60	13	42	106	187	316	.54	33
APR	1823	920	68	2	7	23	34	140	.57	21
MAY	1936	842	75	0	0	0	0	47	.56	11
JUN	1883	821	80	0	0	0	0	14	.53	7
JUL	1748	786	81	0	0	0	0	11	.50	9
AUG	1675	808	81	0	0	0	0	11	.51	17
SEP	1493	930	78	0	0	0	0	22	.52	29
OCT	1318	1174	69	1	5	17	31	122	.56	40
NOV	1008	1168	59	18	55	124	204	344	.53	50
DEC	813	1011	53	65	140	241	376	524	.48	54
YR	1434	969	68	215	501	951	1563	2521	.53	

TAMPA, FLORIDA — ELEV 10 LAT 28.0

	HS	VS	TA	D50	D55	D60	D65	D70	KT	LD
JAN	1011	1148	60	16	46	110	203	316	.52	49
FEB	1259	1156	62	10	31	81	176	253	.54	42
MAR	1594	1051	66	4	14	41	90	185	.57	30
APR	1908	904	72	1	3	11	9	84	.59	18
MAY	1998	839	77	0	0	0	0	32	.57	9
JUN	1847	813	81	0	0	0	0	13	.52	5
JUL	1753	783	82	0	0	0	0	11	.50	7
AUG	1653	775	82	0	0	0	0	10	.50	15
SEP	1492	873	81	0	0	0	0	14	.51	26
OCT	1346	1108	75	0	0	0	0	53	.55	38
NOV	1108	1213	67	3	11	33	71	164	.55	47
DEC	935	1119	62	12	36	93	169	285	.51	51
YR	1493	981	72	47	141	369	718	1421	.54	

WEST PALM BEACH, FLORIDA — ELEV 20 LAT 26.7

	HS	VS	TA	D50	D55	D60	D65	D70	KT	LD
JAN	1000	1075	66	2	8	30	83	178	.50	48
FEB	1233	1075	66	2	6	23	91	150	.52	40
MAR	1556	985	70	1	2	10	25	100	.55	29
APR	1814	851	74	0	0	0	0	43	.56	17
MAY	1845	801	78	0	0	0	0	18	.53	8
JUN	1706	783	81	0	0	0	0	8	.48	3
JUL	1779	789	82	0	0	0	0	5	.51	5
AUG	1663	767	82	0	0	0	0	5	.50	13
SEP	1419	807	82	0	0	0	0	6	.48	25
OCT	1224	946	77	0	0	0	0	20	.49	37
NOV	1060	1087	71	0	2	7	22	80	.51	46
DEC	958	1100	67	1	6	22	78	152	.51	50
YR	1439	922	75	6	24	92	299	765	.51	

ATLANTA, GEORGIA — ELEV 1033 LAT 33.6

	HS	VS	TA	D50	D55	D60	D65	D70	KT	LD
JAN	718	884	42	246	393	546	701	856	.44	55
FEB	969	998	45	161	284	421	560	700	.47	47
MAR	1304	993	51	67	153	283	443	586	.50	36
APR	1686	927	61	3	16	65	144	274	.53	24
MAY	1854	851	69	0	1	7	27	98	.53	15
JUN	1914	842	76	0	0	0	0	19	.53	10
JUL	1812	821	78	0	0	0	0	9	.51	12
AUG	1708	867	78	0	0	0	0	11	.52	20
SEP	1422	966	72	0	0	2	8	49	.51	32
OCT	1200	1165	62	2	11	49	137	246	.54	44
NOV	883	1101	51	61	142	266	408	558	.51	53
DEC	674	881	44	217	360	512	667	822	.44	57
YR	1347	941	61	758	1362	2150	3095	4228	.51	

TABLE III-1 (cont.)

AUGUSTA, GEORGIA ELEV 148 LAT 33.4

	HS	VS	TA	D50	D55	D60	D65	D70	KT	LD
JAN	751	934	46	173	297	443	601	750	.45	55
FEB	1015	1055	48	114	208	333	475	608	.49	47
MAR	1338	1018	55	39	105	199	346	480	.51	36
APR	1728	944	64	3	13	45	90	211	.55	24
MAY	1865	852	72	0	2	6	10	73	.53	14
JUN	1904	838	78	0	0	0	0	15	.53	10
JUL	1803	817	80	0	0	0	0	9	.51	12
AUG	1667	847	80	0	0	0	0	11	.51	20
SEP	1410	951	74	0	0	0	0	41	.50	32
OCT	1220	1182	64	3	12	44	104	211	.55	43
NOV	916	1151	54	46	116	214	344	491	.52	53
DEC	721	962	46	161	281	425	577	732	.47	57
YR	1363	962	63	539	1033	1709	2547	3631	.51	

MACON, GEORGIA ELEV 361 LAT 32.7

	HS	VS	TA	D50	D55	D60	D65	D70	KT	LD
JAN	769	940	48	138	245	384	543	689	.45	54
FEB	1020	1035	50	86	166	280	423	550	.49	46
MAR	1363	1018	57	26	80	162	298	423	.51	35
APR	1736	932	66	2	9	30	66	170	.55	23
MAY	1885	850	74	0	1	4	6	53	.54	14
JUN	1919	838	80	0	0	0	0	12	.53	9
JUL	1785	807	81	0	0	0	0	8	.50	11
AUG	1718	856	81	0	0	0	0	9	.52	19
SEP	1439	953	76	0	0	0	0	30	.51	31
OCT	1247	1186	66	2	9	32	82	178	.55	43
NOV	940	1161	55	35	95	182	304	447	.53	52
DEC	729	950	48	129	233	370	518	673	.46	56
YR	1381	960	65	420	838	1444	2240	3240	.52	

SAVANNAH, GEORGIA ELEV 52 LAT 32.1

	HS	VS	TA	D50	D55	D60	D65	D70	KT	LD
JAN	795	962	50	100	192	323	483	624	.46	53
FEB	1044	1046	52	61	133	236	379	502	.49	46
MAR	1398	1029	58	16	56	131	256	378	.52	35
APR	1761	930	66	2	7	26	63	161	.55	22
MAY	1852	835	73	0	0	0	0	53	.53	13
JUN	1844	817	79	0	0	0	0	12	.51	9
JUL	1783	803	81	0	0	0	0	7	.50	11
AUG	1621	810	81	0	0	0	0	8	.49	19
SEP	1364	885	76	0	0	0	0	25	.48	30
OCT	1217	1125	67	1	5	21	60	147	.53	42
NOV	941	1134	57	20	66	143	253	391	.52	51
DEC	754	972	50	92	181	308	458	608	.47	55
YR	1366	945	66	293	641	1188	1952	2917	.51	

BOISE, IDAHO ELEV 2867 LAT 43.6

	HS	VS	TA	D50	D55	D60	D65	D70	KT	LD
JAN	485	770	29	651	806	961	1116	1271	.44	65
FEB	840	1207	36	408	547	686	826	966	.54	57
MAR	1304	1402	41	288	434	587	741	896	.58	46
APR	1827	1355	49	110	206	337	480	630	.62	34
MAY	2277	1252	57	19	63	141	252	395	.66	25
JUN	2463	1183	65	2	9	35	97	188	.68	20
JUL	2613	1309	75	0	0	0	0	39	.74	22
AUG	2196	1460	72	0	1	5	12	65	.71	30
SEP	1737	1749	63	4	15	52	127	227	.71	42
OCT	1138	1610	52	67	146	261	406	556	.64	54
NOV	628	1034	40	314	458	607	756	906	.52	63
DEC	437	735	32	556	710	865	1020	1175	.44	67
YR	1499	1255	51	2420	3395	4536	5833	7315	.64	

LEWISTON, IDAHO ELEV 1437 LAT 46.4

	HS	VS	TA	D50	D55	D60	D65	D70	KT	LD
JAN	340	523	31	583	738	893	1048	1203	.35	68
FEB	609	849	38	337	474	613	753	893	.43	60
MAR	1020	1108	43	237	379	531	685	840	.48	49
APR	1435	1114	50	84	172	299	441	591	.50	37
MAY	1842	1111	58	13	49	123	232	373	.54	27
JUN	2015	1083	65	2	7	29	84	179	.55	23
JUL	2336	1304	73	0	0	0	0	45	.66	25
AUG	1931	1396	72	0	1	5	17	70	.63	33
SEP	1435	1502	63	3	12	45	124	219	.61	45
OCT	860	1198	52	65	147	267	409	565	.53	56
NOV	413	633	41	293	437	585	735	885	.39	66
DEC	286	454	35	473	627	781	936	1091	.34	70
YR	1214	1024	52	2090	3042	4171	5464	6955	.54	

POCATELLO, IDAHO ELEV 4478 LAT 42.9

	HS	VS	TA	D50	D55	D60	D65	D70	KT	LD
JAN	539	871	23	831	986	1141	1296	1451	.47	64
FEB	882	1259	29	577	717	857	997	1137	.55	57
MAR	1371	1463	35	454	608	763	918	1073	.60	45
APR	1820	1317	45	166	296	442	591	741	.61	33
MAY	2280	1227	54	29	93	194	336	484	.66	24
JUN	2480	1165	62	3	14	56	138	255	.68	20
JUL	2600	1273	72	0	0	0	0	62	.74	22
AUG	2239	1454	70	0	1	6	20	92	.72	29
SEP	1769	1744	59	6	28	90	192	322	.71	41
OCT	1203	1697	48	110	219	363	515	670	.67	53
NOV	689	1151	36	430	579	729	879	1029	.55	62
DEC	477	809	27	716	871	1026	1181	1336	.47	66
YR	1533	1285	47	3322	4413	5666	7063	8651	.65	

CHICAGO, ILLINOIS ELEV 623 LAT 41.8

	HS	VS	TA	D50	D55	D60	D65	D70	KT	LD
JAN	507	756	24	797	952	1107	1262	1417	.42	63
FEB	759	967	27	633	773	913	1053	1193	.46	55
MAR	1107	1054	37	414	565	719	874	1029	.48	44
APR	1459	993	50	98	187	313	453	604	.49	32
MAY	1789	963	60	10	36	99	208	320	.52	23
JUN	2007	973	71	1	2	9	26	90	.55	18
JUL	1944	984	75	0	0	0	0	39	.55	20
AUG	1719	1065	74	0	1	4	8	49	.55	28
SEP	1354	1179	66	2	8	28	57	167	.54	40
OCT	969	1181	55	33	94	183	316	456	.52	52
NOV	566	815	40	299	441	589	738	888	.43	61
DEC	401	593	29	667	822	977	1132	1287	.37	65
YR	1217	960	51	2954	3881	4940	6127	7537	.51	

MOLINE, ILLINOIS ELEV 594 LAT 41.4

	HS	VS	TA	D50	D55	D60	D65	D70	KT	LD
JAN	535	803	22	884	1039	1194	1349	1504	.44	63
FEB	812	1048	26	681	820	960	1100	1240	.48	55
MAR	1119	1055	36	446	599	753	908	1063	.48	44
APR	1459	982	51	81	167	291	436	582	.49	32
MAY	1754	938	61	6	23	77	184	287	.51	22
JUN	1969	952	71	0	1	6	20	79	.54	18
JUL	1939	974	75	0	0	0	0	35	.55	20
AUG	1715	1051	73	0	1	3	11	52	.54	28
SEP	1357	1166	65	2	8	33	79	189	.53	40
OCT	996	1209	54	36	103	200	344	485	.53	51
NOV	595	862	39	330	475	624	774	924	.45	61
DEC	433	651	27	726	880	1035	1190	1345	.39	65
YR	1226	973	50	3191	4117	5178	6395	7786	.51	

TABLE III-1 (cont.)

SPRINGFIELD, ILLINOIS ELEV 614 LAT 39.8

	HS	VS	TA	D50	D55	D60	D65	D70	KT	LD
JAN	585	852	27	723	877	1032	1187	1342	.45	61
FEB	861	1069	30	549	689	829	969	1109	.49	53
MAR	1143	1028	39	337	486	639	794	949	.48	42
APR	1515	979	53	54	126	229	363	508	.50	30
MAY	1865	954	63	4	15	52	132	229	.54	21
JUN	2097	966	73	0	1	4	12	56	.58	16
JUL	2058	984	76	0	0	0	0	28	.58	18
AUG	1806	1058	74	0	1	3	8	42	.57	26
SEP	1454	1205	67	1	5	20	48	142	.56	38
OCT	1068	1254	57	24	76	158	282	419	.55	50
NOV	677	971	42	259	397	544	693	843	.48	59
DEC	490	725	31	605	760	915	1070	1225	.41	63
YR	1304	1003	53	2558	3434	4425	5558	6891	.53	

EVANSVILLE, INDIANA ELEV 387 LAT 38.0

	HS	VS	TA	D50	D55	D60	D65	D70	KT	LD
JAN	574	767	33	541	695	849	1004	1159	.41	59
FEB	823	937	36	398	536	675	815	955	.45	52
MAR	1151	979	44	206	340	489	653	797	.47	40
APR	1501	925	57	22	71	150	263	403	.49	28
MAY	1783	888	66	2	8	30	95	175	.51	19
JUN	1983	903	75	0	1	3	5	37	.55	15
JUL	1920	905	78	0	0	0	0	17	.54	17
AUG	1735	971	76	0	0	0	0	26	.54	25
SEP	1403	1087	69	1	3	12	34	109	.53	36
OCT	1087	1199	58	16	54	128	236	372	.54	48
NOV	682	909	45	186	312	455	603	753	.45	57
DEC	499	688	35	458	611	766	921	1076	.39	61
YR	1264	929	56	1830	2630	3556	4629	5879	.50	

FORT WAYNE, INDIANA ELEV 827 LAT 41.0

	HS	VS	TA	D50	D55	D60	D65	D70	KT	LD
JAN	455	624	25	766	921	1076	1231	1386	.36	62
FEB	698	833	28	627	767	907	1047	1187	.41	55
MAR	982	880	37	421	574	729	884	1039	.42	43
APR	1361	901	49	98	194	326	471	621	.45	31
MAY	1672	893	60	8	32	96	216	328	.48	22
JUN	1842	900	70	0	2	8	23	95	.51	18
JUL	1787	905	73	0	0	0	0	48	.50	20
AUG	1594	966	71	0	1	5	12	71	.50	28
SEP	1274	1064	65	2	8	31	90	189	.50	39
OCT	924	1071	54	41	112	218	363	509	.49	51
NOV	516	690	40	301	446	594	744	894	.38	60
DEC	369	509	29	664	818	973	1128	1283	.33	64
YR	1125	853	50	2927	3874	4963	6209	7650	.46	

INDIANAPOLIS, INDIANA ELEV 807 LAT 39.7

	HS	VS	TA	D50	D55	D60	D65	D70	KT	LD
JAN	496	668	28	685	840	995	1150	1305	.38	61
FEB	747	872	31	541	681	820	960	1100	.42	53
MAR	1037	905	40	328	477	630	784	939	.43	42
APR	1398	897	52	63	139	248	387	532	.46	30
MAY	1688	877	62	5	20	66	159	259	.49	21
JUN	1868	890	72	0	1	6	11	71	.51	16
JUL	1806	891	75	0	0	0	0	35	.51	18
AUG	1643	962	73	0	1	4	5	54	.52	26
SEP	1324	1070	66	2	7	25	63	157	.51	38
OCT	977	1102	56	30	88	175	302	446	.50	50
NOV	579	771	42	264	403	550	699	849	.41	59
DEC	417	572	31	593	747	902	1057	1212	.35	63
YR	1167	873	52	2511	3403	4421	5577	6960	.47	

SOUTH BEND, INDIANA ELEV 774 LAT 41.7

	HS	VS	TA	D50	D55	D60	D65	D70	KT	LD
JAN	416	566	24	806	961	1116	1271	1426	.34	63
FEB	660	791	26	664	804	944	1084	1224	.40	55
MAR	992	911	35	458	611	766	921	1076	.43	44
APR	1387	936	48	122	226	362	507	657	.46	32
MAY	1722	929	58	13	48	122	245	365	.50	23
JUN	1922	940	69	1	3	12	35	114	.53	18
JUL	1852	944	72	0	1	4	6	62	.52	20
AUG	1666	1029	71	0	2	6	24	81	.53	28
SEP	1291	1106	64	3	12	43	98	209	.51	40
OCT	909	1075	53	48	122	226	368	516	.49	52
NOV	497	673	40	319	464	612	762	912	.38	61
DEC	340	467	28	676	831	986	1141	1296	.31	65
YR	1140	864	49	3112	4084	5199	6462	7938	.48	

BURLINGTON, IOWA ELEV 702 LAT 40.8

	HS	VS	TA	D50	D55	D60	D65	D70	KT	LD
JAN	579	878	23	840	995	1150	1305	1460	.46	62
FEB	859	1109	27	636	776	916	1056	1196	.50	54
MAR	1165	1090	37	411	562	716	871	1026	.49	43
APR	1538	1023	51	78	159	275	416	562	.51	31
MAY	1876	981	62	6	23	74	172	271	.54	22
JUN	2121	992	71	0	2	7	16	77	.58	17
JUL	2085	1016	75	0	0	0	0	33	.59	19
AUG	1828	1101	74	0	1	4	8	47	.58	27
SEP	1416	1206	65	2	9	32	70	177	.55	39
OCT	1061	1293	55	34	95	185	320	459	.56	51
NOV	664	987	40	315	458	607	756	906	.49	60
DEC	481	739	28	695	849	1004	1159	1314	.42	64
YR	1308	1034	51	3018	3930	4969	6149	7527	.54	

DES MOINES, IOWA ELEV 965 LAT 41.5

	HS	VS	TA	D50	D55	D60	D65	D70	KT	LD
JAN	581	912	19	949	1104	1259	1414	1569	.48	63
FEB	861	1145	24	723	862	1002	1142	1282	.51	55
MAR	1180	1135	34	502	655	809	964	1119	.51	44
APR	1557	1057	50	109	200	325	465	616	.52	32
MAY	1867	993	61	10	33	91	186	297	.54	22
JUN	2125	1008	71	1	3	10	26	94	.58	18
JUL	2097	1037	75	0	0	0	0	39	.59	20
AUG	1828	1124	73	0	1	5	13	58	.58	28
SEP	1434	1256	64	4	14	45	94	204	.57	40
OCT	1068	1342	54	47	116	211	350	489	.57	51
NOV	658	1005	38	373	518	666	816	966	.50	61
DEC	487	778	25	775	930	1085	1240	1395	.44	65
YR	1314	1065	49	3491	4435	5510	6710	8129	.55	

MASON CITY, IOWA ELEV 1224 LAT 43.1

	HS	VS	TA	D50	D55	D60	D65	D70	KT	LD
JAN	554	917	14	1110	1265	1420	1575	1730	.49	64
FEB	836	1173	19	882	1022	1162	1302	1442	.53	57
MAR	1168	1184	29	651	806	961	1116	1271	.52	46
APR	1519	1078	46	165	288	431	579	729	.51	33
MAY	1895	1046	57	16	58	136	265	394	.55	24
JUN	2114	1041	67	1	4	16	64	135	.58	20
JUL	2084	1074	71	0	1	5	13	74	.59	22
AUG	1833	1183	70	0	2	8	31	94	.59	30
SEP	1405	1294	60	7	28	87	165	302	.57	41
OCT	1010	1322	51	85	174	303	457	605	.56	53
NOV	600	943	34	493	642	792	942	1092	.48	62
DEC	443	731	20	927	1082	1237	1392	1547	.44	66
YR	1291	1081	45	4338	5373	6559	7901	9415	.55	

TABLE III-1 (cont.)

SIOUX CITY, IOWA — ELEV 1102 LAT 42.4

	HS	VS	TA	D50	D55	D60	D65	D70	KT	LD
JAN	569	922	18	992	1147	1302	1457	1612	.48	64
FEB	842	1151	23	745	885	1025	1165	1305	.52	56
MAR	1170	1158	33	523	676	831	986	1141	.51	45
APR	1578	1102	49	109	201	328	474	619	.53	33
MAY	1901	1031	61	9	32	90	189	297	.55	23
JUN	2124	1028	70	1	3	10	33	95	.58	19
JUL	2122	1071	75	0	0	0	0	36	.60	21
AUG	1845	1166	74	0	1	5	10	55	.59	29
SEP	1421	1281	63	5	17	54	113	224	.57	41
OCT	1038	1336	53	59	134	239	378	526	.57	52
NOV	643	1012	36	415	562	711	861	1011	.50	62
DEC	469	769	24	822	977	1132	1287	1442	.45	66
YR	1313	1085	48	3680	4634	5726	6953	8362	.55	

DODGE CITY, KANSAS — ELEV 2582 LAT 37.8

	HS	VS	TA	D50	D55	D60	D65	D70	KT	LD
JAN	827	1303	31	596	750	905	1060	1215	.58	59
FEB	1122	1444	35	417	555	695	834	974	.60	51
MAR	1476	1335	41	289	432	584	738	893	.60	40
APR	1886	1160	54	48	116	210	344	482	.61	28
MAY	2090	997	64	4	15	50	115	218	.60	19
JUN	2358	998	74	0	1	4	21	51	.65	14
JUL	2295	1013	79	0	0	0	0	14	.65	16
AUG	2055	1126	78	0	0	0	0	19	.64	24
SEP	1687	1350	69	1	4	15	41	118	.63	36
OCT	1301	1532	58	20	63	139	247	382	.64	48
NOV	894	1343	43	239	373	518	666	816	.59	57
DEC	732	1202	33	517	670	825	980	1135	.56	61
YR	1562	1232	55	2132	2980	3945	5046	6318	.62	

GOODLAND, KANSAS — ELEV 3688 LAT 39.4

	HS	VS	TA	D50	D55	D60	D65	D70	KT	LD
JAN	789	1310	28	695	850	1004	1159	1314	.59	61
FEB	1056	1414	32	519	658	798	938	1078	.59	53
MAR	1424	1350	36	429	581	735	890	1045	.59	42
APR	1829	1181	49	121	217	347	489	640	.60	30
MAY	2062	1025	59	16	51	122	216	353	.59	20
JUN	2357	1031	69	1	4	14	55	115	.65	16
JUL	2319	1060	76	0	0	0	0	33	.65	18
AUG	2046	1178	74	0	0	0	0	49	.64	26
SEP	1642	1383	64	4	14	45	108	204	.63	38
OCT	1268	1578	53	64	140	247	387	535	.64	49
NOV	857	1357	39	353	497	646	795	945	.60	59
DEC	695	1202	30	618	772	927	1082	1237	.57	63
YR	1531	1255	51	2819	3784	4885	6119	7547	.62	

TOPEKA, KANSAS — ELEV 886 LAT 39.1

	HS	VS	TA	D50	D55	D60	D65	D70	KT	LD
JAN	681	1033	28	683	837	992	1147	1302	.50	60
FEB	941	1181	33	467	605	745	885	1025	.52	53
MAR	1257	1135	41	291	433	584	745	893	.52	42
APR	1642	1045	55	47	113	203	329	468	.54	29
MAY	1915	960	65	5	16	49	118	210	.55	20
JUN	2126	962	74	0	2	6	13	58	.58	16
JUL	2128	993	78	0	0	0	0	21	.60	18
AUG	1910	1094	77	0	0	0	0	27	.60	26
SEP	1516	1239	68	2	6	21	55	134	.58	37
OCT	1147	1350	58	24	71	148	259	392	.58	49
NOV	772	1144	43	240	371	515	663	813	.53	58
DEC	583	907	32	566	720	874	1029	1184	.47	62
YR	1387	1036	54	2325	3175	4137	5243	6527	.56	

WICHITA, KANSAS — ELEV 1339 LAT 37.6

	HS	VS	TA	D50	D55	D60	D65	D70	KT	LD
JAN	784	1191	31	581	735	890	1045	1200	.55	59
FEB	1058	1315	36	389	525	664	804	944	.57	51
MAR	1405	1243	44	231	363	511	671	819	.57	40
APR	1782	1088	57	30	83	161	275	408	.58	28
MAY	2036	974	66	3	11	35	90	176	.59	19
JUN	2264	972	76	0	1	3	7	36	.62	14
JUL	2238	994	81	0	0	0	0	12	.63	16
AUG	2032	1108	80	0	0	0	0	15	.63	24
SEP	1616	1272	71	1	3	12	32	97	.61	36
OCT	1250	1436	60	15	48	115	211	335	.61	48
NOV	871	1280	45	197	319	460	606	756	.57	57
DEC	690	1088	35	484	637	791	946	1101	.53	61
YR	1504	1162	57	1931	2725	3642	4687	5897	.60	

LEXINGTON, KENTUCKY — ELEV 988 LAT 38.0

	HS	VS	TA	D50	D55	D60	D65	D70	KT	LD
JAN	546	714	33	531	685	840	995	1150	.39	59
FEB	779	868	35	414	552	692	832	972	.42	52
MAR	1099	924	44	223	360	510	673	818	.45	40
APR	1479	911	55	31	90	177	302	444	.48	28
MAY	1747	874	65	3	10	37	106	197	.50	19
JUN	1897	877	73	0	1	4	8	54	.52	15
JUL	1850	881	76	0	0	0	0	26	.52	17
AUG	1685	945	75	0	0	0	0	35	.53	25
SEP	1362	1049	69	1	3	13	40	117	.51	36
OCT	1044	1134	58	17	58	134	246	383	.51	48
NOV	657	862	45	192	320	464	612	762	.44	57
DEC	485	660	36	452	605	760	915	1070	.38	61
YR	1221	892	55	1865	2686	3632	4729	6026	.49	

LOUISVILLE, KENTUCKY — ELEV 489 LAT 38.2

	HS	VS	TA	D50	D55	D60	D65	D70	KT	LD
JAN	545	718	33	519	673	828	983	1138	.39	59
FEB	789	890	36	400	538	678	818	958	.43	52
MAR	1102	933	44	214	349	498	661	806	.45	41
APR	1467	908	56	28	83	166	286	426	.48	28
MAY	1720	867	65	3	11	38	105	195	.50	19
JUN	1903	882	73	0	1	4	5	52	.52	15
JUL	1837	880	77	0	0	0	0	22	.52	17
AUG	1680	947	76	0	0	0	0	29	.52	25
SEP	1361	1055	69	1	3	12	35	110	.51	36
OCT	1042	1139	58	16	56	130	241	375	.52	48
NOV	653	861	45	185	310	452	600	750	.44	58
DEC	488	672	36	449	602	757	911	1066	.38	62
YR	1218	896	56	1816	2625	3563	4645	5927	.49	

BATON ROUGE, LOUISIANA — ELEV 75 LAT 30.5

	HS	VS	TA	D50	D55	D60	D65	D70	KT	LD
JAN	785	889	51	90	174	294	451	590	.43	52
FEB	1054	1001	54	46	111	199	335	453	.48	44
MAR	1379	965	60	13	43	109	208	330	.51	33
APR	1681	864	68	1	5	17	33	127	.52	21
MAY	1871	826	75	0	0	0	0	42	.54	11
JUN	1926	831	80	0	0	0	0	11	.54	7
JUL	1746	786	82	0	0	0	0	7	.50	9
AUG	1677	810	82	0	0	0	0	8	.51	17
SEP	1464	914	78	0	0	0	0	22	.51	29
OCT	1301	1159	69	1	5	17	54	129	.55	40
NOV	920	1032	59	17	53	123	208	350	.49	50
DEC	737	883	53	63	139	245	381	532	.44	54
YR	1380	913	67	232	530	1006	1670	2601	.51	

TABLE III-1 (cont.)

LAKE CHARLES, LOUISIANA — ELEV 10 LAT 30.1

	HS	VS	TA	D50	D55	D60	D65	D70	KT	LD
JAN	728	790	52	79	158	265	415	551	.40	51
FEB	1010	934	55	42	100	181	306	422	.45	44
MAR	1313	903	60	15	44	108	200	317	.48	33
APR	1570	810	69	2	6	19	26	127	.49	20
MAY	1849	818	75	0	0	0	0	45	.53	11
JUN	1970	840	81	0	0	0	0	13	.55	7
JUL	1788	795	82	0	0	0	0	9	.51	9
AUG	1657	798	82	0	0	0	0	9	.50	17
SEP	1485	917	78	0	0	0	0	22	.51	28
OCT	1381	1233	70	1	5	16	36	112	.58	40
NOV	917	1012	60	14	44	106	177	310	.48	49
DEC	706	819	54	54	124	217	338	491	.41	53
YR	1366	889	68	207	481	912	1498	2427	.50	

NEW ORLEANS, LOUISIANA — ELEV 10 LAT 30.0

	HS	VS	TA	D50	D55	D60	D65	D70	KT	LD
JAN	835	950	53	73	150	252	403	533	.46	51
FEB	1112	1055	56	39	96	173	299	409	.50	44
MAR	1415	979	61	14	42	105	188	308	.52	32
APR	1780	895	69	2	7	22	29	133	.55	20
MAY	1968	846	75	0	0	0	0	48	.56	11
JUN	2004	846	80	0	0	0	0	15	.56	7
JUL	1813	801	82	0	0	0	0	11	.51	9
AUG	1717	817	82	0	0	0	0	11	.52	17
SEP	1514	933	78	0	0	0	0	24	.52	28
OCT	1335	1176	70	2	5	17	40	118	.56	40
NOV	973	1095	60	16	46	110	179	313	.51	49
DEC	779	936	55	51	118	208	327	476	.45	53
YR	1438	943	68	197	465	887	1465	2399	.53	

SHREVEPORT, LOUISIANA — ELEV 259 LAT 32.5

	HS	VS	TA	D50	D55	D60	D65	D70	KT	LD
JAN	762	920	47	154	264	403	552	707	.45	54
FEB	1038	1052	51	90	169	279	416	547	.49	46
MAR	1341	993	57	28	81	161	291	415	.50	35
APR	1613	870	66	3	9	30	65	163	.51	23
MAY	1886	848	73	0	2	6	5	59	.54	13
JUN	2065	869	80	0	0	0	0	12	.57	9
JUL	2014	864	83	0	0	0	0	6	.57	11
AUG	1877	913	83	0	0	0	0	6	.57	19
SEP	1554	1030	77	0	0	0	0	24	.55	31
OCT	1303	1248	68	2	7	24	70	148	.57	42
NOV	929	1132	56	31	87	167	278	419	.52	52
DEC	731	946	49	120	215	346	490	646	.46	56
YR	1428	973	66	428	832	1415	2167	3152	.53	

CARIBOU, MAINE — ELEV 623 LAT 46.9

	HS	VS	TA	D50	D55	D60	D65	D70	KT	LD
JAN	419	730	11	1218	1373	1528	1683	1838	.45	68
FEB	724	1120	13	1039	1179	1319	1459	1599	.52	61
MAR	1133	1306	24	818	973	1128	1283	1438	.54	49
APR	1414	1111	37	400	549	699	849	999	.50	37
MAY	1578	963	50	84	183	323	474	629	.46	28
JUN	1757	970	60	4	23	82	170	315	.48	24
JUL	1762	1016	65	1	4	21	84	178	.50	26
AUG	1501	1063	62	2	10	46	122	247	.49	33
SEP	1103	1072	54	26	90	192	327	477	.47	45
OCT	688	885	44	206	350	503	657	812	.43	57
NOV	366	544	31	558	708	858	1008	1158	.35	66
DEC	310	524	16	1051	1206	1361	1516	1671	.38	70
YR	1065	941	39	5408	6648	8061	9632	11363	.48	

PORTLAND, MAINE — ELEV 62 LAT 43.6

	HS	VS	TA	D50	D55	D60	D65	D70	KT	LD
JAN	450	689	22	884	1039	1194	1349	1504	.41	65
FEB	682	891	23	759	899	1039	1179	1319	.44	57
MAR	970	941	32	564	719	874	1029	1184	.43	46
APR	1304	920	43	225	370	519	669	819	.44	34
MAY	1567	888	53	32	108	232	381	536	.46	25
JUN	1712	888	62	1	6	37	106	239	.47	20
JUL	1659	894	68	0	1	5	27	102	.47	22
AUG	1461	943	66	0	1	8	55	135	.47	30
SEP	1158	1025	59	3	22	87	200	340	.47	42
OCT	822	1003	49	83	193	339	493	648	.47	54
NOV	459	651	39	343	492	642	792	942	.38	63
DEC	363	559	26	753	908	1063	1218	1373	.37	67
YR	1052	857	45	3648	4758	6039	7498	9142	.45	

BALTIMORE, MARYLAND — ELEV 154 LAT 39.2

	HS	VS	TA	D50	D55	D60	D65	D70	KT	LD
JAN	587	834	33	515	670	825	980	1135	.44	60
FEB	840	1008	35	426	566	706	846	986	.47	53
MAR	1162	1030	43	236	381	534	688	843	.48	42
APR	1488	945	54	33	101	203	340	487	.49	29
MAY	1714	880	64	2	8	36	110	212	.49	20
JUN	1879	887	72	0	0	0	0	49	.52	16
JUL	1823	890	77	0	0	0	0	15	.51	18
AUG	1599	926	75	0	0	0	0	25	.50	26
SEP	1330	1059	69	0	2	8	27	105	.51	37
OCT	998	1114	57	12	50	128	250	393	.50	49
NOV	660	912	46	150	274	418	567	717	.46	59
DEC	499	725	35	457	611	766	921	1076	.41	63
YR	1217	933	55	1831	2662	3623	4729	6042	.49	

BOSTON, MASSACHUSETTS — ELEV 16 LAT 42.4

	HS	VS	TA	D50	D55	D60	D65	D70	KT	LD
JAN	475	706	29	645	800	955	1110	1265	.40	64
FEB	710	900	30	549	689	829	969	1109	.44	56
MAR	1016	961	38	371	524	679	834	989	.44	45
APR	1326	908	49	96	203	344	492	642	.45	33
MAY	1620	893	59	5	29	99	218	355	.47	23
JUN	1817	912	68	0	1	7	27	106	.50	19
JUL	1749	914	73	0	0	0	0	32	.49	21
AUG	1486	932	71	0	0	2	8	57	.47	29
SEP	1260	1098	65	1	4	21	76	180	.50	41
OCT	890	1072	55	16	70	164	301	453	.49	52
NOV	503	705	45	163	297	445	594	744	.39	62
DEC	403	615	33	527	682	837	992	1147	.38	66
YR	1106	884	51	2374	3300	4381	5621	7080	.47	

ALPENA, MICHIGAN — ELEV 689 LAT 45.1

	HS	VS	TA	D50	D55	D60	D65	D70	KT	LD
JAN	362	538	18	998	1153	1308	1463	1618	.35	66
FEB	617	820	18	888	1028	1168	1308	1448	.41	59
MAR	1028	1069	26	738	893	1048	1203	1358	.47	48
APR	1407	1047	40	303	448	597	747	897	.48	35
MAY	1720	1004	51	79	169	301	455	605	.50	26
JUN	1879	989	61	4	19	71	150	280	.52	22
JUL	1885	1035	66	1	5	22	75	169	.53	24
AUG	1583	1071	64	2	7	32	110	200	.51	32
SEP	1156	1074	56	17	65	147	265	413	.48	43
OCT	743	921	47	132	250	396	549	704	.44	55
NOV	382	534	35	454	603	753	903	1053	.34	64
DEC	270	392	23	825	980	1135	1290	1445	.30	68
YR	1089	875	42	4440	5621	6978	8518	10190	.48	

TABLE III-1 (cont.)

DETROIT, MICHIGAN ELEV 627 LAT 42.4

	HS	VS	TA	D50	D55	D60	D65	D70	KT	LD
JAN	417	585	26	760	915	1070	1225	1380	.36	64
FEB	680	847	27	647	787	927	1067	1207	.42	56
MAR	1000	941	35	454	608	763	918	1073	.44	45
APR	1399	963	48	116	223	360	507	657	.47	33
MAY	1716	941	58	10	42	115	238	364	.50	23
JUN	1866	932	69	0	2	8	26	100	.51	19
JUL	1835	951	73	0	0	0	0	43	.52	21
AUG	1575	989	72	0	1	4	11	61	.50	29
SEP	1253	1090	65	2	7	31	80	188	.50	41
OCT	876	1048	54	33	100	199	342	488	.48	52
NOV	478	655	41	276	419	567	717	867	.37	62
DEC	343	487	30	633	787	942	1097	1252	.33	66
YR	1122	869	50	2931	3890	4986	6228	7679	.47	

FLINT, MICHIGAN ELEV 764 LAT 43.0

	HS	VS	TA	D50	D55	D60	D65	D70	KT	LD
JAN	383	532	22	859	1014	1169	1324	1479	.34	64
FEB	636	788	24	734	874	1014	1154	1294	.40	57
MAR	957	906	33	540	694	849	1004	1159	.42	45
APR	1339	932	46	153	280	424	573	723	.45	33
MAY	1658	924	56	19	72	161	306	442	.48	24
JUN	1813	921	66	1	4	18	65	155	.50	20
JUL	1797	946	70	0	1	6	14	89	.51	22
AUG	1555	992	68	0	2	9	36	113	.50	30
SEP	1195	1047	61	4	17	67	147	277	.48	41
OCT	829	992	51	66	152	280	433	583	.46	53
NOV	429	576	38	354	502	651	801	951	.34	62
DEC	309	432	27	719	874	1029	1184	1339	.30	66
YR	1077	832	47	3449	4485	5677	7041	8604	.46	

GRAND RAPIDS, MICHIGAN ELEV 804 LAT 42.9

	HS	VS	TA	D50	D55	D60	D65	D70	KT	LD
JAN	370	504	23	831	986	1141	1296	1451	.32	64
FEB	648	806	25	714	854	994	1134	1274	.40	57
MAR	1014	974	33	525	679	834	989	1144	.45	45
APR	1412	987	47	145	265	407	555	705	.48	33
MAY	1755	971	57	15	58	138	270	403	.51	24
JUN	1956	976	67	1	3	13	44	128	.54	20
JUL	1914	996	72	0	1	4	8	66	.54	22
AUG	1676	1070	70	0	1	6	27	88	.54	29
SEP	1262	1118	62	3	13	52	114	241	.51	41
OCT	858	1037	52	59	140	260	409	559	.48	53
NOV	446	606	39	343	490	639	789	939	.36	62
DEC	311	434	27	701	856	1011	1166	1321	.30	66
YR	1138	873	48	3337	4345	5499	6801	8317	.48	

SAULT STE. MARIE, MICHIGAN ELEV 725 LAT 46.5

	HS	VS	TA	D50	D55	D60	D65	D70	KT	LD
JAN	325	492	14	1110	1265	1420	1575	1730	.34	68
FEB	603	840	15	974	1114	1254	1394	1534	.43	60
MAR	1029	1125	24	806	961	1116	1271	1426	.49	49
APR	1383	1069	38	357	504	654	804	954	.48	37
MAY	1688	1021	49	98	203	344	496	651	.50	27
JUN	1811	988	59	7	33	100	200	342	.50	23
JUL	1835	1045	64	1	7	33	96	208	.52	25
AUG	1523	1068	63	2	8	39	125	224	.50	33
SEP	1049	990	55	20	75	165	291	442	.45	45
OCT	673	844	46	150	280	429	583	738	.41	56
NOV	332	465	33	516	666	816	966	1116	.31	66
DEC	253	382	20	927	1082	1237	1392	1547	.30	70
YR	1044	861	40	4969	6198	7607	9193	10912	.47	

TRAVERSE CITY, MICHIGAN ELEV 630 LAT 44.7

	HS	VS	TA	D50	D55	D60	D65	D70	KT	LD
JAN	311	426	21	905	1060	1215	1370	1525	.30	66
FEB	567	716	21	820	960	1100	1240	1380	.38	58
MAR	1001	1018	29	660	815	970	1125	1280	.46	47
APR	1405	1033	43	231	371	519	669	819	.48	35
MAY	1729	999	53	45	121	236	387	534	.50	26
JUN	1912	995	64	2	8	34	104	205	.52	21
JUL	1910	1037	69	0	2	8	33	105	.54	23
AUG	1609	1078	68	0	2	11	66	126	.52	31
SEP	1165	1071	59	6	28	91	178	322	.48	43
OCT	754	926	50	87	184	321	471	626	.44	55
NOV	377	515	37	395	543	693	843	993	.33	64
DEC	257	360	26	747	902	1057	1212	1367	.28	68
YR	1086	848	45	3899	4997	6257	7698	9284	.47	

DULUTH, MINNESOTA ELEV 1417 LAT 46.8

	HS	VS	TA	D50	D55	D60	D65	D70	KT	LD
JAN	389	650	9	1287	1442	1597	1751	1907	.41	68
FEB	673	1000	12	1061	1201	1341	1481	1621	.48	60
MAR	1034	1145	24	822	977	1132	1287	1442	.49	49
APR	1373	1069	39	345	492	642	792	942	.48	37
MAY	1643	1001	49	91	192	332	484	639	.48	28
JUN	1767	973	59	6	29	94	194	333	.49	23
JUL	1854	1063	66	1	4	18	67	163	.53	25
AUG	1547	1097	64	1	6	29	104	200	.51	33
SEP	1095	1058	54	26	88	186	318	469	.47	45
OCT	725	950	45	170	306	457	611	766	.45	57
NOV	381	574	28	648	798	948	1098	1248	.37	66
DEC	292	477	14	1104	1259	1414	1569	1724	.36	70
YR	1067	921	39	5560	6792	8189	9756	11452	.48	

INTERNATIONAL FALLS, MINNESOTA ELEV 1184 LAT 48.6

	HS	VS	TA	D50	D55	D60	D65	D70	KT	LD
JAN	356	626	2	1491	1646	1801	1956	2111	.42	70
FEB	663	1061	7	1204	1344	1484	1624	1764	.51	62
MAR	1046	1244	21	911	1066	1221	1376	1531	.52	51
APR	1444	1202	38	357	505	654	804	954	.52	39
MAY	1716	1096	50	81	176	312	462	617	.51	30
JUN	1853	1059	60	4	20	74	168	293	.51	25
JUL	1921	1154	66	1	4	18	66	160	.55	27
AUG	1618	1220	63	2	9	39	112	224	.54	35
SEP	1121	1161	53	40	112	222	364	510	.49	47
OCT	704	979	44	216	360	512	667	822	.46	59
NOV	345	541	25	753	903	1053	1203	1353	.36	68
DEC	272	475	9	1280	1435	1590	1745	1900	.38	72
YR	1091	984	37	6341	7580	8983	10547	12241	.50	

MINNEAPOLIS, MINNESOTA ELEV 837 LAT 44.9

	HS	VS	TA	D50	D55	D60	D65	D70	KT	LD
JAN	464	768	12	1172	1327	1482	1637	1792	.45	66
FEB	764	1110	17	938	1078	1218	1358	1498	.51	59
MAR	1103	1169	28	673	828	983	1138	1293	.51	47
APR	1442	1071	45	178	305	449	597	747	.50	35
MAY	1737	1009	57	18	63	143	271	403	.51	26
JUN	1927	1006	67	1	4	18	65	142	.53	22
JUL	1970	1071	72	0	1	5	11	66	.56	24
AUG	1687	1142	70	0	2	8	21	90	.55	31
SEP	1255	1188	60	8	30	90	173	308	.52	43
OCT	860	1126	50	93	186	318	472	620	.50	55
NOV	480	736	32	529	678	828	978	1128	.42	64
DEC	353	572	19	973	1128	1283	1438	1593	.39	68
YR	1172	996	44	4584	5631	6824	8159	9680	.51	

TABLE III-1 (cont.)

ROCHESTER, MINNESOTA — ELEV 1319 LAT 43.9

	HS	VS	TA	D50	D55	D60	D65	D70	KT	LD
JAN	477	762	13	1150	1305	1460	1615	1770	.44	65
FEB	753	1041	17	927	1067	1207	1347	1487	.49	58
MAR	1082	1099	28	688	843	998	1153	1308	.49	46
APR	1410	1013	45	188	320	466	615	765	.48	34
MAY	1696	963	56	20	71	156	292	430	.49	25
JUN	1902	974	66	1	5	20	78	155	.52	21
JUL	1909	1017	70	0	1	6	21	87	.54	23
AUG	1662	1091	69	0	2	10	35	112	.53	30
SEP	1250	1142	59	8	33	97	185	326	.51	42
OCT	870	1100	50	95	192	328	485	633	.50	54
NOV	494	734	33	523	672	822	972	1122	.41	63
DEC	370	583	19	964	1119	1274	1429	1584	.38	67
YR	1158	959	44	4565	5631	6845	8227	9779	.50	

JACKSON, MISSISSIPPI — ELEV 331 LAT 32.3

	HS	VS	TA	D50	D55	D60	D65	D70	KT	LD
JAN	753	898	47	158	268	407	569	710	.44	54
FEB	1026	1029	50	102	184	298	442	567	.48	46
MAR	1369	1011	56	35	94	177	313	436	.51	35
APR	1708	910	66	3	12	37	74	179	.54	23
MAY	1941	861	73	1	2	7	6	71	.56	13
JUN	2024	859	79	0	0	0	0	16	.56	9
JUL	1909	837	82	0	0	0	0	9	.54	11
AUG	1780	873	81	0	0	0	0	11	.54	19
SEP	1509	992	76	0	0	0	0	35	.53	30
OCT	1271	1199	66	3	12	38	91	183	.56	42
NOV	902	1077	55	40	102	187	301	445	.50	52
DEC	709	898	49	128	224	355	504	655	.45	56
YR	1410	953	65	471	898	1506	2300	3318	.53	

MERIDIAN, MISSISSIPPI — ELEV 308 LAT 32.3

	HS	VS	TA	D50	D55	D60	D65	D70	KT	LD
JAN	744	883	47	163	274	413	575	717	.43	54
FEB	1012	1010	50	103	185	298	443	567	.48	46
MAR	1328	976	56	36	95	178	312	437	.50	35
APR	1662	889	65	4	13	41	79	186	.52	23
MAY	1860	839	72	1	2	8	7	76	.53	13
JUN	1963	846	79	0	0	0	0	17	.54	9
JUL	1823	815	81	0	0	0	0	11	.52	11
AUG	1739	858	81	0	0	0	0	13	.53	19
SEP	1454	953	75	0	0	0	0	42	.51	30
OCT	1258	1183	65	5	16	48	111	205	.55	42
NOV	897	1068	54	52	120	211	331	478	.50	52
DEC	699	880	48	146	249	384	530	686	.44	56
YR	1371	933	65	510	955	1582	2388	3434	.51	

COLUMBIA, MISSOURI — ELEV 886 LAT 38.8

	HS	VS	TA	D50	D55	D60	D65	D70	KT	LD
JAN	611	869	29	642	797	952	1107	1262	.45	60
FEB	875	1052	34	461	600	739	879	1019	.48	52
MAR	1179	1035	42	275	417	569	730	877	.48	41
APR	1526	960	55	37	99	187	314	453	.50	29
MAY	1880	940	64	4	13	44	117	207	.54	20
JUN	2089	946	73	0	1	5	11	58	.57	15
JUL	2116	983	77	0	0	0	0	22	.60	17
AUG	1878	1067	76	0	1	2	5	30	.59	25
SEP	1450	1161	68	1	5	17	42	126	.55	37
OCT	1101	1259	58	19	60	135	247	379	.55	49
NOV	703	983	44	212	341	485	633	783	.48	58
DEC	522	760	33	535	689	843	998	1153	.42	62
YR	1330	1001	54	2186	3022	3979	5083	6370	.53	

KANSAS CITY, MISSOURI — ELEV 1033 LAT 39.3

	HS	VS	TA	D50	D55	D60	D65	D70	KT	LD
JAN	648	969	27	710	865	1020	1175	1330	.48	61
FEB	895	1108	32	497	636	776	916	1056	.50	53
MAR	1203	1080	41	303	447	599	753	908	.50	42
APR	1575	1006	54	46	114	206	336	477	.52	30
MAY	1873	948	64	4	15	49	127	216	.54	20
JUN	2080	952	73	0	1	5	15	60	.57	16
JUL	2102	989	78	0	0	0	0	22	.59	18
AUG	1862	1074	77	0	0	0	0	28	.58	26
SEP	1452	1183	68	1	5	19	50	133	.56	37
OCT	1092	1269	58	22	67	144	259	391	.55	49
NOV	737	1077	42	252	387	533	681	831	.51	59
DEC	561	865	31	581	735	890	1045	1200	.46	63
YR	1342	1042	54	2417	3273	4242	5357	6651	.54	

SAINT LOUIS, MISSOURI — ELEV 564 LAT 38.7

	HS	VS	TA	D50	D55	D60	D65	D70	KT	LD
JAN	627	898	31	581	735	890	1045	1200	.46	60
FEB	886	1066	35	421	558	697	837	977	.49	52
MAR	1205	1061	43	237	371	520	682	828	.49	41
APR	1564	982	57	29	83	162	272	410	.51	29
MAY	1871	935	66	3	11	36	103	181	.54	20
JUN	2092	945	75	0	1	4	10	42	.58	15
JUL	2049	959	79	0	0	0	0	19	.58	17
AUG	1816	1032	77	0	0	0	0	26	.57	25
SEP	1459	1166	70	1	4	14	35	110	.55	37
OCT	1100	1252	59	16	52	122	224	349	.55	49
NOV	718	1009	45	191	313	454	600	750	.49	58
DEC	531	776	35	481	633	788	942	1097	.43	62
YR	1329	1006	56	1961	2762	3686	4750	5989	.53	

SPRINGFIELD, MISSOURI — ELEV 1270 LAT 37.2

	HS	VS	TA	D50	D55	D60	D65	D70	KT	LD
JAN	684	956	33	532	686	840	995	1150	.47	58
FEB	926	1071	37	369	505	644	784	924	.49	51
MAR	1235	1042	44	217	350	498	660	806	.50	40
APR	1604	968	57	27	79	158	275	410	.52	27
MAY	1882	912	65	3	11	38	94	192	.54	18
JUN	2075	918	74	0	1	4	10	52	.57	14
JUL	2063	936	78	0	0	0	0	20	.58	16
AUG	1873	1019	77	0	0	2	6	24	.58	24
SEP	1481	1130	69	1	4	13	35	111	.55	35
OCT	1144	1249	59	15	49	120	227	350	.55	47
NOV	775	1058	46	178	298	438	585	735	.50	57
DEC	603	874	36	438	590	744	899	1054	.45	61
YR	1364	1011	56	1779	2573	3501	4570	5828	.54	

BILLINGS, MONTANA — ELEV 3570 LAT 45.8

	HS	VS	TA	D50	D55	D60	D65	D70	KT	LD
JAN	486	863	22	871	1026	1181	1336	1491	.49	67
FEB	763	1153	27	633	773	913	1053	1193	.53	59
MAR	1189	1340	33	540	695	849	1004	1159	.55	48
APR	1526	1177	45	190	319	464	612	762	.53	36
MAY	1913	1134	55	36	102	198	333	482	.56	27
JUN	2174	1139	63	4	15	55	131	238	.60	22
JUL	2384	1303	72	0	1	5	10	68	.68	24
AUG	2022	1441	70	0	2	8	15	92	.66	32
SEP	1470	1517	59	11	41	108	221	339	.62	44
OCT	987	1432	49	105	203	338	487	642	.60	56
NOV	561	972	36	431	580	729	879	1029	.51	65
DEC	421	782	27	719	874	1029	1184	1339	.48	69
YR	1328	1188	46	3541	4630	5878	7265	8835	.59	

TABLE III-1 (cont.)

CUT BANK, MONTANA — ELEV 3839 LAT 48.6

	HS	VS	TA	D50	D55	D60	D65	D70	KT	LD
JAN	402	753	16	1048	1203	1358	1513	1668	.48	70
FEB	688	1122	22	773	913	1053	1193	1333	.53	62
MAR	1128	1385	27	719	874	1029	1184	1339	.56	51
APR	1485	1245	40	318	466	615	765	915	.53	39
MAY	1883	1209	50	85	185	326	477	633	.56	30
JUN	2045	1162	57	12	55	136	267	406	.56	25
JUL	2287	1372	64	1	5	25	82	190	.65	27
AUG	1897	1475	63	2	9	42	125	239	.63	35
SEP	1352	1503	53	34	105	215	368	504	.60	47
OCT	871	1343	44	198	341	493	648	803	.57	59
NOV	480	887	30	609	759	909	1059	1209	.51	68
DEC	334	647	21	887	1042	1197	1352	1507	.46	72
YR	1241	1175	41	4686	5955	7398	9033	10745	.57	

DILLON, MONTANA — ELEV 5210 LAT 45.2

	HS	VS	TA	D50	D55	D60	D65	D70	KT	LD
JAN	526	943	20	924	1079	1234	1389	1544	.51	66
FEB	846	1310	26	686	826	966	1106	1246	.57	59
MAR	1279	1451	30	633	787	942	1097	1252	.59	48
APR	1639	1257	41	276	419	567	717	867	.57	35
MAY	1989	1158	50	84	174	305	453	608	.58	26
JUN	2143	1108	58	14	52	127	238	378	.59	22
JUL	2392	1282	66	1	4	19	54	153	.68	24
AUG	2023	1412	65	2	8	32	85	193	.66	32
SEP	1521	1554	55	30	91	184	325	460	.63	43
OCT	1023	1472	45	183	317	466	620	775	.61	55
NOV	602	1050	32	547	696	846	996	1146	.53	65
DEC	450	838	24	809	964	1119	1274	1429	.50	69
YR	1373	1236	43	4188	5418	6809	8354	10051	.60	

GLASGOW, MONTANA — ELEV 2297 LAT 48.2

	HS	VS	TA	D50	D55	D60	D65	D70	KT	LD
JAN	388	698	9	1265	1420	1575	1730	1885	.45	69
FEB	671	1060	15	974	1114	1254	1394	1534	.51	62
MAR	1105	1324	25	769	924	1079	1234	1389	.54	51
APR	1488	1232	43	234	370	517	666	816	.53	38
MAY	1828	1159	54	40	108	206	344	491	.54	29
JUN	2047	1151	62	5	19	64	151	254	.56	25
JUL	2193	1300	71	0	2	7	15	88	.63	27
AUG	1863	1423	69	1	3	11	30	112	.62	35
SEP	1340	1462	57	18	61	138	263	388	.59	46
OCT	877	1333	46	158	279	425	577	732	.57	58
NOV	479	865	29	630	780	930	1080	1230	.49	68
DEC	334	632	17	1020	1175	1330	1485	1640	.45	72
YR	1221	1137	42	5114	6256	7537	8969	10559	.56	

GREAT FALLS, MONTANA — ELEV 3661 LAT 47.5

	HS	VS	TA	D50	D55	D60	D65	D70	KT	LD
JAN	420	757	21	915	1070	1225	1380	1535	.47	69
FEB	720	1141	27	655	795	935	1075	1215	.53	61
MAR	1170	1398	31	605	760	915	1070	1225	.57	50
APR	1489	1205	43	215	352	499	648	798	.53	38
MAY	1848	1149	53	44	117	225	367	519	.54	28
JUN	2101	1157	61	5	21	74	162	284	.58	24
JUL	2329	1348	69	0	2	8	18	100	.66	26
AUG	1933	1451	67	1	3	14	42	132	.64	34
SEP	1378	1480	57	14	54	130	260	384	.60	46
OCT	925	1400	48	117	225	367	524	673	.59	57
NOV	498	885	35	463	612	762	912	1062	.49	67
DEC	336	612	27	729	884	1039	1194	1349	.43	71
YR	1266	1165	45	3761	4893	6191	7652	9274	.57	

HELENA, MONTANA — ELEV 3898 LAT 46.6

	HS	VS	TA	D50	D55	D60	D65	D70	KT	LD
JAN	419	719	18	989	1144	1299	1454	1609	.44	68
FEB	709	1071	25	689	829	969	1109	1249	.50	60
MAR	1145	1310	31	602	756	911	1066	1221	.54	49
APR	1487	1170	43	232	372	520	669	819	.52	37
MAY	1860	1128	52	56	135	254	401	552	.55	28
JUN	2040	1101	59	8	32	97	194	329	.56	23
JUL	2334	1312	68	1	3	12	33	122	.66	25
AUG	1930	1404	66	1	4	19	57	155	.63	33
SEP	1412	1480	56	23	78	165	304	436	.60	45
OCT	926	1348	45	175	307	457	611	766	.57	57
NOV	521	904	32	549	699	849	999	1149	.49	66
DEC	364	656	23	828	983	1138	1293	1448	.44	70
YR	1266	1134	43	4151	5342	6689	8190	9855	.57	

LEWISTOWN, MONTANA — ELEV 4147 LAT 47.0

	HS	VS	TA	D50	D55	D60	D65	D70	KT	LD
JAN	420	737	19	958	1113	1268	1423	1578	.45	68
FEB	692	1051	24	734	874	1014	1154	1294	.50	61
MAR	1128	1303	28	698	853	1008	1163	1318	.54	49
APR	1444	1143	40	306	449	598	747	897	.51	37
MAY	1807	1107	50	102	198	330	477	633	.53	28
JUN	2059	1121	57	21	70	150	265	405	.57	24
JUL	2288	1304	66	2	8	29	70	177	.65	26
AUG	1901	1399	64	3	10	39	94	202	.62	34
SEP	1372	1444	54	41	109	205	348	482	.59	45
OCT	905	1326	46	177	305	452	605	760	.57	57
NOV	502	872	32	535	684	834	984	1134	.49	66
DEC	363	668	25	791	946	1101	1256	1411	.45	70
YR	1243	1123	42	4367	5617	7026	8586	10289	.56	

MILES CITY, MONTANA — ELEV 2634 LAT 46.4

	HS	VS	TA	D50	D55	D60	D65	D70	KT	LD
JAN	457	811	15	1073	1228	1383	1538	1693	.48	68
FEB	745	1143	22	795	935	1075	1215	1355	.52	60
MAR	1185	1365	30	615	769	924	1079	1234	.56	49
APR	1542	1215	45	190	308	446	591	741	.54	37
MAY	1896	1143	56	37	95	177	288	432	.56	27
JUN	2146	1144	65	5	16	47	117	199	.59	23
JUL	2293	1282	74	1	2	6	9	54	.65	25
AUG	1977	1434	73	1	3	9	16	78	.65	33
SEP	1444	1515	60	16	47	111	217	318	.61	45
OCT	961	1413	49	134	230	359	508	658	.59	56
NOV	551	975	32	531	679	828	978	1128	.52	66
DEC	399	746	22	868	1023	1178	1333	1488	.48	70
YR	1303	1182	45	4265	5334	6544	7889	9378	.58	

MISSOULA, MONTANA — ELEV 3189 LAT 46.9

	HS	VS	TA	D50	D55	D60	D65	D70	KT	LD
JAN	312	473	21	905	1060	1215	1370	1525	.33	68
FEB	574	796	27	638	778	918	1058	1198	.41	61
MAR	981	1069	33	518	673	828	983	1138	.47	49
APR	1382	1081	44	194	335	483	633	783	.48	37
MAY	1782	1089	52	42	122	248	397	552	.52	28
JUN	1933	1057	59	4	24	88	201	335	.53	24
JUL	2327	1321	67	0	2	10	39	134	.66	26
AUG	1881	1377	65	0	3	17	71	172	.62	33
SEP	1358	1418	55	15	66	159	301	441	.58	45
OCT	813	1126	44	195	340	493	648	803	.51	57
NOV	410	641	32	531	681	831	981	1131	.40	66
DEC	267	421	25	784	939	1094	1249	1404	.33	70
YR	1172	990	44	3828	5023	6385	7931	9617	.53	

TABLE III-1 (cont.)

GRAND ISLAND, NEBRASKA — ELEV 1857 LAT 41.0

	HS	VS	TA	D50	D55	D60	D65	D70	KT	LD
JAN	661	1081	22	859	1014	1169	1324	1479	.53	62
FEB	917	1230	28	625	765	904	1044	1184	.54	55
MAR	1265	1223	36	453	606	760	915	1070	.54	43
APR	1692	1141	50	103	192	315	461	604	.56	31
MAY	1972	1028	61	10	35	95	184	303	.57	22
JUN	2242	1035	71	1	3	10	35	92	.62	18
JUL	2216	1068	76	0	1	2	6	30	.63	20
AUG	1939	1175	75	0	1	3	5	41	.61	28
SEP	1509	1317	64	4	14	45	107	203	.59	39
OCT	1138	1442	54	54	127	226	362	508	.60	51
NOV	738	1166	38	362	506	655	804	954	.55	60
DEC	569	960	27	713	868	1023	1178	1333	.51	64
YR	1407	1155	50	3186	4130	5208	6425	7801	.58	

NORTH OMAHA, NEBRASKA — ELEV 1325 LAT 41.4

	HS	VS	TA	D50	D55	D60	D65	D70	KT	LD
JAN	634	1034	20	924	1079	1234	1389	1544	.52	63
FEB	892	1201	26	686	826	966	1106	1246	.53	55
MAR	1222	1185	35	480	633	788	942	1097	.52	44
APR	1558	1055	50	99	187	311	456	601	.52	32
MAY	1873	993	61	9	31	89	186	296	.54	22
JUN	2122	1005	70	1	3	10	33	96	.58	18
JUL	2106	1038	75	0	1	3	7	38	.59	20
AUG	1858	1139	74	0	1	4	10	52	.59	28
SEP	1373	1184	64	3	13	43	99	200	.54	40
OCT	1050	1304	54	44	112	207	342	486	.56	51
NOV	644	969	38	369	515	663	813	963	.48	61
DEC	511	832	26	754	908	1063	1218	1373	.46	65
YR	1323	1078	49	3369	4309	5381	6601	7992	.55	

NORTH PLATTE, NEBRASKA — ELEV 2785 LAT 41.1

	HS	VS	TA	D50	D55	D60	D65	D70	KT	LD
JAN	692	1165	23	825	980	1135	1290	1445	.56	62
FEB	958	1317	28	614	753	893	1033	1173	.57	55
MAR	1333	1316	34	490	643	797	952	1107	.57	44
APR	1724	1169	48	138	241	373	522	667	.57	31
MAY	1988	1037	58	20	60	134	238	371	.58	22
JUN	2266	1045	68	2	6	20	65	135	.62	18
JUL	2277	1093	74	0	1	4	7	49	.64	20
AUG	1989	1209	73	0	2	6	8	64	.63	28
SEP	1565	1385	62	7	24	71	141	253	.61	39
OCT	1177	1521	51	92	176	295	439	590	.62	51
NOV	759	1223	36	419	565	714	864	1014	.56	60
DEC	605	1059	27	720	874	1029	1184	1339	.54	64
YR	1447	1211	49	3326	4325	5473	6743	8206	.60	

SCOTTSBLUFF, NEBRASKA — ELEV 3957 LAT 41.9

	HS	VS	TA	D50	D55	D60	D65	D70	KT	LD
JAN	676	1170	25	778	933	1088	1243	1398	.56	63
FEB	950	1346	30	575	714	854	994	1134	.58	56
MAR	1307	1320	34	489	642	797	952	1107	.57	44
APR	1668	1155	46	163	279	418	564	714	.56	32
MAY	1933	1033	57	27	81	163	280	423	.56	23
JUN	2237	1055	66	2	9	30	91	169	.61	19
JUL	2284	1119	74	0	0	0	0	52	.65	21
AUG	1999	1247	72	1	2	8	8	80	.64	28
SEP	1599	1466	61	8	28	83	160	279	.63	40
OCT	1145	1510	50	100	190	316	459	615	.62	52
NOV	723	1180	36	418	565	714	864	1014	.55	61
DEC	575	1021	28	695	850	1004	1159	1314	.53	65
YR	1427	1218	48	3256	4293	5475	6774	8299	.60	

ELKO, NEVADA — ELEV 5075 LAT 40.8

	HS	VS	TA	D50	D55	D60	D65	D70	KT	LD
JAN	689	1140	23	831	986	1141	1296	1451	.55	62
FEB	1034	1456	29	583	722	862	1002	1142	.61	54
MAR	1463	1476	35	467	620	775	930	1085	.62	43
APR	1900	1288	44	216	350	496	645	795	.63	31
MAY	2303	1161	52	67	147	265	406	562	.67	22
JUN	2534	1115	60	9	34	97	190	319	.70	17
JUL	2623	1197	70	1	2	9	27	102	.74	19
AUG	2316	1394	67	1	5	19	60	145	.73	27
SEP	1893	1754	58	16	55	130	248	376	.74	39
OCT	1322	1782	47	147	265	409	561	716	.70	51
NOV	812	1336	35	458	606	756	906	1056	.60	60
DEC	617	1075	26	747	902	1057	1212	1367	.54	64
YR	1629	1346	45	3541	4696	6018	7483	9117	.67	

ELY, NEVADA — ELEV 6253 LAT 39.3

	HS	VS	TA	D50	D55	D60	D65	D70	KT	LD
JAN	819	1380	24	818	973	1128	1283	1438	.61	61
FEB	1141	1578	28	619	759	899	1039	1179	.64	53
MAR	1606	1580	33	534	688	843	998	1153	.67	42
APR	2009	1301	41	268	412	561	711	861	.66	30
MAY	2311	1115	50	81	178	315	470	620	.67	20
JUN	2513	1068	58	9	42	116	241	371	.69	16
JUL	2447	1094	67	0	2	11	23	130	.69	18
AUG	2230	1274	66	1	4	19	62	166	.70	26
SEP	1935	1698	57	13	55	135	265	400	.74	37
OCT	1408	1832	46	154	285	435	589	744	.71	49
NOV	926	1521	34	481	630	780	930	1080	.64	59
DEC	723	1271	26	738	893	1048	1203	1358	.59	63
YR	1675	1391	44	3716	4922	6291	7814	9500	.68	

LAS VEGAS, NEVADA — ELEV 2178 LAT 36.1

	HS	VS	TA	D50	D55	D60	D65	D70	KT	LD
JAN	978	1553	44	216	346	493	645	800	.65	57
FEB	1339	1739	49	110	197	315	451	586	.69	50
MAR	1823	1646	55	45	111	202	324	475	.72	39
APR	2319	1343	64	5	17	53	126	218	.75	26
MAY	2646	1109	73	0	2	6	10	61	.76	17
JUN	2778	1033	82	0	0	0	0	7	.77	13
JUL	2588	1039	90	0	0	0	0	1	.73	15
AUG	2355	1200	87	0	0	0	0	2	.73	23
SEP	2037	1593	80	0	0	0	0	13	.75	34
OCT	1540	1817	67	2	8	27	74	156	.73	46
NOV	1085	1665	53	59	131	229	357	503	.67	55
DEC	880	1467	45	193	318	463	614	769	.63	59
YR	1866	1431	66	631	1129	1788	2601	3591	.73	

LOVELOCK, NEVADA — ELEV 3904 LAT 40.1

	HS	VS	TA	D50	D55	D60	D65	D70	KT	LD
JAN	804	1396	29	654	809	964	1119	1274	.62	61
FEB	1165	1687	35	416	555	695	834	974	.67	54
MAR	1656	1700	40	316	464	617	772	927	.70	43
APR	2165	1451	49	116	216	350	495	645	.72	30
MAY	2555	1230	58	17	59	137	255	392	.74	21
JUN	2749	1145	66	2	7	27	86	169	.75	17
JUL	2784	1215	74	0	0	0	0	1727	.79	19
AUG	2484	1454	71	0	1	6	17	76	.78	27
SEP	2027	1865	63	4	15	55	126	236	.78	38
OCT	1451	1984	51	77	162	285	428	583	.75	50
NOV	929	1589	38	353	499	648	798	948	.66	59
DEC	714	1299	31	596	750	905	1060	1215	.61	63
YR	1793	1500	50	2551	3538	4688	5990	9166	.73	

TABLE III-1 (cont.)

RENO, NEVADA — ELEV 4400 LAT 39.5

	HS	VS	TA	D50	D55	D60	D65	D70	KT	LD
JAN	800	1345	32	561	716	871	1026	1181	.60	61
FEB	1150	1611	37	363	501	641	781	921	.65	53
MAR	1649	1650	40	305	456	611	766	921	.69	42
APR	2159	1415	47	131	253	397	546	696	.71	30
MAY	2523	1196	55	24	85	186	328	478	.73	20
JUN	2701	1115	62	2	12	55	145	261	.74	16
JUL	2692	1167	69	0	1	5	17	90	.76	18
AUG	2406	1379	67	0	2	11	50	133	.76	26
SEP	1998	1785	60	4	19	73	168	298	.77	38
OCT	1431	1893	50	74	169	305	456	611	.73	49
NOV	912	1500	40	301	448	597	747	897	.64	59
DEC	705	1235	33	527	682	837	992	1147	.58	63
YR	1764	1439	49	2292	3345	4590	6022	7635	.71	

TONOPAH, NEVADA — ELEV 5423 LAT 38.1

	HS	VS	TA	D50	D55	D60	D65	D70	KT	LD
JAN	918	1552	30	614	769	924	1079	1234	.65	59
FEB	1274	1765	35	432	571	711	851	991	.69	52
MAR	1777	1725	40	329	479	633	787	942	.72	41
APR	2251	1404	48	118	224	361	512	657	.73	28
MAY	2577	1161	57	17	63	143	269	409	.74	19
JUN	2788	1089	65	1	6	26	92	171	.77	15
JUL	2703	1121	73	0	0	0	0	48	.76	17
AUG	2438	1325	71	0	1	6	13	80	.76	25
SEP	2043	1734	64	2	10	41	108	214	.77	36
OCT	1520	1945	52	60	140	258	407	556	.75	48
NOV	1031	1694	40	312	457	606	756	906	.69	57
DEC	827	1473	32	562	716	871	1026	1181	.65	61
YR	1848	1497	51	2448	3436	4580	5900	7389	.74	

WINNEMUCCA, NEVADA — ELEV 4340 LAT 40.9

	HS	VS	TA	D50	D55	D60	D65	D70	KT	LD
JAN	690	1148	28	676	831	986	1141	1296	.55	62
FEB	1028	1449	34	446	585	725	865	1005	.60	55
MAR	1472	1494	38	388	540	695	849	1004	.63	43
APR	1967	1343	45	176	304	449	597	747	.65	31
MAY	2362	1188	54	40	110	213	359	503	.68	22
JUN	2569	1127	62	4	18	64	149	260	.71	18
JUL	2678	1217	71	0	1	5	6	76	.76	20
AUG	2348	1418	68	1	3	13	42	127	.74	27
SEP	1907	1778	59	9	35	101	199	329	.75	39
OCT	1322	1789	48	119	226	367	518	673	.70	51
NOV	809	1335	37	384	532	681	831	981	.60	60
DEC	618	1083	30	608	763	918	1073	1228	.55	64
YR	1651	1363	48	2851	3949	5216	6629	8230	.68	

CONCORD, NEW HAMPSHIRE — ELEV 344 LAT 43.2

	HS	VS	TA	D50	D55	D60	D65	D70	KT	LD
JAN	459	696	21	911	1066	1221	1376	1531	.41	64
FEB	686	884	23	767	907	1047	1187	1327	.43	57
MAR	974	933	32	549	704	859	1014	1169	.43	46
APR	1317	920	44	185	326	474	624	774	.45	33
MAY	1582	889	55	14	68	167	315	462	.46	24
JUN	1705	879	65	0	2	16	58	172	.47	20
JUL	1675	894	70	0	0	3	16	75	.47	22
AUG	1455	930	67	0	1	7	45	119	.47	30
SEP	1140	992	60	3	17	75	182	317	.46	41
OCT	817	979	49	81	188	333	487	642	.46	53
NOV	463	648	38	361	510	660	810	960	.38	63
DEC	362	546	25	781	936	1091	1246	1401	.36	67
YR	1055	849	46	3653	4726	5954	7360	8949	.45	

NEWARK, NEW JERSEY — ELEV 30 LAT 40.7

	HS	VS	TA	D50	D55	D60	D65	D70	KT	LD
JAN	552	814	31	577	732	887	1042	1197	.44	62
FEB	793	985	33	488	627	767	907	1047	.46	54
MAR	1109	1019	41	297	447	602	756	911	.47	43
APR	1449	956	52	55	135	257	399	549	.48	31
MAY	1687	894	62	2	13	54	143	260	.49	22
JUN	1795	878	71	0	0	0	0	59	.49	17
JUL	1760	889	76	0	0	0	0	15	.50	19
AUG	1565	941	75	0	0	0	0	26	.49	27
SEP	1273	1052	68	0	2	9	34	115	.50	39
OCT	951	1102	58	10	46	124	243	389	.50	51
NOV	596	838	46	145	271	415	564	714	.44	60
DEC	454	676	35	481	636	791	946	1101	.40	64
YR	1167	920	54	2056	2908	3905	5034	6382	.48	

ALBUQUERQUE, NEW MEXICO — ELEV 5312 LAT 35.0

	HS	VS	TA	D50	D55	D60	D65	D70	KT	LD
JAN	1016	1562	35	459	614	769	924	1079	.65	56
FEB	1342	1664	40	281	420	560	700	840	.67	49
MAR	1768	1515	46	145	287	440	595	750	.69	37
APR	2228	1244	56	7	49	140	282	426	.71	25
MAY	2538	1052	65	0	1	9	58	158	.73	16
JUN	2679	997	75	0	0	0	0	12	.74	12
JUL	2489	995	79	0	0	0	0	2	.70	14
AUG	2290	1132	77	0	0	0	0	5	.70	22
SEP	1972	1468	70	0	0	1	7	59	.72	33
OCT	1547	1745	58	3	22	93	218	366	.71	45
NOV	1134	1692	45	173	316	465	615	765	.68	54
DEC	928	1508	36	428	583	738	893	1048	.64	58
YR	1830	1379	57	1497	2292	3216	4292	5511	.70	

CLAYTON, NEW MEXICO — ELEV 4970 LAT 36.4

	HS	VS	TA	D50	D55	D60	D65	D70	KT	LD
JAN	962	1536	33	524	679	834	989	1144	.64	58
FEB	1241	1576	36	390	529	669	809	949	.64	50
MAR	1652	1466	40	303	454	608	763	918	.65	39
APR	2039	1199	51	67	154	282	431	576	.66	27
MAY	2222	1007	60	4	22	81	172	314	.64	17
JUN	2418	984	69	0	1	6	38	91	.67	13
JUL	2284	981	74	0	0	0	0	34	.64	15
AUG	2097	1100	72	0	0	2	5	47	.65	23
SEP	1802	1390	65	1	4	22	73	172	.67	35
OCT	1433	1660	55	24	84	182	324	472	.68	46
NOV	1028	1556	42	240	383	531	681	831	.65	56
DEC	861	1439	35	463	617	772	927	1082	.63	60
YR	1672	1323	53	2017	2928	3988	5212	6630	.65	

FARMINGTON, NEW MEXICO — ELEV 5502 LAT 36.7

	HS	VS	TA	D50	D55	D60	D65	D70	KT	LD
JAN	944	1513	29	663	818	973	1128	1283	.64	58
FEB	1281	1673	35	420	560	700	840	980	.67	50
MAR	1693	1531	41	294	447	601	756	911	.67	39
APR	2133	1265	50	73	172	311	465	609	.69	27
MAY	2452	1080	60	3	18	79	184	327	.70	18
JUN	2665	1032	68	0	1	5	36	102	.73	13
JUL	2478	1032	75	0	0	0	0	15	.70	15
AUG	2252	1179	73	0	0	1	6	35	.70	23
SEP	1934	1531	65	0	3	17	67	175	.72	35
OCT	1479	1759	53	32	106	227	375	530	.71	47
NOV	1047	1622	39	326	474	624	774	924	.66	56
DEC	837	1398	30	617	772	927	1082	1237	.62	60
YR	1768	1382	51	2428	3371	4465	5713	7130	.69	

TABLE III-1 (cont.)

LOS ALAMOS, NEW MEXICO ELEV 7380 LAT 35.9

	HS	VS	TA	D50	D55	D60	D65	D70	KT	LD
JAN	952	1476	29	651	806	961	1117	1271	.63	57
FEB	1279	1612	32	501	641	781	929	1061	.66	50
MAR	1568	1346	38	379	533	688	856	998	.62	38
APR	1929	1119	46	142	278	426	560	726	.62	26
MAY	2071	951	55	11	62	162	304	459	.59	17
JUN	2132	915	65	0	1	11	87	159	.59	13
JUL	1889	866	68	0	0	3	18	95	.53	15
AUG	1759	934	66	0	1	9	44	145	.54	22
SEP	1656	1235	60	1	10	57	147	290	.61	34
OCT	1267	1368	50	61	160	303	448	611	.60	46
NOV	1037	1540	38	355	504	654	796	954	.64	55
DEC	880	1452	30	608	763	918	1053	1228	.63	59
YR	1535	1232	48	2710	3760	4974	6359	7997	.60	

ROSWELL, NEW MEXICO ELEV 3619 LAT 33.4

	HS	VS	TA	D50	D55	D60	D65	D70	KT	LD
JAN	1046	1516	38	371	524	679	834	989	.63	55
FEB	1373	1607	43	209	341	479	619	759	.66	47
MAR	1807	1464	49	92	195	335	487	642	.69	36
APR	2218	1175	60	5	24	83	185	313	.70	24
MAY	2459	997	69	0	1	7	20	106	.71	14
JUN	2610	962	77	0	0	0	0	11	.72	10
JUL	2441	957	79	0	0	0	0	6	.69	12
AUG	2242	1063	78	0	0	0	0	9	.69	20
SEP	1913	1335	70	0	1	4	17	73	.68	32
OCT	1527	1603	60	5	25	87	195	326	.68	43
NOV	1131	1567	47	131	251	395	543	693	.65	53
DEC	952	1453	39	335	487	642	797	952	.62	57
YR	1812	1306	59	1149	1849	2711	3697	4878	.68	

TRUTH OR CONSEQ., NEW MEXICO ELEV 4859 LAT 33.2

	HS	VS	TA	D50	D55	D60	D65	D70	KT	LD
JAN	1118	1661	40	315	466	620	775	930	.67	54
FEB	1451	1724	45	163	287	424	563	703	.70	47
MAR	1886	1532	50	79	173	309	459	614	.71	36
APR	2338	1220	60	5	26	88	188	319	.74	23
MAY	2557	1010	68	6	27	91	19	330	.73	14
JUN	2650	963	77	0	0	0	0	13	.73	10
JUL	2365	942	79	0	0	0	0	6	.67	12
AUG	2216	1048	77	0	0	0	0	11	.68	20
SEP	1940	1346	72	0	1	3	5	57	.69	31
OCT	1579	1665	61	3	16	63	144	277	.70	43
NOV	1217	1733	49	100	204	342	489	639	.69	53
DEC	1003	1558	41	291	441	595	750	905	.65	57
YR	1861	1364	60	963	1641	2535	3392	4804	.70	

TUCUMCARI, NEW MEXICO ELEV 4039 LAT 35.2

	HS	VS	TA	D50	D55	D60	D65	D70	KT	LD
JAN	1009	1560	37	405	558	713	868	1023	.65	56
FEB	1297	1597	41	256	390	529	669	809	.65	49
MAR	1712	1466	47	141	266	414	567	722	.67	38
APR	2098	1185	57	13	54	133	260	395	.67	25
MAY	2314	1005	66	1	4	20	57	165	.66	16
JUN	2484	974	75	0	0	1	7	23	.69	12
JUL	2349	972	78	0	0	0	0	8	.66	14
AUG	2164	1090	77	0	0	0	0	14	.67	22
SEP	1829	1354	70	0	1	6	20	87	.67	33
OCT	1443	1595	59	8	35	105	217	354	.67	45
NOV	1073	1571	46	147	271	415	564	714	.65	55
DEC	910	1478	39	357	509	664	818	973	.63	59
YR	1725	1319	58	1328	2090	3000	4047	5288	.66	

ZUNI, NEW MEXICO ELEV 6447 LAT 35.1

	HS	VS	TA	D50	D55	D60	D65	D70	KT	LD
JAN	986	1500	30	611	766	921	1076	1231	.63	56
FEB	1297	1590	35	431	571	711	851	991	.65	49
MAR	1688	1434	40	324	478	632	787	942	.66	38
APR	2167	1217	48	96	213	358	507	657	.69	25
MAY	2473	1041	57	7	45	130	264	416	.71	16
JUN	2602	989	65	0	2	11	68	153	.72	12
JUL	2264	952	71	0	0	0	0	47	.64	14
AUG	2078	1052	69	0	0	2	13	77	.64	22
SEP	1895	1406	63	0	4	25	91	208	.69	33
OCT	1496	1671	53	33	111	238	388	543	.69	45
NOV	1088	1596	40	299	447	597	747	897	.66	54
DEC	893	1429	32	558	713	868	1023	1178	.62	58
YR	1745	1321	50	2361	3349	4493	5815	7341	.67	

ALBANY, NEW YORK ELEV 292 LAT 42.7

	HS	VS	TA	D50	D55	D60	D65	D70	KT	LD
JAN	456	674	22	884	1039	1194	1349	1504	.39	64
FEB	688	871	24	742	882	1022	1162	1302	.43	56
MAR	986	933	33	515	670	825	980	1135	.43	45
APR	1335	922	47	136	253	395	543	693	.45	33
MAY	1570	873	58	12	49	125	253	384	.46	24
JUN	1730	882	68	1	3	12	39	125	.47	19
JUL	1725	908	72	0	1	3	9	58	.49	21
AUG	1499	947	70	0	1	7	22	93	.48	29
SEP	1170	1009	62	3	15	58	135	253	.47	41
OCT	817	961	51	66	150	276	422	577	.45	53
NOV	457	623	40	317	463	612	762	912	.36	62
DEC	356	521	26	747	902	1057	1212	1367	.34	66
YR	1068	843	48	3424	4428	5586	6888	8403	.45	

BINGHAMTON, NEW YORK ELEV 1637 LAT 42.2

	HS	VS	TA	D50	D55	D60	D65	D70	KT	LD
JAN	386	520	22	868	1023	1178	1333	1488	.33	63
FEB	576	667	23	762	902	1042	1182	1322	.35	56
MAR	861	772	31	580	735	890	1045	1200	.37	45
APR	1242	840	45	176	312	459	609	759	.42	32
MAY	1496	828	55	18	75	172	320	462	.43	23
JUN	1681	855	65	1	4	20	75	173	.46	19
JUL	1659	871	69	0	1	5	21	91	.47	21
AUG	1425	888	67	0	1	9	40	123	.45	29
SEP	1131	952	60	3	17	71	172	298	.45	40
OCT	779	883	50	72	167	304	456	611	.42	52
NOV	414	531	38	356	504	654	804	954	.32	62
DEC	297	395	25	763	918	1073	1228	1383	.28	66
YR	998	750	46	3598	4657	5875	7285	8863	.42	

BUFFALO, NEW YORK ELEV 705 LAT 42.9

	HS	VS	TA	D50	D55	D60	D65	D70	KT	LD
JAN	349	465	24	815	970	1125	1280	1435	.30	64
FEB	546	636	24	717	857	997	1137	1277	.34	57
MAR	888	820	32	555	710	865	1020	1175	.39	45
APR	1315	911	45	170	306	453	603	753	.44	33
MAY	1596	890	55	17	72	170	321	462	.46	24
JUN	1804	915	66	0	2	13	58	151	.49	20
JUL	1776	935	70	0	0	3	12	72	.50	22
AUG	1513	961	68	0	1	5	33	101	.48	29
SEP	1152	996	62	2	10	49	138	257	.46	41
OCT	784	914	52	52	137	269	419	574	.44	53
NOV	403	526	40	309	456	606	756	906	.32	62
DEC	283	380	28	685	840	995	1150	1305	.28	66
YR	1037	780	47	3322	4363	5551	6927	8468	.44	

TABLE III-1 (cont.)

MASSENA, NEW YORK — ELEV 207 LAT 44.9

	HS	VS	TA	D50	D55	D60	D65	D70	KT	LD
JAN	391	596	15	1101	1256	1411	1566	1721	.38	66
FEB	620	819	17	932	1072	1212	1352	1492	.41	59
MAR	977	992	28	694	849	1004	1159	1314	.45	47
APR	1343	986	42	242	385	534	684	834	.46	35
MAY	1613	939	54	27	93	198	350	493	.47	26
JUN	1779	940	64	1	5	24	78	187	.49	22
JUL	1751	964	64	1	5	25	22	193	.50	24
AUG	1484	993	67	0	2	12	57	136	.48	31
SEP	1124	1029	59	5	25	88	192	327	.47	43
OCT	736	901	49	103	214	359	512	667	.43	55
NOV	388	541	36	424	573	723	873	1023	.34	64
DEC	294	439	20	927	1082	1237	1392	1547	.32	68
YR	1044	845	43	4456	5561	6827	8237	9934	.46	

NEW YORK, NEW YORK — ELEV 187 LAT 40.8

	HS	VS	TA	D50	D55	D60	D65	D70	KT	LD
JAN	500	708	32	552	707	862	1017	1172	.40	62
FEB	721	865	33	465	605	745	885	1025	.42	54
MAR	1037	937	41	282	432	586	741	896	.44	43
APR	1364	898	52	50	127	246	387	537	.45	31
MAY	1636	873	62	2	11	49	137	248	.47	22
JUN	1710	848	72	0	0	0	0	56	.47	17
JUL	1688	861	77	0	0	0	0	14	.48	19
AUG	1483	894	75	0	0	0	0	23	.47	27
SEP	1214	996	68	0	1	7	29	104	.47	39
OCT	895	1017	59	7	33	103	209	353	.47	51
NOV	533	716	47	122	238	380	528	678	.39	60
DEC	404	574	36	451	605	760	915	1070	.36	64
YR	1101	849	55	1931	2759	3737	4848	6177	.45	

ROCHESTER, NEW YORK — ELEV 554 LAT 43.1

	HS	VS	TA	D50	D55	D60	D65	D70	KT	LD
JAN	364	497	24	806	961	1116	1271	1426	.32	64
FEB	559	661	25	706	846	986	1126	1266	.35	57
MAR	903	843	33	528	682	837	992	1147	.40	46
APR	1339	935	46	152	275	419	567	717	.45	33
MAY	1606	899	57	18	66	149	285	421	.47	24
JUN	1817	924	67	1	3	15	46	136	.50	20
JUL	1781	941	71	0	1	4	9	70	.50	22
AUG	1519	970	69	0	2	8	26	99	.49	30
SEP	1160	1011	62	3	13	53	126	243	.47	41
OCT	782	917	52	55	133	251	398	549	.44	53
NOV	404	532	41	292	436	585	735	885	.33	62
DEC	281	380	28	673	828	983	1138	1293	.28	66
YR	1045	793	48	3233	4247	5405	6719	8252	.44	

SYRACUSE, NEW YORK — ELEV 407 LAT 43.1

	HS	VS	TA	D50	D55	D60	D65	D70	KT	LD
JAN	385	538	24	818	973	1128	1283	1438	.34	64
FEB	571	681	25	711	851	991	1131	1271	.36	57
MAR	890	827	33	521	676	831	986	1141	.39	46
APR	1324	923	47	141	263	407	555	705	.45	33
MAY	1578	885	57	14	58	140	272	411	.46	24
JUN	1778	908	67	1	3	13	46	133	.49	20
JUL	1758	931	72	0	1	3	11	62	.50	22
AUG	1504	960	70	0	1	6	18	89	.48	30
SEP	1165	1017	63	2	10	44	120	228	.47	41
OCT	777	909	53	49	126	244	392	543	.43	53
NOV	399	523	41	277	421	570	720	870	.32	62
DEC	285	387	28	679	834	989	1144	1299	.28	66
YR	1037	791	48	3215	4218	5366	6678	8192	.44	

ASHEVILLE, NORTH CAROLINA — ELEV 2169 LAT 35.4

	HS	VS	TA	D50	D55	D60	D65	D70	KT	LD
JAN	722	958	38	377	531	685	840	995	.47	57
FEB	971	1069	39	300	437	577	717	857	.49	49
MAR	1306	1053	46	157	288	438	592	747	.51	38
APR	1668	960	56	17	66	152	279	424	.53	26
MAY	1804	858	64	1	7	33	100	210	.52	16
JUN	1854	839	71	0	1	4	14	70	.51	12
JUL	1776	827	74	0	0	0	0	35	.50	14
AUG	1627	864	73	0	0	0	0	43	.50	22
SEP	1361	969	67	0	3	13	50	135	.50	34
OCT	1147	1171	57	13	56	138	269	411	.53	45
NOV	849	1120	46	143	268	412	561	711	.52	55
DEC	658	918	39	353	506	660	815	970	.46	59
YR	1313	966	56	1362	2162	3112	4237	5609	.51	

CAPE HATTERAS, NORTH CAROLINA — ELEV 7 LAT 35.3

	HS	VS	TA	D50	D55	D60	D65	D70	KT	LD
JAN	686	886	45	167	305	456	611	766	.44	57
FEB	952	1036	46	140	262	398	538	678	.48	49
MAR	1326	1070	51	68	160	296	458	602	.52	38
APR	1774	1015	59	5	27	92	188	335	.57	26
MAY	1962	907	67	0	2	10	47	130	.56	16
JUN	2036	886	74	0	0	0	0	24	.56	12
JUL	1921	869	78	0	0	0	0	7	.54	14
AUG	1705	897	78	0	0	0	0	9	.53	22
SEP	1470	1055	74	0	0	0	0	29	.54	33
OCT	1137	1153	65	1	3	18	76	170	.53	45
NOV	873	1162	56	14	60	146	277	421	.53	55
DEC	659	916	48	116	235	383	536	691	.46	59
YR	1377	987	62	512	1055	1800	2731	3860	.53	

CHARLOTTE, NORTH CAROLINA — ELEV 768 LAT 35.2

	HS	VS	TA	D50	D55	D60	D65	D70	KT	LD
JAN	719	944	42	255	402	555	710	865	.46	56
FEB	971	1061	44	184	311	449	588	728	.49	49
MAR	1317	1057	51	74	164	297	461	602	.51	38
APR	1695	970	61	4	18	69	145	282	.54	25
MAY	1856	872	69	0	1	7	34	102	.53	16
JUN	1921	855	76	0	0	0	0	17	.53	12
JUL	1831	841	79	0	0	0	0	8	.52	14
AUG	1695	891	78	0	0	0	0	10	.52	22
SEP	1416	1007	72	0	0	2	10	52	.52	33
OCT	1173	1198	62	3	14	58	152	266	.54	45
NOV	865	1141	51	66	150	277	420	570	.52	55
DEC	672	938	43	243	390	543	698	853	.47	59
YR	1346	981	61	828	1451	2257	3218	4355	.52	

GREENSBORO, NORTH CAROLINA — ELEV 886 LAT 36.1

	HS	VS	TA	D50	D55	D60	D65	D70	KT	LD
JAN	715	973	39	354	506	660	815	970	.47	57
FEB	970	1096	41	270	405	543	683	823	.50	50
MAR	1313	1085	48	123	237	381	544	688	.52	39
APR	1683	986	59	9	37	106	203	346	.54	26
MAY	1868	889	67	1	3	14	59	136	.54	17
JUN	1953	872	74	0	0	0	0	30	.54	13
JUL	1864	861	77	0	0	0	0	14	.53	15
AUG	1697	910	76	0	0	0	0	19	.52	23
SEP	1418	1037	70	0	1	6	24	88	.52	34
OCT	1141	1193	59	7	33	99	209	339	.54	46
NOV	839	1134	48	111	216	354	501	651	.52	55
DEC	659	949	40	328	478	633	787	942	.47	59
YR	1345	998	58	1202	1916	2797	3825	5047	.52	

TABLE III-1 (cont.)

RALEIGH-DURHAM, NORTH CAROLINA ELEV 440 LAT 35.9

	HS	VS	TA	D50	D55	D60	D65	D70	KT	LD
JAN	694	924	41	300	451	605	760	915	.46	57
FEB	943	1046	42	227	360	499	638	778	.48	50
MAR	1276	1040	49	95	198	338	502	645	.50	38
APR	1644	959	60	5	26	87	180	319	.53	26
MAY	1808	866	67	0	2	11	48	126	.52	17
JUN	1864	846	74	0	0	0	0	26	.51	13
JUL	1776	832	78	0	0	0	0	10	.50	15
AUG	1611	865	77	0	0	0	0	14	.50	22
SEP	1377	996	71	0	1	4	12	71	.51	34
OCT	1105	1134	60	4	21	79	186	309	.52	46
NOV	812	1072	50	79	172	304	450	600	.50	55
DEC	636	893	41	280	429	583	738	893	.45	59
YR	1297	955	59	990	1659	2509	3514	4706	.50	

BISMARCK, NORTH DAKOTA ELEV 1647 LAT 46.8

	HS	VS	TA	D50	D55	D60	D65	D70	KT	LD
JAN	467	857	8	1296	1451	1606	1761	1916	.50	68
FEB	776	1237	14	1022	1162	1302	1442	1582	.55	60
MAR	1168	1358	25	772	927	1082	1237	1392	.56	49
APR	1459	1150	43	230	365	511	660	810	.51	37
MAY	1848	1127	54	40	108	204	339	486	.54	28
JUN	2060	1116	64	3	13	45	122	211	.57	23
JUL	2184	1241	71	0	2	8	18	86	.62	25
AUG	1877	1369	69	1	3	12	35	111	.62	33
SEP	1354	1408	58	18	60	135	252	380	.58	45
OCT	908	1321	47	152	270	413	564	719	.56	57
NOV	507	877	29	633	783	933	1083	1233	.49	66
DEC	373	688	16	1066	1221	1376	1531	1686	.46	70
YR	1251	1145	41	5235	6364	7627	9044	10612	.56	

FARGO, NORTH DAKOTA ELEV 899 LAT 46.9

	HS	VS	TA	D50	D55	D60	D65	D70	KT	LD
JAN	415	720	6	1367	1522	1677	1832	1987	.44	68
FEB	706	1078	11	1100	1240	1380	1520	1660	.51	61
MAR	1098	1249	24	800	955	1110	1265	1420	.52	49
APR	1476	1170	42	243	384	532	681	831	.52	37
MAY	1835	1122	55	31	95	192	334	479	.54	28
JUN	1994	1087	65	1	7	29	97	184	.55	24
JUL	2120	1210	71	0	1	5	13	78	.60	26
AUG	1825	1330	69	0	2	8	33	102	.60	33
SEP	1304	1342	58	12	47	120	234	366	.56	45
OCT	874	1253	47	140	260	406	558	713	.54	57
NOV	457	753	29	642	792	942	1092	1242	.44	66
DEC	337	594	13	1147	1302	1457	1612	1767	.42	70
YR	1206	1075	41	5485	6607	7858	9271	10829	.54	

MINOT, NORTH DAKOTA ELEV 1713 LAT 48.3

	HS	VS	TA	D50	D55	D60	D65	D70	KT	LD
JAN	384	691	8	1305	1460	1615	1770	1925	.45	70
FEB	656	1029	13	1042	1182	1322	1462	1602	.50	62
MAR	1044	1227	24	819	973	1128	1283	1438	.51	51
APR	1461	1208	41	280	420	568	717	867	.52	39
MAY	1846	1174	53	60	136	244	384	535	.55	29
JUN	1975	1116	62	6	21	68	150	257	.54	25
JUL	2098	1249	69	1	4	14	27	119	.60	27
AUG	1800	1371	67	1	6	21	70	147	.60	35
SEP	1277	1373	56	26	79	160	286	418	.56	46
OCT	850	1277	46	168	289	434	586	741	.55	58
NOV	438	759	28	663	813	963	1113	1263	.45	68
DEC	310	567	15	1094	1249	1404	1559	1714	.42	72
YR	1181	1087	40	5465	6633	7943	9407	11025	.54	

AKRON-CANTON, OHIO ELEV 1237 LAT 40.9

	HS	VS	TA	D50	D55	D60	D65	D70	KT	LD
JAN	428	570	26	735	890	1045	1200	1355	.34	62
FEB	649	750	28	625	764	904	1044	1184	.38	55
MAR	964	856	36	430	583	738	893	1048	.41	43
APR	1357	896	49	108	212	349	495	645	.45	31
MAY	1668	889	59	9	38	108	231	354	.48	22
JUN	1839	897	68	0	2	10	33	111	.50	18
JUL	1787	903	72	0	1	4	9	62	.50	20
AUG	1596	964	70	0	1	5	16	83	.50	27
SEP	1272	1058	64	2	9	37	101	207	.50	39
OCT	908	1041	53	42	115	224	369	518	.48	51
NOV	505	666	41	286	431	579	729	879	.37	60
DEC	353	476	29	639	794	949	1104	1259	.31	64
YR	1113	831	50	2876	3839	4951	6224	7705	.46	

CINCINNATI, OHIO ELEV 889 LAT 39.1

	HS	VS	TA	D50	D55	D60	D65	D70	KT	LD
JAN	500	659	31	587	741	896	1051	1206	.37	60
FEB	738	839	33	469	608	748	888	1028	.41	53
MAR	1027	878	42	272	416	568	722	877	.42	42
APR	1398	883	54	44	112	209	341	485	.46	29
MAY	1672	861	63	4	15	53	138	233	.48	20
JUN	1837	872	72	0	1	5	9	65	.50	16
JUL	1771	869	76	0	0	0	0	30	.50	18
AUG	1634	943	74	0	0	0	0	40	.51	26
SEP	1312	1038	68	1	4	17	44	130	.50	37
OCT	990	1098	57	22	72	152	271	413	.50	49
NOV	588	769	44	211	343	488	636	786	.41	58
DEC	432	587	34	507	661	815	970	1125	.35	62
YR	1160	858	54	2117	2973	3951	5070	6419	.47	

CLEVELAND, OHIO ELEV 804 LAT 41.4

	HS	VS	TA	D50	D55	D60	D65	D70	KT	LD
JAN	388	507	27	716	871	1026	1181	1336	.32	63
FEB	601	687	28	619	759	899	1039	1179	.36	55
MAR	922	822	36	433	586	741	896	1051	.40	44
APR	1349	901	48	112	217	355	501	651	.45	32
MAY	1681	904	58	11	43	116	244	366	.49	22
JUN	1843	906	68	1	3	11	40	119	.51	18
JUL	1828	929	71	0	1	4	9	68	.52	20
AUG	1583	969	70	0	1	6	17	88	.50	28
SEP	1239	1041	64	2	9	36	95	202	.49	40
OCT	867	995	54	38	108	212	354	503	.46	51
NOV	466	607	42	262	404	552	702	852	.35	61
DEC	318	420	30	611	766	921	1076	1231	.29	65
YR	1093	808	50	2804	3768	4879	6154	7646	.45	

COLUMBUS, OHIO ELEV 833 LAT 40.0

	HS	VS	TA	D50	D55	D60	D65	D70	KT	LD
JAN	459	606	28	670	825	980	1135	1290	.35	61
FEB	677	769	30	552	692	832	972	1112	.39	54
MAR	980	851	39	339	491	645	800	955	.41	42
APR	1353	873	51	67	150	272	418	564	.45	30
MAY	1647	864	61	4	19	71	176	284	.48	21
JUN	1813	875	70	0	1	5	13	78	.50	17
JUL	1755	876	74	0	0	0	0	39	.50	19
AUG	1641	969	72	0	1	3	8	59	.52	27
SEP	1282	1038	65	1	5	24	76	171	.50	38
OCT	945	1064	54	33	100	201	342	491	.49	50
NOV	538	703	42	259	401	549	699	849	.38	59
DEC	387	521	31	599	753	908	1063	1218	.33	63
YR	112[illegible]	834	52	2524	3438	4491	5702	7111	.46	

TABLE III-1 (cont.)

DAYTON, OHIO — ELEV 1004 LAT 39.9

	HS	VS	TA	D50	D55	D60	D65	D70	KT	LD
JAN	489	660	28	679	834	989	1144	1299	.37	61
FEB	725	843	30	549	689	829	969	1109	.41	54
MAR	1025	898	39	346	497	651	806	961	.43	42
APR	1403	905	51	68	149	268	413	559	.46	30
MAY	1699	886	62	4	19	67	166	272	.49	21
JUN	1874	895	71	0	1	5	13	69	.51	17
JUL	1810	896	75	0	0	0	0	32	.51	19
AUG	1645	968	73	0	1	3	7	49	.52	26
SEP	1318	1071	66	1	5	20	63	150	.51	38
OCT	969	1098	56	25	83	173	307	451	.50	50
NOV	564	749	42	257	399	547	696	846	.40	59
DEC	407	558	31	593	747	902	1057	1212	.34	63
YR	1163	869	52	2523	3422	4453	5641	7009	.47	

TOLEDO, OHIO — ELEV 692 LAT 41.6

	HS	VS	TA	D50	D55	D60	D65	D70	KT	LD
JAN	435	600	25	781	936	1091	1246	1401	.36	63
FEB	680	821	27	641	781	921	1061	1201	.41	55
MAR	997	914	36	442	596	750	905	1060	.43	44
APR	1384	932	48	113	216	352	498	648	.46	32
MAY	1717	925	59	10	40	110	229	352	.50	23
JUN	1878	922	69	0	2	9	32	105	.52	18
JUL	1849	941	72	0	1	4	5	57	.52	20
AUG	1616	994	71	0	1	6	18	79	.51	28
SEP	1276	1087	64	2	10	39	99	206	.50	40
OCT	911	1074	53	49	124	234	379	528	.49	52
NOV	498	672	40	318	463	612	762	912	.38	61
DEC	355	494	28	682	837	992	1147	1302	.32	65
YR	1135	865	49	3040	4007	5120	6381	7852	.47	

YOUNGSTOWN, OHIO — ELEV 1184 LAT 41.3

	HS	VS	TA	D50	D55	D60	D65	D70	KT	LD
JAN	385	499	26	753	908	1063	1218	1373	.31	63
FEB	586	662	27	652	792	932	1072	1212	.35	55
MAR	890	784	35	457	611	766	921	1076	.38	44
APR	1278	848	48	120	231	372	519	669	.43	32
MAY	1586	857	58	12	49	126	258	387	.46	22
JUN	1759	873	67	1	3	14	42	133	.48	18
JUL	1734	888	71	0	1	4	9	75	.49	20
AUG	1506	919	69	0	1	7	22	99	.48	28
SEP	1194	991	63	2	11	46	118	231	.47	39
OCT	851	966	53	49	126	242	384	540	.45	51
NOV	456	586	40	297	442	591	741	891	.34	61
DEC	315	413	29	657	812	967	1122	1277	.28	65
YR	1047	774	49	3001	3988	5130	6426	7963	.43	

OKLAHOMA CITY, OKLAHOMA — ELEV 1302 LAT 35.4

	HS	VS	TA	D50	D55	D60	D65	D70	KT	LD
JAN	801	1114	37	413	565	719	874	1029	.52	57
FEB	1055	1200	41	255	386	524	664	804	.53	49
MAR	1400	1147	48	126	232	371	532	676	.55	38
APR	1725	991	60	8	29	87	180	298	.55	26
MAY	1918	895	68	1	4	14	36	124	.55	16
JUN	2144	912	77	0	0	0	0	20	.59	12
JUL	2128	925	82	0	0	0	0	6	.60	14
AUG	1950	1007	81	0	0	0	0	7	.60	22
SEP	1554	1128	73	0	1	4	12	52	.57	34
OCT	1233	1292	62	5	18	62	148	253	.57	45
NOV	901	1222	49	106	201	331	474	624	.55	55
DEC	725	1059	40	319	467	621	775	930	.51	59
YR	1463	1073	60	1232	1903	2734	3695	4823	.56	

TULSA, OKLAHOMA — ELEV 676 LAT 36.2

	HS	VS	TA	D50	D55	D60	D65	D70	KT	LD
JAN	732	1011	37	419	571	726	880	1035	.49	57
FEB	978	1113	41	258	389	527	666	806	.51	50
MAR	1305	1080	48	126	230	369	528	673	.52	39
APR	1603	943	61	7	28	83	176	287	.52	26
MAY	1822	875	69	1	3	13	28	117	.52	17
JUN	2021	891	77	0	0	0	0	19	.56	13
JUL	2030	911	82	0	0	0	0	5	.57	15
AUG	1865	990	81	0	0	0	0	6	.58	23
SEP	1473	1087	73	0	1	4	10	50	.54	34
OCT	1164	1231	63	4	17	57	143	241	.55	46
NOV	827	1115	49	104	197	326	468	618	.52	56
DEC	659	953	40	325	473	627	781	936	.48	60
YR	1375	1016	60	1245	1910	2731	3680	4796	.54	

ASTORIA, OREGON — ELEV 23 LAT 46.1

	HS	VS	TA	D50	D55	D60	D65	D70	KT	LD
JAN	315	462	41	292	446	601	756	911	.32	67
FEB	545	715	44	183	320	459	599	739	.38	60
MAR	866	879	44	180	329	484	639	794	.41	49
APR	1253	940	48	92	219	366	516	666	.44	36
MAY	1608	963	52	26	106	241	394	549	.47	27
JUN	1626	890	57	4	34	120	255	405	.45	23
JUL	1746	988	60	1	8	57	163	311	.50	25
AUG	1499	1038	60	1	7	52	151	302	.49	33
SEP	1183	1145	58	2	16	81	201	348	.50	44
OCT	713	902	53	21	95	226	378	533	.43	56
NOV	387	568	47	120	256	405	555	705	.36	65
DEC	261	392	43	226	378	533	688	843	.31	69
YR	1003	824	51	1147	2215	3626	5295	7106	.44	

BURNS, OREGON — ELEV 4170 LAT 43.6

	HS	VS	TA	D50	D55	D60	D65	D70	KT	LD
JAN	490	782	25	769	924	1079	1234	1389	.44	65
FEB	792	1107	31	532	672	812	952	1092	.51	57
MAR	1187	1232	36	433	586	741	896	1051	.53	46
APR	1649	1203	44	195	329	475	624	774	.56	34
MAY	2052	1139	52	57	136	254	402	552	.60	25
JUN	2280	1116	59	8	35	101	205	335	.63	20
JUL	2460	1247	68	0	2	10	30	115	.70	22
AUG	2083	1379	66	1	5	20	68	158	.67	30
SEP	1620	1592	58	11	43	114	226	358	.66	42
OCT	1043	1415	47	134	251	396	549	704	.59	54
NOV	593	948	36	428	576	726	876	1026	.49	63
DEC	430	718	28	685	840	995	1150	1305	.44	67
YR	1393	1156	46	3253	4399	5725	7212	8858	.60	

MEDFORD, OREGON — ELEV 1299 LAT 42.4

	HS	VS	TA	D50	D55	D60	D65	D70	KT	LD
JAN	407	565	37	417	571	725	880	1035	.35	64
FEB	737	949	41	251	385	524	664	804	.45	56
MAR	1133	1109	45	184	321	472	626	781	.50	45
APR	1639	1150	50	78	169	299	444	594	.55	33
MAY	2034	1094	57	12	51	130	250	396	.59	23
JUN	2278	1082	64	1	6	29	94	190	.63	19
JUL	2475	1207	72	0	1	3	11	59	.70	21
AUG	2121	1348	70	0	1	4	21	78	.68	29
SEP	1589	1482	64	1	6	28	89	188	.63	41
OCT	982	1233	53	38	110	219	360	515	.54	52
NOV	504	707	44	210	348	496	645	795	.39	62
DEC	337	475	38	384	537	691	846	1001	.32	66
YR	1356	1033	53	1576	2505	3621	4930	6436	.57	

TABLE III-1 (cont.)

NORTH BEND, OREGON — ELEV 16 LAT 43.4

	HS	VS	TA	D50	D55	D60	D65	D70	KT	LD
JAN	438	656	45	202	332	480	632	788	.39	65
FEB	704	925	47	133	244	377	515	655	.45	57
MAR	1058	1048	47	119	254	406	561	716	.47	46
APR	1510	1080	49	77	185	328	477	627	.51	34
MAY	1857	1035	53	26	98	220	369	524	.54	24
JUN	1994	1001	57	4	34	115	243	393	.55	20
JUL	2108	1093	59	2	16	79	188	342	.60	22
AUG	1786	1161	60	2	14	71	168	321	.57	30
SEP	1377	1273	58	3	22	88	201	349	.56	42
OCT	893	1120	55	13	67	170	313	468	.50	53
NOV	525	784	50	61	159	299	447	597	.43	63
DEC	381	594	47	142	270	420	574	729	.38	67
YR	1222	981	52	784	1696	3053	4688	6508	.52	

PENDLETON, OREGON — ELEV 1496 LAT 45.7

	HS	VS	TA	D50	D55	D60	D65	D70	KT	LD
JAN	348	523	32	558	713	868	1023	1178	.35	67
FEB	614	834	39	314	451	591	731	871	.42	59
MAR	1044	1115	44	209	351	503	657	812	.49	48
APR	1503	1152	51	68	153	279	423	573	.52	36
MAY	1925	1138	59	8	37	108	220	360	.56	27
JUN	2144	1122	66	1	4	19	70	160	.59	22
JUL	2396	1305	74	0	0	2	6	37	.68	24
AUG	1994	1413	72	0	1	3	13	62	.65	32
SEP	1502	1557	64	1	7	31	97	197	.63	44
OCT	908	1260	53	47	124	241	384	540	.55	56
NOV	438	668	41	266	410	558	708	858	.40	65
DEC	293	454	36	445	599	753	908	1063	.34	69
YR	1263	1046	52	1917	2848	3957	5240	6710	.56	

PORTLAND, OREGON — ELEV 39 LAT 45.6

	HS	VS	TA	D50	D55	D60	D65	D70	KT	LD
JAN	310	441	38	371	524	679	834	989	.31	67
FEB	554	718	43	210	343	482	622	762	.38	59
MAR	895	903	46	158	293	444	598	753	.42	48
APR	1308	975	51	67	156	286	432	582	.45	36
MAY	1663	984	57	12	53	136	264	413	.49	27
JUN	1772	951	62	2	10	48	128	247	.49	22
JUL	2037	1124	67	0	2	10	48	128	.58	24
AUG	1674	1157	67	0	2	12	56	139	.54	32
SEP	1217	1168	62	2	9	45	119	242	.51	44
OCT	724	905	54	29	98	206	347	503	.43	56
NOV	388	558	45	162	295	442	591	741	.35	65
DEC	260	381	41	293	444	598	753	908	.30	69
YR	1070	856	53	1307	2230	3389	4792	6407	.47	

REDMOND, OREGON — ELEV 3084 LAT 44.3

	HS	VS	TA	D50	D55	D60	D65	D70	KT	LD
JAN	491	811	30	614	769	924	1079	1234	.46	66
FEB	775	1105	36	399	538	678	818	958	.51	58
MAR	1190	1269	39	356	509	663	818	973	.54	47
APR	1683	1260	44	345	493	642	618	942	.58	35
MAY	2079	1177	51	62	148	277	425	580	.61	25
JUN	2287	1140	58	8	37	107	220	356	.63	21
JUL	2446	1270	66	1	4	18	55	161	.69	23
AUG	2069	1403	64	1	7	32	102	208	.67	31
SEP	1584	1586	58	9	42	116	233	371	.65	42
OCT	999	1367	48	107	218	362	515	670	.58	54
NOV	572	929	39	333	481	630	780	930	.49	64
DEC	424	729	33	515	670	825	980	1135	.45	68
YR	1387	1170	47	2750	3913	5274	6643	8517	.60	

SALEM, OREGON — ELEV 200 LAT 44.9

	HS	VS	TA	D50	D55	D60	D65	D70	KT	LD
JAN	332	471	39	348	502	657	812	967	.32	66
FEB	588	760	43	204	340	479	619	759	.39	59
MAR	947	951	45	163	306	459	614	769	.43	47
APR	1370	1009	50	68	168	308	456	606	.47	35
MAY	1738	1009	56	10	57	151	295	444	.51	26
JUN	1849	972	61	1	8	48	133	267	.51	22
JUL	2142	1154	67	0	1	7	43	130	.61	24
AUG	1775	1208	66	0	1	9	53	141	.58	31
SEP	1328	1280	62	1	6	39	120	247	.55	43
OCT	769	959	53	26	97	217	366	521	.45	55
NOV	410	585	45	158	296	444	594	744	.36	64
DEC	277	403	41	285	437	592	747	902	.30	68
YR	1130	897	52	1265	2220	3411	4852	6496	.49	

ALLENTOWN, PENNSYLVANIA — ELEV 384 LAT 40.6

	HS	VS	TA	D50	D55	D60	D65	D70	KT	LD
JAN	527	758	28	688	843	998	1153	1308	.41	62
FEB	763	929	29	577	717	857	997	1137	.44	54
MAR	1078	979	38	372	525	679	834	989	.46	43
APR	1410	926	50	85	178	308	453	603	.47	31
MAY	1637	870	60	6	25	85	190	313	.47	22
JUN	1777	870	70	0	1	7	21	91	.49	17
JUL	1765	890	74	0	0	0	0	33	.50	19
AUG	1546	927	72	0	1	3	6	62	.49	27
SEP	1238	1014	65	1	6	27	85	182	.48	39
OCT	926	1058	54	33	101	203	344	494	.48	51
NOV	568	778	42	242	383	531	681	831	.41	60
DEC	430	622	31	599	753	908	1063	1218	.38	64
YR	1141	885	51	2604	3534	4607	5827	7262	.47	

ERIE, PENNSYLVANIA — ELEV 738 LAT 42.1

	HS	VS	TA	D50	D55	D60	D65	D70	KT	LD
JAN	346	445	25	772	927	1082	1237	1392	.29	63
FEB	577	666	25	694	834	974	1114	1254	.35	56
MAR	920	837	33	530	685	840	995	1150	.40	45
APR	1359	925	45	173	309	457	606	756	.46	32
MAY	1646	900	55	22	83	184	336	478	.48	23
JUN	1847	919	65	1	4	21	80	178	.51	19
JUL	1833	944	69	0	1	5	24	98	.52	21
AUG	1455	905	68	0	1	8	43	119	.46	29
SEP	1201	1023	61	2	12	54	141	263	.48	40
OCT	827	956	52	53	137	266	415	571	.45	52
NOV	416	532	40	300	448	597	747	897	.32	61
DEC	278	359	29	648	803	958	1113	1268	.26	65
YR	1061	785	47	3197	4244	5448	6851	8424	.44	

HARRISBURG, PENNSYLVANIA — ELEV 348 LAT 40.2

	HS	VS	TA	D50	D55	D60	D65	D70	KT	LD
JAN	536	763	30	617	772	927	1082	1237	.41	61
FEB	771	929	32	496	636	776	916	1056	.44	54
MAR	1083	972	41	288	436	589	744	899	.46	43
APR	1410	916	53	48	122	231	370	517	.47	30
MAY	1652	870	63	3	12	46	128	230	.48	21
JUN	1805	875	72	0	0	0	0	58	.50	17
JUL	1764	883	76	0	0	0	0	21	.50	19
AUG	1550	921	74	0	0	0	0	38	.49	27
SEP	1267	1030	67	1	4	16	51	136	.49	38
OCT	934	1055	56	23	77	165	293	442	.48	50
NOV	579	787	44	205	340	487	636	786	.42	60
DEC	447	646	33	540	695	849	1004	1159	.38	64
YR	1152	887	53	2221	3093	4086	5224	6579	.47	

TABLE III-1 (cont.)

PHILADELPHIA, PENNSYLVANIA — ELEV 30 LAT 39.9

	HS	VS	TA	D50	D55	D60	D65	D70	KT	LD
JAN	555	792	32	549	704	859	1014	1169	.42	61
FEB	794	957	34	452	591	731	871	1011	.45	54
MAR	1108	991	42	262	408	562	716	871	.46	42
APR	1434	926	53	46	119	227	367	514	.47	30
MAY	1660	868	63	2	11	44	122	227	.48	21
JUN	1811	873	72	0	0	0	0	53	.50	17
JUL	1758	876	77	0	0	0	0	16	.50	19
AUG	1574	928	75	0	0	0	0	29	.50	26
SEP	1281	1034	68	0	2	11	38	115	.49	38
OCT	958	1081	57	14	53	132	249	393	.49	50
NOV	619	856	46	150	273	416	564	714	.44	59
DEC	470	685	35	460	614	769	924	1079	.40	63
YR	1170	905	55	1935	2775	3749	4865	6190	.48	

PITTSBURGH, PENNSYLVANIA — ELEV 1224 LAT 40.5

	HS	VS	TA	D50	D55	D60	D65	D70	KT	LD
JAN	424	553	28	679	834	989	1144	1299	.33	62
FEB	625	702	29	580	720	860	1000	1140	.36	54
MAR	943	823	38	372	525	679	834	989	.40	43
APR	1317	859	50	80	170	300	444	594	.44	31
MAY	1602	852	60	6	27	89	208	321	.46	21
JUN	1762	863	69	0	2	8	26	105	.48	17
JUL	1689	857	72	0	1	3	7	58	.48	19
AUG	1510	904	70	0	1	5	16	83	.48	27
SEP	1209	981	64	2	8	35	98	203	.47	39
OCT	895	1005	53	42	115	226	372	521	.47	50
NOV	505	656	41	269	413	561	711	861	.37	60
DEC	347	457	31	605	760	915	1070	1225	.30	64
YR	1071	793	50	2635	3574	4669	5930	7399	.44	

SCRANTON, PENNSYLVANIA — ELEV 948 LAT 41.3

	HS	VS	TA	D50	D55	D60	D65	D70	KT	LD
JAN	455	632	26	744	899	1054	1209	1364	.37	63
FEB	689	827	27	636	776	916	1056	1196	.41	55
MAR	991	898	36	435	589	744	899	1054	.42	44
APR	1339	892	49	104	209	348	495	645	.45	32
MAY	1591	860	59	7	32	100	219	347	.46	22
JUN	1760	873	68	0	2	9	28	114	.48	18
JUL	1746	893	72	0	0	2	7	51	.49	20
AUG	1513	923	70	0	1	5	18	82	.48	28
SEP	1199	996	63	2	9	41	116	225	.47	39
OCT	897	1039	53	46	123	241	391	540	.48	51
NOV	490	649	41	282	427	576	726	876	.37	61
DEC	368	513	29	648	803	958	1113	1268	.33	65
YR	1089	833	49	2904	3871	4994	6277	7762	.45	

PROVIDENCE, RHODE ISLAND — ELEV 62 LAT 41.7

	HS	VS	TA	D50	D55	D60	D65	D70	KT	LD
JAN	506	750	28	670	825	980	1135	1290	.42	63
FEB	738	925	29	577	717	857	997	1137	.44	55
MAR	1032	959	37	406	561	716	871	1026	.45	44
APR	1374	927	47	108	235	381	531	681	.46	32
MAY	1655	897	57	5	37	120	259	406	.48	23
JUN	1775	885	66	0	1	6	36	127	.49	18
JUL	1695	878	72	0	0	0	0	35	.48	20
AUG	1499	924	70	0	0	1	10	58	.48	28
SEP	1209	1019	63	0	3	21	93	204	.48	40
OCT	907	1072	54	19	85	202	350	505	.49	52
NOV	538	754	43	207	352	501	651	801	.41	61
DEC	418	627	32	574	729	884	1039	1194	.38	65
YR	1114	884	50	2566	3543	4669	5972	7464	.46	

CHARLESTON, SOUTH CAROLINA — ELEV 39 LAT 32.9

	HS	VS	TA	D50	D55	D60	D65	D70	KT	LD
JAN	744	904	49	120	222	360	521	664	.44	54
FEB	995	1009	51	81	161	275	419	547	.48	47
MAR	1339	1003	57	23	75	157	300	422	.51	35
APR	1732	934	65	2	10	36	69	192	.55	23
MAY	1860	845	72	0	1	5	5	66	.53	14
JUN	1844	820	78	0	0	0	0	16	.51	10
JUL	1799	813	80	0	0	0	0	9	.51	12
AUG	1585	806	80	0	0	0	0	10	.48	19
SEP	1394	926	75	0	0	0	0	31	.50	31
OCT	1193	1127	66	2	7	26	74	165	.53	43
NOV	934	1159	56	24	75	156	271	414	.53	52
DEC	721	943	49	108	205	339	487	642	.46	56
YR	1346	940	65	360	756	1355	2146	3178	.51	

COLUMBIA, SOUTH CAROLINA — ELEV 226 LAT 33.9

	HS	VS	TA	D50	D55	D60	D65	D70	KT	LD
JAN	762	973	45	185	310	456	608	763	.47	55
FEB	1020	1082	48	129	227	353	493	628	.50	48
MAR	1355	1050	54	47	116	212	360	492	.52	36
APR	1747	965	64	4	14	46	83	207	.55	24
MAY	1895	867	72	0	2	7	12	72	.54	15
JUN	1947	852	79	0	0	0	0	15	.54	11
JUL	1842	832	81	0	0	0	0	8	.52	13
AUG	1703	871	80	0	0	0	0	11	.52	20
SEP	1439	987	75	0	0	0	0	42	.52	32
OCT	1211	1192	64	4	14	47	112	212	.55	44
NOV	921	1184	54	49	118	214	341	488	.54	53
DEC	722	985	46	173	294	438	589	744	.48	57
YR	1382	986	64	590	1094	1772	2598	3683	.52	

GREENVILLE, SOUTH CAROLINA — ELEV 971 LAT 34.9

	HS	VS	TA	D50	D55	D60	D65	D70	KT	LD
JAN	730	953	42	248	395	549	704	859	.46	56
FEB	982	1066	44	173	300	437	577	717	.49	49
MAR	1328	1058	51	68	156	288	450	592	.51	37
APR	1697	964	61	3	16	64	144	276	.54	25
MAY	1839	863	69	0	1	6	29	96	.53	16
JUN	1918	852	76	0	0	0	0	16	.53	12
JUL	1830	838	78	0	0	0	0	8	.52	14
AUG	1699	887	78	0	0	0	0	10	.52	21
SEP	1406	990	72	0	0	2	9	55	.51	33
OCT	1180	1195	62	3	13	56	145	265	.54	45
NOV	880	1156	51	64	149	276	420	570	.53	54
DEC	670	922	43	232	377	531	685	840	.46	58
YR	1348	978	61	791	1409	2210	3163	4304	.52	

HURON, SOUTH DAKOTA — ELEV 1289 LAT 44.4

	HS	VS	TA	D50	D55	D60	D65	D70	KT	LD
JAN	488	808	13	1163	1318	1473	1627	1783	.46	66
FEB	745	1047	18	899	1039	1179	1319	1459	.49	58
MAR	1114	1163	29	652	806	961	1116	1271	.51	47
APR	1530	1130	46	169	288	429	576	726	.52	35
MAY	1871	1069	57	22	71	151	273	408	.55	25
JUN	2101	1068	67	1	6	21	72	144	.58	21
JUL	2183	1156	74	0	1	4	9	50	.62	23
AUG	1892	1276	72	0	2	6	13	70	.61	31
SEP	1418	1372	61	8	30	87	169	291	.58	43
OCT	988	1351	50	107	201	332	482	633	.57	54
NOV	577	946	32	529	678	828	978	1128	.49	64
DEC	405	684	19	955	1110	1265	1420	1575	.43	68
YR	1279	1089	45	4506	5549	6735	8054	9537	.55	

TABLE III-1 (cont.)

PIERRE, SOUTH DAKOTA — ELEV 1726 LAT 44.4

	HS	VS	TA	D50	D55	D60	D65	D70	KT	LD
JAN	530	915	16	1068	1222	1377	1531	1686	.50	66
FEB	795	1152	20	832	970	1110	1249	1389	.52	58
MAR	1206	1297	30	641	789	940	1091	1247	.55	47
APR	1614	1203	46	239	332	447	561	720	.55	35
MAY	1966	1120	57	86	151	228	267	434	.57	25
JUN	2195	1106	67	22	44	83	74	221	.60	21
JUL	2278	1199	75	7	15	31	6	112	.65	23
AUG	1992	1350	74	9	18	37	10	129	.64	31
SEP	1496	1474	62	45	86	150	152	315	.62	43
OCT	1052	1483	51	47	89	155	451	326	.61	54
NOV	623	1063	34	512	649	792	936	1087	.53	64
DEC	442	780	22	888	1041	1195	1349	1504	.47	68
YR	1352	1178	46	4396	5405	6544	7677	9171	.59	

RAPID CITY, SOUTH DAKOTA — ELEV 3169 LAT 44.0

	HS	VS	TA	D50	D55	D60	D65	D70	KT	LD
JAN	542	928	22	871	1026	1181	1336	1491	.50	65
FEB	826	1198	26	678	818	958	1098	1238	.53	58
MAR	1229	1312	31	585	738	893	1048	1203	.55	46
APR	1589	1167	45	202	325	466	612	762	.54	34
MAY	1887	1066	55	43	107	196	319	463	.55	25
JUN	2131	1070	64	5	17	52	134	211	.58	21
JUL	2223	1161	73	1	2	8	13	73	.63	23
AUG	1963	1311	72	1	3	10	17	87	.63	31
SEP	1518	1481	61	13	39	100	191	301	.62	42
OCT	1064	1482	50	110	200	325	474	621	.61	54
NOV	647	1104	35	443	589	738	888	1038	.54	63
DEC	476	855	27	729	884	1039	1194	1349	.49	67
YR	1344	1177	47	3681	4749	5965	7324	8837	.58	

SIOUX FALLS, SOUTH DAKOTA — ELEV 1427 LAT 43.6

	HS	VS	TA	D50	D55	D60	D65	D70	KT	LD
JAN	533	887	14	1110	1265	1420	1575	1730	.48	65
FEB	802	1128	19	857	997	1137	1277	1417	.51	57
MAR	1152	1183	30	621	775	930	1085	1240	.51	46
APR	1543	1114	46	165	282	421	567	717	.52	34
MAY	1894	1059	58	20	65	141	259	388	.55	25
JUN	2100	1048	68	2	6	20	65	138	.58	20
JUL	2150	1117	73	0	1	5	10	57	.61	22
AUG	1844	1209	72	1	2	7	18	77	.59	30
SEP	1410	1323	61	9	31	87	165	287	.57	42
OCT	1005	1340	50	100	190	316	465	615	.57	54
NOV	607	982	33	509	657	807	957	1107	.50	63
DEC	441	745	20	930	1085	1240	1395	1550	.45	67
YR	1293	1094	45	4323	5355	6531	7838	9322	.55	

CHATTANOOGA, TENNESSEE — ELEV 689 LAT 35.0

	HS	VS	TA	D50	D55	D60	D65	D70	KT	LD
JAN	630	776	40	314	461	614	769	924	.40	56
FEB	859	891	43	218	344	480	625	759	.43	49
MAR	1176	916	50	102	195	325	483	627	.46	37
APR	1550	887	61	8	30	88	165	296	.49	25
MAY	1732	828	69	1	4	15	51	123	.50	16
JUN	1831	829	76	0	0	0	0	27	.51	12
JUL	1735	810	79	0	0	0	0	13	.49	14
AUG	1630	858	78	0	0	0	0	17	.50	22
SEP	1335	937	72	0	1	6	9	69	.49	33
OCT	1108	1102	61	8	29	87	182	297	.51	45
NOV	773	963	49	113	209	340	483	633	.46	54
DEC	580	753	41	286	431	584	738	893	.40	58
YR	124[illegible]	879	60	1050	1705	2539	3505	4677	.48	

KNOXVILLE, TENNESSEE — ELEV 981 LAT 35.8

	HS	VS	TA	D50	D55	D60	D65	D70	KT	LD
JAN	621	785	41	302	449	602	756	911	.41	57
FEB	863	923	43	219	346	483	630	762	.44	49
MAR	1191	953	50	98	191	322	484	624	.47	38
APR	1599	932	60	8	30	89	173	300	.51	26
MAY	1803	863	68	1	3	14	47	123	.52	17
JUN	1902	856	76	0	0	0	0	29	.52	12
JUL	1804	839	78	0	0	0	0	15	.51	14
AUG	1666	889	77	0	0	0	0	19	.51	22
SEP	1383	998	72	0	1	6	10	71	.51	34
OCT	1121	1152	61	7	27	83	175	294	.53	46
NOV	759	969	49	106	201	331	474	624	.47	55
DEC	569	757	42	277	422	574	729	884	.40	59
YR	1275	909	60	1018	1671	2504	3478	4654	.49	

MEMPHIS, TENNESSEE — ELEV 285 LAT 35.0

	HS	VS	TA	D50	D55	D60	D65	D70	KT	LD
JAN	683	870	41	312	455	606	760	915	.44	56
FEB	945	1015	44	207	324	457	594	734	.47	49
MAR	1278	1013	51	99	182	299	457	591	.50	37
APR	1639	935	63	9	27	75	131	252	.52	25
MAY	1885	879	71	1	4	13	22	100	.54	16
JUN	2045	885	79	0	0	0	0	21	.56	12
JUL	1972	879	82	0	0	0	0	11	.56	14
AUG	1824	943	80	0	0	0	0	14	.56	22
SEP	1471	1046	74	1	2	7	7	61	.54	33
OCT	1204	1232	63	8	25	70	142	248	.56	45
NOV	817	1042	51	97	178	292	423	575	.49	54
DEC	629	845	43	255	391	539	691	847	.43	58
YR	1368	965	62	988	1588	2357	3227	4368	.53	

NASHVILLE, TENNESSEE — ELEV 591 LAT 36.1

	HS	VS	TA	D50	D55	D60	D65	D70	KT	LD
JAN	580	721	38	369	519	673	828	983	.38	57
FEB	824	876	41	263	395	533	672	812	.42	50
MAR	1130	902	49	119	220	357	524	661	.45	39
APR	1544	907	60	9	33	93	176	306	.50	26
MAY	1825	874	69	1	4	14	45	122	.52	17
JUN	1963	875	77	0	0	0	0	22	.54	13
JUL	1891	869	80	0	0	0	0	10	.53	15
AUG	1737	928	79	0	0	0	0	14	.54	23
SEP	1398	1020	72	0	1	5	10	66	.52	34
OCT	1114	1155	61	7	27	84	180	294	.53	46
NOV	711	893	48	120	220	354	498	648	.44	55
DEC	521	676	40	308	455	608	763	918	.38	59
YR	1272	891	59	1195	1874	2720	3696	4856	.49	

ABILENE, TEXAS — ELEV 1752 LAT 32.4

	HS	VS	TA	D50	D55	D60	D65	D70	KT	LD
JAN	924	1208	44	236	364	510	660	816	.54	54
FEB	1183	1253	48	138	230	349	479	620	.56	46
MAR	1576	1197	55	58	127	218	354	486	.59	35
APR	1843	974	65	6	17	49	104	197	.58	23
MAY	2037	887	72	1	4	11	11	84	.58	13
JUN	2209	895	80	0	0	0	0	17	.61	9
JUL	2139	890	84	0	0	0	0	8	.60	11
AUG	1956	939	84	0	0	0	0	8	.60	19
SEP	1598	1058	76	0	0	0	0	41	.57	31
OCT	1315	1259	66	5	15	42	89	185	.58	42
NOV	1008	1267	54	60	129	219	336	482	.56	52
DEC	863	1199	46	180	292	430	577	732	.54	56
YR	1556	1084	65	683	1177	1828	2610	3675	.58	

TABLE III-1 (cont.)

AMARILLO, TEXAS — ELEV 3602 LAT 35.2

	HS	VS	TA	D50	D55	D60	D65	D70	KT	LD
JAN	960	1447	36	437	590	744	899	1054	.62	56
FEB	1243	1502	40	296	430	569	708	848	.63	49
MAR	1631	1379	46	174	302	449	601	757	.63	38
APR	2019	1143	57	22	71	151	275	408	.65	25
MAY	2212	978	66	2	7	28	81	174	.63	16
JUN	2393	959	75	0	1	2	10	35	.66	12
JUL	2280	958	79	0	0	0	0	12	.64	14
AUG	2103	1065	78	0	0	0	0	17	.65	22
SEP	1760	1294	70	1	2	9	20	95	.64	33
OCT	1403	1533	60	10	37	103	206	333	.65	45
NOV	1033	1483	46	155	273	414	561	711	.62	55
DEC	872	1384	39	362	513	667	822	977	.61	59
YR	1661	1259	57	1457	2225	3136	4183	5421	.64	

AUSTIN, TEXAS — ELEV 620 LAT 30.3

	HS	VS	TA	D50	D55	D60	D65	D70	KT	LD
JAN	864	1008	50	116	207	333	483	631	.48	52
FEB	1125	1083	53	58	125	216	344	470	.51	44
MAR	1429	999	60	17	51	119	223	339	.52	33
APR	1605	828	69	2	6	20	44	130	.50	21
MAY	1834	815	75	0	0	0	0	44	.53	11
JUN	2072	860	82	0	0	0	0	10	.58	7
JUL	2105	865	85	0	0	0	0	5	.60	9
AUG	1931	891	85	0	0	0	0	5	.59	17
SEP	1606	999	79	0	0	0	0	18	.56	28
OCT	1333	1187	70	1	4	14	39	109	.56	40
NOV	987	1130	59	18	53	122	205	339	.52	50
DEC	825	1028	52	78	156	264	399	551	.49	54
YR	1478	974	68	289	602	1088	1737	2649	.54	

BROWNSVILLE, TEXAS — ELEV 20 LAT 25.9

	HS	VS	TA	D50	D55	D60	D65	D70	KT	LD
JAN	913	923	60	18	51	116	225	321	.45	47
FEB	1135	944	63	8	24	65	151	220	.47	40
MAR	1458	901	68	3	11	31	89	158	.51	28
APR	1737	816	75	0	0	0	0	53	.53	16
MAY	1927	817	79	0	0	0	0	22	.56	7
JUN	2115	868	83	0	0	0	0	10	.60	3
JUL	2212	867	84	0	0	0	0	7	.63	5
AUG	2027	854	84	0	0	0	0	7	.61	12
SEP	1694	933	82	0	0	0	0	13	.57	24
OCT	1439	1120	76	1	2	5	5	47	.57	36
NOV	1054	1047	68	3	9	28	35	146	.50	45
DEC	862	922	63	10	31	80	145	258	.45	49
YR	1550	917	74	44	127	325	650	1263	.55	

CORPUS CHRISTI, TEXAS — ELEV 43 LAT 27.8

	HS	VS	TA	D50	D55	D60	D65	D70	KT	LD
JAN	898	967	56	36	94	176	304	431	.46	49
FEB	1147	1018	60	16	47	108	199	304	.49	41
MAR	1430	930	65	5	16	49	120	205	.51	30
APR	1642	807	73	0	0	0	0	71	.51	18
MAY	1866	810	78	0	0	0	0	26	.54	9
JUN	2094	861	82	0	0	0	0	9	.59	4
JUL	2186	866	85	0	0	0	0	5	.62	6
AUG	1991	870	85	0	0	0	0	5	.60	14
SEP	1687	978	81	0	0	0	0	12	.57	26
OCT	1416	1172	74	1	2	6	7	59	.57	38
NOV	1043	1108	65	5	16	47	81	198	.51	47
DEC	845	963	59	19	57	128	219	351	.46	51
YR	1522	945	72	82	232	515	930	1676	.55	

DALLAS, TEXAS — ELEV 489 LAT 32.8

	HS	VS	TA	D50	D55	D60	D65	D70	KT	LD
JAN	821	1035	45	189	312	457	608	763	.49	54
FEB	1071	1110	49	106	190	307	437	578	.51	46
MAR	1422	1073	56	35	95	181	314	445	.54	35
APR	1627	882	66	3	9	30	71	163	.51	23
MAY	1888	852	74	0	0	0	0	55	.54	14
JUN	2135	885	82	0	0	0	0	9	.59	9
JUL	2122	890	86	0	0	0	0	3	.60	11
AUG	1950	945	86	0	0	0	0	3	.60	19
SEP	1587	1063	78	0	0	0	0	20	.56	31
OCT	1276	1228	68	2	6	21	55	140	.56	43
NOV	936	1158	56	34	91	173	284	428	.53	52
DEC	780	1052	48	137	239	374	521	676	.50	56
YR	1470	1014	66	505	943	1543	2290	3282	.55	

DEL RIO, TEXAS — ELEV 1027 LAT 29.4

	HS	VS	TA	D50	D55	D60	D65	D70	KT	LD
JAN	958	1125	51	95	180	301	449	596	.51	51
FEB	1206	1149	56	33	87	165	283	404	.54	43
MAR	1580	1087	63	7	23	69	163	254	.57	32
APR	1699	852	72	1	2	8	16	75	.53	20
MAY	1827	808	78	0	0	0	0	20	.52	10
JUN	2024	849	84	0	0	0	0	4	.57	6
JUL	2054	851	87	0	0	0	0	2	.58	8
AUG	1936	878	86	0	0	0	0	3	.59	16
SEP	1584	961	80	0	0	0	0	12	.55	28
OCT	1360	1179	71	1	3	10	34	89	.57	39
NOV	1059	1204	60	14	45	110	184	323	.54	49
DEC	903	1127	52	74	152	261	394	551	.52	53
YR	1518	1005	70	223	492	924	1523	2335	.55	

EL PASO, TEXAS — ELEV 3917 LAT 31.8

	HS	VS	TA	D50	D55	D60	D65	D70	KT	LD
JAN	1125	1572	44	210	355	509	663	818	.65	53
FEB	1480	1671	48	91	194	326	465	605	.69	45
MAR	1909	1472	55	21	81	183	328	478	.71	34
APR	2363	1174	64	1	4	25	89	195	.74	22
MAY	2601	985	72	0	0	0	0	44	.75	13
JUN	2682	946	80	0	0	0	0	3	.75	8
JUL	2450	934	82	0	0	0	0	1	.69	10
AUG	2284	1030	81	0	0	0	0	3	.70	18
SEP	1987	1313	74	0	0	0	0	23	.70	30
OCT	1639	1654	64	1	4	25	92	199	.71	42
NOV	1244	1681	52	50	131	257	402	552	.68	51
DEC	1031	1521	44	188	331	484	639	794	.64	55
YR	1901	1327	63	561	1102	1810	2678	3714	.71	

FORT WORTH, TEXAS — ELEV 538 LAT 32.8

	HS	VS	TA	D50	D55	D60	D65	D70	KT	LD
JAN	805	1007	45	202	329	475	626	781	.48	54
FEB	1069	1107	49	116	205	325	456	597	.51	46
MAR	1409	1062	55	43	107	198	335	469	.53	35
APR	1616	877	65	3	12	39	88	187	.51	23
MAY	1890	853	73	0	0	0	0	71	.54	14
JUN	2153	888	81	0	0	0	0	11	.60	9
JUL	2155	897	85	0	0	0	0	4	.61	11
AUG	1983	957	85	0	0	0	0	4	.61	19
SEP	1621	1088	78	0	0	0	0	22	.58	31
OCT	1293	1249	68	2	7	23	60	147	.57	43
NOV	938	1162	56	34	93	175	287	430	.53	52
DEC	766	1024	48	142	246	383	530	686	.49	56
YR	1477	1014	66	542	999	1618	2382	3410	.55	

TABLE III-1 (cont.)

HOUSTON, TEXAS ELEV 108 LAT 30.0

	HS	VS	TA	D50	D55	D60	D65	D70	KT	LD
JAN	772	852	52	71	150	263	416	556	.42	51
FEB	1034	960	55	31	87	168	294	414	.46	44
MAR	1297	889	61	9	31	89	189	298	.47	32
APR	1522	788	69	1	3	12	23	107	.47	20
MAY	1775	798	76	0	0	0	0	31	.51	11
JUN	1898	824	81	0	0	0	0	8	.53	7
JUL	1828	805	83	0	0	0	0	4	.52	9
AUG	1686	807	83	0	0	0	0	4	.51	17
SEP	1471	906	79	0	0	0	0	13	.51	28
OCT	1276	1110	71	1	2	8	24	88	.54	40
NOV	924	1019	61	8	28	82	155	281	.48	49
DEC	730	855	55	41	108	201	333	480	.43	53
YR	1353	884	69	161	409	825	1434	2285	.49	

LAREDO, TEXAS ELEV 518 LAT 27.5

	HS	VS	TA	D50	D55	D60	D65	D70	KT	LD
JAN	959	1046	57	36	93	174	299	426	.49	49
FEB	1195	1062	61	12	37	92	177	273	.51	41
MAR	1516	981	68	3	9	28	87	154	.54	30
APR	1727	833	76	0	0	0	0	36	.53	18
MAY	1952	827	81	0	0	0	0	12	.56	8
JUN	2073	857	86	0	0	0	0	4	.58	4
JUL	2131	857	88	0	0	0	0	3	.61	6
AUG	2009	870	88	0	0	0	0	3	.61	14
SEP	1705	979	83	0	0	0	0	8	.58	26
OCT	1408	1152	76	0	1	5	8	44	.57	37
NOV	1041	1093	65	5	15	45	74	193	.51	47
DEC	889	1022	59	23	64	138	231	365	.48	51
YR	1552	965	74	79	219	482	876	1522	.56	

LUBBOCK, TEXAS ELEV 3241 LAT 33.6

	HS	VS	TA	D50	D55	D60	D65	D70	KT	LD
JAN	1031	1497	39	343	494	648	803	958	.63	55
FEB	1332	1551	43	218	348	485	624	764	.65	47
MAR	1762	1430	49	108	211	349	508	654	.67	36
APR	2168	1159	60	7	28	87	190	307	.69	24
MAY	2396	989	69	1	2	11	29	115	.69	15
JUN	2544	957	77	0	0	0	0	16	.70	10
JUL	2412	956	80	0	0	0	0	8	.68	12
AUG	2208	1057	78	0	0	0	0	11	.68	20
SEP	1820	1271	71	0	1	5	8	73	.65	32
OCT	1468	1530	61	5	22	76	162	288	.66	44
NOV	1116	1550	49	106	206	341	486	636	.64	53
DEC	934	1425	41	280	427	580	735	890	.61	57
YR	1768	1279	60	1069	1739	2582	3545	4720	.67	

LUFKIN, TEXAS ELEV 315 LAT 31.2

	HS	VS	TA	D50	D55	D60	D65	D70	KT	LD
JAN	794	927	49	125	223	356	509	658	.45	52
FEB	1069	1045	52	66	137	237	371	500	.49	45
MAR	1376	983	58	20	62	138	256	379	.51	34
APR	1624	851	67	2	7	23	56	145	.51	21
MAY	1867	831	74	0	0	0	0	50	.54	12
JUN	2055	860	80	0	0	0	0	11	.57	8
JUL	2006	852	83	0	0	0	0	6	.57	10
AUG	1864	885	83	0	0	0	0	6	.57	18
SEP	1531	976	78	0	0	0	0	22	.54	29
OCT	1349	1245	68	1	5	19	52	134	.58	41
NOV	963	1131	57	24	71	147	256	390	.52	51
DEC	768	962	51	93	179	300	440	596	.47	55
YR	1441	962	67	331	684	1220	1940	2897	.53	

MIDLAND-ODESSA, TEXAS ELEV 2858 LAT 31.9

	HS	VS	TA	D50	D55	D60	D65	D70	KT	LD
JAN	1081	1488	44	225	361	510	663	819	.62	53
FEB	1383	1525	48	124	222	347	482	622	.65	46
MAR	1839	1413	54	44	113	209	349	489	.69	34
APR	2192	1109	64	3	13	43	98	202	.69	22
MAY	2430	961	72	0	0	0	0	68	.70	13
JUN	2562	938	80	0	0	0	0	11	.71	9
JUL	2389	927	82	0	0	0	0	6	.68	11
AUG	2210	1011	82	0	0	0	0	7	.67	18
SEP	1844	1216	75	0	0	0	0	33	.65	30
OCT	1522	1502	66	2	9	31	81	176	.66	42
NOV	1176	1553	53	54	125	225	356	503	.65	51
DEC	1000	1460	46	173	296	441	592	747	.62	55
YR	1804	1257	64	627	1137	1806	2621	3682	.67	

PORT ARTHUR, TEXAS ELEV 23 LAT 29.9

	HS	VS	TA	D50	D55	D60	D65	D70	KT	LD
JAN	800	892	52	78	157	269	420	560	.44	51
FEB	1071	1001	55	37	95	176	302	421	.48	44
MAR	1353	929	60	13	41	105	202	320	.49	32
APR	1609	824	69	1	5	17	33	121	.50	20
MAY	1871	822	75	0	0	0	0	42	.54	11
JUN	2011	847	81	0	0	0	0	10	.56	7
JUL	1846	809	83	0	0	0	0	6	.52	9
AUG	1736	822	83	0	0	0	0	6	.53	16
SEP	1527	939	79	0	0	0	0	17	.53	28
OCT	1321	1157	70	1	4	13	35	108	.55	40
NOV	953	1060	60	12	39	101	184	307	.50	49
DEC	754	891	54	50	120	215	342	493	.44	53
YR	1406	915	69	193	462	897	1518	2412	.51	

SAN ANGELO, TEXAS ELEV 1909 LAT 31.4

	HS	VS	TA	D50	D55	D60	D65	D70	KT	LD
JAN	962	1228	46	178	291	429	577	732	.55	53
FEB	1208	1241	50	100	178	286	413	551	.56	45
MAR	1606	1183	57	33	86	165	287	409	.59	34
APR	1851	954	67	3	10	31	74	158	.58	22
MAY	2031	874	75	0	0	0	0	56	.58	12
JUN	2186	885	82	0	0	0	0	12	.61	8
JUL	2123	877	85	0	0	0	0	6	.60	10
AUG	1966	922	85	0	0	0	0	6	.60	18
SEP	1607	1033	77	0	0	0	0	34	.56	30
OCT	1337	1240	67	3	11	32	73	164	.58	41
NOV	1044	1279	56	45	106	189	298	441	.56	51
DEC	895	1211	48	145	244	375	518	674	.55	55
YR	1570	1076	66	508	927	1509	2240	3244	.58	

SAN ANTONIO, TEXAS ELEV 794 LAT 29.5

	HS	VS	TA	D50	D55	D60	D65	D70	KT	LD
JAN	895	1026	51	93	179	302	451	599	.48	51
FEB	1154	1089	55	39	100	185	310	436	.51	43
MAR	1450	991	61	9	32	91	194	299	.52	32
APR	1612	819	70	1	3	12	31	106	.50	20
MAY	1894	825	76	0	0	0	0	31	.54	10
JUN	2069	857	82	0	0	0	0	6	.58	6
JUL	2121	863	85	0	0	0	0	3	.60	8
AUG	1947	883	85	0	0	0	0	3	.59	16
SEP	1638	996	79	0	0	0	0	13	.56	28
OCT	1350	1172	71	1	3	10	32	96	.56	39
NOV	1009	1129	60	12	40	104	179	319	.52	49
DEC	847	1033	53	58	132	236	373	523	.49	53
YR	1501	973	69	213	490	941	1570	2435	.55	

TABLE III-1 (cont.)

SHERMAN, TEXAS — ELEV 764 LAT 33.7

	HS	VS	TA	D50	D55	D60	D65	D70	KT	LD
JAN	794	1024	42	269	415	568	722	877	.48	55
FEB	1037	1098	46	149	263	397	535	675	.50	47
MAR	1366	1053	52	59	137	253	411	549	.52	36
APR	1610	892	64	2	10	41	114	209	.51	24
MAY	1852	852	71	0	1	5	13	74	.53	15
JUN	2114	888	79	0	0	0	0	9	.59	10
JUL	2077	890	84	0	0	0	0	3	.59	12
AUG	1932	958	84	0	0	0	0	2	.59	20
SEP	1580	1088	76	0	0	0	0	22	.57	32
OCT	1268	1259	66	1	6	24	90	167	.57	44
NOV	919	1171	53	44	114	217	353	499	.53	53
DEC	744	1020	45	190	323	473	626	781	.49	57
YR	1443	1015	64	715	1271	1977	2864	3869	.55	

WACO, TEXAS — ELEV 509 LAT 31.6

	HS	VS	TA	D50	D55	D60	D65	D70	KT	LD
JAN	833	1007	47	156	269	409	558	713	.48	53
FEB	1096	1096	51	84	160	269	401	536	.51	45
MAR	1427	1038	57	25	74	153	280	403	.53	34
APR	1612	853	67	2	7	23	56	146	.51	22
MAY	1774	808	75	0	0	0	0	46	.51	13
JUN	2112	873	82	0	0	0	0	8	.59	8
JUL	2130	880	86	0	0	0	0	3	.60	10
AUG	1958	923	86	0	0	0	0	3	.60	18
SEP	1601	1035	79	0	0	0	0	16	.56	30
OCT	1301	1205	69	1	4	16	51	121	.56	42
NOV	957	1139	58	22	68	143	241	381	.52	51
DEC	803	1043	50	109	201	328	471	627	.49	55
YR	1469	991	67	399	782	1340	2058	3003	.54	

WICHITA FALLS, TEXAS — ELEV 1030 LAT 34.0

	HS	VS	TA	D50	D55	D60	D65	D70	KT	LD
JAN	862	1168	42	284	425	575	729	884	.53	55
FEB	1123	1240	46	164	271	400	535	675	.55	48
MAR	1472	1164	53	74	152	259	409	545	.56	36
APR	1763	975	64	5	17	51	112	209	.56	24
MAY	2017	904	72	1	2	8	13	77	.58	15
JUN	2221	912	81	0	0	0	0	10	.61	11
JUL	2166	914	86	0	0	0	0	4	.61	13
AUG	1969	979	86	0	0	0	0	4	.60	21
SEP	1602	1116	77	0	0	0	0	28	.58	32
OCT	1291	1304	66	3	11	36	92	180	.58	44
NOV	957	1257	53	67	140	241	369	516	.56	53
DEC	799	1148	44	218	347	493	645	800	.53	57
YR	1522	1089	64	816	1365	2063	2904	3931	.58	

BRYCE CANYON, UTAH — ELEV 7588 LAT 37.7

	HS	VS	TA	D50	D55	D60	D65	D70	KT	LD
JAN	914	1511	20	936	1091	1246	1401	1556	.64	59
FEB	1236	1657	23	750	890	1030	1170	1310	.66	51
MAR	1685	1582	29	660	815	970	1125	1280	.68	40
APR	2133	1310	38	370	519	669	819	969	.69	28
MAY	2454	1111	46	141	276	428	583	738	.71	19
JUN	2655	1055	54	137	268	415	330	714	.73	14
JUL	2424	1044	62	1	8	47	128	265	.68	16
AUG	2157	1172	60	2	16	72	176	315	.67	24
SEP	1920	1578	53	31	102	220	363	513	.72	36
OCT	1465	1811	43	230	379	533	688	843	.72	48
NOV	1015	1622	31	579	729	879	1029	1179	.67	57
DEC	818	1419	22	856	1011	1166	1321	1476	.63	61
YR	1742	1404	40	4693	6106	7675	9133	11159	.69	

CEDAR CITY, UTAH — ELEV 5617 LAT 37.7

	HS	VS	TA	D50	D55	D60	D65	D70	KT	LD
JAN	882	1431	29	660	815	970	1125	1280	.62	59
FEB	1180	1548	33	474	613	753	893	1033	.63	51
MAR	1636	1521	38	364	515	670	825	980	.66	40
APR	2092	1284	47	134	249	390	537	687	.68	28
MAY	2467	1115	56	20	71	156	281	430	.71	19
JUN	2706	1064	65	1	6	27	86	177	.74	14
JUL	2503	1064	73	0	0	0	0	45	.71	16
AUG	2241	1213	71	0	1	4	6	70	.70	24
SEP	1968	1627	63	2	11	43	114	220	.74	36
OCT	1460	1802	52	67	150	274	424	574	.71	48
NOV	992	1566	39	340	487	636	786	936	.65	57
DEC	785	1331	31	596	750	905	1060	1215	.60	61
YR	1745	1379	50	2658	3669	4829	6137	7647	.69	

SALT LAKE CITY, UTAH — ELEV 4226 LAT 40.8

	HS	VS	TA	D50	D55	D60	D65	D70	KT	LD
JAN	639	1017	28	683	837	992	1147	1302	.51	62
FEB	989	1363	33	467	605	745	885	1025	.58	54
MAR	1454	1463	40	334	481	633	787	942	.62	43
APR	1894	1284	49	116	208	334	474	625	.63	31
MAY	2362	1184	58	20	61	135	237	371	.68	22
JUN	2561	1122	66	3	9	31	88	167	.70	17
JUL	2590	1186	77	0	0	0	0	29	.73	19
AUG	2254	1356	75	0	1	4	5	47	.71	27
SEP	1843	1693	65	4	13	43	105	195	.72	39
OCT	1293	1724	52	72	150	259	402	548	.68	51
NOV	788	1276	39	337	480	628	777	927	.58	60
DEC	570	953	30	612	766	921	1076	1231	.50	64
YR	1606	1301	51	2648	3612	4725	5983	7409	.66	

BURLINGTON, VERMONT — ELEV 341 LAT 44.5

	HS	VS	TA	D50	D55	D60	D65	D70	KT	LD
JAN	385	572	17	1029	1184	1339	1494	1649	.36	66
FEB	607	782	19	879	1019	1159	1299	1439	.40	58
MAR	940	929	29	648	803	958	1113	1268	.43	47
APR	1296	936	43	220	362	510	660	810	.44	35
MAY	1574	909	55	22	81	180	331	472	.46	25
JUN	1729	911	65	1	3	18	63	165	.47	21
JUL	1721	941	70	0	1	4	20	81	.49	23
AUG	1475	976	67	0	2	9	49	122	.48	31
SEP	1122	1013	59	5	24	87	191	324	.46	43
OCT	740	895	49	98	206	350	502	657	.43	54
NOV	375	507	37	391	540	690	840	990	.32	64
DEC	283	408	23	849	1004	1159	1314	1469	.30	68
YR	1023	815	44	4142	5230	6464	7876	9447	.44	

NORFOLK, VIRGINIA — ELEV 30 LAT 36.9

	HS	VS	TA	D50	D55	D60	D65	D70	KT	LD
JAN	678	932	41	300	450	605	760	915	.46	58
FEB	932	1068	41	247	382	521	661	801	.49	51
MAR	1281	1080	48	112	226	371	532	679	.51	39
APR	1677	1004	58	9	40	114	226	368	.54	27
MAY	1887	909	67	0	2	13	53	139	.54	18
JUN	2000	894	75	0	0	0	0	25	.55	14
JUL	1853	868	78	0	0	0	0	8	.52	16
AUG	1680	918	77	0	0	0	0	12	.52	23
SEP	1396	1044	72	0	0	2	9	53	.52	35
OCT	1083	1145	62	2	13	56	141	265	.52	47
NOV	811	1117	52	55	136	259	402	552	.52	56
DEC	624	907	42	248	395	549	704	859	.46	60
YR	1327	990	59	974	1646	2489	3488	4674	.52	

TABLE III-1 (cont.)

RICHMOND, VIRGINIA ELEV 164 LAT 37.5

	HS	VS	TA	D50	D55	D60	D65	D70	KT	LD
JAN	632	863	38	390	543	698	853	1008	.44	59
FEB	877	1004	39	301	438	577	717	857	.47	51
MAR	1210	1025	47	140	262	408	569	716	.49	40
APR	1566	953	58	11	46	119	226	369	.51	28
MAY	1762	872	67	1	4	17	64	148	.51	18
JUN	1872	864	74	0	0	0	0	32	.52	14
JUL	1774	849	78	0	0	0	0	11	.50	16
AUG	1601	891	76	0	0	0	0	18	.50	24
SEP	1348	1020	70	0	1	6	21	84	.50	36
OCT	1033	1097	59	7	32	98	203	336	.50	47
NOV	733	988	49	100	199	334	480	630	.48	57
DEC	567	810	39	345	497	651	806	961	.43	61
YR	1250	936	58	1296	2021	2909	3939	5170	.49	

ROANOKE, VIRGINIA ELEV 1175 LAT 37.3

	HS	VS	TA	D50	D55	D60	D65	D70	KT	LD
JAN	660	911	36	423	577	732	887	1042	.46	59
FEB	899	1032	38	336	474	613	753	893	.48	51
MAR	1236	1046	45	171	306	457	611	766	.50	40
APR	1581	957	56	17	67	152	283	424	.51	28
MAY	1764	870	64	1	6	28	101	193	.51	18
JUN	1882	865	72	0	0	0	0	56	.52	14
JUL	1796	854	75	0	0	0	0	22	.51	16
AUG	1620	897	74	0	0	0	0	30	.50	24
SEP	1358	1023	68	0	2	9	32	112	.51	35
OCT	1080	1157	58	10	43	119	235	381	.52	47
NOV	765	1043	47	136	257	401	549	699	.50	57
DEC	591	853	37	393	546	701	856	1011	.45	61
YR	1271	958	56	1486	2277	3211	4307	5627	.50	

OLYMPIA, WASHINGTON ELEV 200 LAT 47.0

	HS	VS	TA	D50	D55	D60	D65	D70	KT	LD
JAN	269	385	37	398	552	707	862	1017	.29	68
FEB	503	662	41	255	392	532	672	812	.36	61
MAR	845	876	43	219	367	521	676	831	.40	49
APR	1255	966	48	97	211	355	504	654	.44	37
MAY	1632	999	54	21	86	196	341	496	.48	28
JUN	1693	940	59	3	22	85	197	334	.47	24
JUL	1913	1101	64	1	4	25	89	207	.54	26
AUG	1549	1105	63	1	5	32	103	230	.51	34
SEP	1157	1146	59	4	24	90	198	343	.50	45
OCT	636	795	51	61	155	294	446	601	.40	57
NOV	339	489	43	209	352	501	651	801	.33	66
DEC	221	324	40	327	481	636	791	946	.27	70
YR	1004	816	50	1595	2652	3974	5530	7273	.45	

SEATTLE-TACOMA, WASHINGTON ELEV 400 LAT 47.4

	HS	VS	TA	D50	D55	D60	D65	D70	KT	LD
JAN	262	378	38	367	521	676	831	986	.29	69
FEB	495	657	42	220	356	496	636	776	.36	61
MAR	849	893	44	192	339	493	648	803	.41	50
APR	1293	1013	49	85	197	340	489	639	.46	38
MAY	1714	1061	55	13	68	171	313	468	.50	28
JUN	1802	1004	60	2	14	67	167	308	.50	24
JUL	2248	1299	65	0	2	16	80	181	.64	26
AUG	1616	1174	64	0	3	20	82	200	.53	34
SEP	1148	1150	60	2	15	71	170	313	.50	46
OCT	656	843	52	36	116	246	397	552	.42	57
NOV	337	494	45	173	313	462	612	762	.33	67
DEC	211	309	41	297	450	605	760	915	.27	71
YR	1056	857	51	1386	2393	3662	5185	6903	.48	

SPOKANE, WASHINGTON ELEV 2365 LAT 47.6

	HS	VS	TA	D50	D55	D60	D65	D70	KT	LD
JAN	315	496	25	763	918	1073	1228	1383	.35	69
FEB	606	887	32	499	639	778	918	1058	.45	61
MAR	1041	1190	38	391	543	698	853	1008	.50	50
APR	1495	1215	46	153	276	419	567	717	.53	38
MAY	1918	1197	55	30	94	190	327	476	.57	29
JUN	2083	1151	62	4	18	65	144	265	.57	24
JUL	2357	1368	70	0	2	7	21	94	.67	26
AUG	1942	1465	68	1	3	12	47	122	.64	34
SEP	1435	1572	60	7	30	92	196	318	.62	46
OCT	841	1220	48	125	238	382	533	688	.54	58
NOV	398	635	36	437	585	735	885	1035	.40	67
DEC	255	409	29	651	806	961	1116	1271	.33	71
YR	1227	1068	47	3061	4150	5411	6835	8433	.56	

YAKIMA, WASHINGTON ELEV 1066 LAT 46.6

	HS	VS	TA	D50	D55	D60	D65	D70	KT	LD
JAN	365	585	28	698	853	1008	1163	1318	.38	68
FEB	666	976	36	402	541	680	820	960	.47	60
MAR	1122	1273	42	265	412	565	719	874	.53	49
APR	1598	1277	50	93	188	320	465	615	.56	37
MAY	2008	1218	58	12	48	123	239	378	.59	28
JUN	2169	1161	65	2	7	30	94	188	.60	23
JUL	2358	1324	71	0	1	5	20	78	.67	25
AUG	1975	1442	69	0	2	10	37	111	.65	33
SEP	1483	1583	61	4	18	67	147	270	.63	45
OCT	891	1273	50	86	180	313	462	617	.55	57
NOV	444	711	38	352	499	648	798	948	.42	66
DEC	295	480	31	580	735	890	1045	1200	.36	70
YR	1285	1109	50	2494	3482	4658	6009	7557	.57	

CHARLESTON, WEST VIRGINIA ELEV 951 LAT 38.4

	HS	VS	TA	D50	D55	D60	D65	D70	KT	LD
JAN	498	638	35	483	636	791	946	1101	.36	60
FEB	706	770	37	381	519	658	798	938	.39	52
MAR	1009	841	45	202	334	483	642	791	.41	41
APR	1356	842	56	27	83	165	287	426	.44	29
MAY	1639	837	65	3	11	40	113	202	.47	19
JUN	1776	843	72	0	1	5	10	67	.49	15
JUL	1682	827	75	0	0	0	0	35	.47	17
AUG	1514	863	74	0	0	0	0	49	.47	25
SEP	1272	979	68	1	5	18	46	136	.48	37
OCT	972	1043	57	21	69	149	267	407	.48	48
NOV	613	793	45	176	298	441	588	738	.41	58
DEC	440	585	36	431	584	738	893	1048	.35	62
YR	1125	822	55	1726	2540	3488	4590	5938	.45	

HUNTINGTON, WEST VIRGINIA ELEV 837 LAT 38.4

	HS	VS	TA	D50	D55	D60	D65	D70	KT	LD
JAN	526	689	34	489	642	797	952	1107	.38	60
FEB	757	847	36	393	530	669	809	949	.41	52
MAR	1067	901	44	208	340	489	649	797	.44	41
APR	1448	900	56	30	87	171	293	432	.47	29
MAY	1710	866	65	3	12	41	115	203	.49	19
JUN	1844	865	72	0	1	5	11	64	.51	15
JUL	1769	859	75	0	0	0	0	34	.50	17
AUG	1580	899	74	0	0	0	0	48	.49	25
SEP	1306	1010	68	1	5	18	46	134	.49	37
OCT	1004	1091	57	22	70	149	265	405	.50	48
NOV	638	840	46	175	296	438	585	735	.43	58
DEC	467	636	36	438	590	744	899	1054	.37	62
YR	1178	867	55	1759	2574	3522	4624	5961	.47	

TABLE III-1 (cont.)

EAU CLAIRE, WISCONSIN ELEV 896 LAT 44.9

	HS	VS	TA	D50	D55	D60	D65	D70	KT	LD
JAN	452	738	12	1187	1342	1497	1652	1807	.43	66
FEB	746	1072	15	969	1109	1249	1389	1529	.50	59
MAR	1090	1150	27	704	859	1014	1169	1324	.50	47
APR	1426	1057	45	187	320	466	615	765	.49	35
MAY	1681	977	56	18	68	154	293	430	.49	26
JUN	1872	982	66	1	4	18	65	151	.51	22
JUL	1886	1030	71	0	1	5	14	78	.53	24
AUG	1621	1093	68	0	2	9	37	112	.53	31
SEP	1196	1115	59	8	36	104	202	343	.50	43
OCT	826	1062	49	107	213	354	505	660	.48	55
NOV	450	670	32	540	690	840	990	1140	.39	64
DEC	341	544	18	992	1147	1302	1457	1612	.37	68
YR	1134	957	43	4714	5790	7011	8388	9951	.49	

GREEN BAY, WISCONSIN ELEV 702 LAT 44.5

	HS	VS	TA	D50	D55	D60	D65	D70	KT	LD
JAN	451	722	15	1073	1228	1383	1538	1693	.42	66
FEB	725	1010	18	896	1036	1176	1316	1456	.48	58
MAR	1104	1153	29	664	818	973	1128	1283	.50	47
APR	1439	1056	44	202	339	487	636	786	.49	35
MAY	1719	989	55	28	92	191	338	481	.50	25
JUN	1908	989	65	1	6	27	91	185	.52	21
JUL	1888	1022	69	0	1	7	22	97	.54	23
AUG	1622	1082	68	0	2	11	54	122	.52	31
SEP	1218	1127	59	7	32	98	191	336	.50	43
OCT	821	1037	49	97	199	339	490	645	.48	54
NOV	465	690	34	478	627	777	927	1077	.40	64
DEC	350	554	21	902	1057	1212	1367	1522	.37	68
YR	1145	952	44	4348	5437	6680	8098	9684	.50	

LA CROSSE, WISCONSIN ELEV 673 LAT 43.9

	HS	VS	TA	D50	D55	D60	D65	D70	KT	LD
JAN	481	771	16	1051	1206	1361	1516	1671	.44	65
FEB	765	1065	20	840	980	1120	1260	1400	.49	58
MAR	1101	1125	31	586	741	896	1051	1206	.49	46
APR	1426	1027	48	127	237	375	522	672	.49	34
MAY	1713	972	59	10	38	107	224	346	.50	25
JUN	1905	976	69	1	2	11	39	112	.52	21
JUL	1900	1013	73	0	1	3	10	52	.54	23
AUG	1666	1094	71	0	1	5	17	71	.54	30
SEP	1242	1133	62	4	17	63	130	258	.51	42
OCT	863	1087	52	65	146	266	421	565	.49	54
NOV	494	734	35	440	588	738	888	1038	.41	63
DEC	369	581	22	874	1029	1184	1339	1494	.38	67
YR	1163	964	46	3998	4987	6130	7417	8884	.50	

MADISON, WISCONSIN ELEV 860 LAT 43.1

	HS	VS	TA	D50	D55	D60	D65	D70	KT	LD
JAN	515	822	17	1029	1184	1339	1494	1649	.45	64
FEB	804	1108	20	832	972	1112	1252	1392	.51	57
MAR	1136	1141	30	614	769	924	1079	1234	.50	46
APR	1398	981	45	167	297	442	591	741	.47	33
MAY	1743	969	56	19	70	157	297	436	.51	24
JUN	1948	977	66	1	4	19	72	156	.53	20
JUL	1934	1009	70	0	1	5	14	83	.55	22
AUG	1708	1098	69	0	2	8	39	106	.55	30
SEP	1299	1168	60	6	26	87	173	314	.52	41
OCT	911	1139	50	86	182	318	474	623	.51	53
NOV	504	729	35	460	609	759	909	1059	.41	62
DEC	389	603	22	871	1026	1181	1336	1491	.38	66
YR	1193	978	45	4086	5143	6352	7730	9284	.51	

MILWAUKEE, WISCONSIN ELEV 692 LAT 42.9

	HS	VS	TA	D50	D55	D60	D65	D70	KT	LD
JAN	479	731	19	949	1104	1259	1414	1569	.42	64
FEB	736	967	23	770	910	1050	1190	1330	.46	57
MAR	1089	1071	31	577	732	887	1042	1197	.48	45
APR	1443	1011	45	177	313	460	609	759	.49	33
MAY	1768	977	54	27	92	196	348	490	.51	24
JUN	1977	984	65	1	5	24	90	182	.54	20
JUL	1962	1017	70	0	1	4	15	81	.55	22
AUG	1719	1099	69	0	1	5	36	92	.55	29
SEP	1310	1173	61	3	14	61	140	273	.53	41
OCT	908	1124	51	65	153	285	440	589	.50	53
NOV	525	767	37	406	555	705	855	1005	.42	62
DEC	378	573	24	800	955	1110	1265	1420	.37	66
YR	1194	957	46	3774	4833	6045	7444	8986	.51	

CASPER, WYOMING ELEV 5289 LAT 42.9

	HS	VS	TA	D50	D55	D60	D65	D70	KT	LD
JAN	683	1251	23	831	986	1141	1296	1451	.60	64
FEB	1013	1550	27	650	790	930	1070	1210	.63	57
MAR	1441	1568	31	589	744	899	1054	1209	.64	45
APR	1847	1340	43	234	373	520	669	819	.62	33
MAY	2204	1191	53	53	129	242	388	537	.64	24
JUN	2501	1172	62	4	17	61	147	255	.69	20
JUL	2535	1249	71	0	1	5	13	76	.72	22
AUG	2225	1444	70	0	2	8	17	98	.71	29
SEP	1749	1717	59	10	40	109	229	344	.70	41
OCT	1219	1731	48	129	242	385	536	691	.68	53
NOV	765	1352	34	484	633	783	933	1083	.61	62
DEC	594	1134	26	738	893	1048	1203	1358	.58	66
YR	1568	1390	45	3723	4850	6131	7555	9131	.66	

CHEYENNE, WYOMING ELEV 6142 LAT 41.1

	HS	VS	TA	D50	D55	D60	D65	D70	KT	LD
JAN	766	1362	27	725	880	1035	1190	1345	.62	62
FEB	1068	1548	29	588	728	868	1008	1148	.63	55
MAR	1433	1451	32	571	725	880	1035	1190	.61	44
APR	1770	1203	43	230	371	519	669	819	.59	31
MAY	1995	1040	52	49	127	246	394	546	.58	22
JUN	2258	1043	61	3	15	62	156	268	.62	18
JUL	2230	1076	69	0	1	7	22	98	.63	20
AUG	1966	1195	68	0	2	11	31	123	.62	28
SEP	1667	1502	58	9	38	109	225	357	.65	39
OCT	1242	1645	48	118	233	378	530	685	.66	51
NOV	823	1384	36	436	585	735	885	1035	.61	60
DEC	671	1244	29	645	800	955	1110	1265	.60	64
YR	1493	1306	46	3375	4507	5805	7255	8880	.62	

ROCK SPRINGS, WYOMING ELEV 6745 LAT 41.6

	HS	VS	TA	D50	D55	D60	D65	D70	KT	LD
JAN	735	1311	19	955	1110	1265	1420	1575	.60	63
FEB	1089	1631	23	745	885	1025	1165	1305	.65	55
MAR	1530	1619	29	654	809	964	1119	1274	.66	44
APR	1944	1358	40	299	447	597	747	897	.65	32
MAY	2344	1206	50	62	158	300	453	608	.68	23
JUN	2574	1151	59	3	20	83	198	334	.71	18
JUL	2547	1202	68	0	1	4	18	98	.72	20
AUG	2240	1387	66	0	1	9	49	141	.71	28
SEP	1832	1734	56	8	46	130	269	408	.72	40
OCT	1306	1812	45	176	321	475	629	784	.70	52
NOV	826	1427	31	579	729	879	1029	1179	.63	61
DEC	651	1219	23	849	1004	1159	1314	1469	.60	65
YR	1638	1420	43	4330	5531	6889	8410	10073	.68	

TABLE III-1 (cont.)

SHERIDAN, WYOMING ELEV 3966 LAT 44.8

	HS	VS	TA	D50	D55	D60	D65	D70	KT	LD
JAN	517	900	21	899	1054	1209	1364	1519	.49	66
FEB	788	1157	26	675	815	955	1095	1235	.52	58
MAR	1205	1315	31	590	744	899	1054	1209	.55	47
APR	1537	1150	44	217	349	494	642	792	.53	35
MAY	1883	1087	53	57	131	237	375	526	.55	26
JUN	2156	1102	61	7	27	81	168	280	.59	21
JUL	2329	1237	70	1	2	9	28	94	.66	23
AUG	2006	1379	69	1	3	13	31	113	.65	31
SEP	1502	1504	58	17	57	131	245	369	.62	43
OCT	1005	1409	48	137	245	384	533	689	.59	55
NOV	591	1000	33	500	648	798	948	1098	.51	64
DEC	441	794	26	760	915	1070	1225	1380	.48	68
YR	1333	1170	45	3860	4991	6279	7708	9303	.58	

TABLE III-1 (cont.)

EDMONTON, ALBERTA ELEV 2220 LAT 53.6

	HS	VS	TA	D50	D55	D60	D65	D70	KT	LD
JAN	324	746	4	1421	1574	1728	1883	2038	.55	75
FEB	622	1238	12	1076	1215	1354	1494	1634	.60	67
MAR	1104	1645	21	916	1068	1221	1375	1530	.62	56
APR	1554	1566	38	374	510	655	802	952	.59	44
MAY	1818	1352	51	97	184	307	452	604	.55	35
JUN	1945	1271	57	10	46	121	238	380	.54	30
JUL	1977	1378	62	1	7	38	117	241	.57	32
AUG	1598	1412	60	4	21	83	185	327	.55	40
SEP	1113	1378	51	92	177	297	437	585	.54	52
OCT	689	1185	41	322	458	605	757	911	.55	64
NOV	359	762	23	806	952	1101	1250	1400	.52	73
DEC	235	535	11	1198	1351	1505	1660	1815	.50	77
YR	1114	1205	36	6317	7563	9016	10650	12416	.56	

WINNIPEG, MANITOBA ELEV 820 LAT 49.9

	HS	VS	TA	D50	D55	D60	D65	D70	KT	LD
JAN	461	1011	0	1588	1740	1893	2047	2201	.59	71
FEB	799	1508	4	1303	1440	1578	1717	1857	.65	64
MAR	1240	1676	17	1026	1175	1326	1479	1632	.63	52
APR	1548	1370	38	409	536	674	818	965	.56	40
MAY	1847	1231	51	125	207	314	447	592	.55	31
JUN	2000	1179	62	5	20	68	147	262	.55	27
JUL	2025	1264	67	0	2	9	45	122	.58	29
AUG	1687	1337	66	1	5	21	76	166	.57	36
SEP	1180	1305	55	61	127	214	330	467	.53	48
OCT	726	1082	44	275	387	519	664	814	.50	60
NOV	406	737	24	797	939	1084	1231	1380	.46	69
DEC	336	712	7	1334	1485	1637	1791	1944	.51	73
YR	1190	1199	36	6925	8062	9338	10790	12401	.56	

SUFFIELD, ALBERTA ELEV 2549 LAT 50.3

	HS	VS	TA	D50	D55	D60	D65	D70	KT	LD
JAN	433	937	7	1333	1486	1640	1794	1949	.57	72
FEB	748	1388	17	943	1080	1219	1358	1498	.62	64
MAR	1232	1688	24	815	965	1117	1270	1424	.64	53
APR	1583	1429	41	321	446	584	729	877	.58	41
MAY	1903	1287	53	84	161	262	394	541	.57	31
JUN	2074	1235	61	6	24	78	164	289	.57	27
JUL	2173	1377	67	0	2	11	49	128	.62	29
AUG	1801	1468	65	1	5	23	81	173	.61	37
SEP	1274	1475	55	48	111	195	312	449	.58	48
OCT	822	1329	45	242	355	491	638	790	.57	60
NOV	456	903	28	678	821	968	1116	1265	.53	70
DEC	333	722	17	1029	1181	1334	1487	1642	.52	74
YR	1239	1269	40	5500	6637	7923	9393	11025	.59	

HALIFAX, NOVA SCOTIA ELEV 136 LAT 44.6

	HS	VS	TA	D50	D55	D60	D65	D70	KT	LD
JAN	456	737	26	752	900	1051	1204	1358	.43	66
FEB	695	954	26	682	816	953	1092	1231	.46	58
MAR	1085	1130	32	575	718	868	1020	1174	.49	47
APR	1298	940	41	321	448	588	734	882	.45	35
MAY	1522	882	49	128	219	343	487	638	.44	26
JUN	1738	917	58	10	42	115	225	365	.48	21
JUL	1694	929	65	0	1	8	57	164	.48	23
AUG	1582	1056	65	0	0	5	47	149	.51	31
SEP	1211	1122	60	4	21	76	169	302	.50	43
OCT	812	1024	51	93	176	291	432	583	.47	55
NOV	463	688	42	280	402	540	685	833	.40	64
DEC	332	514	31	612	757	907	1059	1213	.36	68
YR	1076	907	46	3457	4500	5746	7211	8892	.47	

NANAIMO, BRITISH COLUMBIA ELEV 60 LAT 49.2

	HS	VS	TA	D50	D55	D60	D65	D70	KT	LD
JAN	256	398	37	432	579	730	884	1039	.32	70
FEB	538	799	40	304	433	569	708	847	.42	63
MAR	921	1063	42	280	416	564	717	871	.46	52
APR	1409	1189	47	140	246	381	527	676	.51	39
MAY	1860	1215	54	33	100	202	341	493	.55	30
JUN	1956	1132	60	4	20	79	179	316	.54	26
JUL	2092	1279	64	0	2	17	81	197	.60	28
AUG	1727	1344	63	0	2	17	85	211	.58	36
SEP	1234	1354	58	5	28	95	206	348	.55	47
OCT	680	954	50	97	190	319	467	621	.46	59
NOV	333	530	42	256	386	530	678	827	.36	69
DEC	206	327	39	370	513	664	817	972	.30	73
YR	1104	966	50	1921	2915	4169	5691	7417	.51	

MOOSONEE, ONTARIO ELEV 34 LAT 51.3

	HS	VS	TA	D50	D55	D60	D65	D70	KT	LD
JAN	375	800	-4	1696	1848	2000	2154	2308	.53	73
FEB	721	1380	0	1424	1562	1700	1839	1978	.62	65
MAR	1122	1531	11	1237	1387	1538	1691	1845	.59	54
APR	1466	1338	28	692	830	973	1119	1267	.54	42
MAY	1603	1096	42	333	451	586	731	881	.48	32
JUN	1780	1089	53	94	168	258	376	512	.49	28
JUL	1625	1044	60	15	47	113	203	328	.47	30
AUG	1339	1055	58	24	69	144	248	383	.46	38
SEP	956	1026	50	143	227	334	464	606	.44	49
OCT	543	752	40	364	486	624	770	920	.39	61
NOV	297	498	24	795	936	1080	1227	1375	.37	71
DEC	274	571	4	1425	1575	1727	1881	2034	.47	75
YR	1010	1012	30	8243	9584	11078	12702	14436	.49	

VANCOUVER, BRITISH COLUMBIA ELEV 310 LAT 49.3

	HS	VS	TA	D50	D55	D60	D65	D70	KT	LD
JAN	254	395	37	425	572	724	878	1033	.31	71
FEB	495	710	40	287	417	553	693	832	.39	63
MAR	877	997	43	256	392	541	695	849	.44	52
APR	1331	1111	48	127	233	369	516	665	.48	40
MAY	1790	1169	55	24	87	188	329	481	.53	30
JUN	1922	1116	60	4	21	80	179	316	.53	26
JUL	2021	1239	63	0	1	13	82	215	.58	28
AUG	1648	1275	63	0	1	16	92	231	.55	36
SEP	1185	1284	58	4	28	100	218	363	.53	47
OCT	638	874	51	81	171	301	450	604	.43	59
NOV	315	490	43	233	363	508	656	805	.35	69
DEC	201	317	39	351	495	647	800	955	.29	73
YR	1060	916	50	1791	2781	4041	5588	7349	.49	

OTTAWA, ONTARIO ELEV 377 LAT 45.5

	HS	VS	TA	D50	D55	D60	D65	D70	KT	LD
JAN	510	914	13	1169	1320	1473	1627	1782	.51	67
FEB	819	1264	15	979	1116	1255	1394	1533	.56	59
MAR	1203	1347	27	742	889	1040	1193	1347	.56	48
APR	1490	1133	42	287	404	539	682	829	.52	36
MAY	1754	1033	55	61	130	221	344	487	.51	26
JUN	1884	1000	65	1	6	27	87	181	.52	22
JUL	1875	1040	69	0	1	4	23	88	.53	24
AUG	1602	1098	67	0	3	13	57	139	.52	32
SEP	1167	1102	58	23	65	137	231	358	.49	44
OCT	771	986	48	179	275	399	541	690	.46	55
NOV	414	609	35	483	618	760	907	1055	.37	65
DEC	383	667	19	987	1138	1290	1443	1598	.43	69
YR	1158	1015	43	4912	5965	7158	8529	10086	.51	

TABLE III-1 (cont.)

TORONTO, ONTARIO — ELEV 443 LAT 43.7

	HS	VS	TA	D50	D55	D60	D65	D70	KT	LD
JAN	487	777	22	891	1041	1194	1348	1502	.44	65
FEB	747	1021	23	773	910	1048	1187	1327	.48	57
MAR	1100	1116	31	615	761	911	1064	1218	.49	46
APR	1468	1055	44	248	362	495	638	785	.50	34
MAY	1764	994	54	61	132	223	348	492	.51	25
JUN	1950	990	65	2	7	28	86	176	.53	20
JUL	1958	1035	70	0	0	3	18	76	.55	22
AUG	1686	1102	69	0	1	6	32	102	.54	30
SEP	1248	1132	61	10	34	91	172	287	.51	42
OCT	833	1025	50	127	212	326	464	612	.47	54
NOV	428	590	39	371	501	642	788	936	.35	63
DEC	354	542	26	744	893	1044	1198	1352	.36	67
YR	1171	948	46	3842	4853	6013	7343	8865	.50	

NORMANDIN, QUEBEC — ELEV 450 LAT 48.8

	HS	VS	TA	D50	D55	D60	D65	D70	KT	LD
JAN	454	921	0	1564	1719	1873	2028	2183	.55	70
FEB	785	1386	4	1302	1441	1581	1721	1860	.61	62
MAR	1275	1668	17	1042	1195	1349	1504	1658	.63	51
APR	1578	1353	33	532	674	821	970	1119	.56	39
MAY	1652	1059	47	172	277	411	558	711	.49	30
JUN	1831	1052	58	15	54	127	232	367	.50	25
JUL	1707	1031	62	1	7	40	118	242	.49	27
AUG	1471	1097	60	7	29	91	186	320	.49	35
SEP	1018	1026	51	86	165	276	413	558	.45	47
OCT	595	775	41	307	438	584	735	889	.39	59
NOV	375	616	27	689	834	982	1131	1281	.40	68
DEC	343	682	8	1320	1473	1628	1782	1937	.48	72
YR	1092	1053	34	7037	8308	9762	11376	13125	.51	

LOAD COLLECTOR RATIO TABLE

Table III-2 is excerpted from the considerably more detailed LCR table in the *Passive Solar Design Handbook*, Vol. 3, Appendix F. In that reference, systems were categorized partly by their thermal storage capacities and thermal conductivities. What has been done in the following table is to categorize systems by the typical materials and thicknesses used in their construction that correspond closely to thermal capacities and conductivities listed in the *Passive Solar Design Handbook*. It was felt that this would make it easier for the beginning designer to use the table. Values were selected corresponding to the following types of passive systems:

System Abbreviation	*System Description*
WW	18-inch-thick water wall, double-glazed.
WW selective surf	9-inch-thick double-glazed water wall system, with a *selective surface* applied to the collecting surfaces on the wall.
TW concrete	18-inch-thick vented concrete Trombe wall, double-glazed (density of concrete is 143 lb/ft^3).
TW brick	An *average* of the characteristics of vented walls constructed of face brick (with thermal conductivity $k = 0.77$ Btu/ft-°F-hr) and common brick ($k = 0.40$). For greater precision, if the wall is made of face brick, use an LCR value midway between that listed in this table for TW brick and that listed for TW concrete. If common brick is used for the wall, an LCR value midway between TW brick and TW adobe should be employed.
TW adobe	11-inch-thick vented adobe Trombe wall, double-glazed.
TW sel surf concrete	12-inch-thick vented concrete Trombe wall with selective surface applied to it, double-glazed.
TW unvented concrete	18-inch-thick *unvented* Trombe wall, double-glazed.
DG	Direct-gain system; 6 ft^2 of 4-inch-thick concrete thermal storage mass is used for every square foot of aperture.
DG mov insul	Same as above, but with *movable insulation* for all apertures.
SS attchd sngl plane, mov insul	Single-plane attached sunspace with movable insulation. Tilt of glazing is 50° with respect to horizontal. End walls of *all* sunspaces considered in these tables are opaque. Common walls (walls shared with the living space) are of masonry construction.
SS attchd shed roof	Attached shed roof sunspace. Tilt angles of the two glazing planes are 90° (south-wall glazing) and 30° (roof glazing).
SS attchd shed roof, mov insul	Same as above, but with movable insulation for all apertures.
SS semienclosed, mov insul	Semienclosed shed roof sunspace with movable insulation. Angles of glazing planes are the same as for attached shed roof sunspace, 90° (south-wall) and 30° (roof glazing).

Translucent glazing is used in the foregoing systems. This diffuses light more evenly over the collecting and storage mass surfaces than clear glass would. Thermal resistance of the movable insulation is R-9. It is installed in the systems noted above from 5:30 P.M. to 7:30 A.M.

Load collector ratios in this table correspond to solar savings fractions of 30 percent ("solar tempered" structures), 60 percent ("moderately solar heated" buildings), and 80 percent ("heavily solar heated" buildings). As detailed in Chapter 4, after a desired solar savings fraction for a particular system is chosen, the load collector ratio for the location of that system is read from this table, and divided into the building's load coefficient to yield the required area of solar aperture.

TABLE III-2 Load Collector Ratios

	Birmingham, AL			*Mobile, AL*			*Montgomery, AL*			*Phoenix, AZ*		
SSF:	*30%*	*60%*	*80%*	*30%*	*60%*	*80%*	*30%*	*60%*	*80%*	*30%*	*60%*	*80%*
WW	80	29	16	139	53	29	104	39	21	242	95	52
WW selective surf	98	38	21	159	64	36	123	48	27	267	108	60
TW concrete	63	22	11	110	39	21	82	29	15	190	70	37
TW brick	57	18	9	100	33	17	74	24	12	173	59	31
TW adobe	49	15	8	84	28	15	63	20	11	146	50	26
TW sel surf concrete	93	34	18	151	57	31	117	43	23	253	98	53
TW unvented concrete	50	18	9	87	33	18	66	24	13	152	58	31
DG	60	21	7	112	45	24	81	30	14	205	85	48
DG mov insul	86	35	21	135	58	35	106	44	26	225	97	59
SS attchd sngl plane, mov insul	120	39	20	197	67	35	152	50	26	314	107	55
SS attchd shed roof	67	20	10	118	37	19	88	27	13	195	62	32
SS attchd shed roof, mov insul	98	32	17	160	55	29	123	41	22	254	88	46
SS semienclosed, mov insul	115	42	22	192	71	38	147	53	28	308	115	61

	Prescott, AZ			*Tucson, AZ*			*Winslow, AZ*			*Yuma, AZ*		
SSF:	*30%*	*60%*	*80%*	*30%*	*60%*	*80%*	*30%*	*60%*	*80%*	*30%*	*60%*	*80%*
WW	100	40	22	223	89	50	88	34	18	382	148	80
WW selective surf	119	49	28	247	102	58	106	43	24	412	165	91
TW concrete	79	29	16	175	66	36	69	25	13	300	109	58
TW brick	71	24	13	159	56	30	62	21	11	275	93	49
TW adobe	60	21	11	134	47	25	53	18	9	230	78	40
TW sel surf concrete	112	44	24	235	92	51	100	38	21	393	150	81
TW unvented concrete	63	24	13	140	55	30	55	21	11	241	91	49
DG	78	32	15	186	81	47	67	25	11	334	136	79
DG mov insul	101	45	27	206	92	57	91	39	24	350	147	89
SS attchd sngl plane, mov insul	151	51	27	300	104	55	130	43	22	477	161	83
SS attchd shed roof	86	27	14	184	60	31	73	22	11	304	97	49
SS attchd shed roof, mov insul	121	42	22	241	85	46	105	36	19	387	133	69
SS semienclosed, mov insul	144	54	29	293	111	61	125	45	24	471	174	92

Note: Solar tempered: SSF of 30%; Moderately solar heated: SSF of 60%; Heavily solar heated: SSF of 80%; WW, water wall; TW, Trombe wall; DG, direct gain; SS, sunspace.

TABLE III-2 (cont'd)

	Fort Smith, AR			Little Rock, AR			Bakersfield, CA			Daggett, CA		
SSF:	30%	60%	80%	30%	60%	80%	30%	60%	80%	30%	60%	80%
WW	71	26	14	68	25	13	141	52	27	179	69	38
WW selective surf	88	35	19	86	34	19	162	63	33	202	81	45
TW concrete	56	19	10	54	19	10	112	38	19	141	51	27
TW brick	50	16	8	48	15	8	102	32	16	129	43	23
TW adobe	43	14	7	42	13	7	86	27	14	109	36	19
TW sel surf concrete	83	31	17	81	30	16	154	56	30	192	73	40
TW unvented concrete	44	16	9	43	15	8	89	32	16	113	42	23
DG	51	17	—	49	16	—	117	43	20	148	60	33
DG mov insul	77	32	19	75	31	18	140	57	32	170	73	44
SS attchd sngl plane, mov insul	104	34	18	103	34	17	196	62	30	238	79	41
SS attchd shed roof	57	17	8	56	17	8	117	34	16	145	45	23
SS attchd shed roof, mov insul	86	29	15	84	28	15	159	52	26	193	66	34
SS semienclosed, mov insul	100	36	19	99	35	19	190	66	33	232	85	45

	Fresno, CA			Long Beach, CA			Los Angeles, CA			Mt. Shasta, CA		
SSF:	30%	60%	80%	30%	60%	80%	30%	60%	80%	30%	60%	80%
WW	113	39	19	261	102	56	270	104	58	58	20	10
WW selective surf	133	49	26	286	116	65	296	119	66	76	29	15
TW concrete	90	29	14	205	75	41	213	77	42	46	15	7
TW brick	82	25	12	187	64	34	194	66	35	41	12	6
TW adobe	70	21	10	157	53	28	163	55	29	36	11	5
TW sel surf concrete	127	44	22	272	105	58	282	107	59	71	25	13
TW unvented concrete	71	24	12	164	62	34	170	64	35	36	12	6
DG	92	30	11	219	93	54	228	95	55	40	10	—
DG mov insul	117	45	25	239	104	64	248	106	66	68	27	15
SS attchd sngl plane, mov insul	163	49	23	369	125	65	401	135	70	97	29	14
SS attchd shed roof	95	26	11	230	73	37	250	79	40	51	13	5
SS attchd shed roof, mov insul	132	41	20	297	102	54	321	110	57	78	24	12
SS semienclosed, mov insul	156	51	25	359	133	71	388	143	76	91	30	14

Note: Solar tempered: SSF of 30%; Moderately solar heated: SSF of 60%; Heavily solar heated: SSF of 80%; WW, water wall; TW, Trombe wall; DG, direct gain; SS, sunspace.

TABLE III-2 (cont'd)

	Needles, CA			*Oakland, CA*			*Red Bluff, CA*			*Sacramento, CA*		
SSF:	*30%*	*60%*	*80%*	*30%*	*60%*	*80%*	*30%*	*60%*	*80%*	*30%*	*60%*	*80%*
WW	266	102	55	168	63	33	106	37	19	109	38	19
WW selective surf	291	117	64	190	74	40	125	47	25	129	48	25
TW concrete	209	76	40	133	47	24	84	28	14	87	28	14
TW brick	191	64	33	121	39	20	76	23	11	79	24	11
TW adobe	160	54	28	102	33	17	65	20	10	67	20	10
TW sel surf concrete	277	106	57	181	67	36	119	42	22	123	43	22
TW unvented concrete	167	63	33	106	39	20	67	23	11	69	23	11
DG	229	92	52	142	54	28	84	28	10	89	29	10
DG mov insul	248	104	62	165	67	39	109	43	24	114	44	24
SS attchd sngl plane, mov insul	330	111	67	262	81	39	147	45	21	161	48	22
SS attchd shed roof	208	65	33	158	46	21	86	24	11	94	25	11
SS attchd shed roof, mov insul	270	92	48	209	67	33	121	38	18	131	40	19
SS semienclosed, mov insul	327	120	63	251	86	43	142	48	24	154	50	24

	San Diego, CA			*San Francisco, CA*			*Santa Maria, CA*			*Colorado Springs, CO*		
SSF:	*30%*	*60%*	*80%*	*30%*	*60%*	*80%*	*30%*	*60%*	*80%*	*30%*	*60%*	*80%*
WW	302	118	65	161	60	32	175	69	38	70	27	15
WW selective surf	329	134	75	182	71	39	197	81	45	87	35	20
TW concrete	238	87	47	127	44	23	137	51	28	55	20	11
TW brick	217	74	40	115	38	19	124	43	23	49	16	9
TW adobe	182	62	33	97	32	16	105	36	19	42	14	7
TW sel surf concrete	314	121	67	173	64	34	186	73	40	82	32	17
TW unvented concrete	190	72	40	101	37	19	110	42	23	44	16	9
DG	256	109	64	135	51	26	146	61	34	50	19	6
DG mov insul	275	119	74	158	64	38	167	73	45	75	32	20
SS attchd sngl plane, mov insul	442	149	78	254	79	38	283	98	50	105	35	18
SS attchd shed roof	276	88	45	153	44	21	170	55	28	57	17	9
SS attchd shed roof, mov insul	353	121	64	202	65	32	224	79	41	85	29	15
SS semienclosed, mov insul	429	159	85	243	83	42	272	102	54	101	37	20

Note: Solar tempered: SSF of 30%; Moderately solar heated: SSF of 60%; Heavily solar heated: SSF of 80%; WW, water wall; TW, Trombe wall; DG, direct gain; SS, sunspace.

TABLE III-2 (cont'd)

	Denver, CO			*Eagle, CO*			*Grand Junction, CO*			*Pueblo, CO*		
SSF:	*30%*	*60%*	*80%*	*30%*	*60%*	*80%*	*30%*	*60%*	*80%*	*30%*	*60%*	*80%*
WW	72	28	15	46	17	9	67	25	13	76	29	16
WW selective surf	89	36	20	63	25	13	84	33	18	94	38	21
TW concrete	56	20	11	36	12	6	53	18	9	60	22	12
TW brick	50	17	9	32	10	5	47	15	8	53	18	9
TW adobe	43	14	8	28	9	4	41	13	7	46	15	8
TW sel surf concrete	84	32	18	59	22	12	79	30	16	88	34	19
TW unvented concrete	45	17	9	29	10	5	42	15	8	48	18	10
DG	52	19	6	29	—	—	48	15	—	56	21	8
DG mov insul	77	33	20	56	23	13	74	31	18	81	35	21
SS attchd sngl plane, mov insul	105	35	18	78	24	14	97	31	16	109	37	19
SS attchd shed roof	57	18	9	39	10	4	52	15	7	60	18	9
SS attchd shed roof, mov insul	86	29	15	64	20	10	80	26	13	89	31	16
SS semienclosed, mov insul	101	37	20	73	25	12	93	33	17	105	39	21

	Hartford, CT			*Wilmington, DE*			*Washington, DC*			*Apalachicola, FL*		
SSF:	*30%*	*60%*	*80%*	*30%*	*60%*	*80%*	*30%*	*60%*	*80%*	*30%*	*60%*	*80%*
WW	18	4	1	38	13	7	36	12	6	177	68	37
WW selective surf	36	13	7	55	21	11	53	20	11	200	80	44
TW concrete	14	3	—	30	10	5	28	9	4	140	50	27
TW brick	12	1	—	27	8	4	25	7	3	128	43	22
TW adobe	12	2	—	24	7	3	22	6	3	108	36	19
TW sel surf concrete	33	11	5	51	19	10	49	17	9	190	72	39
TW unvented concrete	11	2	—	24	8	4	22	7	3	112	42	23
DG	—	—	—	22	—	—	19	—	—	147	59	33
DG mov insul	34	13	6	50	20	11	48	19	10	169	72	44
SS attchd sngl plane, mov insul	42	12	5	65	20	10	63	19	9	249	85	44
SS attchd shed roof	13	—	—	30	8	3	29	7	3	152	48	25
SS attchd shed roof, mov insul	35	10	5	54	17	9	52	16	8	201	70	37
SS semienclosed, mov insul	38	12	5	61	21	10	59	20	10	242	90	49

Note: Solar tempered: SSF of 30%; Moderately solar heated: SSF of 60%; Heavily solar heated: SSF of 80%; WW, water wall; TW, Trombe wall; DG, direct gain; SS, sunspace.

TABLE III-2 (cont'd)

	Daytona Beach, FL			Jacksonville, FL			Miami, FL			Orlando, FL		
SSF:	30%	60%	80%	30%	60%	80%	30%	60%	80%	30%	60%	80%
WW	287	114	67	192	75	42	1140	444	243	359	143	80
WW selective surf	314	129	73	215	88	49	1204	485	269	389	161	91
TW concrete	226	84	46	151	56	30	897	328	176	282	106	58
TW brick	206	72	38	138	47	25	821	281	148	258	90	48
TW adobe	173	60	32	116	40	21	686	233	123	216	75	40
TW sel surf concrete	299	117	65	204	79	44	1150	441	240	371	145	81
TW unvented concrete	181	70	39	121	46	25	719	273	148	226	88	48
DG	243	105	62	158	67	38	1018	426	259	308	133	80
DG mov insul	261	115	72	179	79	49	1019	429	264	325	143	89
SS attchd sngl plane, mov insul	388	137	73	262	91	49	1534	532	282	484	172	92
SS attchd shed roof	243	81	43	161	52	27	990	324	169	305	102	54
SS attchd shed roof, mov insul	314	112	60	212	75	40	1231	432	230	390	141	76
SS semienclosed, mov insul	382	147	81	257	98	53	1512	570	309	477	185	101

	Tallahassee, FL			Tampa, FL			West Palm Beach, FL			Atlanta, GA		
SSF:	30%	60%	80%	30%	60%	80%	30%	60%	80%	30%	60%	80%
WW	160	62	34	357	143	79	760	298	163	74	27	14
WW selective surf	181	73	41	387	160	90	808	327	182	92	36	20
TW concrete	126	46	25	280	105	57	598	220	118	58	20	10
TW brick	115	39	20	256	90	48	547	188	99	52	17	8
TW adobe	97	32	17	214	75	40	457	156	82	45	14	7
TW sel surf concrete	172	66	36	368	145	80	771	297	162	86	32	17
TW unvented concrete	101	38	21	225	87	48	480	183	99	46	16	9
DG	130	53	29	307	133	80	678	282	171	54	18	—
DG mov insul	152	66	40	324	143	89	687	289	179	80	33	19
SS attchd sngl plane, mov insul	223	76	40	482	172	92	1015	354	187	113	37	19
SS attchd shed roof	135	43	22	304	102	54	653	215	112	62	18	9
SS attchd shed roof, mov insul	181	63	33	389	141	75	818	289	153	92	30	16
SS semienclosed, mov insul	217	81	44	476	185	101	1004	380	206	108	38	20

Note: Solar tempered: SSF of 30%; Moderately solar heated: SSF of 60%; Heavily solar heated: SSF of 80%; WW, water wall; TW, Trombe wall; DG, direct gain; SS, sunspace.

TABLE III-2 (cont'd)

	Augusta, GA			*Macon, GA*			*Savannah, GA*			*Boise, ID*		
SSF:	*30%*	*60%*	*80%*	*30%*	*60%*	*80%*	*30%*	*60%*	*80%*	*30%*	*60%*	*80%*
WW	96	36	19	108	41	22	126	48	26	52	16	7
WW selective surf	115	45	25	128	50	28	146	58	32	71	25	12
TW concrete	76	27	14	86	30	16	100	35	19	42	12	5
TW brick	68	22	11	78	25	13	90	30	16	37	9	4
TW adobe	58	19	10	66	21	11	77	25	13	33	9	3
TW sel surf concrete	109	41	22	121	45	25	139	52	28	67	22	11
TW unvented concrete	60	22	12	68	25	13	79	29	16	33	10	4
DG	74	27	12	85	32	15	101	39	20	36	—	—
DG mov insul	99	41	25	109	46	27	125	52	32	66	24	12
SS attchd sngl plane, mov insul	139	46	24	155	51	27	178	60	31	83	23	10
SS attchd shed roof	80	24	12	90	28	14	106	33	17	42	9	3
SS attchd shed roof, mov insul	113	38	20	126	43	22	145	49	26	69	20	9
SS semienclosed, mov insul	134	49	26	150	55	29	174	64	34	78	23	10

	Lewiston, ID			*Pocatello, ID*			*Chicago, IL*			*Moline, IL*		
SSF:	*30%*	*60%*	*80%*	*30%*	*60%*	*80%*	*30%*	*60%*	*80%*	*30%*	*60%*	*80%*
WW	31	5	—	45	14	6	24	6	2	24	7	3
WW selective surf	51	16	7	63	23	11	41	15	8	41	15	8
TW concrete	26	3	—	36	10	4	19	5	2	19	5	2
TW brick	22	—	—	32	8	3	16	3	—	16	3	—
TW adobe	21	3	—	28	7	3	15	3	1	15	3	1
TW sel surf concrete	47	14	6	59	20	10	38	13	6	38	13	6
TW unvented concrete	20	2	—	28	8	4	15	4	1	15	4	1
DG	—	—	—	29	—	—	—	—	—	—	—	—
DG mov insul	50	16	7	58	22	11	39	14	7	39	14	7
SS attchd sngl plane, mov insul	59	14	5	75	21	9	48	14	6	48	14	6
SS attchd shed roof	24	—	—	36	8	2	18	—	—	18	2	—
SS attchd shed roof, mov insul	49	12	5	61	18	8	40	12	5	40	12	6
SS semienclosed, mov insul	53	13	5	70	21	10	44	13	6	43	13	6

Note: Solar tempered: SSF of 30%; Moderately solar heated: SSF of 60%; Heavily solar heated: SSF of 80%; WW, water wall; TW, Trombe wall; DG, direct gain; SS, sunspace.

TABLE III-2 (cont'd)

	Springfield, IL			*Evansville, IN*			*Ft. Wayne, IN*			*Indianapolis, IN*		
SSF:	*30%*	*60%*	*80%*	*30%*	*60%*	*80%*	*30%*	*60%*	*80%*	*30%*	*60%*	*80%*
WW	33	11	5	37	12	6	15	—	—	22	6	2
WW selective surf	50	19	10	55	20	11	34	12	6	40	14	7
TW concrete	26	8	4	30	9	4	12	—	—	18	4	1
TW brick	23	6	3	26	7	3	9	—	—	15	3	—
TW adobe	21	6	3	23	6	3	10	—	—	14	3	1
TW sel surf concrete	47	16	8	51	18	9	31	10	5	37	12	6
TW unvented concrete	21	6	3	23	7	4	9	—	—	14	3	1
DG	15	—	—	20	—	—	—	—	—	—	—	—
DG mov insul	46	18	10	50	19	10	33	12	6	38	14	7
SS attchd sngl plane, mov insul	59	18	9	66	20	10	41	11	5	48	14	6
SS attchd shed roof	26	6	2	30	7	3	11	—	—	17	2	—
SS attchd shed roof, mov insul	49	15	7	54	17	8	34	9	4	40	12	6
SS semienclosed, mov insul	55	18	9	61	20	10	36	10	4	43	13	6

	South Bend, IN			*Burlington, IA*			*Des Moines, IA*			*Mason City, IA*		
SSF:	*30%*	*60%*	*80%*	*30%*	*60%*	*80%*	*30%*	*60%*	*80%*	*30%*	*60%*	*80%*
WW	11	—	—	30	10	4	28	9	4	23	6	2
WW selective surf	31	10	5	48	18	9	46	17	9	41	15	7
TW concrete	9	—	—	24	7	3	23	6	3	18	4	2
TW brick	5	—	—	21	5	2	19	5	2	15	3	—
TW adobe	8	—	—	19	5	2	18	5	2	14	3	1
TW sel surf concrete	29	9	4	44	15	8	42	15	7	37	12	6
TW unvented concrete	6	—	—	19	6	3	18	5	2	14	4	1
DG	—	—	—	12	—	—	—	—	—	—	—	—
DG mov insul	32	10	5	44	17	9	43	16	9	39	14	7
SS attchd sngl plane, mov insul	38	9	4	55	16	8	52	15	7	46	13	6
SS attchd shed roof	—	—	—	23	5	2	21	4	1	16	—	—
SS attchd shed roof, mov insul	32	8	4	46	14	7	44	13	6	39	11	5
SS semienclosed, mov insul	33	8	3	51	16	8	48	15	7	42	13	6

Note: Solar tempered: SSF of 30%; Moderately solar heated: SSF of 60%; Heavily solar heated: SSF of 80%; WW, water wall; TW, Trombe wall; DG, direct gain; SS, sunspace.

TABLE III-2 (cont'd)

	Sioux City, IA			Dodge City, KS			Goodland, KS			Topeka, KS		
SSF:	30%	60%	80%	30%	60%	80%	30%	60%	80%	30%	60%	80%
WW	28	8	4	68	25	14	59	22	12	47	17	8
WW selective surf	45	16	9	85	34	19	76	30	17	65	25	13
TW concrete	22	6	3	53	19	10	47	16	9	37	12	6
TW brick	19	4	2	48	15	8	41	13	7	33	10	5
TW adobe	17	4	2	41	13	7	36	12	6	29	9	4
TW sel surf concrete	42	14	7	80	30	16	72	27	15	60	22	11
TW unvented concrete	17	5	2	42	15	8	37	13	7	30	10	5
DG	—	—	—	48	17	—	41	13	—	30	—	—
DG mov insul	42	16	8	74	31	18	67	28	17	58	23	13
SS attchd sngl plane, mov insul	51	15	7	100	33	17	90	29	15	75	23	12
SS attchd shed roof	20	3	—	54	16	8	47	14	7	37	10	4
SS attchd shed roof, mov insul	43	13	6	82	27	14	74	25	13	62	20	10
SS semienclosed, mov insul	47	15	7	96	34	18	86	31	16	71	24	12

	Wichita, KS			Lexington, KY			Louisville, KY			Baton Rouge, LA		
SSF:	30%	60%	80%	30%	60%	80%	30%	60%	80%	30%	60%	80%
WW	64	23	12	34	11	5	35	11	6	133	50	27
WW selective surf	81	32	17	51	19	10	52	19	10	153	61	33
TW concrete	50	17	9	27	8	4	28	8	4	105	37	20
TW brick	45	14	7	23	6	3	24	7	3	96	31	16
TW adobe	39	12	6	21	6	3	22	6	3	81	26	14
TW sel surf concrete	76	28	15	47	16	8	49	17	9	146	54	30
TW unvented concrete	40	14	7	21	6	3	22	7	3	84	31	16
DG	45	14	—	16	—	—	18	—	—	107	42	21
DG mov insul	71	29	17	47	18	10	48	18	10	130	55	33
SS attchd sngl plane, mov insul	95	31	16	62	18	9	63	19	9	190	64	33
SS attchd shed roof	50	15	7	28	6	3	28	7	3	114	35	18
SS attchd shed roof, mov insul	78	26	13	51	16	8	52	16	8	155	53	27
SS semienclosed, mov insul	91	32	17	57	19	9	59	19	10	185	68	36

Note: Solar tempered: SSF of 30%; Moderately solar heated: SSF of 60%; Heavily solar heated: SSF of 80%; WW, water wall; TW, Trombe wall; DG, direct gain; SS, sunspace.

TABLE III-2 (cont'd)

	Lake Charles, LA			*New Orleans, LA*			*Shreveport, LA*			*Caribou, ME*		
SSF:	*30%*	*60%*	*80%*	*30%*	*60%*	*80%*	*30%*	*60%*	*80%*	*30%*	*60%*	*80%*
WW	133	50	27	162	62	34	109	41	22	9	—	—
WW selective surf	153	61	33	184	73	41	129	51	28	30	10	4
TW concrete	105	37	20	128	46	24	86	30	16	8	—	—
TW brick	96	31	16	117	39	20	78	25	13	—	—	—
TW adobe	81	26	14	99	33	17	67	22	11	7	—	—
TW sel surf concrete	146	54	29	175	66	36	122	45	25	27	8	3
TW unvented concrete	84	31	16	102	38	20	69	25	13	—	—	—
DG	107	42	21	132	53	29	86	32	15	—	—	—
DG mov insul	130	55	33	155	66	40	110	46	27	30	10	4
SS attchd sngl plane, mov insul	192	65	33	229	77	41	156	52	27	34	8	3
SS attchd shed roof	115	36	18	139	44	22	91	28	14	—	—	—
SS attchd shed roof, mov insul	156	53	28	185	64	34	128	43	23	29	7	3
SS semienclosed, mov insul	187	69	36	223	82	44	152	55	29	29	7	2

	Portland, ME			*Baltimore, MD*			*Boston, MA*			*Alpena, MI*		
SSF:	*30%*	*60%*	*80%*	*30%*	*60%*	*80%*	*30%*	*60%*	*80%*	*30%*	*60%*	*80%*
WW	16	3	—	41	14	7	27	8	4	—	—	—
WW selective surf	34	12	6	58	22	12	44	16	9	27	8	3
TW concrete	13	2	—	32	10	5	22	6	3	—	—	—
TW brick	10	—	—	29	8	4	19	5	2	—	—	—
TW adobe	10	1	—	25	7	4	17	4	2	—	—	—
TW sel surf concrete	31	10	5	54	20	10	41	14	7	25	6	2
TW unvented concrete	10	—	—	26	9	4	17	5	2	—	—	—
DG	—	—	—	24	—	—	—	—	—	—	—	—
DG mov insul	33	12	6	52	21	12	41	16	8	29	9	3
SS attchd sngl plane, mov insul	41	11	5	68	21	11	52	15	7	34	7	2
SS attchd shed roof	12	—	—	33	8	4	21	4	1	—	—	—
SS attchd shed roof, mov insul	34	10	4	56	18	9	43	13	6	28	6	2
SS semienclosed, mov insul	36	10	5	64	22	11	48	15	7	28	5	—

Note: Solar tempered: SSF of 30%; Moderately solar heated: SSF of 60%; Heavily solar heated: SSF of 80%; WW, water wall; TW, Trombe wall; DG, direct gain; SS, sunspace.

TABLE III-2 (cont'd)

	Detroit, MI			Flint, MI			Grand Rapids, MI			Sault Ste. Marie, MI		
SSF:	30%	60%	80%	30%	60%	80%	30%	60%	80%	30%	60%	80%
WW	15	—	—	—	—	—	—	—	—	—	—	—
WW selective surf	34	12	6	28	9	4	29	9	4	24	6	2
TW concrete	12	—	—	—	—	—	—	—	—	—	—	—
TW brick	9	—	—	—	—	—	—	—	—	—	—	—
TW adobe	10	—	—	—	—	—	5	—	—	—	—	—
TW sel surf concrete	31	10	5	25	7	3	27	7	3	21	4	—
TW unvented concrete	9	—	—	—	—	—	—	—	—	—	—	—
DG	—	—	—	—	—	—	—	—	—	—	—	—
DG mov insul	33	11	5	29	9	4	30	9	4	27	7	2
SS attchd sngl plane, mov insul	40	11	5	34	8	3	35	8	3	29	4	—
SS attchd shed roof	10	—	—	—	—	—	—	—	—	—	—	—
SS attchd shed roof, mov insul	34	9	4	29	7	3	30	7	3	25	4	—
SS semienclosed, mov insul	36	10	4	29	7	2	30	7	2	23	—	—

	Traverse City, MI			Duluth, MN			International Falls, MN			Minneapolis-St. Paul, MN		
SSF:	30%	60%	80%	30%	60%	80%	30%	60%	80%	30%	60%	80%
WW	—	—	—	—	—	—	—	—	—	12	—	—
WW selective surf	24	6	2	26	8	3	23	6	2	32	11	5
TW concrete	—	—	—	—	—	—	—	—	—	10	—	—
TW brick	—	—	—	—	—	—	—	—	—	7	—	—
TW adobe	—	—	—	—	—	—	—	—	—	9	—	—
TW sel surf concrete	22	5	1	23	6	2	20	4	—	29	9	4
TW unvented concrete	—	—	—	—	—	—	—	—	—	7	—	—
DG	—	—	—	—	—	—	—	—	—	—	—	—
DG mov insul	27	7	2	27	8	3	26	7	2	32	11	5
SS attchd sngl plane, mov insul	31	5	1	30	6	2	25	3	—	36	9	4
SS attchd shed roof	—	—	—	—	—	—	—	—	—	—	—	—
SS attchd shed roof, mov insul	26	5	1	25	5	2	22	4	—	30	8	3
SS semienclosed, mov insul	24	—	—	24	3	—	19	—	—	31	8	3

Note: Solar tempered: SSF of 30%; Moderately solar heated: SSF of 60%; Heavily solar heated: SSF of 80%; WW, water wall; TW, Trombe wall; DG, direct gain; SS, sunspace.

TABLE III-2 (cont'd)

	Rochester, MN			*Jackson, MS*			*Meridian, MS*			*Columbia, MO*		
SSF:	*30%*	*60%*	*80%*	*30%*	*60%*	*80%*	*30%*	*60%*	*80%*	*30%*	*60%*	*80%*
WW	12	—	—	101	38	20	97	36	19	39	13	6
WW selective surf	31	11	5	120	47	26	115	45	25	56	21	11
TW concrete	10	—	—	80	28	15	76	27	14	31	9	5
TW brick	7	—	—	72	23	12	69	22	11	27	8	3
TW adobe	8	—	—	62	20	10	59	19	10	24	7	3
TW sel surf concrete	29	9	4	114	42	23	109	41	22	52	18	10
TW unvented concrete	7	—	—	64	23	12	61	22	12	24	8	4
DG	—	—	—	78	29	13	74	27	12	21	—	—
DG mov insul	31	11	5	104	43	26	100	41	24	51	20	11
SS attchd sngl plane, mov insul	36	9	4	148	49	25	142	47	24	66	20	10
SS attchd shed roof	—	—	—	86	26	13	81	25	12	31	8	3
SS attchd shed roof, mov insul	30	8	4	120	40	21	115	39	20	55	17	8
SS semienclosed, mov insul	31	8	3	143	52	27	137	49	26	62	20	10

	Kansas City, MO			*St. Louis, MO*			*Springfield, MO*			*Billings, MT*		
SSF:	*30%*	*60%*	*80%*	*30%*	*60%*	*80%*	*30%*	*60%*	*80%*	*30%*	*60%*	*80%*
WW	41	14	7	44	15	8	51	18	9	37	12	5
WW selective surf	58	22	12	62	23	12	68	26	14	55	20	10
TW concrete	33	10	5	35	11	5	40	13	7	30	9	4
TW brick	29	8	4	31	9	4	36	11	5	26	7	3
TW adobe	25	7	4	27	8	4	31	9	5	23	6	3
TW sel surf concrete	55	20	10	58	20	11	64	23	12	51	18	9
TW unvented concrete	26	9	4	28	9	4	32	11	6	23	7	3
DG	24	—	—	27	—	—	33	7	—	20	—	—
DG mov insul	53	21	12	56	22	12	61	24	14	51	19	10
SS attchd sngl plane, mov insul	68	21	10	72	22	11	81	26	13	62	18	8
SS attchd shed roof	32	8	4	35	9	4	41	11	5	28	5	2
SS attchd shed roof, mov insul	56	18	9	60	19	9	67	22	11	51	15	7
SS semienclosed, mov insul	64	22	11	68	23	12	77	26	14	58	18	8

Note: Solar tempered: SSF of 30%; Moderately solar heated: SSF of 60%; Heavily solar heated: SSF of 80%; WW, water wall; TW, Trombe wall; DG, direct gain; SS, sunspace.

TABLE III-2 (cont'd)

	Cut Bank, MT			Dillon, MT			Glasgow, MT			Great Falls, MT		
SSF:	30%	60%	80%	30%	60%	80%	30%	60%	80%	30%	60%	80%
WW	28	6	—	40	13	6	15	—	—	32	8	2
WW selective surf	46	16	8	57	21	11	35	11	5	50	17	8
TW concrete	23	4	—	32	10	4	12	—	—	26	6	2
TW brick	20	2	—	28	8	3	10	—	—	23	4	—
TW adobe	18	3	—	25	7	3	11	—	—	21	4	1
TW sel surf concrete	43	14	6	53	19	9	32	9	4	47	15	7
TW unvented concrete	18	3	—	25	8	3	9	—	—	20	5	1
DG	—	—	—	23	—	—	—	—	—	14	—	—
DG mov insul	45	15	7	52	20	11	36	11	5	48	17	8
SS attchd sngl plane, mov insul	52	13	5	67	19	9	38	9	3	57	15	6
SS attchd shed roof	20	—	—	31	7	2	—	—	—	24	—	—
SS attchd shed roof, mov insul	44	12	5	55	17	8	32	8	3	47	13	6
SS semienclosed, mov insul	48	13	5	62	20	9	33	8	3	52	15	6

	Helena, Mt			Lewiston, MT			Miles City, MT			Missoula, MT		
SSF:	30%	60%	80%	30%	60%	80%	30%	60%	80%	30%	60%	80%
WW	28	6	1	28	7	2	28	7	2	—	—	—
WW selective surf	46	16	8	46	16	8	46	16	8	34	10	4
TW concrete	22	5	—	23	5	1	22	5	2	9	—	—
TW brick	19	2	—	20	4	—	19	3	—	—	—	—
TW adobe	18	3	—	18	4	1	18	4	1	9	—	—
TW sel surf concrete	43	14	6	43	14	7	43	14	7	31	8	2
TW unvented concrete	17	3	—	18	4	1	17	4	1	—	—	—
DG	—	—	—	—	—	—	—	—	—	—	—	—
DG mov insul	44	15	7	44	16	8	44	15	8	36	10	4
SS attchd sngl plane, mov insul	53	14	6	54	14	6	51	13	6	41	8	2
SS attchd shed roof	21	—	—	21	—	—	19	—	—	—	—	—
SS attchd shed roof, mov insul	44	12	5	45	13	5	43	12	5	34	7	2
SS semienclosed, mov insul	48	13	5	49	14	6	46	13	6	35	6	—

Note: Solar tempered: SSF of 30%; Moderately solar heated: SSF of 60%; Heavily solar heated: SSF of 80%; WW, water wall; TW, Trombe wall; DG, direct gain; SS, sunspace.

TABLE III-2 (cont'd)

	Grand Island, NE			North Omaha, NE			North Platte, NE			Scottsbluff, NE		
SSF:	30%	60%	80%	30%	60%	80%	30%	60%	80%	30%	60%	80%
WW	42	14	7	33	11	5	47	17	9	49	18	9
WW selective surf	59	22	12	50	19	10	64	25	14	66	26	14
TW concrete	33	10	5	26	8	4	37	12	6	38	13	7
TW brick	29	8	4	23	6	3	33	10	5	34	10	5
TW adobe	26	8	4	20	6	3	29	9	4	30	9	5
TW sel surf concrete	55	20	10	47	16	9	60	22	12	61	23	12
TW unvented concrete	26	9	4	21	7	3	29	10	5	30	11	5
DG	24	—	—	15	—	—	30	—	—	31	6	—
DG mov insul	53	21	12	46	18	10	57	23	13	59	24	14
SS attchd sngl plane, mov insul	67	21	10	57	17	8	74	23	12	76	24	12
SS attchd shed roof	32	8	4	25	6	2	36	10	4	38	10	5
SS attchd shed roof, mov insul	56	18	9	48	15	7	61	20	10	62	20	10
SS semienclosed, mov insul	63	21	11	53	17	9	70	24	12	72	25	13

	Elko, NV			Ely, NV			Las Vegas, NV			Lovelock, NV		
SSF:	30%	60%	80%	30%	60%	80%	30%	60%	80%	30%	60%	80%
WW	54	19	10	59	22	12	163	63	34	82	30	16
WW selective surf	72	28	15	76	30	17	184	74	41	100	39	21
TW concrete	43	14	7	46	16	9	128	46	25	64	22	12
TW brick	38	12	6	41	13	7	117	39	21	58	19	9
TW adobe	33	10	5	35	12	6	98	33	17	49	16	8
TW sel surf concrete	67	25	13	71	27	15	175	67	37	94	35	19
TW unvented concrete	34	12	6	37	13	7	103	39	21	51	18	10
DG	37	9	—	41	13	—	133	54	29	62	21	7
DG mov insul	64	26	14	66	28	17	156	67	41	87	36	21
SS attchd sngl plane, mov insul	89	27	13	95	31	16	212	71	37	119	38	19
SS attchd shed roof	46	12	5	50	15	7	128	40	20	66	19	9
SS attchd shed roof, mov insul	72	23	11	77	26	13	172	59	31	97	32	16
SS semienclosed, mov insul	83	28	14	90	32	17	207	76	40	114	40	20

Note: Solar tempered: SSF of 30%; Moderately solar heated: SSF of 60%; Heavily solar heated: SSF of 80%; WW, water wall; TW, Trombe wall; DG, direct gain; SS, sunspace.

TABLE III-2 (cont'd)

	Reno, NV			Tonopah, NV			Winnemucca, NV			Concord, NH		
SSF:	30%	60%	80%	30%	60%	80%	30%	60%	80%	30%	60%	80%
WW	84	32	17	86	33	18	64	23	12	14	—	—
WW selective surf	102	41	22	105	42	24	82	32	17	32	11	6
TW concrete	66	23	12	68	25	13	51	17	9	11	—	—
TW brick	59	19	10	61	20	11	45	14	7	9	—	—
TW adobe	51	17	8	52	17	9	39	12	6	9	—	—
TW sel surf concrete	96	36	20	99	38	21	77	28	15	30	10	5
TW unvented concrete	53	19	10	54	20	11	40	14	7	8	—	—
DG	64	23	8	65	25	10	46	14	—	—	—	—
DG mov insul	89	37	22	90	39	23	72	30	17	32	11	6
SS attchd sngl plane, mov insul	128	41	20	129	42	22	100	31	15	38	10	5
SS attchd shed roof	71	21	10	72	22	11	53	15	7	9	—	—
SS attchd shed roof, mov insul	103	34	17	104	35	18	81	26	13	32	9	4
SS semienclosed, mov insul	122	43	22	123	45	23	95	33	16	34	10	4

	Newark, NJ			Albuquerque, NM			Clayton, NM			Farmington, NM		
SSF:	30%	60%	80%	30%	60%	80%	30%	60%	80%	30%	60%	80%
WW	36	12	6	97	38	21	83	32	18	75	29	16
WW selective surf	52	20	11	115	47	27	101	41	23	93	38	21
TW concrete	28	9	4	76	28	15	65	24	13	59	21	11
TW brick	25	7	3	68	23	12	59	20	11	53	18	9
TW adobe	22	6	3	58	20	11	50	17	9	46	15	8
TW sel surf concrete	49	17	9	109	42	23	95	37	20	88	34	18
TW unvented concrete	22	7	3	61	23	13	52	20	11	47	18	10
DG	19	—	—	74	30	14	62	24	10	56	20	7
DG mov insul	48	19	10	98	43	26	87	38	23	80	35	21
SS attchd sngl plane, mov insul	61	19	9	137	47	25	123	42	22	113	38	19
SS attchd shed roof	28	7	3	78	25	13	69	21	11	62	19	9
SS attchd shed roof, mov insul	51	16	8	111	39	20	100	34	18	91	31	16
SS semienclosed, mov insul	57	19	10	133	50	27	118	44	24	108	39	21

Note: Solar tempered: SSF of 30%; Moderately solar heated: SSF of 60%; Heavily solar heated: SSF of 80%; WW, water wall; TW, Trombe wall; DG, direct gain; SS, sunspace.

TABLE III-2 (cont'd)

	Los Alamos, NM			Roswell, NM			Truth or Consequences, NM			Tucumcari, NM		
SSF:	30%	60%	80%	30%	60%	80%	30%	60%	80%	30%	60%	80%
WW	67	26	14	107	42	23	126	50	28	103	40	22
WW selective surf	84	34	19	125	51	29	146	60	34	121	50	28
TW concrete	52	19	10	84	31	17	99	37	20	81	30	16
TW brick	47	16	8	75	26	14	89	31	16	73	25	13
TW adobe	40	14	7	64	22	12	76	26	14	62	21	11
TW sel surf concrete	79	31	17	118	46	25	138	54	30	114	44	25
TW unvented concrete	42	16	9	67	25	14	79	30	17	64	24	13
DG	47	18	5	83	33	16	100	42	22	79	32	16
DG mov insul	72	32	19	107	47	28	123	54	33	103	45	28
SS attchd sngl plane, mov insul	106	36	19	151	52	27	176	60	32	144	49	26
SS attchd shed roof	57	18	9	87	28	14	103	33	17	83	26	13
SS attchd shed roof, mov insul	85	30	16	122	42	22	141	50	26	117	41	22
SS semienclosed, mov insul	101	37	20	146	54	29	170	64	35	139	52	28

	Zuni, NM			Albany, NY			Binghamton, NY			Buffalo, NY		
SSF:	30%	60%	80%	30%	60%	80%	30%	60%	80%	30%	60%	80%
WW	74	29	16	13	—	—	—	—	—	—	—	—
WW selective surf	92	37	21	32	11	5	23	7	3	24	7	3
TW concrete	58	21	11	11	—	—	—	—	—	—	—	—
TW brick	52	18	9	8	—	—	—	—	—	—	—	—
TW adobe	45	15	8	9	—	—	—	—	—	—	—	—
TW sel surf concrete	86	33	18	29	9	4	20	5	2	21	5	2
TW unvented concrete	46	18	10	8	—	—	—	—	—	—	—	—
DG	54	20	7	—	—	—	—	—	—	—	—	—
DG mov insul	79	34	21	31	11	5	24	8	3	26	8	3
SS attchd sngl plane, mov insul	118	39	20	37	10	4	28	6	2	30	6	2
SS attchd shed roof	64	20	10	8	—	—	—	—	—	—	—	—
SS attchd shed roof, mov insul	94	32	17	31	9	4	24	6	2	25	6	2
SS semienclosed, mov insul	111	41	22	33	9	4	23	5	—	24	4	—

Note: Solar tempered: SSF of 30%; Moderately solar heated: SSF of 60%; Heavily solar heated: SSF of 80%; WW, water wall; TW, Trombe wall; DG, direct gain; SS, sunspace.

TABLE III-2 (cont'd)

	Massena, NY			*(Central Park) NYC, NY*			*Rochester, NY*			*Syracuse, NY*		
SSF:	*30%*	*60%*	*80%*	*30%*	*60%*	*80%*	*30%*	*60%*	*80%*	*30%*	*60%*	*80%*
WW	—	—	—	29	9	4	—	—	—	—	—	—
WW selective surf	24	7	3	46	17	9	25	8	3	25	8	4
TW concrete	—	—	—	23	7	3	—	—	—	—	—	—
TW brick	—	—	—	20	5	2	—	—	—	—	—	—
TW adobe	—	—	—	18	5	2	—	—	—	—	—	—
TW sel surf concrete	21	6	2	43	15	8	22	6	2	23	6	3
TW unvented concrete	—	—	—	18	5	2	—	—	—	—	—	—
DG	—	—	—	10	—	—	—	—	—	—	—	—
DG mov insul	26	8	3	43	16	9	27	8	3	27	8	3
SS attchd sngl plane, mov insul	28	6	2	54	16	8	31	7	2	31	7	3
SS attchd shed roof	—	—	—	23	5	2	—	—	—	—	—	—
SS attchd shed roof, mov insul	24	5	2	45	14	7	26	6	2	26	6	3
SS semienclosed, mov insul	23	4	—	50	16	8	25	5	1	26	6	2

	Asheville, NC			*Cape Hatteras, NC*			*Charlotte, NC*			*Greensboro, NC*		
SSF:	*30%*	*60%*	*80%*	*30%*	*60%*	*80%*	*30%*	*60%*	*80%*	*30%*	*60%*	*80%*
WW	62	23	12	92	34	18	78	29	15	67	25	13
WW selective surf	79	31	17	111	43	24	96	37	20	85	33	18
TW concrete	49	17	9	73	25	13	61	21	11	53	18	9
TW brick	43	14	7	66	21	11	55	17	9	47	15	8
TW adobe	38	12	6	56	18	9	47	15	8	41	13	7
TW sel surf concrete	74	28	15	105	39	21	90	33	18	80	30	16
TW unvented concrete	39	14	7	58	21	11	49	17	9	42	15	8
DG	44	13	—	71	25	10	58	20	6	48	16	—
DG mov insul	70	29	17	96	39	23	84	34	20	75	30	18
SS attchd sngl plane, mov insul	98	32	16	136	44	23	114	37	19	101	33	17
SS attchd shed roof	52	15	7	78	23	11	64	19	9	55	16	8
SS attchd shed roof, mov insul	80	26	14	111	37	19	94	31	16	83	27	14
SS semienclosed, mov insul	93	33	17	131	47	25	110	39	21	97	34	18

Note: Solar tempered: SSF of 30%; Moderately solar heated: SSF of 60%; Heavily solar heated: SSF of 80%; WW, water wall; TW, Trombe wall; DG, direct gain; SS, sunspace.

TABLE III-2 (cont'd)

	Raleigh-Durham, NC			Bismarck, ND			Fargo, ND			Minot, ND		
SSF:	30%	60%	80%	30%	60%	80%	30%	60%	80%	30%	60%	80%
WW	68	25	13	19	—	—	—	—	—	—	—	—
WW selective surf	86	34	18	37	13	6	29	9	4	31	9	4
TW concrete	54	18	10	15	—	—	—	—	—	6	—	—
TW brick	48	15	8	12	—	—	—	—	—	—	—	—
TW adobe	42	13	7	12	—	—	5	—	—	7	—	—
TW sel surf concrete	81	30	16	34	11	5	26	7	3	28	8	3
TW unvented concrete	43	15	8	11	—	—	—	—	—	—	—	—
DG	50	16	—	—	—	—	—	—	—	—	—	—
DG mov insul	76	31	18	37	13	6	30	9	4	32	10	4
SS attchd sngl plane, mov insul	103	33	17	41	10	4	31	7	3	34	7	2
SS attchd shed roof	56	16	8	11	—	—	—	—	—	—	—	—
SS attchd shed roof, mov insul	84	28	14	34	9	4	27	6	3	29	7	3
SS semienclosed, mov insul	99	35	18	36	10	4	26	6	2	28	6	2

	Akron-Canton, OH			Cincinnati, OH			Cleveland, OH			Columbus, OH		
SSF:	30%	60%	80%	30%	60%	80%	30%	60%	80%	30%	60%	80%
WW	12	—	—	26	8	3	—	—	—	17	3	—
WW selective surf	32	11	5	43	16	8	29	9	4	36	12	6
TW concrete	10	—	—	21	6	2	5	—	—	14	2	—
TW brick	7	—	—	18	4	1	—	—	—	11	—	—
TW adobe	9	—	—	16	4	2	6	—	—	11	2	—
TW sel surf concrete	29	9	4	40	14	7	27	8	3	33	11	5
TW unvented concrete	7	—	—	16	4	2	—	—	—	11	2	—
DG	—	—	—	—	—	—	—	—	—	—	—	—
DG mov insul	32	11	5	41	15	8	30	10	4	35	12	6
SS attchd sngl plane, mov insul	40	10	4	52	15	7	37	9	4	44	12	5
SS attchd shed roof	9	—	—	21	4	1	—	—	—	14	—	—
SS attchd shed roof, mov insul	33	9	4	44	13	6	31	8	3	36	10	5
SS semienclosed, mov insul	34	9	4	48	15	7	31	8	3	39	11	5

Note: Solar tempered: SSF of 30%; Moderately solar heated: SSF of 60%; Heavily solar heated: SSF of 80%; WW, water wall; TW, Trombe wall; DG, direct gain; SS, sunspace.

TABLE III-2 (cont'd)

	Dayton, OH			Toledo, OH			Youngstown, OH			Oklahoma City, OK		
SSF:	30%	60%	80%	30%	60%	80%	30%	60%	80%	30%	60%	80%
WW	21	5	2	14	—	—	—	—	—	73	27	15
WW selective surf	39	14	7	33	11	5	26	8	4	91	36	20
TW concrete	17	4	1	11	—	—	—	—	—	58	20	11
TW brick	14	2	—	8	—	—	—	—	—	52	17	9
TW adobe	13	3	—	9	—	—	—	—	—	45	14	7
TW sel surf concrete	36	12	6	30	9	4	24	7	3	86	32	17
TW unvented concrete	13	3	—	8	—	—	—	—	—	46	17	9
DG	—	—	—	—	—	—	—	—	—	53	19	5
DG mov insul	37	13	7	33	11	5	28	9	4	79	33	20
SS attchd sngl plane, mov insul	47	13	6	40	10	4	33	8	3	108	36	18
SS attchd shed roof	16	—	—	10	—	—	—	—	—	59	18	9
SS attchd shed roof, mov insul	39	11	5	33	9	4	28	7	3	88	30	15
SS semienclosed, mov insul	42	13	6	35	10	4	28	6	2	104	37	20

	Tulsa, OK			Astoria, OR			Burns, OR			Medford, OR		
SSF:	30%	60%	80%	30%	60%	80%	30%	60%	80%	30%	60%	80%
WW	64	24	13	47	14	5	39	12	5	45	12	4
WW selective surf	82	32	18	65	23	11	57	21	10	63	22	10
TW concrete	51	18	9	38	10	4	31	9	4	36	9	3
TW brick	45	14	7	34	8	2	28	7	2	32	7	2
TW adobe	39	12	6	29	7	3	24	6	2	28	7	2
TW sel surf concrete	77	29	15	61	20	9	53	18	9	59	19	9
TW unvented concrete	40	14	8	30	8	3	25	7	3	28	7	2
DG	46	15	—	32	—	—	23	—	—	29	—	—
DG mov insul	72	30	17	61	22	11	53	20	10	60	21	10
SS attchd sngl plane, mov insul	96	31	16	84	23	9	69	19	8	79	21	8
SS attchd shed roof	52	15	7	42	8	—	32	6	1	38	6	—
SS attchd shed roof, mov insul	79	26	14	69	19	8	57	17	7	65	18	7
SS semienclosed, mov insul	92	33	17	79	23	10	64	19	9	73	21	9

Note: Solar tempered: SSF of 30%; Moderately solar heated: SSF of 60%; Heavily solar heated: SSF of 80%; WW, water wall; TW, Trombe wall; DG, direct gain; SS, sunspace.

TABLE III-2 (cont'd)

	North Bend, OR			Pendleton, OR			Portland, OR			Redmond, OR		
SSF:	30%	60%	80%	30%	60%	80%	30%	60%	80%	30%	60%	80%
WW	82	30	15	34	6	—	38	8	1	52	17	8
WW selective surf	100	39	21	53	17	7	57	18	8	69	26	13
TW concrete	65	22	11	27	4	—	31	6	—	41	13	6
TW brick	58	18	9	24	—	—	27	4	—	36	10	4
TW adobe	50	16	8	22	3	—	25	5	—	32	9	4
TW sel surf concrete	94	35	18	50	14	6	54	16	7	65	23	11
TW unvented concrete	51	18	9	21	3	—	24	5	—	32	10	5
DG	64	21	—	13	—	—	21	—	—	35	—	—
DG mov insul	89	36	20	52	16	7	55	18	8	63	24	13
SS attchd sngl plane, mov insul	133	42	19	62	15	6	71	17	7	85	25	11
SS attchd shed roof	75	21	9	26	—	—	32	—	—	43	10	4
SS attchd shed roof, mov insul	107	35	16	51	13	5	58	15	6	70	21	10
SS semienclosed, mov insul	127	43	21	56	14	5	64	17	6	80	25	12

	Salem, OR			Allentown, PA			Erie, PA			Harrisburg, PA		
SSF:	30%	60%	80%	30%	60%	80%	30%	60%	80%	30%	60%	80%
WW	42	11	3	26	8	4	—	—	—	30	10	5
WW selective surf	60	20	9	44	16	8	26	7	3	47	18	9
TW concrete	34	8	2	21	6	3	—	—	—	24	7	3
TW brick	30	6	—	18	4	2	—	—	—	20	5	2
TW adobe	27	6	2	16	4	2	—	—	—	19	5	2
TW sel surf concrete	57	18	8	40	14	7	23	6	2	43	15	8
TW unvented concrete	26	6	2	16	5	2	—	—	—	19	6	3
DG	26	—	—	—	—	—	—	—	—	12	—	—
DG mov insul	57	20	9	41	15	8	27	8	3	44	17	9
SS attchd sngl plane, mov insul	75	19	8	52	15	7	33	7	2	55	17	8
SS attchd shed roof	36	5	—	21	4	1	—	—	—	23	5	2
SS attchd shed roof, mov insul	62	17	7	43	13	6	28	6	2	46	14	7
SS semienclosed, mov insul	69	19	8	48	15	7	27	5	—	51	17	8

Note: Solar tempered: SSF of 30%; Moderately solar heated: SSF of 60%; Heavily solar heated: SSF of 80%; WW, water wall; TW, Trombe wall; DG, direct gain; SS, sunspace.

TABLE III-2 (cont'd)

	Philadelphia, PA			Pittsburgh, PA			Wilkes-Barre-Scranton, PA			Providence, RI		
SSF:	30%	60%	80%	30%	60%	80%	30%	60%	80%	30%	60%	80%
WW	37	12	6	12	—	—	16	2	—	27	8	4
WW selective surf	54	20	11	32	11	5	35	12	6	44	17	9
TW concrete	29	9	4	10	—	—	13	1	—	22	6	3
TW brick	25	7	3	7	—	—	10	—	—	18	5	2
TW adobe	23	7	3	9	—	—	11	1	—	17	4	2
TW sel surf concrete	50	18	9	29	9	4	32	10	5	41	14	7
TW unvented concrete	23	7	4	7	—	—	10	—	—	17	5	2
DG	20	—	—	—	—	—	—	—	—	—	—	—
DG mov insul	49	19	11	32	11	5	34	12	6	41	16	8
SS attchd sngl plane, mov insul	63	19	10	40	10	4	42	11	5	52	16	7
SS attchd shed roof	29	7	3	9	—	—	12	—	—	21	4	1
SS attchd shed roof, mov insul	52	17	8	33	9	4	35	10	4	44	13	6
SS semienclosed, mov insul	59	20	10	35	9	4	37	11	5	48	16	8

	Charleston, SC			Columbia, SC			Greenville-Spartanburg, SC			Huron, SD		
SSF:	30%	60%	80%	30%	60%	80%	30%	60%	80%	30%	60%	80%
WW	111	42	23	98	36	20	79	29	15	18	3	—
WW selective surf	131	51	28	116	46	25	97	38	21	37	13	6
TW concrete	88	31	16	77	27	14	62	21	11	15	2	—
TW brick	80	26	13	69	23	12	56	18	9	12	—	—
TW adobe	68	22	11	59	19	10	48	15	8	12	2	—
TW sel surf concrete	124	46	25	110	41	22	91	34	18	34	11	5
TW unvented concrete	70	25	14	61	22	12	49	18	9	11	1	—
DG	88	33	16	75	28	12	59	20	6	—	—	—
DG mov insul	113	46	28	100	42	25	85	34	20	36	12	6
SS attchd sngl plane, mov insul	159	53	27	139	46	24	116	38	19	41	11	5
SS attchd shed roof	93	28	14	80	24	12	65	19	9	11	—	—
SS attchd shed roof, mov insul	130	44	23	114	38	20	95	32	16	35	10	4
SS semienclosed, mov insul	154	56	30	135	49	26	112	40	21	37	10	4

Note: Solar tempered: SSF of 30%; Moderately solar heated: SSF of 60%; Heavily solar heated: SSF of 80%; WW, water wall; TW, Trombe wall; DG, direct gain; SS, sunspace.

TABLE III-2 (cont'd)

	Pierre, SD			*Rapid City, SD*			*Sioux Falls, SD*			*Chattanooga, TN*		
SSF:	*30%*	*60%*	*80%*	*30%*	*60%*	*80%*	*30%*	*60%*	*80%*	*30%*	*60%*	*80%*
WW	28	8	3	38	12	6	23	6	2	55	19	10
WW selective surf	46	16	8	56	21	11	41	15	7	72	28	15
TW concrete	22	6	2	30	9	4	18	4	2	43	14	7
TW brick	19	4	1	27	7	3	15	3	—	39	12	6
TW adobe	18	4	2	24	7	3	15	3	1	34	10	5
TW sel surf concrete	42	14	7	52	18	9	38	12	6	68	24	13
TW unvented concrete	17	5	2	24	7	4	14	3	1	34	12	6
DG	—	—	—	21	—	—	—	—	—	37	9	—
DG mov insul	43	16	8	51	19	11	39	14	7	65	25	15
SS attchd sngl plane, mov insul	51	14	6	64	19	9	46	13	6	88	28	14
SS attchd shed roof	20	3	—	29	6	2	16	—	—	46	13	6
SS attchd shed roof, mov insul	43	12	6	53	16	8	39	11	5	72	23	12
SS semienclosed, mov insul	47	14	7	60	19	9	42	12	6	84	29	15

	Knoxville, TN			*Memphis, TN*			*Nashville, TN*			*Abilene, TX*		
SSF:	*30%*	*60%*	*80%*	*30%*	*60%*	*80%*	*30%*	*60%*	*80%*	*30%*	*60%*	*80%*
WW	57	20	11	67	24	13	46	16	8	115	44	24
WW selective surf	75	29	16	85	33	18	63	24	13	134	54	30
TW concrete	46	15	8	53	18	9	36	11	6	90	33	17
TW brick	41	12	6	48	15	7	32	9	4	82	27	14
TW adobe	35	11	5	41	13	6	28	8	4	69	23	12
TW sel surf concrete	71	25	13	80	29	16	59	21	11	127	48	27
TW unvented concrete	36	12	6	42	15	8	28	9	5	72	27	15
DG	39	10	—	48	15	—	28	—	—	90	36	18
DG mov insul	67	26	15	75	30	18	57	22	12	114	49	30
SS attchd sngl plane, mov insul	91	29	14	102	33	17	76	24	12	163	55	29
SS attchd shed roof	48	13	6	55	16	8	38	10	5	95	30	15
SS attchd shed roof, mov insul	75	24	12	84	27	14	63	20	10	132	45	24
SS semienclosed, mov insul	86	30	15	98	34	18	74	24	12	158	58	31

Note: Solar tempered: SSF of 30%; Moderately solar heated: SSF of 60%; Heavily solar heated: SSF of 80%; WW, water wall; TW, Trombe wall; DG, direct gain; SS, sunspace.

TABLE III-2 (cont'd)

	Amarillo, TX			*Austin, TX*			*Brownsville, TX*			*Corpus Christi, TX*		
SSF:	*30%*	*60%*	*80%*	*30%*	*60%*	*80%*	*30%*	*60%*	*80%*	*30%*	*60%*	*80%*
WW	92	36	20	140	54	29	294	114	63	223	88	48
WW selective surf	110	45	25	160	64	36	321	129	72	247	101	56
TW concrete	72	26	14	110	40	21	232	84	45	175	65	35
TW brick	65	22	12	100	33	17	212	72	38	160	55	29
TW adobe	55	19	10	85	28	15	178	60	32	134	46	24
TW sel surf concrete	104	40	22	152	58	32	306	117	64	235	91	50
TW unvented concrete	58	22	12	88	33	18	185	70	38	141	54	29
DG	70	27	12	113	45	24	249	105	61	189	79	45
DG mov insul	94	41	25	136	58	35	269	116	71	209	91	55
SS attchd sngl plane, mov insul	132	45	23	199	67	35	421	145	76	311	110	57
SS attchd shed roof	75	23	12	119	37	19	263	85	44	192	64	33
SS attchd shed roof, mov insul	107	37	19	161	55	29	338	118	62	251	90	47
SS semienclosed, mov insul	128	47	25	193	72	38	412	155	83	306	117	63

	Dallas, TX			*Del Rio, TX*			*El Paso, TX*			*Fort Worth, TX*		
SSF:	*30%*	*60%*	*80%*	*30%*	*60%*	*80%*	*30%*	*60%*	*80%*	*30%*	*60%*	*80%*
WW	108	41	23	169	65	36	147	58	32	103	39	21
WW selective surf	127	51	28	191	77	43	168	68	38	122	49	27
TW concrete	85	31	16	133	48	26	116	42	23	81	29	15
TW brick	77	26	13	121	41	21	105	36	19	73	24	13
TW adobe	65	22	11	103	34	18	89	30	16	62	21	11
TW sel surf concrete	120	46	25	182	69	38	159	62	34	115	44	24
TW unvented concrete	68	25	14	107	40	22	93	35	19	65	24	13
DG	84	33	16	140	56	31	119	49	27	80	31	14
DG mov insul	108	46	28	161	69	42	142	62	38	104	44	27
SS attchd sngl plane, mov insul	153	52	27	235	80	42	203	69	36	147	50	26
SS attchd shed roof	89	28	14	143	45	23	121	38	20	85	27	13
SS attchd shed roof, mov insul	125	43	23	190	66	35	163	57	30	120	41	22
SS semienclosed, mov insul	149	55	30	229	85	46	196	73	40	143	53	28

Note: Solar tempered: SSF of 30%; Moderately solar heated: SSF of 60%; Heavily solar heated: SSF of 80%; WW, water wall; TW, Trombe wall; DG, direct gain; SS, sunspace.

TABLE III-2 (cont'd)

	Houston, TX			*Laredo, TX*			*Lubbock, TX*			*Lufkin, TX*		
SSF:	*30%*	*60%*	*80%*	*30%*	*60%*	*80%*	*30%*	*60%*	*80%*	*30%*	*60%*	*80%*
WW	139	53	29	249	96	52	109	42	23	121	46	25
WW selective surf	160	64	35	274	110	61	128	52	29	141	56	31
TW concrete	110	39	21	196	71	38	85	31	17	95	34	18
TW brick	100	33	17	179	61	32	77	26	14	86	29	15
TW adobe	85	28	15	150	51	26	65	22	12	73	24	13
TW sel surf concrete	152	57	31	261	100	54	120	47	26	133	50	27
TW unvented concrete	88	33	17	157	59	32	68	26	14	76	28	15
DG	113	45	23	215	87	49	85	34	17	96	38	19
DG mov insul	136	58	35	235	98	60	109	47	29	119	51	31
SS attchd sngl plane, mov insul	198	68	35	344	118	61	154	53	28	173	58	30
SS attchd shed roof	119	38	19	214	69	35	89	28	14	102	32	16
SS attchd shed roof, mov insul	161	56	29	278	97	51	124	43	23	141	48	25
SS semienclosed, mov insul	194	72	38	338	126	67	149	56	30	168	62	33

	Midland-Odessa, TX			*Port Arthur, TX*			*San Angelo, TX*			*San Antonio, TX*		
SSF:	*30%*	*60%*	*80%*	*30%*	*60%*	*80%*	*30%*	*60%*	*80%*	*30%*	*60%*	*80%*
WW	140	55	30	145	55	30	133	52	28	154	59	32
WW selective surf	160	65	37	166	66	37	153	62	35	175	70	39
TW concrete	110	40	22	114	41	22	105	38	21	121	43	23
TW brick	100	34	18	104	34	18	95	32	17	110	37	19
TW adobe	84	29	15	88	29	15	80	27	14	93	31	16
TW sel surf concrete	152	59	33	157	59	32	145	56	31	166	63	34
TW unvented concrete	88	34	18	91	34	18	84	32	17	97	36	20
DG	112	47	25	117	47	25	107	43	23	125	50	27
DG mov insul	135	59	36	140	60	36	130	56	34	148	63	38
SS attchd sngl plane, mov insul	194	67	35	207	70	37	187	64	34	217	74	39
SS attchd shed roof	115	37	19	125	39	20	111	35	18	131	42	21
SS attchd shed roof, mov insul	157	55	29	168	58	30	152	52	28	176	61	32
SS semienclosed, mov insul	188	71	38	202	75	40	182	68	37	211	79	42

Note: Solar tempered: SSF of 30%; Moderately solar heated: SSF of 60%; Heavily solar heated: SSF of 80%; WW, water wall; TW, Trombe wall; DG, direct gain; SS, sunspace.

TABLE III-2 (cont'd)

	Sherman, TX			Waco, TX			Wichita Falls, TX			Bryce Canyon, UT		
SSF:	30%	60%	80%	30%	60%	80%	30%	60%	80%	30%	60%	80%
WW	86	32	17	118	45	25	97	37	20	54	21	11
WW selective surf	104	41	23	138	55	31	116	46	26	70	29	16
TW concrete	68	24	13	93	33	18	77	27	15	42	15	8
TW brick	61	20	10	85	28	15	69	23	12	37	12	6
TW adobe	52	17	9	72	24	12	59	19	10	32	11	6
TW sel surf concrete	98	37	20	131	49	27	110	42	23	66	25	14
TW unvented concrete	54	20	11	74	28	15	61	23	12	34	12	7
DG	65	24	9	94	37	18	75	29	13	36	11	—
DG mov insul	90	38	22	117	50	30	99	42	26	62	26	16
SS attchd sngl plane, mov insul	125	42	22	168	57	30	138	46	24	94	31	16
SS attchd shed roof	71	22	11	99	31	16	79	24	12	48	14	7
SS attchd shed roof, mov insul	102	35	18	137	47	25	113	38	20	75	25	13
SS semienclosed, mov insul	121	44	23	164	60	32	134	49	26	88	32	17

	Cedar City, UT			Salt Lake City, UT			Burlington, VT			Norfolk, VA		
SSF:	30%	60%	80%	30%	60%	80%	30%	60%	80%	30%	60%	80%
WW	71	27	15	58	20	10	—	—	—	71	26	14
WW selective surf	89	36	20	76	29	15	23	7	3	89	35	19
TW concrete	56	20	11	46	15	7	—	—	—	56	19	10
TW brick	50	17	9	41	12	6	—	—	—	50	16	8
TW adobe	43	14	7	36	11	5	—	—	—	43	14	7
TW sel surf concrete	83	32	17	72	25	13	21	6	2	83	31	17
TW unvented concrete	45	17	9	37	12	6	—	—	—	44	16	8
DG	52	19	5	41	9	—	—	—	—	52	17	—
DG mov insul	77	33	20	69	27	15	25	8	3	78	32	19
SS attchd sngl plane, mov insul	110	36	18	91	27	13	28	6	2	105	34	17
SS attchd shed roof	60	18	9	47	12	5	—	—	—	58	17	8
SS attchd shed roof, mov insul	89	30	15	74	23	11	24	5	2	87	29	15
SS semienclosed, mov insul	104	38	20	86	28	14	22	4	—	101	36	19

Note: Solar tempered: SSF of 30%; Moderately solar heated: SSF of 60%; Heavily solar heated: SSF of 80%; WW, water wall; TW, Trombe wall; DG, direct gain; SS, sunspace.

TABLE III-2 (cont'd)

	Richmond, VA			*Roanoke, VA*			*Olympia, WA*			*Seattle-Tacoma, WA*		
SSF:	*30%*	*60%*	*80%*	*30%*	*60%*	*80%*	*30%*	*60%*	*80%*	*30%*	*60%*	*80%*
WW	55	20	10	54	19	10	30	—	—	34	—	—
WW selective surf	72	28	15	71	28	15	50	15	6	54	16	7
TW concrete	43	14	7	43	14	7	25	—	—	28	—	—
TW brick	39	12	6	38	12	6	22	—	—	25	—	—
TW adobe	34	10	5	33	10	5	20	—	—	22	—	—
TW sel surf concrete	68	25	13	67	25	13	47	13	5	50	14	5
TW unvented concrete	34	12	6	34	12	6	19	—	—	21	—	—
DG	37	9	—	36	9	—	—	—	—	14	—	—
DG mov insul	65	26	15	63	26	15	49	15	6	53	16	7
SS attchd sngl plane, mov insul	85	27	14	85	27	14	63	14	5	67	15	5
SS attchd shed roof	44	12	6	44	12	6	27	—	—	29	—	—
SS attchd shed roof, mov insul	70	23	12	70	23	12	52	12	5	55	13	5
SS semienclosed, mov insul	81	28	15	81	28	15	57	13	4	60	14	4

	Spokane, WA			*Yakima, WA*			*Charleston, WV*			*Huntington, WV*		
SSF:	*30%*	*60%*	*80%*	*30%*	*60%*	*80%*	*30%*	*60%*	*80%*	*30%*	*60%*	*80%*
WW	22	—	—	34	4	—	30	9	4	35	11	5
WW selective surf	43	12	5	53	16	7	48	17	9	52	19	10
TW concrete	18	—	—	28	3	—	24	7	3	28	8	4
TW brick	16	—	—	25	—	—	21	5	2	24	6	3
TW adobe	15	—	—	22	3	—	19	5	2	22	6	3
TW sel surf concrete	40	10	4	50	14	6	44	15	8	49	17	9
TW unvented concrete	14	—	—	21	—	—	19	5	2	22	7	3
DG	—	—	—	13	—	—	11	—	—	17	—	—
DG mov insul	44	13	5	53	16	7	45	16	9	48	18	10
SS attchd sngl plane, mov insul	50	10	3	62	14	5	58	17	8	63	19	9
SS attchd shed roof	16	—	—	26	—	—	25	5	2	28	6	3
SS attchd shed roof, mov insul	42	9	3	51	12	5	48	14	7	52	16	8
SS semienclosed, mov insul	44	9	2	56	13	5	53	17	8	59	19	9

Note: Solar tempered: SSF of 30%; Moderately solar heated: SSF of 60%; Heavily solar heated: SSF of 80%; WW, water wall; TW, Trombe wall; DG, direct gain; SS, sunspace.

TABLE III-2 (cont'd)

	Eau Claire, WI			*Green Bay, WI*			*La Crosse, WI*			*Madison, WI*		
SSF:	*30%*	*60%*	*80%*	*30%*	*60%*	*80%*	*30%*	*60%*	*80%*	*30%*	*60%*	*80%*
WW	9	—	—	13	—	—	15	—	—	19	4	—
WW selective surf	30	10	5	33	11	5	34	12	6	37	13	6
TW concrete	7	—	—	11	—	—	13	—	—	15	3	—
TW brick	—	—	—	8	—	—	10	—	—	12	—	—
TW adobe	7	—	—	9	—	—	10	—	—	12	2	—
TW sel surf concrete	27	8	4	30	9	4	31	10	5	34	11	5
TW unvented concrete	4	—	—	8	—	—	9	—	—	11	—	—
DG	—	—	—	—	—	—	—	—	—	—	—	—
DG mov insul	30	10	4	33	11	5	33	12	6	35	13	6
SS attchd sngl plane, mov insul	33	8	3	38	10	4	38	10	4	42	12	5
SS attchd shed roof	—	—	—	—	—	—	9	—	—	13	—	—
SS attchd shed roof, mov insul	28	7	3	32	8	4	32	9	4	36	10	5
SS semienclosed, mov insul	29	7	2	33	9	3	34	9	4	38	11	5

	Milwaukee, WI			*Casper, WY*			*Cheyenne, WY*			*Rock Springs, WY*		
SSF:	*30%*	*60%*	*80%*	*30%*	*60%*	*80%*	*30%*	*60%*	*80%*	*30%*	*60%*	*80%*
WW	18	3	—	55	20	11	58	22	12	51	19	10
WW selective surf	36	13	6	72	28	15	75	30	17	68	27	15
TW concrete	14	2	—	43	15	8	46	16	9	40	14	7
TW brick	12	—	—	38	12	6	41	13	7	36	11	6
TW adobe	12	2	—	33	11	5	35	12	6	31	10	5
TW sel surf concrete	33	11	5	67	25	13	71	27	15	64	24	13
TW unvented concrete	11	—	—	34	12	6	37	13	7	32	11	6
DG	—	—	—	37	10	—	40	13	—	34	8	—
DG mov insul	35	12	6	64	26	15	66	28	17	60	25	14
SS attchd sngl plane, mov insul	43	12	5	84	27	13	90	30	15	82	26	13
SS attchd shed roof	13	—	—	43	12	5	47	14	7	41	11	5
SS attchd shed roof, mov insul	36	10	4	69	22	11	73	25	13	67	22	11
SS semienclosed, mov insul	38	11	5	79	28	14	86	31	16	77	27	14

Note: Solar tempered: SSF of 30%; Moderately solar heated: SSF of 60%; Heavily solar heated: SSF of 80%; WW, water wall; TW, Trombe wall; DG, direct gain; SS, sunspace.

TABLE III-2 (cont'd)

	Sheridan, WY			Edmonton, Alberta			Suffield, Alberta			Nanaimo, British Columbia		
SSF:	30%	60%	80%	30%	60%	80%	30%	60%	80%	30%	60%	80%
WW	36	11	5	16	—	—	28	6	1	38	—	—
WW selective surf	53	19	10	38	11	4	47	16	8	57	17	7
TW concrete	28	8	4	14	—	—	23	5	—	31	—	—
TW brick	25	6	3	11	—	—	20	3	—	27	—	—
TW adobe	22	6	3	12	—	—	18	3	—	25	—	—
TW sel surf concrete	49	17	9	35	9	3	43	14	6	54	15	5
TW unvented concrete	22	7	3	10	—	—	18	4	—	24	—	—
DG	18	—	—	—	—	—	—	—	—	21	—	—
DG mov insul	49	18	10	39	12	4	45	16	7	56	17	7
SS attchd sngl plane, mov insul	61	18	8	39	7	2	49	12	5	69	15	5
SS attchd shed roof	27	5	2	—	—	—	18	—	—	31	—	—
SS attchd shed roof, mov insul	51	15	7	33	7	2	41	11	5	57	13	5
SS semienclosed, mov insul	57	18	8	34	5	—	45	12	5	63	14	4

	Vancouver, BC			Winnipeg, Manitoba			Halifax, Nova Scotia			Moosonee, Ontario		
SSF:	30%	60%	80%	30%	60%	80%	30%	60%	80%	30%	60%	80%
WW	35	—	—	16	—	—	25	7	2	—	—	
WW selective surf	55	16	7	35	12	6	43	16	8	26	7	3
TW concrete	29	—	—	13	—	—	20	5	2	—	—	—
TW brick	25	—	—	11	—	—	17	4	—	—	—	—
TW adobe	23	—	—	11	—	—	16	4	1	—	—	—
TW sel surf concrete	52	14	5	32	10	5	39	14	7	23	6	2
TW unvented concrete	22	—	—	10	—	—	16	4	1	—	—	—
DG	16	—	—	—	—	—	—	—	—	—	—	—
DG mov insul	54	16	6	34	12	6	40	15	8	27	8	3
SS attchd sngl plane, mov insul	66	14	5	36	9	3	52	15	7	27	4	—
SS attchd shed roof	29	—	—	—	—	—	20	—	—	—	—	—
SS attchd shed roof, mov insul	55	13	4	31	8	3	43	13	6	23	5	1
SS semienclosed, mov insul	60	13	4	32	8	3	47	15	7	22	—	—

Note: Solar tempered: SSF of 30%; Moderately solar heated: SSF of 60%; Heavily solar heated: SSF of 80%; WW, water wall; TW, Trombe wall; DG, direct gain; SS, sunspace.

TABLE III-2 (cont'd)

	Ottawa, Ontario			*Toronto, Ontario*			*Normandin, Quebec*		
SSF:	*30%*	*60%*	*80%*	*30%*	*60%*	*80%*	*30%*	*60%*	*80%*
WW	22	6	1	24	7	3	15	—	—
WW selective surf	40	14	7	41	15	8	33	11	5
TW concrete	18	4	—	19	5	2	12	—	—
TW brick	15	2	—	16	4	—	9	—	—
TW adobe	14	3	—	15	4	1	10	—	—
TW sel surf concrete	37	12	6	38	13	7	31	10	4
TW unvented concrete	14	3	—	15	4	1	9	—	—
DG	—	—	—	—	—	—	—	—	—
DG mov insul	38	14	7	39	15	8	33	11	5
SS attchd sngl plane, mov insul	45	12	5	48	14	6	37	9	3
SS attchd shed roof	15	—	—	18	2	—	—	—	—
SS attchd shed roof, mov insul	37	11	5	40	12	6	31	8	3
SS semienclosed, mov insul	40	12	5	44	14	6	32	8	3

Note: Solar tempered: SSF of 30%; Moderately solar heated: SSF of 60%; Heavily solar heated: SSF of 80%; WW, water wall; TW, Trombe wall; DG, direct gain; SS, sunspace.

Source: J. Douglas Balcomb et al., *Passive Solar Design Handbook*, Vol. 3, Los Alamos National Laboratory, Los Alamos, NM, for U.S. Department of Energy, DOE/C5-0127/3, 1982, Appendix F. The LCR tables in this reference are more extensive and detailed than those above. Also included are methods of determining month by month solar savings performances of a building. The *Passive Solar Design Handbook* is a strongly recommended reference for all designers of solar systems.

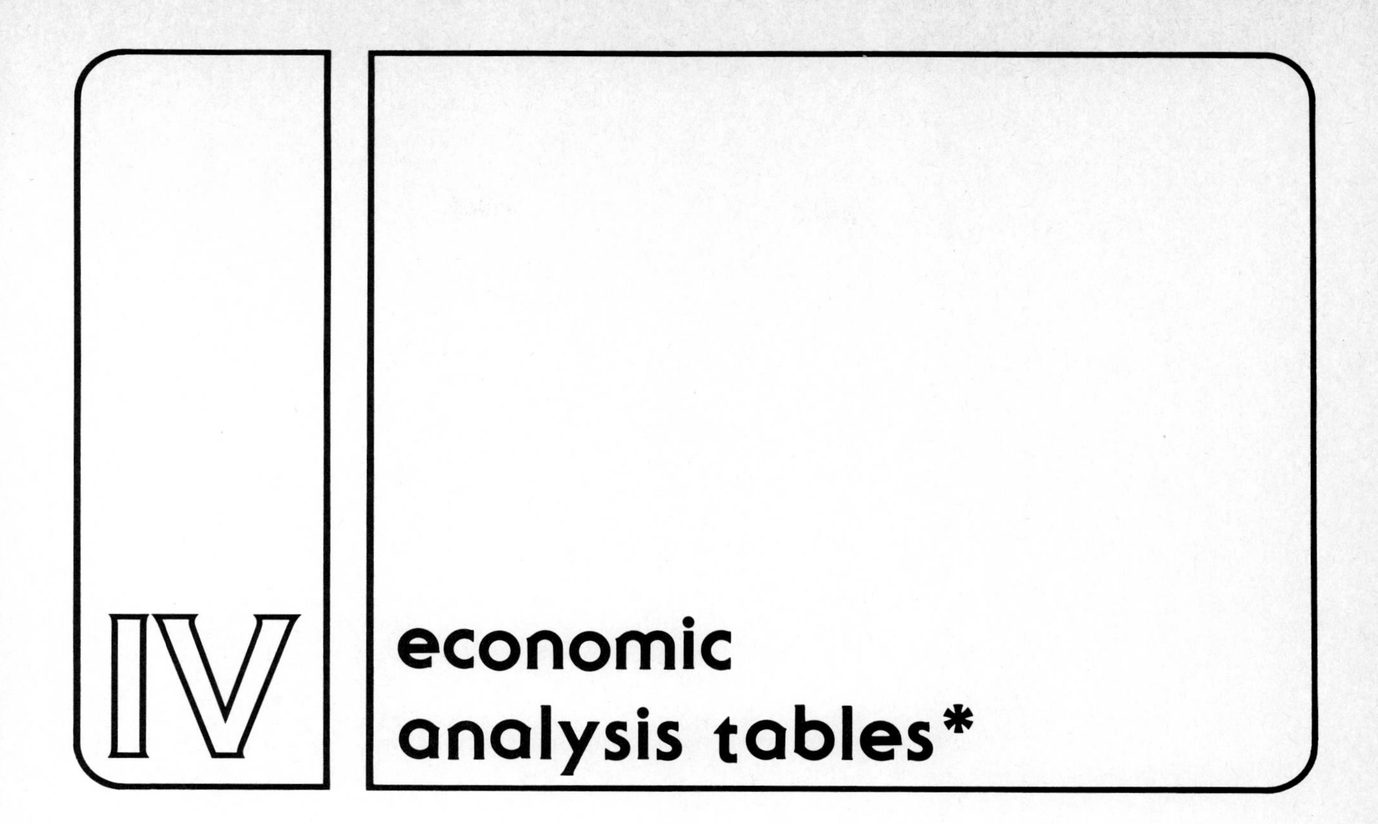

*Tables IV-1 through IV-3 are excerpted from Harold E. Marshall and Rosalie T. Ruegg, *Simplified Energy Design Economics*, National Bureau of Standards Special Publication 544, U.S. Department of Commerce, Washington, DC, 1980.

TABLE IV-1 Uniform Present-Value Factors

DISCOUNT FACTORS

	DISCOUNT RATES						
N	6%	8%	10%	12%	15%	20%	25%
1	0.943	0.926	0.909	0.893	0.870	0.833	0.800
2	1.833	1.783	1.736	1.690	1.626	1.528	1.440
3	2.673	2.577	2.487	2.402	2.283	2.106	1.952
4	3.465	3.312	3.170	3.037	2.855	2.589	2.362
5	4.212	3.993	3.791	3.605	3.352	2.991	2.689
6	4.917	4.623	4.355	4.111	3.784	3.326	2.951
7	5.582	5.206	4.868	4.564	4.160	3.605	3.161
8	6.210	5.747	5.335	4.968	4.487	3.837	3.329
9	6.802	6.247	5.759	5.328	4.772	4.031	3.463
10	7.360	6.710	6.145	5.650	5.019	4.192	3.571
11	7.887	7.139	6.495	5.938	5.234	4.327	3.656
12	8.834	7.536	6.814	6.194	5.421	4.439	3.725
13	8.853	7.904	7.103	6.424	5.583	4.533	3.780
14	9.295	8.244	7.367	6.628	5.724	4.611	3.824
15	9.712	8.559	7.606	6.811	5.847	4.675	3.859
16	10.106	8.851	7.824	6.974	5.954	4.730	3.887
17	10.477	9.122	8.022	7.120	6.047	4.775	3.910
18	10.828	9.372	8.201	7.250	6.128	4.812	3.928
19	11.158	9.604	8.365	7.366	6.198	4.843	3.942
20	11.470	9.818	8.514	7.469	6.259	4.870	3.954
21	11.764	10.017	8.649	7.562	6.312	4.891	3.963
22	12.042	10.201	8.772	7.645	6.359	4.909	3.970
23	12.303	10.371	8.883	7.718	6.399	4.925	3.976
24	12.550	10.529	8.985	7.784	6.434	4.937	3.981
25	12.783	10.675	9.077	7.843	6.464	4.948	3.985
26	13.003	10.810	9.161	7.896	6.491	4.956	3.988
27	13.211	10.935	9.237	7.943	6.514	4.964	3.990
28	13.406	11.051	9.307	7.984	6.534	4.970	3.992
29	13.591	11.158	9.370	8.022	6.551	4.975	3.994
30	13.765	11.258	9.427	8.055	6.566	4.979	3.995
31	13.929	11.350	9.479	8.085	6.579	4.982	3.996
32	14.084	11.435	9.526	8.112	6.591	4.985	3.997
33	14.230	11.514	9.569	8.135	6.600	4.988	3.997
34	14.368	11.587	9.609	8.157	6.609	4.990	3.998
35	14.498	11.655	9.644	8.176	6.617	4.992	3.998
40	15.046	11.925	9.779	8.244	6.642	4.997	3.999
45	15.456	12.108	9.863	8.283	6.654	4.999	4.000
50	15.762	12.233	9.915	8.304	6.661	7.999	4.000

TABLE IV-2 Present Value Factors Modified for Energy Price Escalation

a) 8% Discount Rate

N	RATE OF ENERGY PRICE ESCALATION 1%	2%	3%	4%	5%	6%	7%	8%	9%	10%
1	.935	.944	.954	.963	.972	.981	.991	1.000	1.009	1.019
2	1.810	1.836	1.863	1.890	1.917	1.945	1.972	2.000	2.028	2.056
3	2.628	2.679	2.731	2.783	2.836	2.890	2.945	3.000	3.056	3.112
4	3.393	3.474	3.558	3.643	3.730	3.818	3.908	4.000	4.093	4.189
5	4.108	4.226	4.347	4.471	4.598	4.729	4.863	5.000	5.141	5.285
6	4.777	4.936	5.099	5.268	5.443	5.623	5.809	6.000	6.197	6.401
7	5.402	5.606	5.817	6.036	6.264	6.500	6.745	7.000	7.264	7.538
8	5.987	6.239	6.501	6.776	7.062	7.361	7.674	8.000	8.341	8.696
9	6.534	6.837	7.154	7.488	7.838	8.207	8.593	9.000	9.427	9.876
10	7.046	7.401	7.777	8.173	8.593	9.036	9.505	10.000	10.524	11.077
11	7.525	7.935	8.370	8.834	9.326	9.850	10.407	11.000	11.630	12.301
12	7.972	8.438	8.936	9.469	10.039	10.649	11.302	12.000	12.747	13.547
13	8.391	8.914	9.476	10.082	10.733	11.434	12.188	13.000	13.875	14.817
14	8.782	9.363	9.991	10.671	11.407	12.203	13.066	14.000	15.012	16.110
15	9.148	9.787	10.483	11.239	12.062	12.905	13.936	15.000	16.161	17.426
16	9.490	10.188	10.951	11.786	12.699	13.700	14.797	16.000	17.320	18.768
17	9.810	10.567	11.398	12.312	13.319	14.428	15.651	17.000	18.489	20.134
18	10.110	10.924	11.824	12.819	13.921	15.142	16.497	18.000	19.670	21.525
19	10.390	11.262	12.230	13.307	14.507	15.843	17.335	19.000	20.861	22.942
20	10.651	11.580	12.618	13.777	15.076	16.532	18.165	20.000	22.063	24.386
21	10.896	11.881	12.987	14.230	15.629	17.207	18.988	21.000	23.277	25.856
22	11.125	12.166	13.340	14.666	16.167	17.870	19.802	22.000	24.502	27.353
23	11.339	12.434	13.676	15.086	16.690	18.520	20.610	23.000	25.738	28.878
24	11.539	12.688	13.996	15.490	17.199	19.159	21.410	24.000	26.985	30.431
25	11.727	12.928	14.302	15.879	17.694	19.785	22.202	25.000	28.244	32.013

b) 10% Discount Rate

N	RATE OF ENERGY PRICE ESCALATION 1%	2%	3%	4%	5%	6%	7%	8%	9%	10%
1	0.918	0.927	0.936	0.945	0.955	0.964	0.973	0.982	0.991	1.000
2	1.761	1.787	1.813	1.839	1.867	1.892	1.919	1.946	1.973	2.000
3	2.535	2.584	2.634	2.684	2.735	2.787	2.839	2.892	2.946	3.000
4	3.246	3.324	3.403	3.483	3.566	3.649	3.734	3.821	3.910	4.000
5	3.899	4.009	4.123	4.239	4.358	4.480	4.605	4.734	4.865	5.000
6	4.498	4.645	4.797	4.953	5.115	5.281	5.453	5.630	5.812	6.000
7	5.048	5.234	5.428	5.628	5.837	6.053	6.277	6.509	6.750	7.000
8	5.553	5.781	6.019	6.267	6.526	6.796	7.078	7.372	7.680	8.000
9	6.017	6.288	6.572	6.871	7.184	7.512	7.858	8.220	8.601	9.000
10	6.443	6.758	7.090	7.441	7.812	8.203	8.616	9.053	9.513	10.000
11	6.834	7.194	7.575	7.981	8.411	8.868	9.354	9.870	10.418	11.000
12	7.193	7.598	8.030	8.491	8.983	9.510	10.072	10.672	11.314	12.000
13	7.523	7.972	8.455	8.973	9.530	10.127	10.770	11.460	12.202	13.000
14	7.825	8.320	8.853	9.429	10.051	10.723	11.449	12.233	13.082	14.000
15	8.102	8.642	9.226	9.860	10.549	11.296	12.109	12.993	13.954	15.000
16	8.357	8.941	9.576	10.268	11.268	11.849	12.752	13.738	14.818	16.000
17	8.592	9.218	9.903	10.653	11.477	12.382	13.377	14.470	15.674	17.000
18	8.807	9.475	10.209	11.018	11.910	12.896	13.984	15.189	16.523	18.000
19	9.004	9.713	10.496	11.362	12.323	13.390	14.576	15.895	17.363	19.000
20	9.186	9.934	10.764	10.688	12.718	13.867	15.151	16.588	18.196	20.000
21	9.351	10.139	11.015	11.996	13.094	14.326	15.711	17.268	19.022	21.000
22	9.504	10.329	11.251	12.287	13.454	14.769	16.255	17.936	19.840	22.000
23	9.645	10.505	11.471	12.562	13.796	15.196	16.784	18.591	20.650	23.000
24	9.774	10.668	11.678	12.822	14.124	15.607	17.299	19.235	21.454	24.000
25	9.892	10.819	11.871	13.068	14.437	16.003	17.800	19.867	22.250	25.000

TABLE IV-2 (cont.)

c) 12% Discount Rate	RATE OF ENERGY PRICE ESCALATION									
N	1%	2%	3%	4%	5%	6%	7%	8%	9%	10%
1	.902	.911	.920	.929	.938	.946	.955	.964	.973	.982
2	1.715	1.740	1.765	1.791	1.816	1.842	1.868	1.894	1.920	1.947
3	2.448	2.495	2.543	2.591	2.640	2.690	2.740	2.791	2.842	2.894
4	3.110	3.183	3.258	3.335	3.413	3.492	3.573	3.655	3.739	3.825
5	3.706	3.810	3.916	4.025	4.137	4.252	4.369	4.489	4.612	4.738
6	4.244	4.380	4.521	4.666	4.816	4.970	5.129	5.293	5.462	5.636
7	4.729	4.900	5.078	5.262	5.452	5.650	5.856	6.068	6.289	6.517
8	5.166	5.373	5.589	5.814	6.049	6.294	6.550	6.816	7.094	7.383
9	5.561	5.804	6.060	6.328	6.609	6.903	7.213	7.537	7.877	8.234
10	5.916	6.197	6.492	6.804	7.133	7.480	7.846	8.232	8.639	9.069
11	6.237	6.554	6.890	7.247	7.625	8.026	8.451	8.902	9.381	9.889
12	6.526	6.880	7.256	7.658	8.086	8.542	9.029	9.549	10.103	10.694
13	6.787	7.176	7.593	8.039	8.518	9.031	9.581	10.172	10.805	11.486
14	7.022	7.446	7.902	8.394	8.923	9.494	10.109	10.773	11.489	12.263
15	7.234	7.692	8.187	8.723	9.303	9.931	10.613	11.352	12.155	13.026
16	7.426	7.916	8.449	9.028	9.659	10.346	11.095	11.911	12.802	13.775
17	7.598	8.120	8.689	9.312	9.993	10.738	11.555	12.450	13.433	14.512
18	7.754	8.306	8.911	9.575	10.306	11.109	11.994	12.970	14.046	15.234
19	7.894	8.475	9.114	9.820	10.599	11.461	12.414	13.471	14.643	15.945
20	8.020	8.629	9.302	10.047	10.874	11.793	12.815	13.954	15.224	16.642
21	8.134	8.769	9.474	10.258	11.132	12.108	13.199	14.420	15.789	17.337
22	8.237	8.897	9.632	10.454	11.374	12.405	13.565	14.869	16.340	18.000
23	8.330	9.013	9.778	10.636	11.600	12.687	13.914	15.302	16.875	18.660
24	8.414	9.119	9.912	10.805	11.813	12.954	14.249	15.720	17.396	19.309
25	8.489	9.216	10.035	10.961	12.012	13.207	14.568	16.123	17.904	19.947

TABLE IV-3 Single Present-Value Factors

DISCOUNT FACTORS

N	6%	8%	10%	12%	15%	20%	25%
			DISCOUNT	RATES			
1	0.943	0.926	0.909	0.893	0.870	0.833	0.800
2	.890	.857	.826	.797	.756	.694	0.640
3	.840	.794	.751	.712	.658	.579	0.512
4	.792	.735	.683	.636	.572	.482	0.410
5	.747	.681	.621	.567	.497	.402	0.328
6	.705	.630	.565	.507	.432	.335	0.262
7	.665	.584	.513	.452	.376	.279	0.210
8	.627	.540	.467	.404	.327	.233	0.168
9	.592	.500	.424	.361	.284	.194	0.134
10	.558	.463	.386	.322	.247	.162	0.107
11	.527	.429	.351	.288	.215	.135	0.086
12	.497	.397	.319	.257	.187	.112	0.069
13	.469	.368	.290	.229	.163	.094	0.055
14	.442	.341	.263	.205	.141	.078	0.044
15	.417	.315	.239	.183	.123	.065	0.035
16	.394	.292	.218	.163	.107	.054	0.028
17	.371	.270	.198	.146	.093	.045	0.022
18	.350	.250	.180	.130	.081	.038	0.018
19	.331	.232	.164	.116	.070	.031	0.014
20	.312	.215	.149	.104	.061	.026	0.012
21	.294	.199	.135	.093	.053	.022	0.009
22	.278	.184	.123	.083	.046	.018	0.007
23	.262	.170	.112	.074	.040	.015	0.006
24	.247	.158	.102	.066	.035	.013	0.005
25	.233	.146	.092	.059	.030	.010	0.004
26	.220	.135	.084	.053	.026	.009	0.003
27	.207	.125	.076	.047	.023	.007	0.002
28	.196	.116	.069	.042	.020	.006	0.002
29	.185	.107	.063	.037	.017	.005	0.002
30	.174	.099	.057	.033	.015	.004	0.001
31	.164	.092	.052	.030	.013	.004	0.001
32	.155	.085	.047	.027	.011	.003	0.001
33	.146	.079	.043	.024	.010	.002	0.001
34	.138	.073	.039	.021	.009	.002	0.000
35	.130	.068	.036	.019	.008	.002	0.000
40	.097	.046	.022	.011	.004	.001	0.000
45	.073	.031	.014	.006	.002	.000	0.000
50	.054	.021	.009	.004	.001	.000	0.000

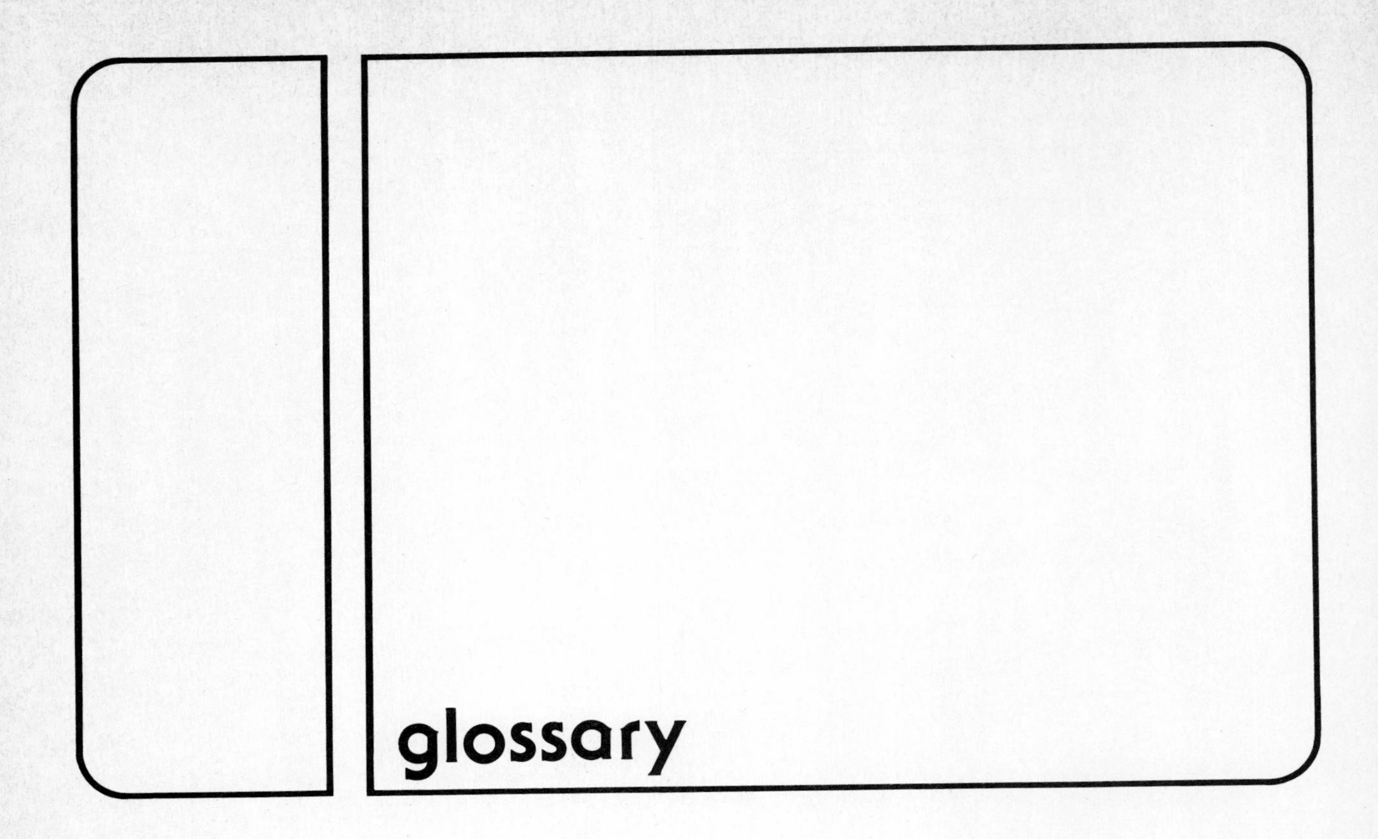

Terms in italic type are defined elsewhere in the Glossary.

absorber plate A component of a *flat-plate collector* that is generally black in color and whose purpose is to receive radiant solar energy and transform it into heat.

absorptivity A surface's ability to absorb solar energy, calculated by dividing the amount of energy absorbed by the amount of energy striking the surface.

active solar system A system that utilizes externally powered mechanical devices such as pumps or blowers to aid in collecting, distributing, and/or storing solar energy.

ambient temperature Outside air temperature.

American Society of Heating, Refrigerating and Air-Conditioning Engineers (ASHRAE) An organization that has set standards for the testing of solar collectors and heat storage systems. ASHRAE also publishes extensive reference data on air conditioning, heating, ventilation, and refrigeration technology, much of which is of great value to solar designers.

angle of incidence The angle between the sun's rays and a line perpendicular to the surface those rays are striking.

ASHRAE See *American Society of Heating, Refrigerating and Air-Conditioning Engineers.*

backdraft damper Used in air ducts and in the vents of Trombe walls and other vented walls to allow air to flow in one direction only.

batch water heater A type of passive solar water heater that combines a collection function with a storage function. It is typically constructed of a dark-colored water tank enclosed in a glazed and insulated box. Also called an *integral passive water heater.*

batt-type insulation Insulation generally made of glass or rock wool fibers held loosely together and precut into the shape of long rectangles, made to fit snugly between studs, rafters, or joists.

bread box water heater A type of *batch water heater* with doors that may be closed over the glazing at night. Generally insulated, the doors dramatically cut heat loss. Their interior surfaces are often highly reflective, and when opened, bounce additional light into the collector.

British thermal unit The amount of heat necessary to raise 1 pound of water 1 degree Fahrenheit.

Btu See *British thermal unit.*

building envelope The exterior shell of a building. Includes all the layers of the exterior walls, windows, roof, doors, and lowest floor of the conditioned space.

building load coefficient (BLC) The BLC expresses the amount of thermal energy needed per day to increase a building's temperature 1 degree Fahrenheit and hold it at that temperature.

centrifugal pump A pump that functions by rotating the fluid inside it to a sufficient angular speed to discharge it from the pump body.

cfm Cubic feet per minute.

clerestory Vertical windows located high in a wall, often between an upper and a lower roof.

closed loop A system in which heat transfer fluid such as silicone oil, Freon, or water circulates in sealed plumbing between the solar collection array and the thermal storage containers. *Heat exchangers* are employed to transfer the fluid's thermal energy to the fluid in the storage containers. There is thus no contact between the two fluids.

conduction Thermal energy transport through a substance by means of motions and vibrations of adjacent molecules and atomic particles.

convection Thermal energy transport by means of fluid flows. In "forced" convection, the currents are generated by an external energy source. In "natural" convection, they are generated by spatial temperature differences in the fluid itself.

dead air space A sealed air space that acts as an insulating layer by inhibiting both convection and conduction of thermal energy.

degree-days A unit for measuring the severity of a climate. The degree-day employed in this text is often called "heating degree-day" and is generally determined by the following formula:

$$\begin{matrix}\text{degree-days accrued} \\ \text{in a 24-hour period}\end{matrix} = 65^\circ\text{F} - \begin{pmatrix}\text{average temperature} \\ \text{during that period}\end{pmatrix}$$

Average degree-day data for various locations is tabulated on monthly and annual bases (see Table III-1).

DHW system See *domestic hot water system.*

differential thermostat An electric switch that responds to the temperature difference between two locations. In solar systems it is often used to automatically control pumps and blowers.

direct gain A passive system in which sunlight heats a living space by entering into it through *solar apertures* and warming the surfaces of the space. These surfaces give off *radiant heat*, as well as initiating *convection* currents that help distribute thermal energy.

diurnal Daily; happening each day.

domestic hot water system A system whose function is to supply the hot water requirements for a residence.

drain-back In some solar systems, the heat transfer fluid drains out of the collectors and into a storage vessel whenever the circulating pump is not running. In swimming pool systems, the storage vessel can be the pool itself.

drain-down A design in which the water is drained from the collectors whenever there is a freezing danger. This design is often used in *open-loop, active DHW systems.*

efficiency of a solar collector The percentage of *normally* incident solar energy that is collected by a solar panel.

electromagnetic spectrum The full range of *electromagnetic waves*, including x-rays, gamma rays, ultraviolet, visible, and infrared light and radio waves.

electromagnetic wave A disturbance that is transmitted from one location to another by means of changing electric and magnetic fields.

EM waves See *electromagnetic wave.*

emissivity A measure of the propensity of a material to give off thermal radiation.

energy The capacity to do work. It is encountered in various forms, such as thermal, kinetic, chemical, or electrical energy. One form may be transformed into another.

equation of time Of aid in converting "clock time" to "solar time."

equinox The two times during the year at which the length of day and length of night are approximately equal. The autumnal equinox occurs around September 22 and the vernal equinox around March 22.

***f*-chart** A method for predicting the percentage of a building's heating needs that will be supplied by a certain solar system.

flat-plate collector A type of solar collector employing a flat absorbing surface.

float glass Glass manufactured by pouring molten glass over molten metal. It is relatively inexpensive because no polishing of surfaces is required.

fluid A liquid or a gas.

galvanic corrosion A type of corrosion that occurs when different metals are in contact with each other.

Glauber's salt A variety of *phase-change* material whose melting point is approximately 90°F.

glazing Glass or plastic coverings that may be either clear or translucent.

greenhouse effect The situation in which a medium such as *glazing* admits visible light into a building but does not permit infrared (heat) radiation to exit. The result is heating of the building interior on sunny days.

header pipe The pipes in a solar collector that carry the heat transfer fluid to and from the grid of pipes that run across the absorber surface.

heat The energy that flows into a body by contact with, or radiation from, a warmer body.

heat capacity The amount of heat necessary to raise a unit volume of a substance 1 degree. In British units, heat capacity is measured in terms of the energy required to raise 1 pound of a material 1 degree Fahrenheit. Heat capacity differs from substance to substance.

heat exchanger A device used to transfer thermal energy from one fluid to another without the two fluids contacting one another.

heating load The rate of thermal energy gain necessary to maintain a building at a comfortable interior temperature.

heat pump A device that transports thermal energy from one medium to another, in the process cooling the first medium and heating the second.

horizon The line at which the sky appears to meet the earth.

infiltration The entering of outside air into a building through cracks and gaps in the building envelope.

infrared radiation Electromagnetic radiation with wavelengths longer than visible light but shorter than radio waves.

insolation The amount of solar energy received by a unit area of surface per unit time.

insulation Material used to minimize the passage of heat from one area into another.

integral passive water heater See *batch water heater.*

latent heat The amount of thermal energy required to change the state of a substance from solid to liquid or liquid to gas. In the British system, latent heat is measured in terms of the number of Btu per pound required. For instance, 144 Btu is needed to melt 1 pound of ice.

latitude The angular distance north or south of the equator, measured in degrees.

LCR See *load collector ratio.*

life-cycle-cost analysis A method of economic analysis that sums the energy costs of a building or piece of equipment with the net costs of purchase, construction or installation, maintenance, replacement, and all other costs attributable to the investment over the predicted lifetime of the building.

load collector ratio Used in passive solar design methodology developed by Los Alamos National Laboratory, the LCR is the single variable which most influences the solar heating performance of the buildings. It is calculated by dividing the *building load coefficient* by the area of solar collection surfaces.

longitude The angular distance east or west of the standard meridian line passing through Greenwich, England.

magnetic declination The angular deviation between true north and magnetic north. Magnetic declination varies depending on location.

mass The quantity of matter in a body. It differs from weight in that mass is independent of the gravitational force. In other words, an object has the same mass whether it is on earth, in outer space, or on another planet.

movable insulation Also called "night insulation," it is insulation that is installed over windows (usually at night) to prevent heat loss, but removed during daytime in order to allow solar heating, daylighting, venting, and a view.

night sky radiation Radiation of excess heat into the night sky can be used as a means of cooling off a building or thermal storage medium.

normal Perpendicular.

open loop A solar water heating system in which the water that is used in a pool or in a building (for showers, bathing, or washing, for instance) is the same water that has flowed through the solar collector array.

outgassing Many materials emit gases under conditions such as high temperatures. Paint, glue, or insulation in poorly designed solar collectors sometimes do this and can lead to fogging of the glazing.

passive solar heating system A system in which solar energy is collected, stored, and distributed without using pumps, blowers, or other equipment requiring external energy sources such as electricity.

phase change A substance's transformation between its solid, liquid, and gaseous forms.

plate glass Window glass manufactured by pouring molten glass into flat, very smooth molds and then polishing the surfaces to produce low-distortion light transmission.

plenum A large duct whose purpose is to carry air to or from a system of smaller ducts.

power Rate of energy use. Common units of power: Btu per hour, watt (equal to 1 joule of energy per second), or horsepower (equal to 746 watts).

radiant heat Heating by means of infrared waves emitted by bodies warmer than their surroundings.

radiation or radiant energy Energy flow by means of *electromagnetic waves.*

reflectance The fraction of light incident on a surface that is reflected from that surface.

relative humidity The amount of water vapor in the air divided by the amount of water vapor the air is capable of holding at that temperature and pressure.

retrofit Installation of a solar system into an existing building.

riser pipes The pipes in a solar collector that run across the absorbing surface.

rock storage bin A thermal energy storage container that uses gravel or other rock as the medium of storage.

R-value Useful for comparing the insulating values of various materials.

selective surface A surface with a high absorptance of incident solar energy, but a low emittance of infrared radiation.

skyline The line along which the sky appears to touch the earth.

solar savings fraction (SFF) A measure of the energy savings due to a passive solar system.

solar time A system of time based on the movements of the sun, relative to an observer on earth. It is solar time that is used in the design of solar systems.

solar wall Solar energy collection surfaces located between the glazing of a solar aperture and the interior of the building. Solar walls are typically within several inches of the glazing and massive enough to store as well as collect energy.

solstices, summer and winter The solstices occur during the time of year when the sun is farthest north or south of the equator and are characterized by the longest and shortest days of the year.

specific heat The amount of thermal energy necessary to raise a unit mass of a substance 1 degree. In the British system of units, specific heat is expressed in terms of Btu per pound per degree Fahrenheit.

stagnation temperature The temperature a solar collector will attain in full sunlight, with no fluid flowing through it.

sunspace An enclosed area whose primary function is to collect solar energy for the purpose of heating an adjacent building.

thermal storage mass Materials that are used to store heat. Materials often used include concrete, brick, water, adobe, and gravel.

thermosyphon Natural convection.

translucent Allowing light to pass through, but diffusing it sufficiently so that objects on the other side cannot be distinguished.

transmittance The fraction of radiant energy incident on the surface of a substance that is transmitted through that substance.

transparent Transmitting light rays so that objects on the other side may be distinctly seen.

Trombe wall A *solar wall* constructed of solid thermal storage mass.

ultraviolet radiation Electromagnetic radiation of wavelengths just shorter than those at the violet end of the visible-light spectrum.

vapor barrier A layer of material that is resistant to the flow of water vapor, used to prevent condensation within a building envelope.

visible light Light that is perceptible by the eye.

water wall *Solar wall* employing water as the heat storage medium.

work The transference of force from one body or system to another.

zenith (1) The highest point in the sky reached by the sun (or by some other celestial body. (2) The point in the sky directly above the observer.

index